Advanced Information Technologies for Industrial Material Flow Systems

NATO ASI Series

Advanced Science Institutes Series

A series presenting the results of activities sponsored by the NATO Science Committee, which aims at the dissemination of advanced scientific and technological knowledge, with a view to strengthening links between scientific communities.

The Series is published by an international board of publishers in conjunction with the NATO Scientific Affairs Division

A Life Sciences B Physics	Plenum Publishing Corporation London and New York
C Mathematical and Physical Sciences D Behavioural and Social Sciences E Applied Sciences	Kluwer Academic Publishers Dordrecht, Boston and London
F Computer and Systems Sciences G Ecological Sciences H Cell Biology	Springer-Verlag Berlin Heidelberg New York London Paris Tokyo

Series F: Computer and Systems Sciences Vol. 53

Advanced Information Technologies for Industrial Material Flow Systems

Edited by

Shimon Y. Nof
Colin L. Moodie

School of Industrial Engineering,
Purdue University, W. Lafayette, IN 47907, USA

Springer-Verlag
Berlin Heidelberg New York London Paris Tokyo
Published in cooperation with NATO Scientific Affairs Division

Proceedings of the NATO Advanced Research Workshop on Advanced Information Technologies for Industrial Material Flow Systems held in Grenoble, France, June 13–17, 1988.

ISBN-13: 978-3-642-74577-5 e-ISBN-13: 978-3-642-74575-1
DOI: 10.1007/978-3-642-74575-1

Softcover reprint of the hardcover 1st edititon 1989

2145/3140-543210 – Printed on acid-free paper

PREFACE

This book contains the results of an Advanced Research Workshop that took place in Grenoble, France, in June 1988. The objective of this NATO ARW on Advanced Information Technologies for Industrial Material Flow Systems (MFS) was to bring together eminent research professionals from academia, industry and government who specialize in the study and application of information technology for material flow control. The current world status was reviewed and an agenda for needed research was discussed and established. The workshop focused on the following subjects:

The nature of information within the material flow domain.

Status of contemporary databases for engineering and material flow.

Distributed databases and information integration.

Artificial intelligence techniques and models for material flow problem solving.

Digital communications for material flow systems.

Robotics, intelligent systems, and material flow control.

Material handling and storage systems information and control.

Implementation, organization, and economic research-issues as related to the above.

Material flow control is as important as manufacturing and other process control in the computer integrated environment. Important developments have been occurring internationally in information technology, robotics, artificial intelligence and their application in material flow/material handling systems. In a traditional sense, material flow in manufacturing (and other industrial operations) consists of the independent movement of work-in-process between processing entities in order to fulfill the requirements of the appropriate production and process plans. Generally, information, in this environment, has been communicated from processors to movers. In modern, computerized environments, planning as well as sensing and communication are accomplished more or less automatically, and in a manner which synchronizes and integrates materials movement, material processing and information flow.

Such systems, if properly controlled, offer substantial advantages for low cost production of high variety, low to medium order quantity items. As inferred above, information is a key factor in the operation of any production system, particularly a computer integrated one. This includes information sensing, communication, storage and processing: all vital to the effective planning and control of such systems.

While readily recognizing the importance of information to material flow in computer integrated environments, it is evident that more research and development are necessary to improve the information advantage. For example, concepts of knowledge-bases to augment data-bases, object-oriented programming versus procedural programming and intelligent control systems need to be investigated further with respect to effective information control and utilization. Another example is how to generate, represent and communicate production information and process plans within the material processing and handling network. These issues, focusing on advanced information technology, are important to the future development of computer integrated material flow systems.

We would like to express our gratitude and appreciation to the NATO Division of Scientific Affairs, who sponsored this workshop, and in particular to Dr. Craig Sinclair, Director of the ARW Programme, for his invaluable support in the organization of the meeting; the co-sponsors of the workshop: NSF Program on Computer Integrated Engineering, especially its director, Tony Woo; the European Research Office of the U.S. Air Force, especially Major Mel Townsend; FACTROL, Inc., particularly its president, Hank Grant; The National Polytechnic Institute, Grenoble, France; and the School of Industrial Engineering, Purdue University.

We would also like to express our thanks to our distinguished colleagues and friends who participated in the Workshop and contributed so creatively to its success; the members of our organizing committee, Guy Doumeingts, Bob Hollier, Francois Roubellat, and Jocelyn Traccaz. Special thanks go to Louis Balme from INPG, who joined the organizing committee and, as our local liaison, shared with us the best of beautiful Grenoble; Nava Nof, for organizational support; Joyce Hinds and Barb Ford, for administrative support, and Barbara Kester from the NATO Publication Coordination Office for helping to bring this book to print.

The book is organized in seven parts: The results of the research panels; Nature of information in MFS; Data and knowledge methodologies in MFS; Robotics and communication; Modeling issues of specific MFS functions; Integration models and techniques; and Implementation and Management issues.

This Advanced Research Workshop was arranged because its unique focus has not been emphasized at other national and international conferences which consider artificial intelligence or information aspects of manufacturing. The output of the Workshop is expected to benefit those researchers who believe too much in information technology and neglect other aspects of the material flow control problem, and also, those researchers who have been so far disappointed in the rate of transfer of information technology research to actual applications.

West Lafayette, Indiana
November 1988

Shimon Y. Nof
Colin L. Moodie

Table of Contents

Part 6. Implementation and Management Issues

Nature of Information in Material Flow Systems

MANAGING INFORMATION COMPLEXITY IN MATERIAL FLOW SYSTEMS

James J. Solberg
Joseph A. Heim
Engineering Research Center for Intelligent Manufacturing Systems
Purdue University
West Lafayette, IN 47907

I. Introduction.

Few people seem to realize how extraordinarily complex the everyday activities involved in making products really are. Even those who are directly involved in these activities ordinarily see only a limited view of the entire process. In this paper, we would like to first call attention to some facts about the character of the information involved in manufacturing, emphasizing how difficult it is to manage. This perspective is essential background to the real purpose of the paper, which is to examine alternative strategies for managing manufacturing information. Four strategies will be assessed, their merits and deficiencies will be compared, and a few comments will be added about hybrid methods.

We would like to emphasize from the outset that we are not addressing issues of networking, standards, protocols, or database design. Instead, we are focusing on more fundamental issues related to the organization of information, taking into account the specific nature of manufacturing needs.

II. Characteristics of Manufacturing Information.

Before considering the strategic options for dealing with information, we should attempt to understand the nature of the problem independently of the means that people use to cope with it. That is, we should separate the raw problem in its natural state from any formulation or simplification that may be imposed upon it. With this thought in mind, we will begin with some informal discussion of the challenges that must be faced.

The most obvious aspect of manufacturing information is the vast quantity of it. Just as an exercise, let us imagine trying to inventory the total

NATO ASI Series, Vol. F53
Advanced Information Technologies for
Industrial Material Flow Systems
Edited by Shimon Y. Nof, Colin L. Moodie

contents of a typical factory at a fixed point in time. The number of distinct item types--workpieces, machines, people, tools, fixtures, dies, carts, containers, purchased components, repair items, and so forth--would undoubtedly number in the tens or even hundreds of thousands. We are thinking now just of giving a unique identifier to each type; in addition, there may be several or many instances of each type. Usually simple items such as nuts or washers are not distinguished in practice. Since any one of the class is completely replaceable by any other, only a count is considered necessary. In many cases, however, it is important to identify (e.g., with a serial number) each individual instance of a class. Furthermore, many of the items are complex assemblies, made up of possibly many components which are themselves items. Hence, if we are to be really complete, we will be dealing with on the order of hundreds of millions of item identifiers.

Now each identifier is only a "pointer" to the information necessary to describe the named item. A complete description would include all of the geometric dimensions, materials, specifications, tolerances, and other information of the kind you would find on an engineering drawing. Even this would not be complete, for an engineering drawing makes full sense only to someone who is familiar with local practices. It leaves a great deal to the reader to interpret and to supply default values. For example, the description of an ordinary, single diameter hole can require as many as 32 attributes, including radius, depth, concentricity, angles, and so forth. Functioning mechanisms may have kinematic specifications, strength requirements, load capacities, and other engineering constraints. All together, the complete description of any one item may take, typically, on the order of one million bytes. Hence, a complete catalogue of information about all items in a factory could take in the neighborhood of 10^{15} bytes. To put this in perspective, a modern compact optical disk can store on the order of one gigabyte; it would take a million of these to record all of this data.

Obviously, no one attempts to treat the information in this way. Instead, various aspects of the information are structured, simplified, cross-referenced, or in some cases ignored. But it is important to realize that the sheer quantity of information that is there to be dealt with in some manner is enormous. Thus, the greatest challenge in managing manufacturing information is not to ensure access, but rather to control access, so that the decision makers will not be swamped by irrelevant details. Fortunately, not all information is of equal importance.

Next we come to information about the relationships among items. Some items may be components of other items; some may be substitutable for others. There could be location relationships, such as"next to", "within", and the like. Although only a small fraction of the total number of possible relationships may have any practical meaning, it is nevertheless clear that the combinatorial nature of the count of interaction possibilities presents us with astronomical numbers.

What we have mentioned so far is information about *things*. There is another kind of vital information having to do with description of methods and procedures. This would include operating instructions, process plans, production standards, safety rules, time standards, maintenance procedures, inspection plans, record keeping procedures, and all the other standards, conventions, and practices which are followed in a factory. Those which are written down commonly occupy many pages of small print in shelves of handbooks. For example, a typical handbook of machining data has hundreds of pages of tables. One should also realize that most of the information used to carry out manufacturing tasks is not written down anywhere. We rely upon the common sense knowledge of people to attend to details which are obvious. For example, if an instruction is written to "insert bolt in hole", it is assumed that anyone reading the instruction understands what a bolt is, how to pick one up, how to orient it to insert it in a hole, what to do if it is dropped, and so forth. One of the greatest difficulties in automating such tasks is to make explicit what every human already knows so well that it is never even consciously considered.

So far, we have considered only static information. But manufacturing is, by its very nature, a dynamic process. The changes occur over many different time scales. Some phase changes within materials undergoing processing can take only fractions of a second; some machine control requirements demand responses in milliseconds. At the opposite extreme, issues of factory design and equipment selection, can produce effects that last for decades. Between these extremes lie dynamic issues that are expressed in seconds, minutes, days, weeks, months, and years. Of course, the time scale for material flow is normally in the range of minutes to hours.

The usual approach toward these changes is to select a particular timescale as "important" for the issues that one has in mind and then ignore all changes that occur over much shorter or much longer intervals. For example, in considering a scheduling problem, we normally assume that the factory resources are fixed, when in actuality we know that they change over time, and we ignore the details of what happens during an operation. One can easily argue for the logic of doing this; in fact, it is difficult to imagine any alternative. The point we want to make is that this selective perception is only a human construct; in the real world, the dynamic changes are linked.

For the purposes of this paper, we particularly wish to emphasize the inevitability of what might be considered long term changes. Usually, when we engineer a hardware or software system, we do so with the expectation that our system design will last for the foreseeable future. In practice, we find that our systems are obsolete almost as soon as they are working. New products, engineering changes to current products, customer driven changes, responses to competition, new equipment or software, new vendors, legislative actions, labor negotiations, new management policies, technological advances--all these and more contribute to the general turmoil. While an outside observer may see a factory as a rather stable system engaged in routine, repetitive activity, those on the inside invariably mention the difficulty of keeping up with externally imposed changes. The implication for our choice of strategy in managing the complexity associated with manufacturing information is that we must anticipate the need to change whatever systems we build.

This discussion has hardly touched upon the difficulties in managing manufacturing information. We have not mentioned uncertainty, exceptions, failures, organizational divisions, and long delays between actions and consequences--all important aspects of the general problem. But we have said enough to have made the point that a consciously chosen strategy for dealing with this tangled web of information is worthwhile.

II. Strategies for Complexity Management.

In this section, we will introduce four strategic approaches to the management of manufacturing information. The advantages, disadvantages, and additional comments pertaining to each will be expanded in subsequent sections.

The first strategic approach--which is perhaps better described as an absence of strategy--is to select a portion of the problem and to manage that portion by ignoring everything else. We will call this approach ***subsystem isolation***. This approach is by far the most common in current practice, for reasons that will be mentioned later. It is often criticized for producing "islands of automation" that preclude the coordination that is essential to global system efficiency.

The second approach, which we will call ***total integration***, represents the opposite extreme. Many people have advocated this approach to ensure consistency of information handling throughout the entire manufacturing enterprise. The idea is to design a complete information system at one time, allowing for all possible uses of the information. Usually this takes the form of a single database.

The third strategy involves a more structured view which we will call ***hierarchical decomposition***. This approach emphasizes the coordination of activities through a control structure which is conceptually similar to (but need not match) the organizational chart of a company.

The fourth strategy rejects hierarchy in favor of a more "democratic" form of coordination which relies upon communication and negotiation among relatively independent units. We will call this strategy ***heterarchical decomposition***. It is consistent with the "object-oriented" concepts which are achieving popularity among software engineers.

Figure 1 portrays, in simplified form, some of the distinctions among these approaches. In figure 1a, the circles represent subsystems which have been separately constructed, and the lines indicate the (potential) interfaces among them. Figure 1b portrays a large central database, with various applications linked to it. The hierarchy is represented in obvious fashion in figure 1c. In figure 1d, the circles represent the independent objects which communicate through some bus or channel indicated by the upper rectangle.

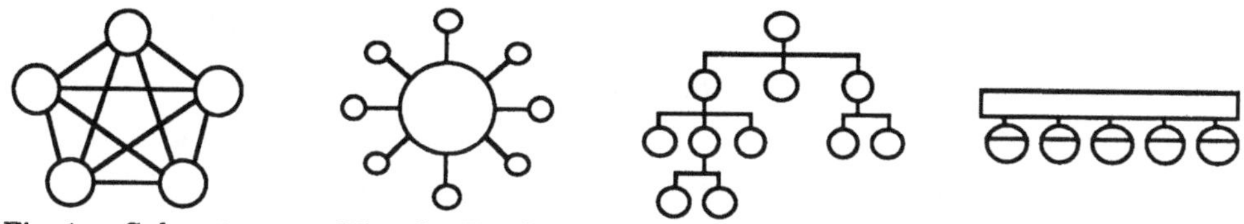

Fig. 1a. Subsystem Optimization — Fig. 1b. Total Integration — Fig. 1c. Hierarchical Decomposition — Fig. 1d. Heterarchical Decomposition

Figure 1. Strategy Alternatives.

It is worth mentioning at this point, although it will become obvious later, that these approaches are not mutually exclusive. Furthermore, we do not claim that they exhaust the possibilities for coping with the complexity of manufacturing information.

III. Subsystem Isolation.

The subsystem isolation approach is the oldest and still the most followed approach. Those who follow this strategy often do so unconsciously, or without recognition that there is any alternative. Sometimes, because of various constraints on budget, time, or jurisdiction, there really is no alternative. However, our position is that the choice of strategy should be made consciously, even when the choice is limited, so that the consequences may be understood in advance.

Whenever a company purchases a software package from an outside vendor, it is almost always following a strategy of subsystem isolation. Certainly the so-called "turn-key" systems that manufacturing companies seem to like are invariably isolated subsystems. For example, a company may purchase from one source an MRP system, from another an AGV system, from another a real time production control system, and from still another a factory monitoring system. As purchased, none of these may be able to communicate with any other. Whether purchased externally or developed internally, most software that is in use today addresses only portions of the manufacturing information needs and essentially ignores everything else. Sometimes claims are made that a "fully integrated" system has been developed, but what is really meant is that the subsystem dealt with is larger than normal.

Despite the criticism that this approach has received, there are many reasons why it is still attractive. It may be the quickest and/or least expensive way to solve an immediate problem; it may avoid capital investment; it may provide the only access to externally owned patented or copyrighted methods; the company may lack the capacity or capability to build its own system; and so forth. Perhaps the most common reason that the strategy is followed is related to organizational jurisdiction. One individual division needs to solve its problem and either cannot or will not wait for other units of the company to express their related concerns.

The principle objection to the approach is that it leaves coordination gaps. Without linkage, the separately developed subsystems can operate at cross-purposes. Getting information from one to another by manual means usually penalizes system performance through delays and costs. On the other hand, trying to achieve unity from the pieces involves many two-way interfaces that are usually specific to the pair. That is, the technical details of the programs at each end of the link usually matter in connecting the two for communication purposes. As a result, building a software bridge between two pre-existing systems, neither of which was designed with interface considerations in mind, is often just about as difficult as starting over. Moreover, the number of potential interface pairs grows as the square of the number of units that have to communicate. Thus, the subsystem isolation strategy, while understandably popular, leaves formidable barriers to achieving real system coordination.

IV. Total Integration.

Recognizing the deficiencies mentioned above, many authors have advocated the total integration approach. Throughout the literature of Computer Integrated Manufacturing (CIM) are scattered charts of concentric circles or linked boxes which attempt to express a "top-down" view of the total generic manufacturing information system. Depending upon the level of abstraction chosen, these can be quite simple or very complex in structure, but an effort is always made to be complete. In attempting to be comprehensive and recognizing the limits of conventional usage of terms, these authors spend great efforts to find the right words. For example, "inspection" is clearly too narrow; "quality assurance" is better.

If carried to a ridiculous extreme, the concept of total integration is clearly unworkable. In some ultimate sense, everything in the universe depends on everything else. But to be fair to the concept, no reasonable person would take it that far. Usually, what is meant by "total" is "complete at the level of a corporate enterprise". Even by this definition, we question the practicality of attempting to build a monolithic system that is capable of dealing with the kind of complexity described above. Even if it were possible, the continual changes that occur in manufacturing would soon render portions of the totally integrated system ineffective or worse.

To our knowledge, there have been no large scale (i.e., total enterprise) successes of the total integration approach to date. Well meaning people continue to advocate efforts in this direction, and certainly there is a managerial purpose served by the concept of a completely coordinated organization. But, in view of the quantity and quality of information involved in conducting manufacturing operations (as discussed above), we believe that the total integration strategy as an approach to engineering practical information systems is just too simplistic to work. A more structured approach, based upon decomposition (i.e., separation of concerns) and abstraction (i.e., deliberate suppression of details), offers better opportunities for coping with this extreme complexity.

V. Hierarchical Decomposition.

Hierarchical decomposition [1] is based upon a consistent, easily understood concept which can be recursively applied at different levels. It offers the advantage of a simple, consistent organizational structure which supports the notion of abstraction (the ability to suppress details). At the lowest levels, physical devices such as motors and sensors carry out simple actions under command of software modules at the next higher layer. An organizational unit at any intermediate level is responsible for receiving status information from lower level units and issuing commands to them, while passing status information upwards and carrying out commands from above. Finally, a single unit at the top of the hierarchy bears ultimate responsibility for fulfilling the goals of the whole system.

The hierarchical structure, because it is a familiar form in human organizations, is both acceptable and intuitively understandable to many people. The importance of understandability cannot be overestimated for the maintainability of the system over its full life.

A pure implementation of hierarchical decomposition, requiring strict adherence to the concept, would forbid communication of a unit with any unit not directly superior or subordinate to it. That is, all communication must go through the channels of the hierarchy. Maintaining such a discipline promotes tight control, with all of its positive and negative connotations. In cases where tight control is desirable (e.g., coordinating a robot with a lathe for part loading), the hierarchy makes eminently good sense. In other cases, where the associations are loose (e.g., machines in different departments of a factory), the advantages can be outweighed by the burden of overhead. A very large hierarchy can become highly inefficient, because of the "distance" between units. For example, to carry out an action which requires the coordination of physical devices which are at the same (lowest) level, but are remote from one another in the hierarchy, may require passing messages through many layers. This organizational overhead can adversely affect overall system performance, and is particularly detrimental to time-critical responses.

In practice, most implementations allow some deviations from the pure hierarchy (dotted line connections) where experience indicates that "shortcuts" are needed. As long as these are confined to instances where the need is clear, they do not violate the spirit of the hierarchical concept. It is the levels of control that distinguish hierarchical decomposition from the other strategies.

The major deficiencies of the approach relate to its structural rigidity. Properly designed, a hierarchical system can deal with many anticipated variations, so it would be unfair to label it as inflexible. However, a hierarchical system requires relatively complete design prior to implementation and, once completed, is not easily changed. In most cases, there is no single, obviously correct decomposition of authority into units, so it is quite likely that any first attempts could be improved upon if modifications were not so difficult. This "resistance to evolution" is what we mean by structural rigidity.

We will illustrate both the strengths and weaknesses of the hierarchical decomposition strategy by means of an example. We will use a typical small batch manufacturing factory with modern automated material handling as our test case. It is not intended to be either complete or fully realistic, but will suffice for expository purposes.

The numerically controlled machine tools are organized as two group technology (GT) cells with material handling robots loading and unloading the NC machines. Products are manufactured in large batches and automated guided vehicles (AGV's) are used to move pallets of work-in-process, raw materials, and tools between the group technology cells. The shop controller is the entry point for new orders, controls the release of orders to manufacturing, and may coordinate activity of the GT-cells. The amount of control and direction which the shop controller provides will depend upon the control strategy adopted.

Figure 2 illustrates the manufacturing facility as a 3-level hierarchy: shop level control, cell control, and direct machine control. Of course the full hierarchical decomposition would be extended to the level of motors and switch closures, but this example will display only these three levels.

The shop controller accepts production orders from outside the shop and uses its process planning functions to decompose the orders into more elementary components to be distributed to the appropriate GT-cell controllers. The shop controller incorporates each of these smaller tasks into its shop-level schedule which it uses to coordinate and direct the facility. Feedback and status information from the cells is used by the shop controller to coordinate movement of jobs/tasks between cells and to determine when production orders are complete.

The cell controllers continue the hierarchical decomposition strategy by dividing cell-level tasks, received from the shop controller, into an agenda of functions that can be performed by the NC machines and robots under the cell's control. The cell controller relies upon information feedback from the direct machine controllers to monitor, coordinate operations between machines within the cell, and to determine when assigned operations have been

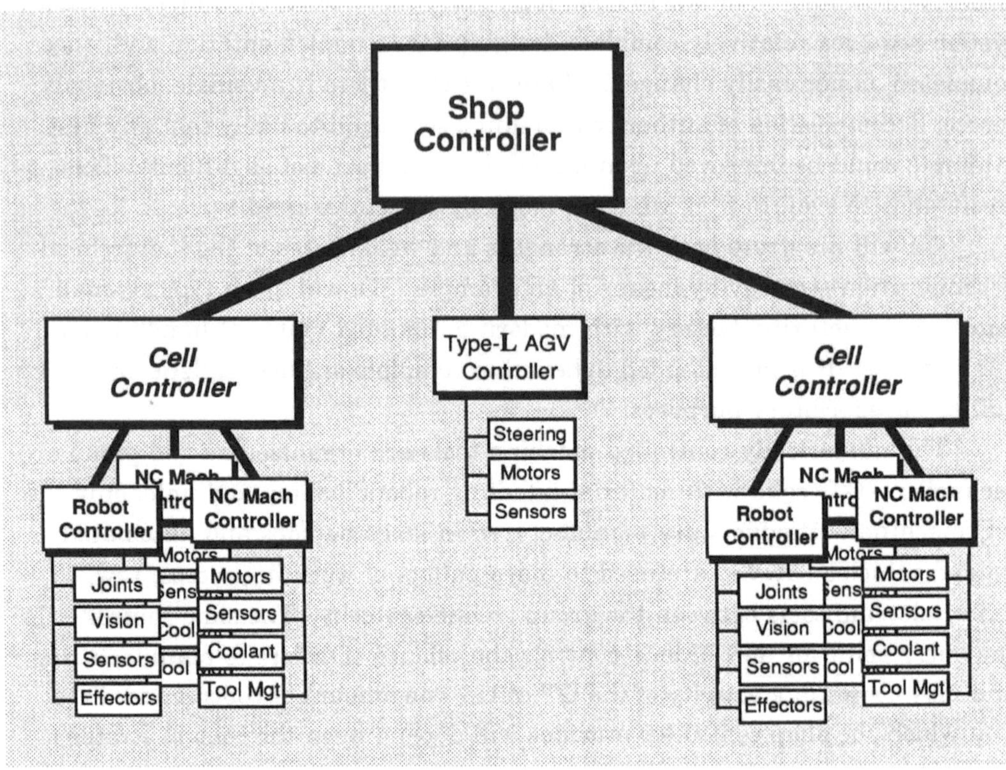

Figure 2. Hierarchical Decomposition Example.

completed. The NC machine controllers and robots use sensory feedback to control and coordinate their operations.

The cell controller activates jobs on its task schedule by requesting from the shop controller the resources (e.g., raw materials, part programs, tools, fixtures) needed to begin a machine-level operation. The shop controller dispatches an AGV to pickup and deliver the requested materials to the cell. Depending upon the capabilities of the AGV, there may be a significant amount of communications between the shop controller and the transport vehicle as it makes its pickup and delivery but eventually the AGV must notify the shop controller when it arrives at its cell destination. The information received from the robots, AGV's, and machine tools is used by the cell controller to coordinate unloading operations and placement of the raw materials in the NC machine. During each step of the operations the cell controller remains an intermediary for the machines, robots, and AGV's involved.

When all feedback data indicates that the NC machine has been loaded, and is ready to begin operations, the cell controller sends a "start executing control program" command to the NC machine controller. Status information is fed back to the cell controller throughout the machining process to indicate current conditions and to alert the cell controller to error situations. After the NC machine signals completion of the assigned operations, the cell controller will instruct a robot to transfer the work to another machine within the cell, place the component in a buffer stock location, or load the work onto an AGV requested by the cell controller and dispatched by the shop controller.

It is evident that there is a very tight coupling between the levels of the hierarchy. The shop controller coordinates and mediates the relationships of the cell controllers and AGV's while the cell controllers provide coordinating functions for the machines within the cells. Changes in machine tools, robots, or even the AGV system would most likely require modifications to be made to the cell controllers as well as the shop controller.

Let us now consider one significant system modification, to observe its impact upon the already functioning hierarchical system. In the original system, large, slow moving AGV's were used since the size of the production batches required relatively infrequent movement of work-in-process and other materials between machining cells. Let us say that the primary goal of the modified manufacturing system is to reduce work-in-process through just-in-time production control. The standard batch will be changed from pallet -size quantities to batches of one, although some orders may still be produced as large batches. The new system will use autonomous roving vehicles, a type of AGV capable of operating without a guide-path and controlled by an on-board system that provides the same functions as Type-L ARV off-vehicle controllers. These small vehicles are unit load transporters and will be used to move single parts between machine cells. The robots will still provide cell-level material handling functions.

Figure 3 illustrates one possible way in which the new vehicles could be incorporated into the hierarchical strategy. Two new control levels have been added: an AGV Dispatcher and a Type-S AGV Dispatcher. The AGV Dispatcher accepts transport commands from the shop controller and determines which size vehicle should be dispatched. This dispatching had previously been performed by the shop controller. The Type-S Dispatcher receives the small-vehicle task assignments from the AGV Dispatcher and decides which one of the Type-S vehicles to assign based upon vehicle proximity to the pickup location and availability.

These additional layers of "management" increase the distance between the controllers and the NC machines, robots, and ARV's, that must be coordinated. The sensory feedback that was used by the shop controller will now be

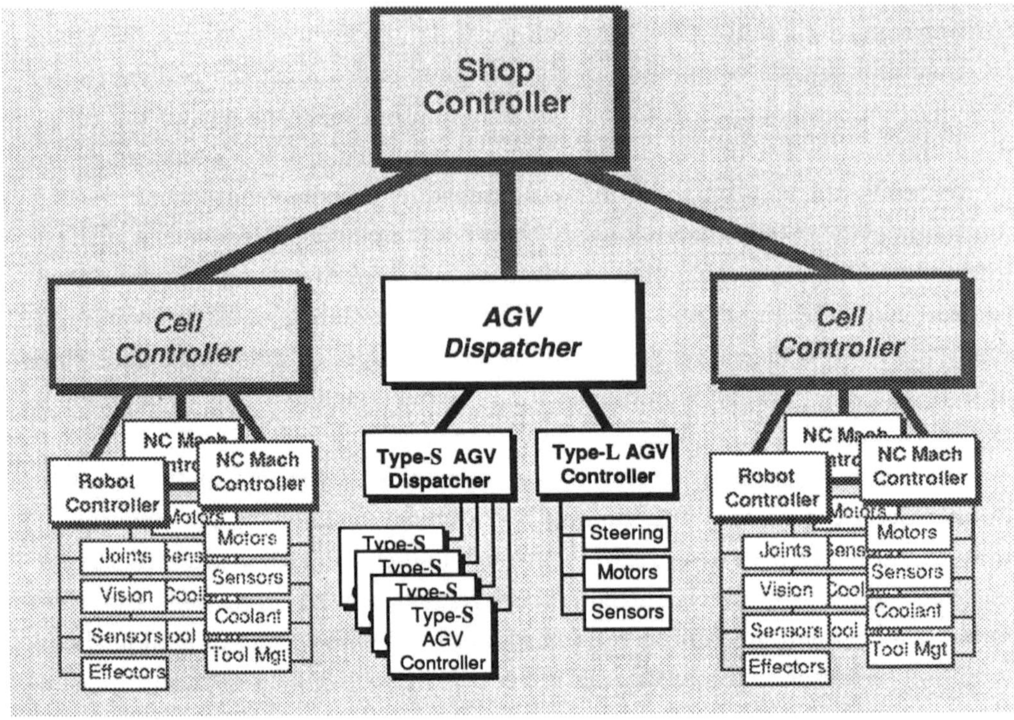

Figure 3. Modified Hierarchical Decomposition Example.

returned to the AGV Dispatcher. This can create problems because some of the functions that depend upon this feedback channel are still being performed by the shop controller. Therefore, AGV Dispatcher must propagate some of the status information received from the vehicle controllers (but not all of it) so that the shop controller can continue to perform a subset of its previous coordinating tasks. A clean hierarchical decomposition becomes difficult to change because of the tight coupling.

In summary, the hierarchical system strategy will require numerous changes throughout the controller hierarchy as well as incorporation of new controller mechanisms.

VI. Heterarchical Decomposition.

Heterarchical decomposition is the most recent, least familiar approach, and therefore deserves an extended discussion. The concept emphasizes modularity by encapsulating information into relatively autonomous "objects."

An "object", in this context, is a software module that operates in a certain fashion in relation to other modules. It is analogous in many respects to a physical component, and in fact some authors refer to objects as "software components" [2, 3]. Like physical components, a good software object will (1) have some simple, well-defined function, (2) be usable for many different final products, and (3) present a simple, standardized interface to any present or future system. Internally, an object may carry out its function in ways which are "hidden" from other objects in the system. These properties ensure that objects will be portable (i.e., reusable in other systems), modular (i.e., parts can be replaced with minimal side effects), and maintainable.

We should mention in passing that there are issues of object classes and inheritance of methods which are normally considered to be essential aspects of object-oriented programming. These do induce a different kind of hierarchy--one related to economy of representation, rather than control. However, these issues are not germane to the present discussion, and are therefore ignored here.

In contrast to hierarchical decomposition which requires that communications between system elements follow the "chain of command", heterarchical decomposition adopts a communications structure that allows components of a system to exchange information as peers. Heterarchically decomposed systems function as independent cooperating processes or agents without a centralized or explicitly directed control. Bargaining between among agents is one method used to coordinate and balance the interests of the system components [4,5]. Mutual agreement of the participants determines each required control decision.

One mechanism for implementing heterarchical systems is the bulletin board (BB) paradigm [6]. A BB provides a neutral medium for collection and exchange of data and information with all participants following a strict communications protocol. The concept has been tested in a few manufacturing cell control experiments [7].

Let us examine the same example as was used in the description of hierarchical decomposition, to see how the implementation according to heterarchical principles would differ. In figure 4 the heterarchical model of a manufacturing control system is represented by the individual system components loosely clustered about a common bulletin board.

Orders still enter the heterarchical system via the shop controller for process planning, but the capabilities and responsibilities of the shop controller have been reduced. Instead of assigning the component operations to particu-

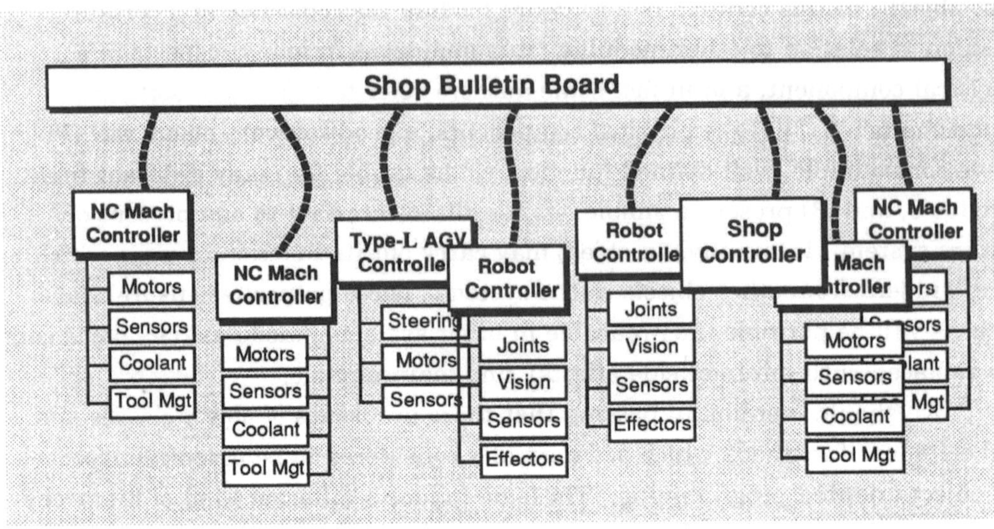

Figure 4. Heterarchical Decomposition Example.

lar machining cells, the shop module simply "posts" the production requirements on the BB as outstanding tasks to be performed. Since there are no cell-level controllers, each of the direct equipment controllers must assume those functions which were provided by cell controllers in the hierarchical example.

NC machines recognize the posted task operations which they can perform and bargain/bid to perform them. There are a number of techniques that can be considered for determining which machine will be awarded the task, but the point is that the BB is used as a passive medium that permits the independent manufacturing components to operate in a cooperative manner rather than by hierarchical assignment. This is the "democratic" element of heterarchical decomposition.

Tasks are removed from the BB by the "winning" bidder. The "winner" determines the resources and coordination activities that will be needed and posts the requests on the BB. These requests are just another type of task and they are cooperatively assigned and negotiated by the same bidding process as the machine operation tasks. As tasks are completed they are returned to the BB with an updated process status so that their next machining operation or material handling request may be bid upon. After an order has completed all production operations, its posting will be recognized as such by the shop controller, and a task for removal (e.g.,to finished goods) will be posted.

When additional vehicles are added to the heterarchical configuration (figure 5) the impact would be expected to be minimal since it just means that there are more elements using the BB for cooperative control. However, it is possible that the bidding algorithm might have to be changed slightly, because there is now more than one type of vehicle to be considered in the cooperative decision process. This may not have been a consideration that was anticipated in the original design. The tasks that are posted by the NC machine must now be able to indicate the type of AGV required so that the bids will correspond to the capabilities of the vehicles bidding.

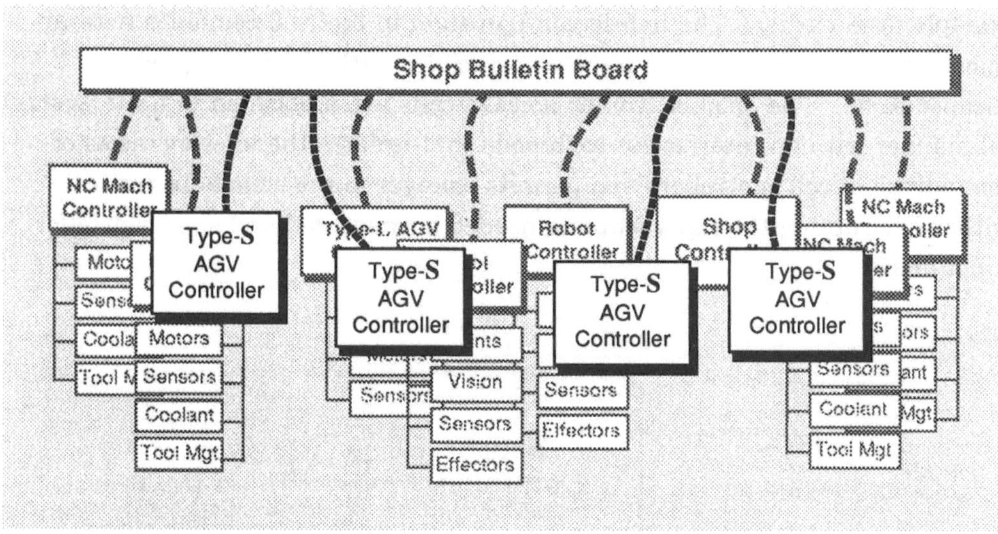

Figure 5. Modified Heterarchical Decomposition Example.

In addition, the number of tasks that must be cooperatively allocated will increase significantly and it is difficult to predict whether the same BB will be able to support the increased activity. It may be necessary to increase the resources for the BB mechanism. There is no control abstraction mechanism such as the Type-S AGV Dispatcher so that when even more vehicles are added to the system the BB mechanism becomes more and more loaded, and this can directly impact the response seen by other users of the BB.

VII. Hybrid Methods.

A comparison of hierarchical and heterarchical decomposition indicates that each strategy for managing information complexity and providing control flow has both strengths and weaknesses subject to the context in which they are examined. The hierarchical approach is appropriate for tightly coupled systems with strong interactions, while the heterarchical approach offers advantages where the coupling is more loose. Therefore, combining the two strategies appropriately should provide the most flexible design methodology and opportunity for incorporating changes that are certain to occur. [8] At the same time, it must be acknowledged that something is lost when the approaches are mixed, namely, the simplicity of a consistent philosophy.

To see how a hybrid system might be organized, we will revisit the shop example used earlier. The hybrid configuration in figure 6 combines heterarchical and hierarchical decomposition principles with object-oriented encapsulation. The shop controller and GT-cells are associated in a hierarchical manner with an interface at each node that isolates the idiosyncrasies of the individual cell controllers and permits changes to the cells to be made without disrupting the tight coupling needed between cell and shop controller functions.

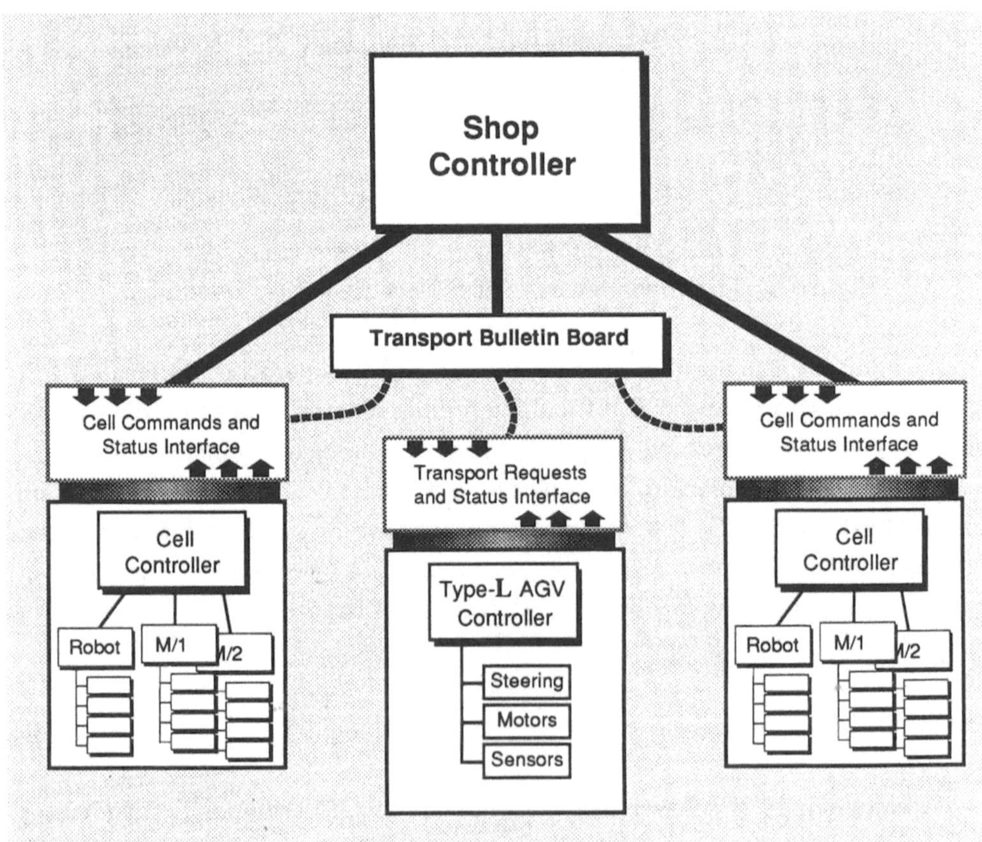

Figure 6. Hybrid Example.

The transport functions have also been isolated with an interface that remains constant and independent of any changes that might be made to the contents of the "transport cell". Instead of the shop controller providing coordination for inter-cell movements, the transportation services are available to the cell controllers and the shop controller via a bulletin board. This decouples the transport requests and makes them directly available (i.e., not hierarchically controlled) to new functions that might be added to the system. This mechanism was used, as shown in figure 7, to incorporate the new vehicle type in the example.

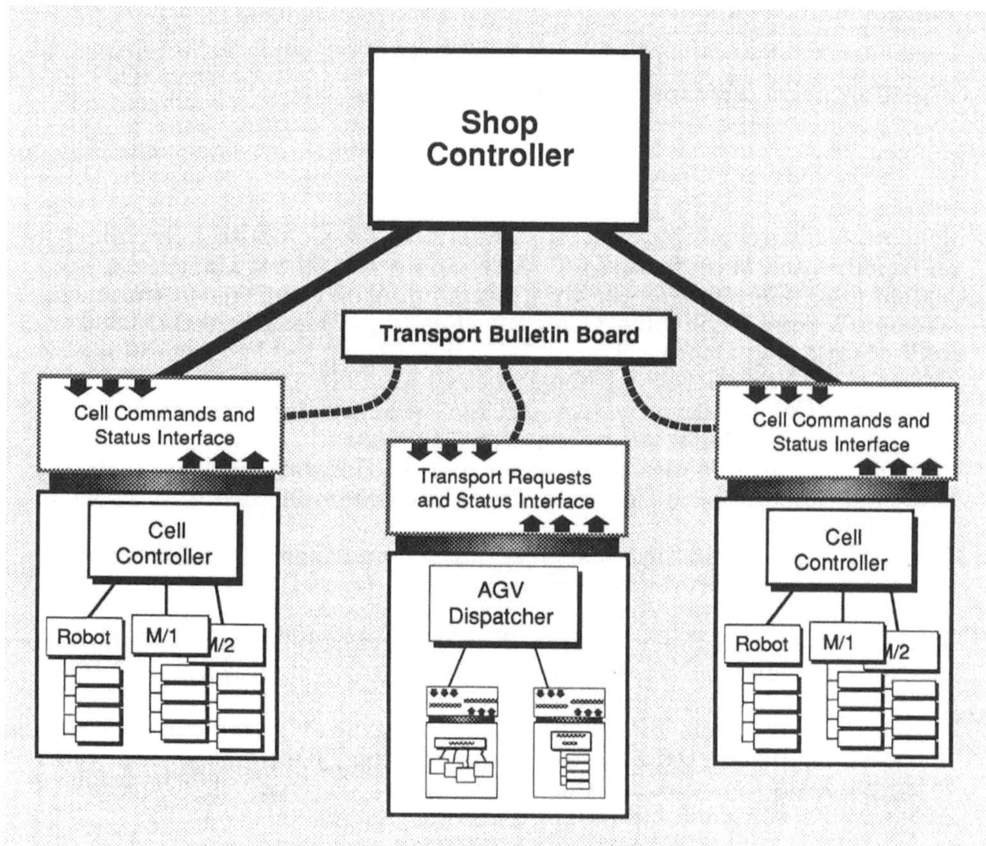

Figure 7. Modified Hybrid Example.

VIII. Issues for Future Work.

The only conclusion supported by the comparison we have made is that there is no single best strategy for managing the information associated with manufacturing systems. Instead of debating the merits of one or the other, we would prefer to see research focusing on the *conditions* under which one might prove superior to another. The possibilities for hybrid approaches, of which we offered only one, provide a rich opportunity for exploration. The compatibility or incompatibility of systems which are organized according to different principles must be faced, because it is inevitable that they will come into contact with one another. Questions related to deliberate evolution from one approach (such as subsystem isolation) to another (such as heterarchical decomposition) also offer intriguing issues for research.

References.

1. McLean, C., Mitchell, M., and Barkmeyer, E., "A Computing Architecture for Small Batch Manufacturing," IEEE Spectrum, 59-64, May, 1983.
2. Cox, Brad J., Object Oriented Programming: An Evolutionary Approach, Addison-Wesley Publishing Company, Reading, Massachusetts, 1986.
3. Stefik, Mark, and Bobrow, Daniel G., "Object-Oriented Programming: Themes and Variations," AI Magazine, 40-62, Fall, 1985.
4. Vamos, T., "Co-operative Systems-An Evolutionary Perspective", IEEE Control Systems Magazine, v. 3, n. 2 , 9-14, 1983.
5. Hatvany, J., "Intelligence and Cooperation in Heterarchic Manufacturing Systems," Robotics and Computer Integrated Manufacturing, v. 2, No. 2, 101-104, 1985.
6. Nii, H.P., "Blackboard Systems: The Blackboard Model of Problem Solving and the Evolution of Blackboard Architecture (Part I)," AI Magazine, 38-53, Summer 1986.
7. Duffie, Neil A., and Piper, Rex S., "Nonhierarchical Control of Manufacturing Systems," Journal of Manufacturing Systems, vol. 5, No. 2, 137-139, 1986.
8. Naylor, Arch W., and Volz, Richard, A., "Integration and Flexibility of Software for Integrated Manufacturing Systems," Design and Analysis of Integrated Manufacturing Systems, Compton, W. Dale, ed., National Academy Press, Washington, D.C., 1988.

INFORMATION DYNAMICS FOR COMPUTER INTEGRATED PRODUCT REALIZATION

Charles A. Fritsch
AT&T Bell Laboratories
6200 E. Broad St.
Columbus, Ohio
U.S.A.

ABSTRACT

Shop floor disruptions are a major concern in today's manufacturing environment. These disruptions can be minimized by a second generation system for computer aided product realization that integrates product development engineering activity with manufacturing engineering activity. The naturally occurring design changes, trade-offs, and iterations are kept in an information preparation environment where such instabilities can be managed and controlled. Stable information is then drawn onto the shop floor just when it is needed to drive shop processes. This is accomplished by managing and controlling the engineering information through a network of precedence controlled activities in a product realization *process*. Thus, product functionality can be maintained while manufacturing and marketing concerns are addressed. Data consistency is guaranteed through the management of the data generating process. Post-activity editing is not allowed and design iterations are controlled from a broad perspective of product development goals.

1. *INTRODUCTION*

In the approach of the 1980's to computer aided product realization, a common database is placed at the core and activities read from and write to the database [1]. In this structure, activities are carried out in an unspecified sequence. Thus, the coordination and control of the product design is established through management committees, procedures, job definitions, and audits on the design data. These control structures make the overall process centralized and sequential as naturally distributed activities are coalesced into a single stream.

A common database centered, sequential, product realization process is adequate for products with long life cycles. However, in the electronics market, the demand for flexible products based on the newest technologies causes a high rate of change activity and short product life cycles [2]. These factors require a product realization process where change is the way of life. Such a process must also support the iterative nature of design where each pass through the development life cycle produces more understanding of the customer requirements and the product functionality. Additionally, each pass must allow for the incremental addition of manufacturing information. In such an environment where many changes must be supported, the centralized sequential process is complex, slow, and costly. Hence, ad hoc methods have arisen to circumvent it.

In the past few years, we have worked on devising a radically different form of computer integrated product realization. The new form is a distributed information-flow system with a control network logically separate from the data flow network. In such a system, parallel processing and change management is supported and controlled through a precedence hierarchy of information. Activities are decoupled into independent streams, to the extent possible. Thus, the effects of changes are localized to subprocesses

NATO ASI Series, Vol. F53
Advanced Information Technologies for Industrial Material Flow Systems
Edited by Shimon Y. Nof, Colin L. Moodie

in contrast to a blind broadcasting of change information. Additionally, the precedence ordering of tasks supports multiple passes through the development life cycle where product definition and manufacturing information is incrementally added and validated. In the resulting second generation computer integrated system, a design (information preparation) environment feeds stable information to shop floor (execution) processes.

In this paper, the concepts that underlie the new form are developed. In Section 2, the traditional organizational structures are reviewed in terms of the product life cycle. Hierarchical and lateral network control models are compared. It is argued that production is achieved through lateral networks. In Section 3, information generation processes that facilitate lateral networks are described. The concepts that underlie the second generation computer integrated system are stated and explained. Data management issues are also addressed. In Section 4, an example of the information flow in flexible manufacturing of electronic equipment assembly is given. Finally, in Section 5, organizational changes required to implement these concepts are suggested. All these issues come under our term, *Information Dynamics*.

2. *ORGANIZATIONAL STRUCTURES*

2.1 *Corporate Divisions That Align With the Product Life Cycle*

The product life cycle consists of customers' needs and expectations being interpreted into requirements that are then translated into functional specifications that are then transformed into designs and implementations.

NEEDS --> REQ'MTS --> FUNC. SPECS --> DESIGN -->IMPLEMENTATIONS

The organizational structure of corporations has evolved to support the life cycle approach to product realization. Systems planners, marketing, product development, product design, documentation, manufacturing engineering, process planning, testing, materials management, shop floor control, sales, and maintenance are some of the names associated with traditional divisions in a corporation.

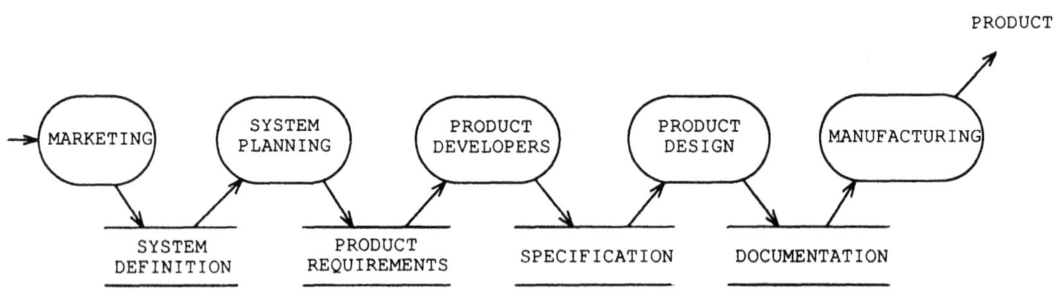

Each of these divisions has a collection of activities that constitute the role of that division.

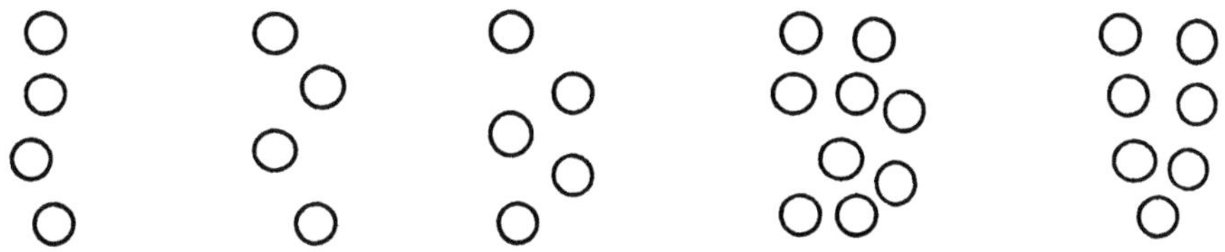

Classically, each division has defined interfaces to the neighboring divisions in the flow of product realization. These interfaces are often defined in terms of a knowledge store, a set of documents, drawings, or volumes of text description and referenced standards.

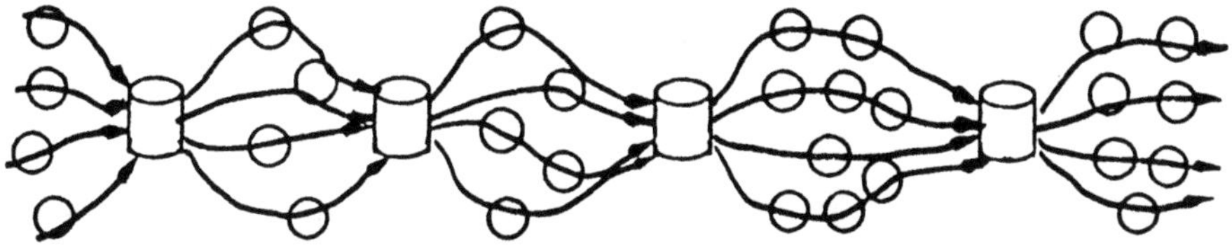

A top-down definition of production follows the organization chart. The knowledge store at the boundary between divisions becomes the output of that division. Once that store is written, the work of that division is complete. Contracts are specified, commitments are met, and rewards are given based on that writing. The alignment of corporate divisions with stages in the product life cycle supports the hierarchical arrangement of corporations and adds a supposed dimension of control.

In this sense, hierarchical means authoritarian control through decomposition of work. Additionally, policies are set and resources are distributed through the organizational hierarchy. We will show that the decomposition of work that follows the organizational hierarchy is often inadequate to cause products to be effectively realized.

2.2 *Network Based Production*

The corporation can be thought of as a cloth where the threads running the length of the bolt (the warp) are analogous to the divisions within the corporation. The cross threads are called the weave. They are the essence (the life) of the cloth since they contain the color and the character. These threads are analogous to the unofficial networks in a corporation where people from various divisions work together to achieve the objectives of the corporation. In this section, we will describe the unofficial networks that function across divisions in the corporation. In subsequent sections we will show how these networks instantiate the process by which products are realized.

A half step towards network based production is the project (matrix) organization. The project or matrix organization has grown out of a movement towards decentralization. By the mid-1920's, organizations had grown in complexity and in the number of products being manufactured.

Hence, product oriented divisions were created with the associated redundancies of duplicated capabilities. The ever growing number of products, specialists, and specialized facilities caused resource scarcity and coordination problems. The answer to this condition was the introduction of the matrix organization.

In the matrix form of organization, the major portion of the firm is organized into functional departments. Superimposed on the functional organization is a Project (Program) Manager with staff assigned to the project from the functional departments. Once the project is completed, the staff return to their respective departments to be reassigned to the next project. The project manager has no direct authority over the staff but must get the job done by *lateral relationships* that replace authority and power with negotiating, trading, and compromise.

In the matrix organization, the lateral relationships are exercised between the project manager and the conventional department management. The vertical, superior-subordinate relationship is shifted to lateral (horizontal) relationships. This emphasizes the behavior of managers with different functions but interdependent goals. These consist of trading resources within the respective departments, securing stability of operations through a negotiated scheduling of inputs and outputs, and developing open communications to avoid interactions up and down the authority structure.

Lateral networks are the full step to network based production. Lateral networks in production differ from matrix management structures outlined above in that they operate at the worker and lower manager levels and are usually informal. However, even though they are unofficial, they are essential.

In a Ph.D dissertation [3] by William Stephans of the Department of Sociology at the University of Kansas, the relationships in and organization of routine production in enterprises were studied. Two pervasive assumptions about "capitalist production" were reviewed and critiqued:

- Top down production--Production is generated through a set of hierarchical relations, with the beginning point at the top of the hierarchy. From this point specific directions and coordination efforts move down the hierarchy until they reach the direct producers at the lowest levels.

- Managerial indispensability--Managers, exercising mental activities at various levels in the hierarchy, are necessary for the completion of any production process.

The consequence of these assumptions is that when production fails "there is an all out attempt to 'fix management.'" This has only served to mask fundamental problems facing industry today.

Stephans argues that hierarchical organizations promote domination. If domination produces obedience rather than product, then some other process must be responsible for the production of goods and services. Stephens hypothesizes that production is achieved through extensive social relations in a network that extends *across* the divisions of a corporation. "these lateral relations often take the form of unofficial organizations within the formal structure of the enterprise." Such organizations are "complete with elaborate social structures and normative regulations...(that) include transactions, information flows, exchange, conflict, and coercion."

Stephans points out that Karl Marx spoke of the role of managers and foreman in the command of the labor process "in the name of capital". Furthermore, "...all labour in which many individuals co-operate necessarily requires a commanding will to co-ordinate and unify...the total activity of the workshop, much as that of the orchestra conductor." Marx assumed that production cannot take place without the presence of the commanding will. The experience of many of us is counter to this assumption, at least as we observe today's production culture. But, it is worth adding, Marx was emphasizing the capitalist-worker relationship whereas we are interested in the worker-worker relationship. That is, we distinguish between the relationship of management and control *to* production from the relationship of managing and control *in* production.

In the book by Majchrzak, et al., [4] the authors note that the introduction of CAD technology supports cooperation within organizations in that the interdependence of drafters, design engineers, and manufacturing engineers is encouraged. With CAD, input from all is now feasible through increased access and communication. Hence, concurrently engineered products (i.e., products engineered with concurrently considered design and manufacturing issues,) can be developed to meet the higher quality expected by customers.

Although CAD allows the parties involved to view the various aspects of a design at will, the engineers are left to other devices to bring agreement to the activity. Hence new skills are required for problem solving. There must be a change from independently creating alternatives to selecting the best solution from a broad (concurrently engineered) perspective. This requires negotiation that must be formalized by the mechanics of the communication process. Majchrzak, et al., note further that a common conceptual basis across department managers is needed for the decision making. This formalization is expressed as an increased concern for the accuracy of data and a systemization of procedures by which work is performed. In short, an integration pattern is needed: integration must be achieved horizontally (lateral networks), as well as vertically.

In summary, from Majchrcak, et al., computer aided processes can be expected to bring the following changes in organizational structures:

- Increased communication and formalization will bring integrative and coordination mechanisms among the units of a firm.

- Increased use of integrative mechanisms will decrease the need to keep organizational units differentiated. For example, design engineering and manufacturing could be combined.

- Increased integration will encourage decentralization of decision making. That is, to ensure optimal designs, the design decisions can be made by those at the level most knowledgeable about the design process. Since managers can be assured that design decisions are made from a broad perspective, they will make fewer decisions themselves causing a decentralization of the decision making authority.

Stephans points out that while lateral relations operate within a single hierarchical level, other levels actively participate in the construction and maintenance of such lateral networks. In short, a manager expects the workers to cooperate across departments and to do "whatever it is that you are supposed to do, and do it fast." That is, in much of today's industrial environment, lateral relations are not only encouraged but they are also expected.

In building computer integrated networks for product realization, the distinction must be made between *formally* rationalized working relations and *substantively* rationalized working relations. Stephans suggests developing working relationships from the actual work (the substance) rather than imposing a controlled set of relations (the form) on the work. He gives examples where those who impose form are in contradiction to their own best interests. Lateral networks are naturally derived from the work and are therefore substantively rational.

Following Stephans we assert:

> Real organizational behavior is evidenced by the lateral networks. These networks are informal and unofficial. Hence, organizations are not the end result of some social process in a static sense, but rather, organizations are in the process of becoming.

Lateral networks support the concept of Total Quality Control (TQC). Ishikawa [5] states that quality is realized when the product or service conforms to the expectations of the customer. Implementation of a process and corporate procedures to bring about TQC requires recognizing the following:

- The ideal state of TQC is when control no longer calls for inspection. Inspection as an after-the-fact occurrence is punitive in character. Quality control is an in-process activity that calls for human responsibility.

- The basic notion of control is the prevention of recurrence of errors. This requires the discovery of the causes of the errors and the relationship of these causes to the result.

- In TQC, the next task in the process is the customer. This requires a definition of the tasks and their relationships (process) to identify the customer so that one may determine their expectations (process goals).

- Expectations must be translated into true quality characteristics. That is, TQC uses the functions and capabilities of products and services expressed in understandable, measurable, and testable forms.

Ishikawa introduces a "fish-bone diagram" that relates defects to possible causes. (Note, as will be discussed later, this is similar to defining the process by which an output is achieved.) The diagram of cause/effect relationships helps to identify the actors involved in the process. These actors sit together in a quality circle to discover the source of errors and to agree to a modification of the process to prevent the recurrence of errors. The concept of quality circles is another expression of interdepartment activity that draws on the lateral networks.

In the following sections, we will show ways that *Information Dynamics* and computer integrated processes can be structured and used to support the naturally occurring lateral networks within an organically grown organization.

3. *CONCEPTS FOR PROCESS BASED PRODUCT REALIZATION*

As noted earlier, in building a second generation computer integrated product realization system, several major issues must be addressed:

- The demand for customized products in a short life cycle requires rapid throughput in design and manufacturing.

- The demand for high quality products requires that design and manufacturing errors be eliminated before any products are shipped and that high quality be maintained in the face of changes.

These factors require a product realization environment that has not only a fast response time but also one that maintains data integrity while allowing the concurrent consideration of design and manufacturing engineering issues.

A fast response time requires that we increase our effectiveness by supporting the way people really work. That is, our experience tells us that the rigid controls in a sequential hierarchically organized process causes people to work around these mechanisms. Because of such "end-runs", the real process is ad hoc and unofficial and thus uncontrolled. In short, we must recognize and rationalize the process embedded in the lateral networks.

To maintain data integrity and support concurrent engineering, we must recognize the limitations of the concept of a common repository of information. It is important to understand that this concept arises from a data management world where the data to be accessed is fixed and supreme. In the product realization world, the *goal* of a set of activities is supreme and the data is generated to achieve the goal. This turns the emphasis around so that the process is managed and controlled rather than the data. That is, the data is the output of the process and is subservient to the goal. The result of this view is that the introduction of a design or manufacturing change can be efficiently managed through the management of the process.

The differences between goal centered and data centered views of the world can be illustrated by the following examples. In a software product, an executable process is a computer program. Inputs of data and operation sequences are given and results are data outputs. The software product is described as data such as the program source code, fixed data and parameters, structure charts, etc. In a hardware product, an executable product (a car) takes inputs (passengers and destinations) and provides outputs. The hardware product is described in its data world which is given to a shop floor which "compiles" the executable product. In both cases, a goal centered activity sits outside of these data centered worlds. That activity is the information preparation portion of the product realization process which takes functional requirements, performance constraints, resources, technologies and ways of working, and uses all of these to produce the data representations.

In a goal centered view, the information preparation process serves the execution process. The execution process places demands on the information process so that when the information process meets these demands it has met its goal. In this view, the roles of customer and supplier are clearly delineated. Metrics can be devised to monitor the quality of the process since the output can be well defined. In our work, this view is used to partition the product realization process. Furthermore, it is proposed that the information generation process should serve as the basis for managing and controlling product information quality and consistency.

To address all these concerns, we introduce the following concepts:

- The generation of information is *precedence controlled*. That is, the process of information generation and maintenance is laid out as related tasks that have been strung together to achieve a particular output. The effects of changes are controlled through the selective reactivation of only those tasks affected by the change.

- Multiple sequences are set up as *streams of information flows* to allow parallel development to the extent possible.

- An *information generation process* is used to provide a minimally complex data model. Data consistency is guaranteed in that the generation of each data item is the result of a specific task and only that task.

- In the process, *control flow is separated from data flow* for the management of iterations and changes. Thus, the iterative nature of design is supported straightforwardly without a loss of efficiency.

- Product realization is partitioned into two parts: An *information preparation environment* which feeds *stable information* to the shop floor processes in an *execution environment*.

In the following sections, these concepts are explained.

3.1 *Precedence Controlled Activities*

Goal centered product realization is managed through the establishment of the process that produces the outputs required to achieve the goal. The network of activities that make up the process are related through temporal and data dependencies. There is an additional relationship that conveys a sense of authority because of what has been stated as the result of higher level and upstream activity. These two aspects of process relationships, goal centered perspective and the authority of preceding activity, we lump into one concept, *precedence control*. The term is borrowed from the legal sense of precedence. Decisions made at some past time carry weight on the present decision. In product realization, this comes from the life cycle where corporate objectives and market goals have been transformed into product functional requirements. These requirements are transformed into a product architecture, design, and finally implementation. At every step, the preceding concern cannot be violated. For example, designs must meet the market expectations of functionality, and manufacturing must preserve this functionality regardless of how nicely the product might otherwise be made. Consequently, preceding decisions cannot be violated by lower level and down stream activity.

The precedence of decisions contains the sense of overall objectives. There is an authority in the high level view that cannot be violated by local decisions. Hence, there is a (classic) tension between authority exercised locally (subsidiarity) and authority contained in the high level view. In a process that is precedence controlled, the broad perspective must be captured in the activities and outputs of the tasks. This means that the decomposition of a task into subtasks is more than strict reductionism. That is, a task can be divided into subtasks but aspects of the organic whole must be recognized and maintained during the decomposition. In our experience, we manage this decomposition through two attributes:

- The subtasks are defined as functions that produce certain outputs given specified inputs. Decisions internal to the tasks are locally determined.

- The operation of subtasks are subjected to *constraints* that are stated to reflect the objectives of the organic whole as it exists at the higher levels.

Thus, the two senses of precedence control are introduced. The individual tasks that have the authority to output a specific item may proceed subject to constraints. If a task cannot run to completion, negotiation is required, but, under no circumstances, can a task output an item outside of its authority or violate a constraint.

3.2 *Streams of Information Flows*

In the traditional view of the product life cycle, activities are coalesced onto a single stream where the common repository of data is passed from division to division along that stream. Our experience is that such a centralized, sequential process is clogged. Controlled parallel processing can be introduced through the definition of *streams* of information flows. This means that multiple concerns within the product realization process can be identified and the tasks associated with these concerns can be divided into streams of separate concerns.

This concept arises from the author having worked as a part of lateral networks in AT&T. A network of knowledgeable people becomes known to the engineering level where tasks can be accomplished through interactions *across* organizational boundaries. A reflective analysis of these interactions led to the realization that such networks are organized along subjects such as component selection and procurement, assembly, and testing. Consequently, a natural division of the product realization process can be devised so that parallel processing can occur through streams of information flows across divisions.

We have made some effort to quantify the act of creating streams with only limited success [6]. Our experience has been that these streams are best identified by forming teams of people from multiple divisions with the assignment of "baselining" the current process used to achieve a specific goal. Concerns can be naturally separated when enhancements to the baselined process are considered. More work in this area is needed.

3.3 *Information Generation Processes*

The costs of generating and maintaining consistent data is a major concern. Earlier we have pointed out that a goal centered approach to product realization causes us to set down precedence related and streamed processes that can be managed and controlled. In this section, we will state that such processes also produce well formed information models as defined by Flavin [7]. That is, "objects" are defined that are "all that are needed/only those that are needed" and hence the set of objects is "minimally complex."

The process chart (domain model) relates object types to object types, not relationships to relationships or attributes to attributes. Hence, in the Flavin sense, the information model is a first-order system in that it contains no compounded structures. Additionally, the process chart defines each component in terms of the policies, laws, rules, and conventions that govern the system being modeled. Thus, all component definitions are mutually consistent and free from contradictions. Hence, the information model is *fundamentally consistent*. This consistency is maintained in the product realization process through precedence control because each data item is the result of a specific task and only that task.

3.4 *Separation of Control Flow From Data Flow*

It has been proposed that the throughput, observability, and control of the product realization process can be enhanced through the recognition of a data-flow like structure with a logically separate control network and data flow network [8]. The precedence control aspects of the process are defined in the network of tasks that generate the flow of data. Adding precedence to data flow forms a "data generation network". In a perfect world, this would be activated by the presence of data on the inputs to each task much like a data-flow machine. In the real world, resources have to be allocated to each task before execution can occur. Additionally, many tasks cannot run to completion. For example, a particular design will fail a verification test, perhaps by failing to meet a particular constraint. The negotiation and arbitration of the relaxation of a constraint or a change in a particular input item must take place outside of the data generation network. This is where the concept of a control network comes into play.

Again, this concept arises from observing the way people really work. Individuals interact and meetings are called to negotiate and arbitrate particular concerns. Experienced people know where to go in their lateral network of co-workers to most effectively redo a particular data item or constraint. This led to the proposal of a control network much like the separation of the telephone network into a call setup and a transmission network. In such a separation, the response of the data-generation network can be readily monitored. Control information, such as the decision to redo a particular task, can be quickly sent to the people responsible for the affected task. The ability for particular process configurations to respond to the exceptions to task completions can also be monitored and compared.

3.5 *Information Environment Feeding Execution Systems*

This section deals with the relationship of the product information preparation environment to an execution environment that supports shop floor activities. The rationale for the partitioning of activities is based on separating unstable from stable information. This responds to the fact that a major source of manufacturing difficulties arises from unstable information causing shop floor disruptions.

An information preparation environment is needed where both development and manufacturing engineers work to deal with product design and manufacturing changes that affect product information. Concurrent engineering is employed so that the output of the information environment is a jointly approved product definition/manufacturing package. Thus, the classical shop floor management techniques can be used since only stable information is loaded on material selection and assembly processes.

We have said we want a goal centered product realization process which is demand driven. That is, it starts with a given objective and defines a process by which that objective is achieved. While the manufacturing shop floor operations may be driven by design data, the shop floor operations have goals which form processes which accept design data as inputs. The shop floor operations place a demand on the design (information preparation) environment. The goal of the design information preparation environment is to meet that demand with *stable information*.

The nature and time distribution (frequency) of changes depends upon the type of product realization process used. In a traditional sequential process products are defined and the design is "frozen", and sent to manufacturing engineers, who create their own changes, then "vault" the drawings for manufacturing. This official information is used to drive a volume manufacturing shop. Since the customer delivery date is set by the market, the traditional process is started early to meet that date. Consequently, design decisions are made with a premature understanding of the product. The resulting design errors are superimposed on the modifications required for new product features. This all produces a change distribution with the need for changes arising during "volume manufacturing". Consequently, products coming out of volume manufacturing shops are patched at great expense.

In an integrated flexible system, products and manufacturing processing steps are defined concurrently, and the resulting information package is used to drive a highly responsive, flexible manufacturing system. The time distribution of changes is somewhat different in that the customer ship date can be met by having the product information on the shop floor for only a short time. Additionally, the extra time can be used for maturing the product definition and validating the design with prototypes *before* releasing the information to the shop floor. The information is stable with reference to the manufacturing interval.

4. *AN APPLICATION TO ASSEMBLY*

The concepts described above are being applied to the assembly of electronic circuit packs in a flexible manufacturing shop. This application builds on the existence of a lateral network of development and manufacturing engineers. It also builds on a decade of experience with procedures that require early review of proposed designs for manufacturability and testability. The procedures include joint approval by manufacturing and development engineers of a documentation package for circuit pack assembly.

The steps in that application are as follows:

1). Identify a strategic goal.

- Minimize shop floor disruptions through improving the accuracy and timeliness of information.

2). Establish a specific goal.

- Derive the information for circuit pack assembly operations.

3). Establish a multi-division cross organization team.

- The team should consist of shop floor engineers and foremen, manufacturing documentation staff, and product development engineers.

4). Develop an information based understanding of the process by which the goal is achieved today.

- Develop a baseline description of the tasks, the data and the control dependencies in the process.

5). Clean up the shortcomings.

- For example, add partitioning of the circuit pack at layout time to enhance machine insertion.

6). Partition and implement the process.

- Obtain commitments from the functional organizations. Establish the resources, procedures, and the training required for the process.

7). Execute the process.

- Monitor the execution and apply on-going improvements.

In this work, a data flow representation of the process was first generated. These diagrams showed the data dependencies among tasks and defined all the tasks that must be performed to output the desired shop instructions. Such diagrams are fine for the perfect instance where a single pass through a process produces the desired output. Thus, the creation of a data flow representation is a good starting point for process representation. This view can be considered a statement of the process data requirements.

A high level data flow view is given in Figure 1. The circles contain actions which generate data items (named on data flow arcs) and data files (named between horizontal lines). Actions external to this view of the process are shown in rectangular boxes.

Once the process data flow requirements are understood, control relations can be addressed. Some of the control flow aspects are presented in Figure 2. Here, the graphical representation is an adaptation of the System Specification Language in the Distributed Computing Design System [9]. In this diagram, the solid lines represent the flow of the *behavior* of the process. The dashed lines show data flows between tasks (actions) described in the rectangular boxes. Data items are labeled on the dashed lines while data files are shown inside flattened ovals. The various join symbols represent the concurrent actions that can be managed in a distributed working environment. For example, the "and" symbol represents multiple activities that can take place in parallel. The "plus" symbol represents the choice of one or another activity. Iterations are represented by the "at" symbol. These symbols are in nodes that act as synchronization points before and after branching of behavior. Note, the temporal aspects of control are not included in this view.

Once the requirements of the process are understood by definition of the data flow and control flow dependencies, the next step is to design the process by partitioning it to various parts of an organization and assigning resources to the functions. Activation and termination conditions must be agreed to and ownership (commitments) established. (The temporal and ownership aspects can be represented in the traditional forms of PERT and Gantt charts.) The process is then implemented and execution can begin. Thus, both process control and on-going process improvement can be achieved provided the organization structures are in place to support process-based product realization. The organizational issues are addressed in the following section.

5. *ORGANIZATIONAL CHANGES*

The traditional operation of a corporation follows the sequential execution of tasks outlined in Section 2. This model of the organization is built on two assumptions:

- The process by which a product is realized is fully known and fixed. It is a linear sequential pass through the phases of a life cycle where the output of each preceding phase is the input to the next phase and the *control* of the data is passed along with the data.

- Production is achieved by management generating a set of hierarchical relations, with the beginning point at the top of the hierarchy. From this point specific directions and coordination efforts move down the organizational levels until they are executed by the producers at the lowest levels.

We have argued here that both of these assumptions are no longer valid. In the following sections, organizations that support process-based product realization and lateral networks of cooperating individuals are suggested. Furthermore, we note that such organizations are in keeping with the principle of Total Quality Control as described in Section 2.

5.1 *Organizing Process-Based Teams*

It has been said recently that the two most significant organizational weaknesses in the American corporation are the failure to encourage teamwork and excessive layers of management. The entire organization should be structured to create team identity, team score keeping, team problem solving, and team victory. We suggest the formation of teams based on the process.

Numerous corporations have dealt with the issue of integration and concurrent engineering by forming teams that contain staff from all the disciplines involved. Sometimes these teams are collected into separate divisions and separate buildings and given clear charters for the responsibility of a particular product. This is an improvement over the matrix management approach mentioned in Section 2.2 in that rewards and control are clearly defined. However, teams so organized may not be flexible enough to meet the dynamics of the product realization process.

Here, it is proposed that implementation teams be formed based on the *process* that will be used to realize the product. Once the requirements for the process have been defined and the process designed, the resources required to implement the process can be identified. These resources include staff members who will be assigned to specific tasks. These tasks are activated as needed by the control mechanism in the Control Network of Section 3. The thought here borrows from the methods for management of a flexible manufacturing system. Staff members are "dynamically" allocated and virtual organizations exist similar to virtual production lines in a flexible shop floor. Load balancing and resource contention problems will have to be managed but the flexibility that results may be worth the effort. At least, the conventional empires will be deemphasized.

CAD and CAM systems have grown up in support of the linear sequential model of organizations. Particular systems have been built to service particular divisions within the corporation. Ponderous organizations have been established in support of the sequentially phased realization paradigm. As we move to process-based product realization, the roles of such CAD and CAM development organizations will have to change. Furthermore, CAD and CAM systems that were developed with embedded processes (hard-wired) will become anachronisms. These systems often contain the vendors' views of the process which haven't always agreed with the process in the intended application. Consequently, many of the CAD and CAM systems provided omnipotent editors in them. These editors make it impossible to control the influence of changes except by post activity audits. Hence, computer aided *tools* must be devised to support flexible processes. These tools must support distributed control and distributed data. Knowledge integration environments will have to be devised to augment the use of flexible and distributed tools in process-based product realization.

5.2 *Supporting vs. Controlling Managers*

It is becoming recognized that the era of the controlling manager is over. The image of the organization as the organization chart with the top levels providing the control is one of domination and heavy-handed action. We propose a different image of the organization that is an inversion of the organization chart: It looks more like a pedestal table where the management sits beneath the working layer in support of the action on the "table-top". The image in this case is one of being lifted up and supported. This view is at least two-thousand years old. It is also the view supported in the recent book by Peters [10].

In the inverted view, the controlling manager is replaced by the supporting manager. The purpose of the corporation, e.g.,to make and sell products, is the primary reason for existence. Management supports this purpose by setting the objectives in accordance with the corporate goals. Resources are then allocated for the transformation of these goals into objectives, plans, and finally actions appropriate for each level.

Managers should monitor the *processes* of their staff and not the detailed technical decisions. Following the proposals in this paper, the control should be in the process and not the data. As noted in Section 2.2, the studies quoted by Majchrzak, et al., [4], indicate that managers are most concerned that design decisions are made from a broad enough perspective. Hence, we suggest that managers approve the process used by their staff and leave the action-level decisions to those at the level most knowledgeable about the action item. This approach should greatly enhance the productivity of an organization for a number of reasons:

- The focus is placed on the process that can be controlled, measured, and managed. Every working element is required to define their actions in terms of what they are expected to accomplish and their interfaces to both upstream and downstream functions.

- Every working element is motivated by having the responsibilities for their actions and the autonomy to make their own decisions within their sphere of authority. A closer match between control (responsibility) and knowledge will exist so that one can effectively control what one knows well.

- Management is relieved from being expected to review and make action (e.g., design) decisions. Instead, they are given the freedom to concentrate on setting goals and implementation plans in terms of their understanding of the technology issues.

These changes are in keeping with the spirit of Total Quality Control [5]. The definition and emphasis on process allows quality to be "inspected in" by everyone in the corporation. The identification of customers as the next one in the line of the process is clearly supported. The finding of errors and exceptions by those closest to the action (quality circles) is also supported. The sense of management as supportive and facilitating might even make Western corporations "good places to work" by "establishing true 'industrial democracy'".

6. *CONCLUSIONS*

To respond to today's market for customized products yet minimize shop floor disruptions, we must change how our corporations operate. The traditional view of linearly sequenced activity moving through a set of centrally controlled phase-based divisions must be replaced by organizations that support distributed process-based product realization. We suggest that the naturally occurring lateral networks be recognized and supported by policies and rewards. Furthermore, we propose that if people really do work effectively in lateral networks, then support these networks by computer aids that promote distributed but coordinated activity and decision making.

In a cross-organization integrated design environment instabilities due to product design and manufacturing changes that affect product and assembly information (including the specification of parts) are dealt with by both development and manufacturing engineers. That is, the concept of concurrent engineering is used and the output of the information environment is a jointly approved product definition/manufacturing package. Manufacturing engineering activities are partitioned to deal with design changes in the information environment. Shop floor disruptions are minimized by loading only stable information on the shop floor activities.

The effects of such a definition of activity on present organizational structures will be extensive. Organizational structures that were set up to cause production through a hierarchical decomposition of tasks will become passe. That is, they will be replaced by structures that support laterally networked integration. Additionally, organizations that were chartered to develop interfacing CAD and CAM "systems" will have to be reoriented to support integrated computer aided product realization. Specifically, the data centered approach to product realization must be replaced by goal centered, process-based realization. These changes must be made if we are to move from systems that interface to systems that integrate.

7. *ACKNOWLEDGEMENT*

The ideas presented in this paper have been formed through interaction with many people in AT&T. However, I would like to especially acknowledge the members of the Advanced Design Technology Group, both past and present. I would also like to acknowledge the people at Purdue University, particularly Prof. James Solberg and Prof. Fredinand Leimkuhler, who have given me the opportunity to test these ideas in the highly challenging and stimulating course we call *Information Dynamics*.

REFERENCES

1. Mike Kutcher, et al.,"Data-Driven Automation", *IEEE Spectrum*, May 1983.

2. David Ben-Arieh, Charles A. Fritsch, and Kostia Mandel, "Competitive Product Realization in Today's Electronics Industries," *Industrial Engineering*, Vol. 18, No. 2, Feb. 1986, pp. 34-42.

3. William Richard Stephans, "Theory L: An Investigation of Lateral Relations in Production," Ph.D Thesis, University of Kansas, January 1985.

4. Ann Majchrzak, Tien-Chien Chang, W. Barfield, R. Eberts, and G. Salvendy, *Human Aspects of Computer-Aided Design*, Taylor and Francis Ltd, Philadelphia PA, 1987.

5. K. Ishikawa, *What is Total Quality Control? The Japanese Way* Prentice-Hall, Inc., Englewood Cliffs NJ, 1985.

6. R. Venugopal, D. Ben-Arieh, C. A. Fritsch, and K. Mandel, "Strategy for Change Management in Computer Integrated Manufacturing", *IEEE 1986 International Conference on Systems, Man and Cybernetics*, October, 1986, Atlanta, GA.

7. Matt Flavin, *Fundamental Concepts of Information Modeling*, Yourdon Press, New York, 1981, pg.96 ff.

8. C. A. Fritsch, "A Distributed Data-Flow Structure for Computer Integrated Manufacturing--Some Concepts", AT&T Bell Laboratories Internal Report, May, 16, 1985.

9. Mack Alford, "SREM at the Age of Eight: The Distributed Computing Design System", *Computer*, vol. 18, No.4, April 1985.

10. Tom Peters, *Thriving on Chaos*, *Handbook for a Management Revolution*, Alfred A. Knopf, Inc., New York, 1987.

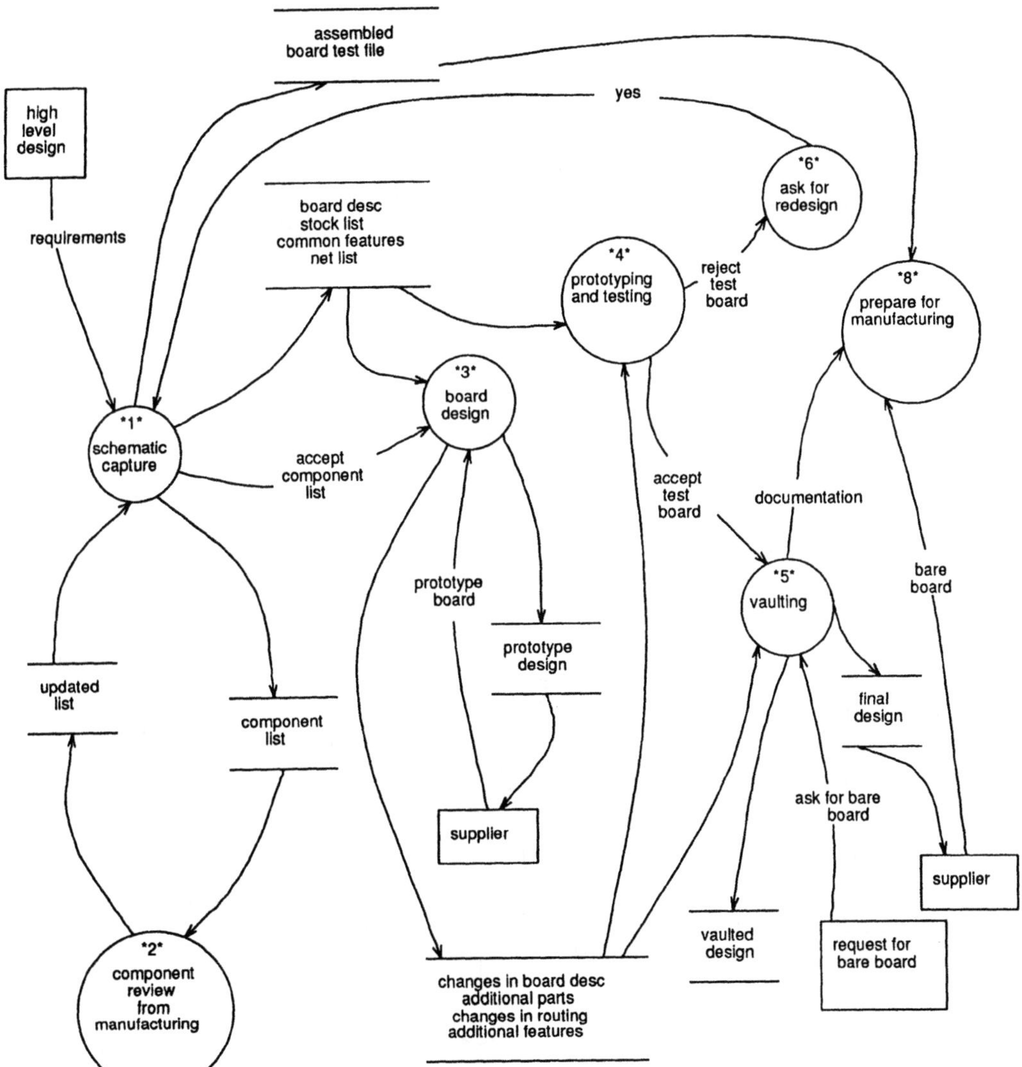

Figure 1. Data Flow View of Electronic Circuit Pack Assembly.

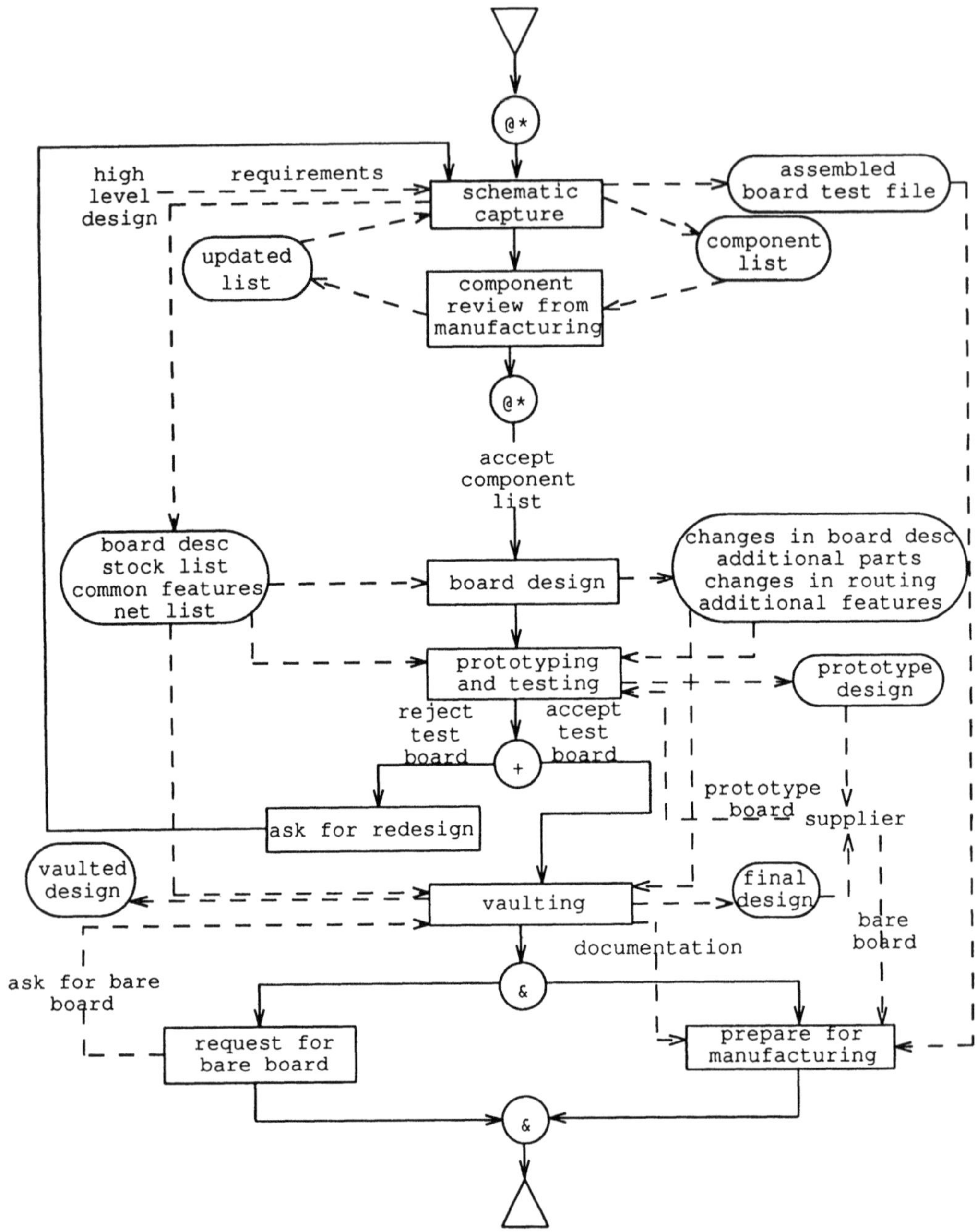

Figure 2. Behavioral View of Electronic Circuit Pack Assembly Process.

Some Interactions of Information and Control in Integrated Automation Systems

Edward J. Barkmeyer
U.S. National Bureau of Standards
Bldg 220 Room A127
Gaithersburg, MD 20899

The ready availability of inexpensive and standard communications hardware in the late 1980s, coupled with sufficient standardization of intermediate-layer protocols to make reliable machine-to-machine communication through any combination of standard devices a reality, has laid the groundwork for integrated automation on a grand scale. The contribution of the Manufacturing Automation Protocols (MAP) effort [MAP87], thus far, has been to speed delivery of this powerful communications capability to the industrial community. We must realize, however, that the delivery of the capability has now outstripped our ability to use it. On the factory floor, we have no devices which are prepared to control automation on a grand scale, and even worse, we have few controllers which are capable of contributing their local automation to any kind of integrated whole. The stumbling block is the lack of common languages or even a common experience base in which to communicate about automation tasks. We are now in the midst of a flurry of intellectual activity, on both sides of the Atlantic, developing languages, architectures, control techniques, data systems, communications systems and ultimately controllers to meet this challenge. In such a burst of activity, it is to be expected that experts in individual areas will hasten to solve the problems with which they are most conversant, and that certain interdisciplinary problems will fail to get timely attention. It is the purpose of this paper to draw attention to several such issues now, in the hope that they may get due consideration as the process advances. The issues discussed here are:

- separation of data flow from control flow,
- the impact of perfect communication on real-time control,
- consistency of data with physical reality,
- the significance of standard data models,
- the identification of shared and private data.

What all of the issues have in common is that their solution requires common attention and joint consensus by the experts in communications, data systems and control.

The presentation of these issues, and the recommendations for approach to solutions, where they appear, draw on seven years of experience in integrated automation in the

NATO ASI Series, Vol. F53
Advanced Information Technologies for Industrial Material Flow Systems
Edited by Shimon Y. Nof, Colin L. Moodie

NBS Automated Manufacturing Research Facility (AMRF) [Nanz84], a laboratory for joint government/industry/university research in flexible automation and in-process measurement supported by the Bureau of Standards and the Navy Manufacturing Technology Program.

Issue 1: Separation of Data Flow from Control Flow

Before 1984, industrial organizations who were linking control systems together were limited by the available communications technologies of the time, which were characteristically point-to-point. Thus any station which needed to communicate with more than one other station needed multiple physical links and interfaces to accomplish the communication. As a consequence, for reasons of efficiency and reliability, it became common practice, particularly for direct equipment controllers, to make a single communications interface provide the entire spectrum of external services. In particular, it became the norm to integrate the commands of a supervisory controller with the data needed to execute them, because the data had to come through the same physical channel anyway.

In the last five years, it has become inexpensive to use physical interfaces which permit a broad spectrum of logical connections through a single physical interface, e.g. IEEE 802 networks [IEEE83a,IEEE83b,IEEE84,IEEE85]. More importantly, compliments of the MAP-enforced acceptance of the Open Systems Interconnection model [ISO84], we now have software to support distinction between physical links and service "sessions" in such a way that there is no required relationship between them. Since the original motivation for integration of data and control functions has become obsolete, it is time to re-examine this integration from the pure design point-of-view.

The primary advantage of integrating control functions with the related data is that it minimizes the complexity of the subordinate process: the subordinate is not "burdened" with two logical interfaces, one for commands and the other for supporting information, like machine control programs. This allows the use of relatively simple systems for equipment control, but it has two corresponding disadvantages. First, it maximizes complexity of the supervisory system, because the supervisory system has to handle, package and communicate all of the data necessary for the functions of all of its subordinates, in addition to all of the data necessary for its own functions. Secondly, it enforces a single control architecture, in which commands and data must come together, and therefore presumably "down from the top" in a fixed hierarchy. Among other things, it makes necessary the continuing revision of control interfaces and supervisory control programs to support equipment controllers which are better designed for integration. The rigid simplicity of the interface prevents the equipment controller from operating on information units not envisioned by its supervisor and from making other choices of data repository, even local ones, not supported by the supervisory interface.

Choosing to separate data from control is an application of the design principle of *separation of concerns*, and realizes the primary benefit of that principle - flexibility, the ability to make independent optimal choices for the subsystems, in this case control services and data services. It is accomplished by reducing the data present in the commands to the keys to the major information units and the parameters of the control function itself, and by providing a separate data service function and protocol in the subordinate controllers [Albu81].

Now the choice of control architectures becomes arbitrary; hierarchical, cooperative, reactive and hybrid architectures are all possible, because the method of accessing data in the controller is independent of the method of determining the functions which use it. Similarly, the choice of data services may be made optimally. The control system can take advantage of whatever local data caches and databases it may have, and can obtain or modify data from whatever sites in the integrated facility its data service may access. The external control functions still direct the use of the data, and may provide direction for early acquisition or planned usage, but the supervisor (if there is one) does not have to handle the data itself, or in many cases, even be aware of all the data which is accessed. Where the communications are still point-to-point, the physical flow of the data will be the same, whether data and control are integrated or logically distinguished. But here the engineering cost of providing the separate data service path will be more than justified by the simplification and increased longevity of supervisory control programs. And when the communications are upgraded, into a broadband facility, for example, the equipment controllers can obtain workpiece descriptions and machine control programs directly from engineering or inventory systems, and the need for copying such data down the hierarchy vanishes, *without change to the control programs*. And even when the equipment controller is upgraded to use and provide more information in its operations, the supervisory controller may still survive unchanged. The flexibility achieved by separating data from control is a mandatory step to future experimentation in integrated systems; and the resulting opportunities for improved control schemes and elimination of "middlemen" in the movement of engineering data from producer to consumer, can result in significant improvements in facility performance.

Issue 2: The impact of perfect communication on real-time control

One of the fundamental concepts in computer-based automation is that of interprocess communication, i.e. communication between separate programs engaged in different aspects of the same larger function. Forgetting all paradigms in human communication, we intuitively expect communication among computer processes to be *perfect*: we demand that every utterance produced by a sending process be heard by the designated receiver. In the automation of systems divorced from dynamic physical environments, especially those who have infrequent, weighty communications, this requirement is appropriate and can be met without unacceptable overhead. But in the automation of systems tightly coupled to rapidly changing physical environments, the consequences of this requirement are dramatic. In such an environment, a control process cannot always afford to wait for the intended recipient to get and acknowledge the messages. But, using the perfect communication model, the control process has no choice: it must at least wait for acknowledgement, and occasionally wait for recoveries of failed transmissions; otherwise information may be lost. The conventional solution to the sender wait problem, borrowed from environments unburdened by physical coupling, is *message queuing*. Message queuing, supported by some class of multiprogramming for communications, appears to allow the sender process to drop a message in a queue and go about its business, with the assurance that the message will ultimately arrive at the intended destination. Unfortunately, message queues were invented solely to insulate the sender process from the vagaries in communications timing, always assuming that the recipient process can, on average, read its mail faster than the sender can generate it. When this assumption is untrue, or untrue for a long enough period of time, messages will gradually back up, first in the receiver's inbound queue and then when it is full, in the sender's outbound queue, until at some inopportune time, the sender control process will come by with its regular message, and find that there is no place to drop it. Then, queuing notwithstanding, it will be faced with the inevitable dilemma: *wait* until there is room in the queue, or *throw the message away*. Formulated concisely, *perfect communication requires the sender to be prepared to wait*. If the sender *refuses* to wait, then some of its communications must be *imperfect*: some messages must be thrown away.

In designing control systems which are tightly coupled to real-world equipment, and which also have to communicate with others, this rule must be taken into account. Controllers must be designed either to tolerate unpredictable waits at the message enqueuing point, or to deliver all relevant information at the next messaging opportunity. Waiting is not normally a problem in issuing commands or supplying needed data to a subordinate process, but it can be a serious problem in reporting status *from* a real-time control or sensory process. If the status reports are designed to be incremental, i.e. separate messages for each interesting event, then the controller cannot deliver all relevant information at one opportunity, and it must be prepared to experience processing delays caused by communication waits. Optimistic engineers may believe that the solution is to connect such a controller only to supervisory systems which will always be able to consume the status reports of all subordinate controllers with time to spare, or conversely, to greatly limit the messaging volume permitted of subordinate controllers. Such a solution is intrinsically inflexible and closed-ended, and a controller so designed is obsolescent the day the prototype is built.

The alternative approach is the one we have always used in interfacing such systems to human operators: systems are designed to produce a *total* status report on each "cycle". In the human case, the presentation method is an array of gauges, displays, lights and alarms, but regardless of how busy he might otherwise be, when the operator examines the status, he gets the complete picture. This presentation method is chosen, because the communication is known to be *imperfect* - the operator can miss any number of changes in the state of the panel, but he can't miss anything important, because when he looks at the panel he gets the up-to-the-minute information on all aspects. This is the paradigm which should be used for designing status reports from real-time controllers. Each status report message should contain all of the information describing the subordinate controller which may be useful to the supervisor. The subordinate does not have to distinguish and report changes per se; the supervisor detects changes, in those information units which it considers important, between the status reports it actually sees [Barb82].

This approach has three side-affects. First, just as the human operator of a complex machine must be trained to recognize events from, and find relevant information in, the status display, supervisory control processes must be programmed to recognize events needing attention and find interesting information in the status report. This is not easy, but the second effect makes it worth the investment: there is now no need for any relationship between the rate of production of status reports by the subordinate controller and the rate of consumption of those reports by the supervisory controller, other than that required for responsiveness to changes in the physical environment, i.e. that required for the task at hand. Each controller processes at whatever rate it can, and the supervisory process can get as much data as it can absorb. The interface is open-ended and simultaneously upward and downward compatible. The third effect is that there is no need for deep queuing of messages from the subordinate; in fact, deep queuing only serves to slow down the communication of current information. "Double-buffering" is ideal: for the sender, one buffer is the report being transmitted and the other is the next report to go;

for the receiver, one buffer is the last report received and the other is the next arriving. The control process rewrites the "next to go" buffer on each controller cycle or event, and whenever a communications opportunity arises, the communications process copies the "next to go" into the transmission buffer. Any further queuing simply introduces out-of-date status reports into the path, delaying the arrival of current information. The definition of "communications opportunity" is a protocol issue: it can be a request from the supervisory process, or the combination of local change, acknowledgement of the previous status report and available time on the communications channel. The former protocol is incorporated in the current draft of the Manufacturing Messaging Service [EIA87], while the latter protocol has been in use in the AMRF since 1983 [Mitc84]. The MMS method is simpler, but it is slightly less efficient and it makes exception-driven control more difficult. Either protocol permits the coupling of supervisory and subordinate systems with arbitrary reporting and consumption rates, and with arbitrary communications equipment.

Issue 3: Consistency of data with physical reality

In business automation systems, a process's only source of knowledge about the outside world is in the databases available to it. When there is more than one such database, or when information in two elements of a single database is interrelated, there is the potential for problems in consistency. The universal rule is: *When a process can see two views of the world, it is vital that the overlap between those views be consistent.*

Because the only views available to a process in a business automation environment are of the available databases, this rule has been interpreted to require management and/or software services which maintain consistency between multiple databases. This interpretation is still necessary, but no longer sufficient, in the case of materials flow systems. In automated materials flow systems, much of the active data mirrors physical reality in a much more intimate way than in business systems automation. In such systems, it is entirely possible, and often the case, that a process has both access to the databases describing the physical realities of the workplace and sensory access to other measures of the same physical reality as well. Take, for example, the workstation which can detect the arrival of an automatic guided vehicle (AGV), or the robot whose clearance depends on the current location of a transport element or the position of a window. In such cases, whenever databases are used to determine or support the physical circumstances, we must provide software services which maintain consistency between the databases and the real world. That is, *the database must be kept consistent with the physical reality.*

The alternative is to deprive some controllers of one of the two views, thereby avoiding the potential inconsistency. This is sometimes the best solution, applying a reliable engineering rule-of-thumb: "Always use the best data available." Unfortunately, it is not always obvious what the "best data" is. The usual interpretation of this rule for the real-time controller is: "Don't use the database for information you detect and measure directly." And in our experience, this interpretation is frequently unsatisfactory. Consider the case in which a workstation controller detects the arrival of an AGV and the unload of the tray/palette it carries. The only part of the physical reality that the controller can detect locally is the unload. The true nature of the load, that which is on the tray/palette, its configuration, contents, identification, is contained in carefully controlled databases which are guaranteed to be consistent by a global data system implementing consistency controls. If the controller uses the tray arrival as an indicator, it has no way to associate that arrival reliably with the contents data. In other words, the best data available, the local detection of the tray arrival, is inadequate when the controller needs to know more than that it arrived! On the other hand, the database update is not something that the controller can readily detect; so in this example, we are left with the consistency problem.

There are three common methods of maintaining consistency among databases, so we will consider them as alternatives for maintaining consistency with physical reality. They are: time-stamping, two-phase commitment and scheduled release.

One of the most flexible techniques for accommodating the cooperation of separate processes acting on the same data units is the time-stamp. The idea is that the control process simultaneously issues the database update and the command for the physical change; and the database update is time-stamped. Then any retrieval of the data prior to the change will get the data preceding the change and any later retrieval will get the data matching the real-world after the change. Using the AGV unload example, if the workstation detects the arrival and starts the contents retrieval operation after the transport system starts the update operation, even if the update has not yet completed, the data system must recognize the chronological dependence and stall the retrieval until it produces the correct (i.e. consistent) answer. But, while the time-stamp is a reliable method of enforcing proper subtask serialization *within* a data system or control system, it is a very unreliable mechanism for serialization of operations *at the interface*. If the data system assigns the time-stamps to the operations when they arrive at the data system interface, so that the time-stamps are based on a common clock, it is almost impossible to avoid "race" conditions between the two client processes. The network services, the operating system services, and even the sampling algorithm used by the data service itself, are all capable of making two messages with clearly staggered start times by-the-clock appear to arrive out of order at some point in the message management process. If, on the other hand, each controller assigns the time-stamp to its own request, then global clock synchronization is necessary to guarantee that the time-stamps are meaningful. But global clock synchronization is subject to tolerances, as a result of the medium of distribution and the variation in response time in the recipient systems, and as soon as more than a few systems are involved, these tolerances may exceed those of the operations we are trying to synchronize, The bottom line is that transaction synchronization by time-stamping is difficult, at best, and some other means of consistency control must be found.

The two-phase commitment technique appears at first glance much simpler and more reliable. Using this technique, a control process simultaneously requests the data system to *commit* to the update and the subordinate or device to *commit* to the physical change. The *commitment request* essentially says: "Here's what I want you to do; don't do it yet, but tell me if you can and when you are ready." Each recipient responds "yes" or "no" and the data service locks out any further transactions until it gets the second "phase" of the request. When both recipients respond "yes", the controlling process then sends the second phase *execute* to both to effect the changes. If either responds "no", the controlling process then sends a second phase *abort* to both, and deals with the error recovery. In practice, many data systems cannot determine a priori whether to commit to an update or not, so they save the affected records, attempt to execute the update, report the results and lock out further transactions until they get the second phase of the request. If the second phase is *execute*, they then simply release the locks. But if the second phase is *abort*, they must undo or *rollback* the update they have done, before releasing the locks. The application of this method to consistency with physical reality has similar characteristics. In our example, the transport controller could conceivably direct a commitment request to the global data manager to update the tray/palette location at the same time it directs a commitment request to the AGV to unload. If they both respond "yes", the transport controller issues the *execute* to both and the workstation controller only sees the arrival after the database is updated. If the AGV can truly commit to the unload before attempting it, this method will successfully prevent consistency problems in the workstation views. But if the AGV has to *try* the unload in order to commit, the purpose of the commitment protocol is defeated outright, because the workstation, with its arrival detector, may see the physical change and initiate the contents retrieval while there is as yet no confirmation that the data system has seen the update transaction. So we must revise the approach as follows: The control process first requests the data system to commit to the update. When the data system responds "yes", then the control process commands the physical change. And when the physical change completes, the control process issues the second phase *execute* to the data system. This is a workable technique, which avoids the previous difficulty. But it does require the global data service to support two-phase commit protocol for the *user process*, not just internally, and, since commitments are subject to timeout, it requires the commitment holding time to be long enough to complete the physical change. These are somewhat strong requirements, which may affect overall data service performance adversely.

This leaves us with the scheduled release mechanism. In conventional systems, this is a management technique, rather than a software technique, in which database B is not permitted to be used until it has been reconciled with database A. Fortunately, this management technique translates directly to the operation-gating control technique: some coordinating process must officially "release" the information describing the physical reality, so that the control process which detects the change in the real-world must also detect the information release before it can start the dependent operation. That is, some control process has to be responsible for making the data and the physical reality consistent *at some point in time* and that process must post the command or indication that marks that point in time. In our example, the transport controller, or some higher-level coordinator, must post an indicator that the arrival and unload of the AGV has been made consistent with the global databases, so that the workstation may retrieve whatever contents information is appropriate. And the workstation controller must follow the rules, so that its use of the tray arrival indicator, and the corresponding access to data, occurs *after* the consistency has been established. The actual algorithm assumes the existence of some agreed-upon consistency indicator for this particular physical situation. The coordinating control process first clears the consistency indicator, then orders both the database update and the physical change, and finally, when both are complete, sets the consistency indicator. The affected process gates the dependent operations with the logical *and* of the change-detect and the consistency indicator. This mechanism follows well-known control techniques, and works without extraordinary changes to existing systems. It is recommended as a method of resolving potential conflicts between the databases and the physical reality in the design of controllers in materials flow systems.

Issue 4: Significance of standard data models

In its work on standard database definition languages [ISO82], the ISO working group formulated what is now called the Helsinki Principle: "Any meaningful exchange of utterances depends upon the prior existence of an accepted set of semantic and syntactic rules. The recipients of the utterances must use *only* these rules to interpret the received utterances, if the received meaning is to be the same as that which was meant by the utterer."

The first sentence of this principle is intuitively obvious, but only recently has the computer applications community begun to understand the ramifications of the second sentence. Until very recently, automation systems tended to operate in closed communities - a particular organization or a single vendor's equipment - so that the "accepted set of ... rules" contained many unwritten elements implicitly used by members of that community. What we have found as we try to design for integration across vendors and organizational elements is that these implicit rules vary substantially from community to community even for the same application. The Initial Graphics Exchange Specification (IGES) is a case in point. The intention of IGES 1.0 [IGES80] was to facilitate the transport of designs from one computer-aided-design (CAD) system to another. But the carefully crafted data exchange rules, in the light of the Helsinki principle, permitted exactly the exchange of the drawing: the only interpretation of the information provided by the rules of IGES was what the picture should look like, not, except superficially, the geometry or features of the object(s) being depicted. Several IGES implementations attempted to extend the drawing information to deliver geometry, but their experiences proved the pessimistic assessment to be correct. More recent versions of the standard [IGES88], therefore, have been extended to express geometric information specifically.

Historically, interchange standards and shared databases have been developed by getting the practitioners into a room to hammer out the list of information units which they all had in common and then devise a data organization or presentation scheme for those units. This technique is fraught with exactly the problems which motivated the formulation of the Principle. The fact that we agree on the syntax of the data, its name, position and formation rules, does not necessarily mean that we agree on the meaning of the data, exactly what it describes and how it is intended to be used. Without a more complete specification of the semantics, the shared data is often inadequate to the purpose, as in the IGES case, and the information units which are shared are often misinterpreted or misused. In order to get real meaning out of the data, we must also have formulated, and agreed on, a model of the world the data describes. We now understand that this actually involves two different kinds of model [Brod84]: the static associations between the data and the real-world physical and conceptual objects it describes, called the *information model*, and the rules for the use and modification of the data, which are derived from the dynamic characteristics of the objects themselves, called the *functional model*. The significance of these models to data interchange for manufacturing and materials flow was recognized early in the Air Force Integrated Computer Aided Manufacturing (ICAM) Project and gave rise to the IDEF formal modeling project [ICAM84]. IDEF produced a specification for a formal functional modeling approach (IDEF0) and an information modeling language (IDEF1)

[ICAM81]. The more recent Product Data Exchange Specification (PDES) project in the U.S. [PDES88], the related ISO Standard for the exchange of product model data (STEP) [Kall88] and the Computer Integrated Manufacture Open Systems Archictecture (CIMOSA) [ISO87] project in the European Economic Community have wholeheartedly accepted the notion that useful data sharing is not possible without formal semantic models of the context the data describes. Within their respective spectra of efforts, each of these projects has a panoply of information models for manufactured objects, materials and product characteristics, and for manufacturing and assembly processes. Each also has a commitment to detailed functional models of the various phases of product life cycle. The object of all of these recent efforts is to standardize the interchange of information in many aspects of product design, manufacture, delivery and support. And following the Helsinki Principle, the standardization of information and functional models in these areas is seen as a necessary means to that end.

But, while all of these efforts have concentrated on the problem of data exchange, the specification of functional and information models for the related processes has significant impact on the design of control systems as well. Consider that a functional model must describe all the significant changes that the objects described by the data can undergo, within some process context. If a standard functional model for some process context is adopted, then clear boundaries have been defined for the functions of a controller operating within that context, at least with respect to the perceived effects of that controller on the outside world. The controller can do some or all of those functions, but it cannot do other functions without participating in other contexts, and it is not permitted to have nonstandard side-effects on the modelled context(s) in which it participates. It is likely that the standardization of a functional model for a fairly sweeping context, like transport, will in fact become a hard boundary on controllers operating in that context, that is, such a controller must not participate in any other context. What will inevitably arise out of standardization of functional models, therefore, is a kind of standard modularization of various materials flow processes, into which nearly "plug-compatible" controllers can be fitted. The externally distinguishing features of controllers will be reduced to the degree of "completeness" in their implementations of the functional model, although there may be numerous internally distinguishing characteristics - actual technique, speed, power consumption, footprint, hardiness, etc. - which are incidental to the external perception of the process performed.

Lest we take these last implications as pie-in-the-sky, let me paraphrase the second sentence of the Principle: "Unless we can agree on what can happen to the objects described, we cannot meaningfully exchange the data describing them!" Integrated automation requires controllers to exchange data on a rather large scale. The Principle tells us, and experience has verified, that non-trivial exchange of data requires prior common functional models. It follows that successful steps toward integrated automation will inevitably lead to standardization of controller functionality in many aspects of materials flow. And the current strong desire for integrated systems suggests that such standardization will occur rather sooner than later. It is important to assure that control architectures are studied now, and that control system designers participate in the standardization efforts, with a view to imposing sufficient flexibility in the functional models to support the desirable control architectures. Otherwise we may find that our haste to standardize interchanges results in the proscription of potentially valuable system architectures.

Issue 5: Shared Information and Private Information

In attempting to modify existing "islands of automation" into integrated systems, the first realization that some of the data from SystemA must be used and perhaps modified by SystemB begets *information resource automation*, in which a common model, a common vocabulary and a shared data dictionary are derived, and each system is modified to import and export the relevant elements of the common databases for its own use. This "refitting-for-integration" produces a loose coupling between subsystems rather than a robust integration, but it is incrementally practicable. It also produces a distinction between "private" information, that data which is of use to only one subsystem, and "public" information, that data which shared among different subsystems. Experience in the AMRF indicates that such divisions tend to be arbitrary. The public information units are identified by committee, based on some consensus perception of current information needs and those of the near future, and the result is not a complete model of the common context, but rather a kind of greatest-common-factor of the existing individual models.

The recommended design-for-integration rule for data sharing is that all data handled by any subsystem is public, except that which is meaningless outside of the subsystem and that which is treated *by the subsystem itself* as duplicate, uninterpreted or unreliable. This rule works comfortably with the principle of modeling the objects before you define the interchanges, because anything which is externally comprehensible about the objects and processes with which the subsystem deals must be part of the information model, and therefore part of the public information base. Properly applied, the rule allows the manufacturer to protect certain private information units which are intimately involved with the method by which a controller carries out a generic function, while requiring it to make available information units which may be important to outside systems working on coordination, planning and control.

Not necessarily are these information units completely distinct. For example, it may be important to report the position or envelope of moving elements of the controlled apparatus so as to avoid collisions with other moving objects in the area. However, the public position data must be phrased in some standardized workspace coordinates and dimensions, which are globally meaningful, rather than reporting local joint or turntable positions which are meaningless outside the controller. Similarly, it is important to report such things as coolant levels and available power in terms of the ability of the subsystem to execute its nominal processes, rather than in absolute measures, which mean nothing to any system which doesn't know the consumption profiles of the device. What is hidden in this recommendation, therefore, is the further requirement that controllers must, perhaps continuously, *render certain private information units into meaningful public information.*

Part of the problem with the consensus agreements is that they are often colored by a second distinction: Access to public information must use *global data systems*, while access to private information can use the local data system, with which the authors of the local control programs are comfortable. The more public information a controller uses, the more involved it becomes with, and more importantly the more *dependent* it becomes on, the global data systems, which are *out of its control* (and out of the control of the engineers who built it). Fundamentally, no one wants his system to be dependent on elements out of his control, so left to his own devices, he designs and negotiates for the absolute minimum retrieval and insertion of public data by his controller. There is, in effect, a built-in characteristic of human engineers to design *against* integration! The imposition of standards by a higher authority is a path to solution, but standards reflecting a strong design-for-integration approach will only result from a standards development effort dominated by users and system integrators rather than a consensus of implementors.

Moreover, the adherence even to limited public data utilization standards is often implemented by a technique known as the "shadow database". The controller maintains a section of the private local databases which is logically identical to the segments of the global databases it is required to share. The internal workings of the controller always use only the local databases, and at certain key points in time, the public data is copied from the global databases into the local (shadow) databases, and at other key points, the updated data is copied in the other direction. From the engineer's point of view, this encapsulates the global data dependence in the copy-in/copy-out functions, so that the controller can proudly run with "local data" whenever "necessary". This form of "design for dis-integration" has one serious drawback: it moves the responsibility for maintenance of database consistency out of the global data services and into the distributed control processes. Unsurprisingly, the various control processes are rarely up to the task. In our experience, the single most common cause of integration test failures has been inconsistency of the local databases. And this is simply because too many decisions in the management of the shared data have been left to the individual control systems. The recommended design principle here is that *consistency of shared data is a vital part of integrated control* - the rules for maintenance of public data consistency, whether global systems are used directly or shadow databases are constructed, must be embodied in the control standards and protocols.

Summary:

This paper identifies five issues for interdisciplinary attention in the automation of materials flow. My contentions are:

- That one must logically separate data flow paths from control paths, so that data systems architectures and control systems architectures can be chosen independently for optimal performance in a given facility.

- That the conventional perfect communication model may cause delays in real-time control systems and that either the control systems must be prepared for the delays or an imperfect communication model, necessitating total status reporting, must be adopted.

- That consistency between databases and the physical world can be an acute problem in real-time control environments, and that consistency indicators and gating, among other choices, may represent the most practicable solution to such problems.

- That data sharing means standard information and process models, and standard models will mean standards for controller functionality.

- That the identification of shared versus private data cannot be left to the consensus of the controller builders, nor can the management of shared data within the controllers, but both must be part of a concerted design-for-integration approach.

Although this paper provides approaches to some of these problems, it does not mean to solve them so much as to draw attention to the need for interdisciplinary attention to problems such as these in the automation of materials flow.

References:

[Albu81] Albus, J.S., Barbera, A.J., Nagel, R.N., "Theory and Practice of Hierarchical Control", Proceedings of the 23rd International Conference of the IEEE Computer Society, September, 1981.

[Barb82] Barbera, A.J., Fitzgerald, M.L., Albus, J.S., "Concepts for a Real-Time Sensory-Interactive Control System Architecture", Proceedings of the 14th Southeastern Symposium on System Theory, April, 1982.

[Brod84] Brodie, M., Mylopoulos, J., Schmidt, J.W., editors, "On Conceptual Modeling", Springer-Verlag, New York, 1984.

[EIA87] Electronic Industries Association, Standardization Project 1393A: "Manufacturing Messaging Standard Service Specification and Protocol", draft 7, August 1987, unpublished.

[ICAM81] "ICAM Architecture Part 2, Volume 5: Information Modeling Manual (IDEF1)", AFWAL TR-81-4023, Air Force Materials Laboratory, Wright Aeronautical Laboratories, USAF Systems Command, Wright-Patterson Air Force Base, OH, June, 1981.

[ICAM84] "ICAM Conceptual Design for Computer Integrated Manufacturing Framework Document", Air Force Materials Laboratory, Wright Aeronautical Laboratories, USAF Systems Command, Wright-Patterson Air Force Base, OH, 1984.

[IEEE83a] IEEE Standards Project 802.2: "Local Area Networks - Logical Link Control", Institute of Electrical and Electronics Engineers, New York, NY, September, 1983.

[IEEE83b] IEEE Standards Project 802.3: "Local Area Networks - Carrier Sense Multiple Access with Collision Detection (CSMA/CD) Access Method and Physical Layer Specifications", Institute of Electrical and Electronics Engineers, New York, NY, September, 1983.

[IEEE84] IEEE Standards Project 802.4: "Local Area Networks - Token-Passing Bus Access Method and Physical Layer Specification", Institute of Electrical and Electronics Engineers, New York, NY, September, 1984.

[IEEE85] IEEE Standards Project 802.5: "Local Area Networks - Token-Passing Ring Access Method and Physical Layer Specification", Institute of Electrical and Electronics Engineers, New York, NY, September, 1985.

[IGES80] Nagel, R., Braithwaite, W., Kennicott, P., "International Graphics Exchange Specification (IGES) - Version 1.0", NBSIR 80-1978R, National Bureau of Standards, Washington, DC, January, 1980.

[IGES88] Smith, B., Rinaudot, G., Wright, T., "International Graphics Exchange Specification (IGES) - Version 4.0", National Bureau of Standards, Gaithersburg, MD, in publication.

[ISO82] "Concepts and Terminology for the Conceptual Schema and the Information Base", ISO-TR-6007, International Organization for Standardization, Geneva, 1982.

[ISO84] International Standard ISO 7498, "Information Processing Systems - Open Systems Interconnection - Basic Reference Model", International Organization for Standardization, Geneva, 1984.

[ISO87] "Reference Model for a Computer-Integrated-Manufacture Open System Architecture", ISO/TC184/SC4/WG1/N95, March, 1987, unpublished.

[Kall88] Kallel, M., "Standards for Data Exchange in an Integrated Environment: A Methodological Approach", Master's Thesis, Sloan School of Management, Massachusetts Institute of Technology, January, 1988.

[MAP87] "Manufacturing Automation Protocol Specification, Version 3.0", Society of Manufacturing Engineers, Detroit, MI, July, 1987.

[Mitc84] Mitchell, M. and Barkmeyer, E., "Data Distribution in the NBS AMRF", Proceedings of the IPAD II Conference, Denver, CO, April, 1984.

[Nanz84] Nanzetta, P., "Update: NBS Research Facility Addresses Problems In Set-ups for Small Batch Manufacturing," Industrial Engineering, pp 68-73, June 1984.

[PDES88] Proceedings of the SME Conference on Product/Process Definition Data, R. A. Carringer, Ed., Dallas, 1988.

Data and Knowledge Methodologies

in Material Flow Systems

DISTRIBUTED, INTELLIGENT INFORMATION SYSTEM FOR AUTOMATED, INTEGRATED MANUFACTURING SYSTEMS*

Detlef M. Weber and Colin L. Moodie
School of Industrial Engineering
Purdue University
West Lafayette, Indiana, U.S.A. 47907

ABSTRACT

As manufacturing systems have gravitated toward the Computer Integrated Manufacturing (CIM) ideal, they have become more dependent on capturing relevant data, storing it, operating on it and passing it between the members of the manufacturing system. As the systems become larger, this means increasing amounts of data to manage, higher complexity of operations on these data, higher number of requests and a need for significantly higher reliability and availability.

A model is presented here which strives to handle these highly demanding data management tasks for integrated manufacturing systems. A distributed system of cooperating, consistent, knowledge base sites is proposed that uses a semi-intelligent mechanism to predict the user processes in timing and data need. Data need is anticipated and the prepared information is stored in object frames, ready for quick access by the user. Test runs of a prototype version of this data storage and retrieval strategy demonstrate its utility.

INTRODUCTION

The emergence of CIM as a viable strategy for programmable, flexible, low production volume automation, in recent years, has stimulated much interest in the role of information management in manufacturing systems[1,2,3,4,5,20] Information, and its control, present the strongest integrating influence on the system, and currently available, and used data management tools cannot provide the capabilities that are required by manufacturing systems of the future.

Hsu and Skevington[20,21] suggest that true integration has to go beyond interfacing software packages and transmitting data between manufacturing functions and between data bases. An underlying information model is required to integrate the "differing functional views and knowledge of the system components on the conceptual level" and to "drive the various functions of the manufacturing

The research described here was supported by Purdue University's Engineering Research Center for Intelligent Manufacturing Systems with funding from NSF Grant CDR 8500022.

NATO ASI Series, Vol. F53
Advanced Information Technologies for Industrial Material Flow Systems
Edited by Shimon Y. Nof, Colin L. Moodie

system[20]." We present a conceptual model here, as well as a prototype information system, that has potential for meeting the needs of data intensive CIM systems.

INFORMATION MODEL FOR MANUFACTURING

Current Information Management Tools

An elementary data management strategy which is found in contemporary integrated manufacturing system would be to have several local and unconnected data bases which are specialized on the needs of the local user with respect to data, structure and in the implementation of the management system. The cooperation between data base and local application software can be described as a Master-Slave-Relationship. The data bases only act on request by their local application software, process the request and quit processing until the next request. The problems and shortcomings are the cost of re-entering or transferring data between manufacturing functions, the limited accessibility to knowledge of other functions, the time and cost intensive maintenance and inconsistencies between the data.

As an improvement on the local data base model, some researchers and practitioners have proposed a centralized and independent data base server as an alternative[1,7]. This data base is implemented on a dedicated processor and connected to all users in the system. The physical as well as logical independence from user modules improves the performance and thus allows the storage of more data and the ability to service more users with the same efficiency. The main advantage of the centralized model is the sharing of data. The possibility of two user modules using different versions of one data item and subsequently making inappropriate decisions should not occur with this configuration.

A centralized data base server is not, however, an adequate solution for a large, flexible and highly integrated manufacturing system. In times of high access users have to queue in front of the data base; in the case of a crash of the centralized data base the whole manufacturing system is blocked completely and finally there is a negative trade-off between efficiency and the amount of data that can be stored (the more data the worse the performance).

Manufacturing systems of the future require an efficient and distributed information management system. Neither of these two conventional approaches is suitable. An independent information management system is proposed that could provide intelligent, reliable and efficient support.

Proposed Model

A double structure of two parallel, but independent, distributed systems is proposed. One system is called a decision processing and control system, and performs most of the tasks of the traditional control system. The second is called a knowledge processing system and is responsible for all functions of information handling, manipulation and interpretation.

Figure 1 is a schematic view of this double structure. The three level hierarchy of shop control, cell controllers and workstations (in the center of the drawing) define the decision processing and control hierarchy. The four knowledge bases (on the outside of the figure) compose the knowledge processing system.

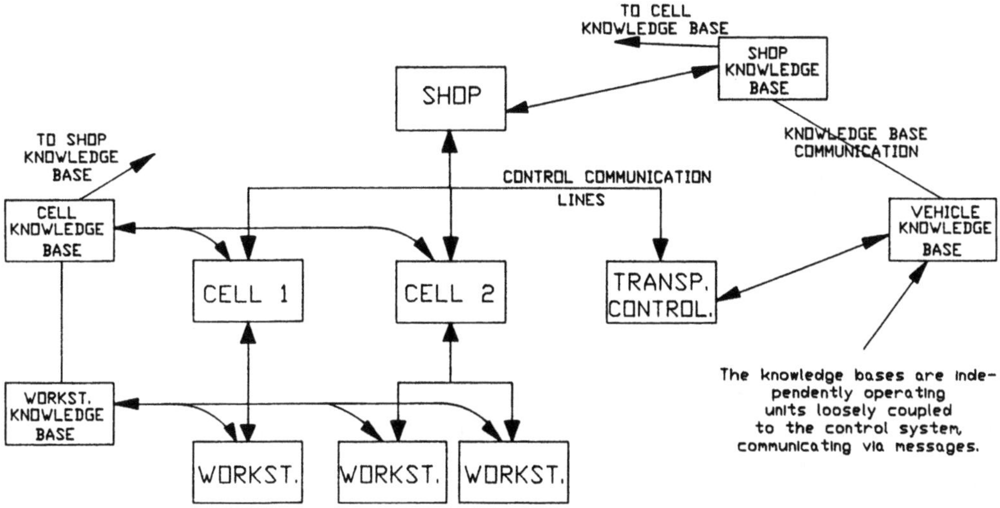

Figure 1. Double Structure

A comparison of this system with the local data base and centralized data base server strategies reveals the major differences. The local data bases are directly attached to their users and are not connected to the other data bases. The knowledge processing structure in figure 1 is independent from, but connected to, the modules of the decision making and control system; the knowledge base sites in the knowledge processing system are connected to each other. The centralized data base server will serve all user modules; however, as seen in figure 1 several connected knowledge base sites perform this task together with each knowledge base site being dedicated to one or a set of users. The sharing of data is maintained through the connections between these knowledge bases.

The main features of this proposed knowledge processing model are summarized as follows:

- Each knowledge base site serves one user or a group of users with similar tasks. The knowledge base maintains all information that are routinely accessed by its user or group of users. The queuing that was observed in the centralized server approach is reduced to a minimum.

- Site recovery and network partition protocols minimize the impacts of component failure and thus provide a high degree of availability and reliability.

- The users have global access to all information in the entire system. If a data item is not available at the local knowledge base the request is transmitted to a knowledge base site that maintains this data.

- The global consistency is maintained by the concurrency control mechanism. An update is only committed (accepted) if the update operation has been successfully processed by all sites that keep a copy of this data[14,15,16].

Manufacturing System with Information Function

In figure 2 the generic picture of the "Double Structure" is extended to a detailed and complete model of a manufacturing system consisting of a conventional hierarchical control structure and the proposed parallel, distributed information management system.

The decision making and control system is represented in a five level hierarchical structure using the model proposed by the National Bureau of Standards[1,7,18]. The facility and the shop levels each contain one module, the cell level is composed of two manufacturing cells, the inventory, the assembly cell and the vehicle controller. The work station level distinguishes handling and machining or assembly work stations and on the equipment level vehicles, machines, and other equipment are displayed.

The knowledge processing system is grouped around this manufacturing system. Six local knowledge base sites serve the system. Each knowledge base site is assigned to a particular user module or to a group of very similar user modules (e. g. Note the cell control knowledge base for the manufacturing cells). These assignments are displayed through 'view' links shown in figure 2. The views represent the part of the knowledge in the knowledge base site that is currently relevant for the user module.

AN INTEGRATED KNOWLEDGE BASE
IN AN
INTELLIGENT MANUFACTURING SYSTEM

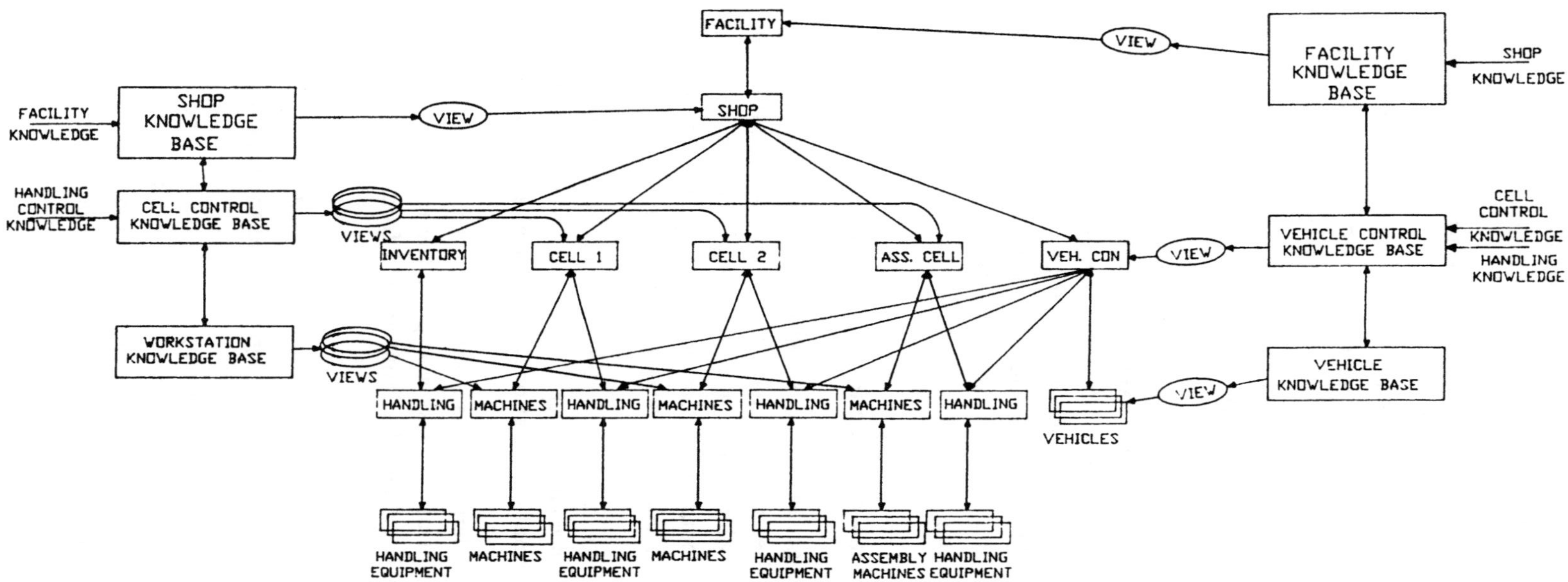

Figure 2. Integrated Manufacturing System

The knowledge base sites are connected with a communication system that operates independently from the manufacturing communication system.

As noted above, the decision processing system proposed by the National Bureau of Standards is used; this does not mean, however, that the decision processing and control system necessarily needs to be hierarchically structured. Heterarchical systems as they were proposed by Hatvany[4], Shaw and Whinston[6] and others, could also profit from the information system that will be described.

SYSTEM ARCHITECTURE

The six knowledge base sites of Figure 2 are identical in most of their components: they have the same software structure, use the same knowledge representation schemes and communication protocols, and share the larger part of the implemented software. We will describe, in detail, *one* of these local knowledge base sites here. This site is used for presenting the structure, capabilities and features that are shared by *all* sites and serves also as an example for the features of prediction-planning-preparation (ppp) functions and object frame base.

The Modules and Components of a Knowledge Base Site

One knowledge base site is shown in figure 3. It can represent *any* of the knowledge base sites that are part of the knowledge processing system displayed in figure 2. Each knowledge base site consists of the five main modules: Input Module, Output Module, Coordinator Module, Interface Module and Knowledge Module. The modules themselves consist of several components.

The knowledge module is the central module of the local knowledge base system. It executes control over the local knowledge base system, is the main local repository of information (relational data base) and incorporates the ppp-function.

Control is implemented as a repetitive cycle of calls to the modules of the site. Each module performs its tasks and posts the results at a predefined location. Then the next module is called, the input loaded from this location and the task of this module is executed.

A diagram of the control cycle is presented in Figure 4. The blocks represent the modules as they are shown in figure 3 and the arcs represent the flow of control.

The cycle begins with the input module reading all current inputs in the mailboxes and identifying their headers. The information of the headers is passed to the coordinator. One transaction is selected, loaded through the input module and analyzed by the coordinator. The transaction is sent to another site for remote processing at this site or to the interface for local processing.

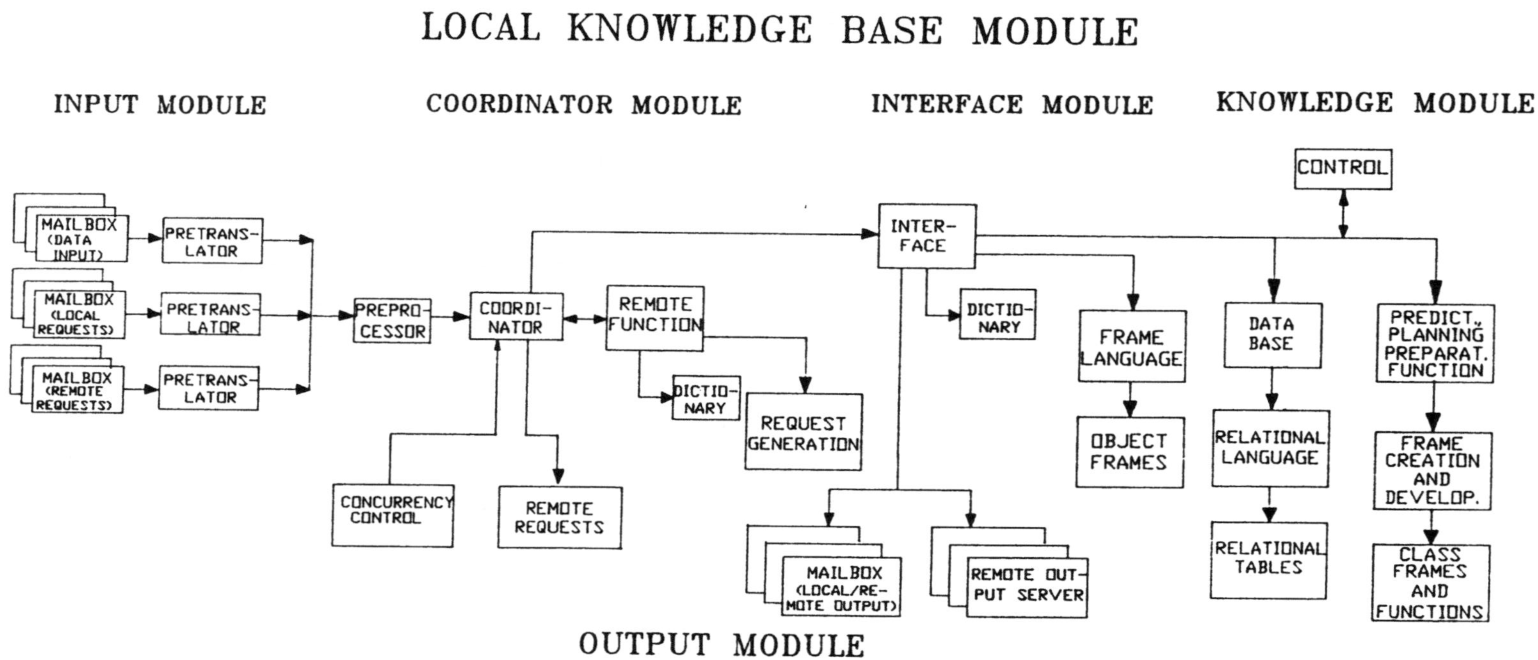

Figure 3. Knowledge Base Site

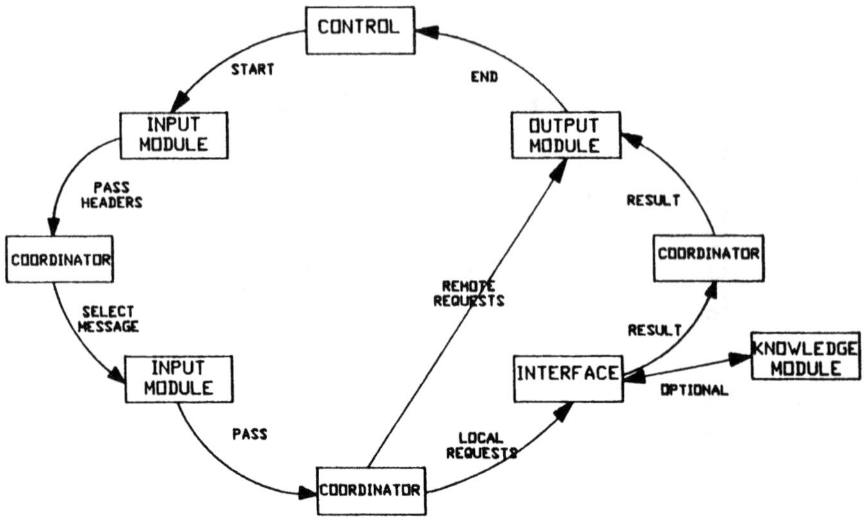

Figure 4. Control Cycle

The interface module and, if needed, the knowledge module process the request by either updating some information and object frames or retrieving some requested information.

The coordinator generates messages to other sites for updating their copies or the messages to the user providing the requested information, and distributes these messages via the output module. Then the next cycle is started.

Knowledge Representation

A mix of two knowledge representation schemes is proposed for the knowledge base sites. The relational data base scheme is used for representing the general shop and product information. This information includes all data that are independent of product order or state of the manufacturing system. In figure 3 the relational data base is part of the knowledge module.

The frame representation scheme[9,10,11,12,13] is used here to build and operate the user interface. Object frames represent order-related product information, and process state information in an semantically organized manner. They model the situation in the shop. In figure 3 the object frames are a component of the interface module. A frame is displayed in figure 5 in the format that is used. The mix of the two representation methodologies overcomes the shortcomings of the conventional relational data base scheme. The frame scheme is used for applications that are not suited for the relational scheme because of its slow operation, and difficulty in representing intelligent and object-oriented information. It makes use of the advantageous features of the relational scheme in other applications that need high flexibility and low redundancy of data.

```
process(
             [  [id,[value,[1]]],
             [  [trigger,[trigger,[17,209]]],
             [  [alarm,[alarm,[209]]],
             [  [parent,[isa,[33,component]]],
             [  [subordinate,[include,[]]]],
             [  [class,[isa,[process]]],
             [  [process-name,[value,[turning]]],
                        [priority,[value,[1]]],
                        [time-estimate,[value,[2]]],
                        [process-finish,[plan,[211]]],
                        [cell_name,[value,[[turning_cell]]]],
                        [nc-programm,[value,[[[nc294k]]]]],
                        [material,[value,[[oil34j]]]],
                        [tools,[value,[[turninghead3d]]]],
                        [machine,[plan,[[[machine2]]],
                           [value,[]]]],
                        [t-arrival,[plan,[208]]],
                        [t-departure,[plan,[212]]],
                        [process_begin,[plan,[209]]],
                        [entry-next-cell,[value,[]]]],
                        [nr_of_intransp,[plan,[17]]],
                        [in-transports,[value,[[]]]],
                        [out-transports,[value,[]]]],
                        [cell-transport,[value,[]]]],
                        [nc-handling,[value,[]]]],
                        [status,[default,[open]]],
                        [delay-status,[value,[]]]],
                        [next-cell,[value,[drilling]]],
                        [process-plan,[value,[processplan00001]]],
                        [number_of_outtr,[value,[]]]],
                        [problems,[value,[]]]],
     ]).
```

Figure 5. Object Frame

Two special features of frames are in particular useful: procedural attachment and inheritance. Procedural attachments are used to represent abstract information about data in place of the data. This abstract information is a method or a retrieval instruction that provides the concrete data when executed by an actual user request. The mechanism of inheritance provides the capability of linking two frames. Inheritance can be used in three ways for the proposed model:

- representing unavailable data,

- linking frames of different classes,

- linking frames with their class frames.

The object frames are organized in a hierarchy that is defined globally for the entire information system. This means the connecting inheritance pointers relate object frames that reside at different sites. This hierarchy represents the information about the products and the manufacturing processes they will undergo.

One scheme is to have the top level object frame contains the product order that was issued by the marketing department to the facility. The product order describes a system that is manufactured and assembled in this plant (robot gripper, for exlample). The next lower level contains the object frames for the components of the product (arm, hand, fingers and control module). On the level of manufacturing operations (turning, drilling or milling) the stations to which a component must go for processing are represented by the process object frames. Then the part needs to be transported to the cell where the manufacturing process takes place. The transport information is represented in the transportation frame. Figure 6 exhibits the hierarchy.

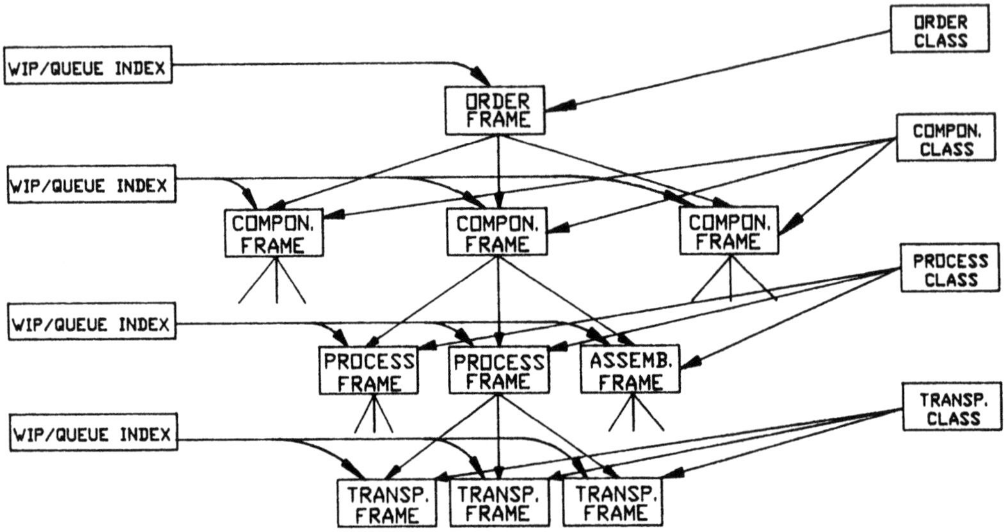

Figure 6. Frame Hierarchy

The object frames reside in the interface modules of different knowledge base sites. This can be seen in Figure 2. The order object frame resides in the Facility knowledge base site, the component object frames in the Shop knowledge base site, the process object frames in the Cell control knowledge base site and the transport frames in the Vehicle control knowledge base site.

The frames in the hierarchy are connected to their class frames and to two types of queue frames by additional inheritance pointers. The class frames contain meta information that is particularly useful for correctness checking. The two queue frames represent the waiting queue and the in-process queue for a manufacturing cell. They relate the manufacturing cells to the current state of the component of a product order.

The frame approach thus provides a very efficient and very flexible tool for combining product information, manufacturing data and the state of the manufacturing situation in a highly useful, efficient and semantically expressive product model.

Interface

The interface receives the requests from its user modules in the manufacturing control system (or from other sites) through the coordinator module and manages the retrieval and update operations. The incoming request is interpreted and it is decided, whether relational data base, object frame base or both are required by this request. Then the necessary operations are initialized.

The interface also includes the frame language and the object frame base. The frame language is a retrieval and manipulation instrument for the management of the object frames [13]. Fifteen operations on frames have been implemented that provide the necessary features for this environment. Figure 7 lists these 15 operations and provides information about their tasks.

The object frame base of a site contains object frames that store the information that is necessary for the operation and decision making of the connected user modules. For example, the site that is serving the manufacturing cells contains the process object frames.

FRAME LANGUAGE OPERATIONS		
PROCEDURE NAME	OPERATION	TYPE
get_operation(A,B,C,D)	retrieve of basic set	frame retrieve
get_extended(A,B,C)	retrieve of extended set	frame retrieve
standard_set(A,b.C)	retrieve of standard set	frame retrieve
get_operation(A,B,C,Con)	constraint retrive	frame retrieve
get((K,N),(S),Z)	retrieve of data	slot retrieve
const_get((K,N),S,F,Z)	constraint retrieve	slot retrieve
feed_back(X,Y,Z)	join of two frames	frame update
trigger_frame(...)	trigger a subframe	frame update
finish_frame(...)	final report and remove	frame update
frameupdate(...)	add slots to frame	frame update
frameupdate(...)	insert facets	slot update
frameupdate(...)	insert data	slot update
frameupdate(...)	delete data	slot update
frameupdate(...)	replace data	slot update
frameupdate(...)	modify data	slot update

Figure 7. Frame Language

Prediction, Planning and Preparation (PPP)

The goal of the prediction, planning and preparation process (ppp-mechanism) is to transfer the available information from a data-oriented relational structure into an object-oriented structure that represents the information for a particular product order. The data have to be organized in such a way that they respond directly to the predicted user requests and thus permit a very efficient retrieval and usage. This product-order-oriented structure is called product model.

The key assumption for the ppp-mechanism is that the behavior of the user modules in the manufacturing control system (decision making system) is predictable in its operations that are associated with the manufacturing of a particular product order or one of its components. The prediction has to cover three aspects: the requested information, the time of the request and the requesting user (location).

A manufacturing system is a well-planned (production plan, schedule and process plans), and well-organized system (celluar organization, integrated control and hierarchical structure). Furthermore it is well-documented (data, history, statistic and experience), well-informed (integrated information system, computerized communication system and inter-human communication) and highly repetitive (a product is manufactured several times and different products share some identical components). The prerequisites for an exact prediction seem to be fulfilled.

But reality in manufacturing systems is not that easy. The manufacturing system has to be able to control a constantly changing product mix every manufacturing period. The complexity of the control problem and the combinatorial explosion of a planning process for such a problem makes it impossible to plan the operation with the exactness that would be necessary for the prediction process. Furthermore, reality in manufacturing systems is characterized by breakdowns, varying product sequences and other unexpected and unpredictable events. An off-line prediction process is not capable of providing predictions with the required certainty and accuracy.

The ppp-mechanism is an on-line prediction and planning tool that includes the capability to react flexibly to short term events. The ppp-mechanism predicts and plans only one process ahead for a product order. The ppp-process is triggered by an alarm when the production of the product order has reached a state that allows it to start the process that will be predicted, planned and processed. The interval between the ppp-process and the use of the prepared data by the user request is minimized. The important information is used by the ppp-mechanism in its prediction and planning, and it is not likely that the data will loose validity before the actual user request.

Even if all required data items are clearly specified and the time of the request is precisely predicted, not all problems of the ppp-process are solved. It is likely that some data items that are needed for answering the predicted user request are not available at the time of preparation. This can mean that these data are subject to frequent change (a data value is available but likely to change before the user request arrives) or that this data has not been created. These data are not represented as concrete information but as virtual information, e.g., as pointer to their location or formula.

The ppp-process is started with the entry of a new order into the system. The order entry contains a basic information set which includes the customer, the requested system, the amount of these, the order date and the due date are supplied. Based on these data the order object frame is created.

The prediction component of this process has two parts. One part is implemented by the systems designer and specifies which information, in what format, are supplied to a particular user process. The other part is a procedure that identifies the temporal and organizational relationships between the objects on the four levels of the frame hierarchy and uses this to build a time frame for the execution of the manufacturing task.

The planning component takes this time frame and relates it to the required operations. It develops a schedule for the creation and generates the instructions that execute these operations. The schedule (alarms) as well as the instructions (triggers) are implemented in the object frames.

The process of creating a frame begins with an alarm in a superior object frame. A time bound was matched or the preceding process has sent the termination signal as an alarm. The trigger is initiated that calls the creation routine for the next object frame.

The class frame is used as a template. This reduces the creation operation to deleting unwanted information from a frame and inserting necessary information into it. The preparation procedure scans step by step through the class frame, interprets the information in the slots and executes the appropriate operations. A class frame is exhibited in figure 8.

```
process (class,[
         [process-name,[build,[process]]],
         [priority,[build,[priority,process]]],
         [time-estimate,[build,[process_data]]],
         [process-begin,[build,[process_begin]]],
         [entry-next-cell,[value,[]]],
         [number-of-intr,[value,[]]],
         [status,[default,[open]]],
         [delay-status,[value,[]]],
         [next_cell,[build,[next_cell]]],
         [process-plan,[build,[process_plan]]],
         [time-estimate,[bound,[11]]],
         [priority,[type,[integer]]],
         [status,[type,[string]]],
         [number_of_transports,[type,[integer]]],
         [time-estimate,[type,[integer]]],
         [process-begin,[type,[integer]]]
              ]).
```

Figure 8. Class Frame

When a preparation process has finished its operation, control is passed back to the prediction and planning function. The header for the frame is implemented. This header contains inheritance pointers to the class frame, the superior object, and the supporting object frames (at creation time an empty list), the identification slot of the frame and in particular the alarms and triggers for the next operations.

```
process(
        [  [id,[value,[1]]],
        [  [trigger,[trigger,[17,209]]],
        [  [alarm,[alarm,[209]]],
        [  [parent,[isa,[33,component]]],
        [  [subordinate,[include,[]]]],
        [  [class,[isa,[process]]],
        [  [process-name,[value,[turning]]],
                [priority,[value,[1]]],
                [time-estimate,[value,[2]]],
                [process-finish,[plan,[211]]],
                [cell_name,[value,[[turning_cell]]]],
                [nc-programm,[value,[[[nc294k]]]]],
                [material,[value,[[oil34j]]]],
                [tools,[value,[[turninghead3d]]]],
                [machine,[plan,[[[machine2]]],
                    [value,[]]]],
                [t-arrival,[plan,[208]]],
                [t-departure,[plan,[212]]],
                [process_begin,[plan,[209]]],
                [entry-next-cell,[value,[]]]],
                [nr_of_intransp,[plan,[17]]],
                [in-transports,[value,[[]]]],
                [out-transports,[value,[]]]],
                [cell-transport,[value,[]]]],
                [nc-handling,[value,[]]]],
                [status,[default,[open]]],
                [delay-status,[value,[]]]],
                [next-cell,[value,[drilling]]],
                [process-plan,[value,[processplan00001]]],
                [number_of_outtr,[value,[]]]],
                [problems,[value,[]]]],
  ]).
```

Figure 9. Created Object Frame

Finally the superior frame is updated by inserting the inheritance link to the new supporting frame. The result of this creation procedure is displayed in Figure 9.

The advantages of this prediction, planning and preparation scheme can be summarized as follows:

- Reduced Average Retrieval Time - only one search is required for identifying and retrieving all data,

- Decoupling of User and Information System - the information system can execute many of the operations for a user request in advance,

- Higher Informational Power - the retrieved information are supplied in the required format and in the most usable form for the user,

- Correctness - the supplied information has a higher degree of correctness,

The limitations of this ppp-operations can be defined as follows:

- lack of knowledge for the planning and prediction - the available knowledge of the systems designer is not sufficient for the prediction of many user processes and their data requests,

- time limitation - the execution of ppp-functions requires some time. The user processes needs to be predicted some time in advance.

- processing cost - too many ppp-functions will overload the CPU of the knowledge base machine and slow the service efficiency down again.

Distribution

The problem of distribution can be separated into three components: communication, distribution/coordination and information management. The communication problem is involved with providing the facilities for exchanging messages and includes such aspects as communication protocols, synchronization of communication and the sending and receiving operations. Distribution and coordination focus on issues like replication of information, concurrency control, component failure handling, providing global access and transparency etc. Several software packages are available with these capabilities[14,15,16,17,21].

For this work the information management aspect was of interest. A model of six sites was chosen for the manufacturing control system. In figure 2 these six sites were displayed in the context of the entire system parallel to the decision making and control system. In Figure 10 the six sites are shown with their major information contents. The configuration is user oriented. User modules with similar data needs are serviced by the same knowledge base. This permits specialization of the local knowledge base manager and reduces the number of copies in the systems. Hence the update and control effort is reduced and so is the amount of required communication between knowledge bases for a retrieval execution.

A data dictionary is used to determine the location of the required data and to generate the message. With this the coordinator module provides the transparency that is required in distributed systems.

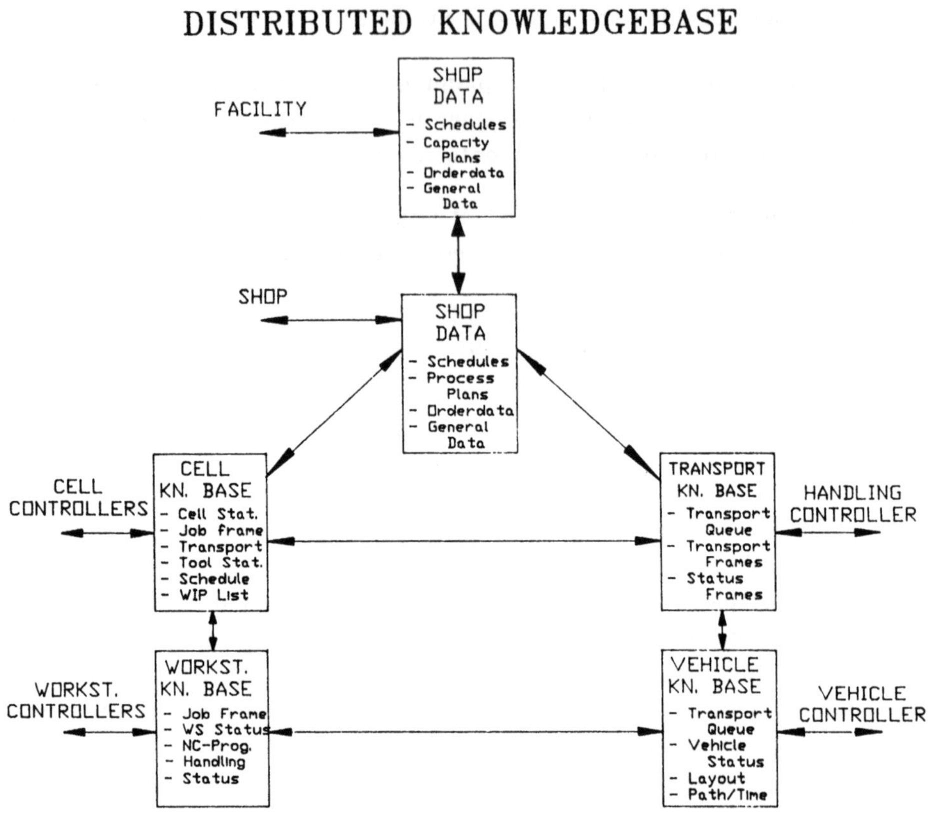

Figure 10. Six Sites of the Information System

IMPLEMENTATION

Software

Prototype software was developed using PROLOG as the implementation language.
The built-in inference machine frees the programmer from work on the control aspect of programming. The implementation work can be concentrated almost entirely on the logic design of the software. Besides these general advantages, PROLOG turned out to be very suitable for an implementation of a knowledge base using a structured representation scheme.

On the other side, PROLOG has serious drawbacks as far as efficiency is concerned. The execution of each subgoal requires a linear search over the data base until a matching and unification process is successful.

The prototype system was operated on three SUN 3/50 work stations using the X-window software package by DEC/MIT. Four knowledge bases are operated in four windows thus implementing four logical sites. The knowledge bases are the facility knowledge base, the shop knowledge base, the cell knowledge base and the transportation knowledge base. The communication between the four sites was simply achieved through the shared file system in the file server. The operations are executed in a serial manner assuming that there are no other interfering requests.

An Example

The following scenario describes the life cycle of a process object frame from its creation to its end. The life cycle exhibits the stages of a typical cooperation between the manufacturing cell and the cell knowledge base site during the preparation, execution and termination of the manufacturing of a component of a particular product order at this cell.

Eventually the alarm is triggered in a component object frame that resides at the shop knowledge base site. The current system date and the manufacturing start date of this component match. A message is sent from the shop to the cell knowledge base site to initiate the creation of the process object frame for the first manufacturing process.

At the cell knowledge base site the class frame for processes is loaded and the creation is started. The result of this creation process is displayed in figure 11. The finished process object frame is posted in the frame base of the cell knowledge base site. A pointer is appended to the queue frame of the manufacturing cell that is scheduled to execute the necessary manufacturing processes. The queue is the list of product orders waiting for processing in this cell.

Sometime later the cell controller decides that it is time to load the next job to replace a job that is close to termination. This job is processed on machine_5 of the cell therefore a new job is needed that can be manufactured on this machine. A request is issued to the cell knowledge base site for the next available job that can be manufactured with machine-5.

The knowledge base interface initiates the frame language routines that load the object frames, one by one, via the queue index, and checks in slot machine_options to find a job for machine_5. The first object frame that satisfies this constraint is selected and the base set of information is loaded from the object frame and sent to the cell controller. The pointer is removed from the queue and inserted into the work-in-process (wip) queue.

```
process(
            [  [id,[value,[2]]],
            [  [trigger,[trigger,[9,219]]],
            [  [alarm,[alarm,[219]]],
            [  [parent,[isa,[33,component]]],
            [  [subordinate,[include,[]]]],
            [  [class,[isa,[process]]],
            [  [process-name,[value,[grinding]]],
                       [priority,[value,[1]]],
                       [time-estimate,[value,[2]]],
                       [process-finish,[value,[223]]],
                       [cell_name,[value,[[grinding_cell]]]],
                       [nc-programm,[value,[[[nc68k]]]]],
                       [material,[value,[[oil4j]]]],
                       [tools,[value,[grindinghead14]]]],
                       [machine_option,[default,[machine4,machine_7]]],
                       [machine,[plan,[[[machine4]]],
                          [value,[machine_7]]]],
                       [t-arrival,[value,[219]]],
                       [t-departure,[plan,[224]]],
                       [process_begin,[value,[222]]],
                       [entry-next-cell,[value,[]]]],
                       [nr_of_intransp,[value,[9]]],
                       [in-transports,[value,[[]]]],
                       [out-transports,[value,[]]]],
                       [cell-transport,[value,[]]]],
                       [nc-handling,[value,[]]]],
                       [status,[value,[in_planning]]]],
                       [delay-status,[value,[]]]],
                       [next-cell,[value,[drilling]]],
                       [process-plan,[value,[processplan00001]]],
                       [number_of_outtr,[value,[]]]],
                       [problems,[value,[]]]],
      ]).
```

Figure 11. Process Object Frame

Based on the content of this base set of information, the cell controller executes its planning operations for this job. Among these defined planning decisions are the transportation path to use, select the tools to use etc. The results of these planning operations are returned as updates to the cell knowledge base site and are inserted into the process object frame. These updates also initialize some ppp-functions that plan the necessary transportation. Messages are sent to the vehicle knowledge base site to trigger the creation of transport object frames. Figure 12 displays such a message. At the vehicle knowledge base site the transport object frames are created, posted in the frame base and placed in the queue for the vehicle controller. Eventually the transport object frames are scheduled and executed. When a transport arrives at the cell, the vehicle knowledge base site terminates the transport object frames and reports to the cell knowledge base site to update the status of the process object frame. When all transports are terminated the job is ready for manufacturing.

At the cell the preceding job on machine_5 has terminated, and the cell controller requests the prepared job for machine_5. The cell knowledge base site loads the frame via wip-index, executes the pointers and methods and sends the information to the cell controller. Figure 13 exhibits such an information chunk.

The cell controller should now have all information required for the processing of the job on machine_5. If this is not the case, an additional request is sent to the knowledge base site to extract these data from the relational data base or remote

```
transport([
        [[id,[value,[17]]]],
        [[trigger]],
        [[alarm]],
        [parent,[isa,[15,process]]]],
        [[subordinate,[include,[]]]],
        [[class,[isa,[transport]]]],
        [[pickup-place,[value,[inventory]]],
         [pick-up-time,[value,]]],
         [pick-up-status,[default,[open]],[value,[]]],
         [drop-place,[value,[[turning_cell]]]],
         [drop-time,[value,[]]],
         [drop-status,[default,[open]]],
         [process-name,[value,[turning]]],
         [weight,[value,[turning]]],
         [status,[value,[]]],
         [priority,[value,[]]],
         [trans-time,[value,[]]],
         [delay-status,[default,[]]],
         [vehicle,[value,[]]],
         [path,[value,[]]],
         [transp_bat,[value,[0]]],
         [capacity_load,[value,[0]]],
         [t_object,[value,[control]]],
         [process-plan,[value,[procxessplan00001]]],
         [distance,[value,[]]]
      ]).
```

Figure 12. Transport Object Frame

```
input(turning_cell,frame,retrieval,immediate,return,
      [],[],[[
      [process-name,[value,[turning]]],
      [priority,[value,[]]],
      time-estimate,[value,[11]]],
      [process_finish,[plan,[211]]],
      [cell_name,[value,[turning_cell]]],
      [nc_program,[value,[nc294k]]],
      [material,[value,[oi134j]]],
      [tools,[value,[turninghead3d]]],
      [t-arrival,[plan,[208]]],
      [t-departure,[plan,[212]]],
      [process_begin,[plan,[p109]]],
      [machine_option,[default,[[mcahine_2]]]],
      [nr_of_intransp,[plan,[7]]],
      out-transports,[value,[2]]],
      [status,[value,[in_process]]],
      [next_cell,[value,[drilling]]],
      [process-plan,[value,[processplan00001]]] ]]).
```

Figure 13. Information Set

locations. Figure 14 displays a request and the response of the cell knowledge base site.

```
input(turning_cell,frame,retrieval,immediate,request,slot,
                                          fixture).

input(turning_cell,frame,retrieval,immediate,return,slot,
                                          [fixture,[fixture56j8]]).
```

Figure 14. Additional Request

During the execution of the job the cell controller regularly sends updates to the knowledge base site updating the process object frame. Finally the job at machine_5 is finished. The cell controller sends the information about scrap, finished parts etc. to the cell knowledge base site. The knowledge base again initializes the ppp-functions and triggers the transports required to remove scrap and tools from the cell.

A report is sent to the parent component object frame at the shop knowledge base site. Figure 15 displays the report message. The shop controller updates the component object frame setting the status for this process to 'ready'. This has the effect that the creation of the next process object frame is triggered. The life cycle of the next process object frame begins. The old process object frame is deleted after all transports are executed.

```
input(turning_cell,frame,update,immediate,terminate,[],
                                [process],[[status],[finished]]).
```

Figure 15. Report Message

Comparison

The alternatives to the proposed model are the centralized data base server concept and the local data base concept. The interest of the following comparison is to highlight the advantages and capabilities of the proposed concept and to demonstrate its achievements in relation to these two conventional alternative models.

This comparison includes an evaluation of features and capabilities and an analysis of the performance characteristics of the three alternative models. The performance analysis is not a comparison of three softwares using predefined measurement parameters. This would have required comparable and appropriate implementations of the two alternative models. The performance analysis is based on the available knowledge about the different methods and components of the three concepts. An informative picture of the potential performance differences is drawn.

Figure 16 displays the comparison of features and capabilities in six issues. Figure 17 visualizes the main aspects of the performance analysis: waiting time, response time, number of requests and operation time of the knowledge base site.

FEATURES AND CAPABILITIES	PROPOSED SYSTEM	CENTRAL. SYSTEM	LOCAL SYSTEM
SHARING OF DATA	++	++	--
RELIABILITY	+	--	++
RANGE OF KNOWLEDGE	++	+	--
INTEGRATION SUPP.	++	+	--
FLEXIBLE KNOWLEDGE REPRESENTATION	++	--	0
PREDICTION AND PLANNING	++	--	--

Figure 16. Features and Capabilities

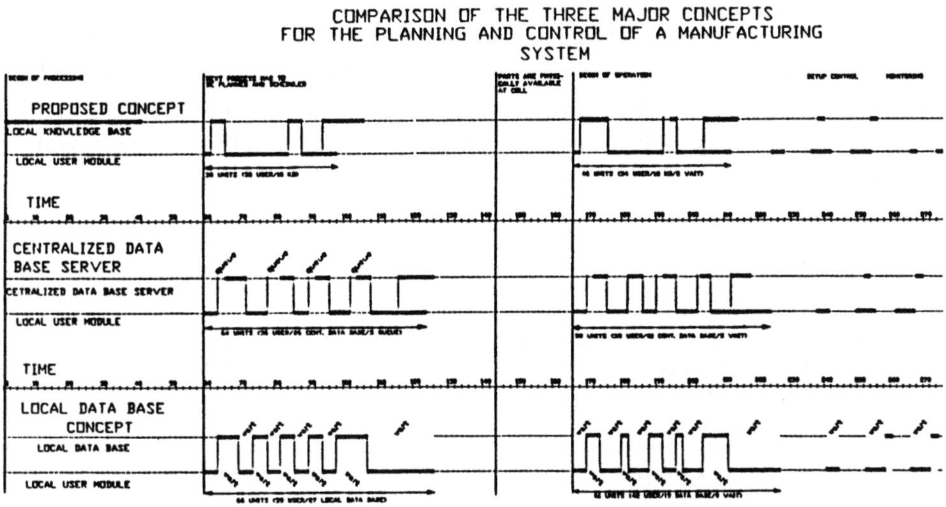

Figure 17. Performance Comparison

CONCLUSION

A concept for an information system for an automatic manufacturing environment has been presented and test results of a prototype software were outlined.

The information system is characterized by its potential for high efficiency of operation (low response time for routine requests, low waiting times), high quality of retrieved information (correctness, usability, and informational power of data) and its active contribution to the decision making of the user (defaults, special computations for data protection) instead of passive responding to user requests.

The information system consists of several distributed but cooperating knowledge base sites that operate independently from any control by the user system. The sites include a relational data base, the distributed communication and synchronization facilities, a scheduler, mailboxes and a user interface.

Furthermore, a semi-intelligent module, the ppp-mechanism, provides the capabilities for planned operation and active participation in the decision making. The ppp-mechanism anticipates the routine behavior of the users and predicts their data needs in timing and content. Based on these predictions the operations of the site are planned and scheduled. The retrieval, correctness checking and other data base operations are executed at the scheduled time, before the predicted user request arrives. Also some defaults for standard decisions are created and some data handling operations are executed in order to reduce the work load of the user.

The results of these operations are stored in an object frame which is then posted in the user interface ready for use by the user request that eventually will arrive. The advantages of the object frames are their semantic organization of the data, high degree of usability and understandability, comprehensive representation and ease of access.

REFERENCES

1. Barkmeyer, E., M. Mitchell, K.P. Mikkilineni, S.Y.W. Su and H. Lam: An architecture for distributed data management in computer integrated manufacturing, 1986.

2. Lillehagen, F.: The product information system and data bases in mechanical CAE, Robotics & Computer Integrated Manufacturing, vol. 1, no. 3/4, pp. 255-260, 1984.

3. Spur, G.: Growth, crises and future of the factory, Robotics & Computer Integrated Manufacturing, vol. 1, no. 1, pp. 21-37, 1984.

4. Hatvany, J.: Intelligence and cooperation in heterarchic manufacturing systems, Robotics & Computer Integrated Manufacturing, vol. 2, no. 2, pp. 101-104, 1985.

5. Solberg, J.J., D.C. Anderson, M.M. Barash and R.P. Paul: Factories of the future: Defining the target. Report to the National Science Foundation, Purdue University, January 1985.

6. Shaw, M.J. and A.B. Whinston: Task bidding and distributed planning in flexible manufacturing, 2nd Conference on Artificial Intelligence Applications, Miami Beach, Florida, pp. 184-189, 1985.

7. McLean, C.R., M. Mitchell and E. Barkmeyer: Distributed computing architecture for small batch manufacturing, IEEE Spectrum, May 1983.

8. Spur, G.: Rechnerunterstuetzte zeichnungserstellung und arbeitsplanung, ZWF-Lehrblaetter, Zeitschrift fuer wirtschaftliche Fertigung, pp. 1-98, 1979-1980.

9. Nau, D.S. and T.C. Chang: Hierarchical representation of problem-solving knowledge in a frame-based process, Journal of Intelligent Systems, February 1986.

10. Ray, S.R.: A knowledge representation scheme for processes in an automated manufacturing environment, Technical Report, National Bureau of Standards, 1986.

11. Minsky, M.: A framework for representing knowledge, in: The psychology of computer vision, (P. Winston, ed.), McGraw-Hill, Inc., pp. 211-277, 1975.

12. Fikes, R. and T. Kehler: The role of frame-based representation in reasoning. Communications of ACM, vol. 28, no. 9, pp. 904-920, September 1985..

13. Finin, T.: Understanding frame languages, Implementing PFL, AI Expert, pp. 51-56, December 1986.

14. Garcia-Molina, H.: Performance of udate algorithms for replaced data in a distributed data base. Ph.D. Dissertation, Stanford University, 1979.

15. Bernstein, P.A. and N. Goodman: Fundamental algorithms for concurrency control in distributed database systems. Technical Report, February 1980.

16. Bhargava, B.: Resilient concurrency control in distributed data base systems. IEEE Transactions on Reliability, vol. R-32, no. 5, December 1983.

17. Liskov, B.: On linguistic support for distributed programs. Proceedings 9th ACM SIGAT - SIGPLAN Symph. Principles of Programming, Albuquerque, NM, pp. 7-19, January 1982.

18. Albus, J.S., A.J. Barbera and R.N. Nagel: Theory and practice of hierarchical control. Twenty-third IEEE Computer Society International Conference, pp. 18-15, 1981.

19. Popek, Gerald J. and Bruce J. Walker: The LOCUS distributed system architecture. The MIT Press, Computer Systems Series, 1984.

20. Hsu, Chen and Craig Skevington: Integration of data and knowledge in manufacturing enterprises: A conceptual framework. Journal of Manufacturing Systems, vol. 6, no. 6, pp. 277-285, 1987.

21. Hsu, Cheng and Craig Skevington: Manufacturing architecture for integrated systems. Proceedings of the International Conference on the Manufacturing Science and Technology of the Future, MSTF'87, Boston, Ma, pp. 165-168, 1987.

APPLYING DISTRIBUTED ARTIFICIAL INTELLIGENCE TO FLEXIBLE MANUFACTURING

Michael J. Shaw
Department of Business Administration
University of Illinois at Urbana-Champaign
350 Commerce West
Champaign, IL 61820

Andrew B. Whinston
Krannert Graduate School of Management
Purdue University
West Lafayette, IN 47907

ABSTRACT

The information system for flexible manufacturing has become increasingly intelligent and distributed, primarily due to the technological advancements in the areas of artificial intelligence and computer networking. This paper presents a framework applying a distributed artificial intelligence (DAI) method for flexible manufacturing, where a manufacturing cell is treated as a problem solving agent. We will show the application of this DAI approach to the dynamic scheduling and configuring/reconfiguring of the manufacturing system.

1. Introduction

This paper describes a method incorporating distributed artificial intelligence (DAI) in flexible manufacturing systems (FMSs); the objective is to create adaptive organizations for flexible manufacturing, with the capability to improve the system performance by learning from manufacturing experience. The method is composed of two processes: (1) the bidding process, as the one used in the contract-net framework (Smith [1980]), for introducing an element of competition in the FMS and for recording a cell's (i.e., an agent's) performance; and (2) a genetic transformation process, for searching for more efficient configurations.

The bidding process has been shown to be an effective coordination mechanism for multi-agent problem solving (Davis and Smith [1983], Shaw and Whinston [1985], Parunak [1987]). Based on the contract-net framework, each agent in the loosely-coupled system bids for announced tasks and the best bidder would be selected to be the contractor for task sharing. By treating each cell in an FMS as a problem solving agent, our method extends this framework in the following manner: As a task is awarded to the best

NATO ASI Series, Vol. F53
Advanced Information Technologies for Industrial Material Flow Systems
Edited by Shimon Y. Nof, Colin L. Moodie

bidder, a hypothetical "payment" equivalent to the bid is paid by the manager of the task to the contractor, which in turn will affect the "strength" of the agents involved. The strength possessed by an agent reflects its ability to bid for tasks in the future.

Furthermore, an FMS is viewed as an adaptive system capable of learning to improve its performance. The learning is achieved by the Genetic Algorithm (Holland [1987]). It uses a strength parameter as the indication of fitnesses to discover desirable characteristics of those agents who have been successful; weaker agents will then be replaced by new agents inheriting these desirable characteristics, so that the overall performance can be improved. This competitive learning process would help the DAI system to adapt to its environment by the familiar "reproduction to fitness" evolution.

To illustrate the DAI approach, this method is applied to the scheduling of FMSs which have been treated as a multi-agent decision-making process in our previous work (Shaw and Whinston [1985, 1988a]).

2. Distributed Artificial Intelligence

The study of DAI is concerned with the cooperative solution of problems by a decentralized group of agents (Hughs [1987]). Several cooperation strategies have been discussed. For example, Davis and Smith [1983), working on the contract net, produced a cooperative approach using the negotiation metaphor to model the assignment of tasks to processors. Shaw and Whinston [1985, 1988a] have applied the contract-net method to intelligent manufacturing. Cammarata [1983] described strategies of cooperation that groups require to solve shared tasks effectively. In dealing with multi-agent planning, Georgeff [1983] discussed the problem of assuring noninterference among the plans for different agents; synchronization techniques were used to resolve potential conflicts between plans. He also developed a general theory of action for these plans. Rosenschein and Genesereth [1984] outlined several communication strategies and they presented a description and analysis of information passing between intelligent agents. In addition, Genesereth et al. [1986] considered strategies for cooperation in situations where communication between the agents is impossible. Shaw and Whinston [1988b] describes a method for incorporating the learning capability in a DAI system.

Research in distributed artificial intelligence, consistent with most decision theoretic research, assumes that each intelligent agent is self-interested and attempts to maximize its local rewards. To display coordinated behavior, self-interested agents must recognize that cooperation is to their benefit. This recognition can be achieved by sharing high-level goals (Durfee et al. [1985]) or by sharing decision matrices (Rosenschein and Genesereth [1985]).

Most approaches to distributed artificial intelligence have circumvented the self-interest problem by assuming that the agent responsible for decomposing the original problem into more manageable smaller ones does so in a way which ensures that the subtasks do not interact--although in many situations this may be too constraining. Alternatively, the agents can communicate their intentions to one another (Cammarata et al. [1983]). However, seeking to maximize their individual payoff, the agents do not have incentives to communicate honestly and are subject to moral hazard. Also, once we decide to design the multi-agent system in such a way that each agent is asynchronous, autonomous, and with distributed control, it becomes impossible to incorporate "commitment binding" as part of the agent's knowledge base (Rosenschein [1982]; Genesereth et al. [1986]).

Therefore, the information-exchange protocol and coordination strategies are critical design issues in distributed artificial intelligence (Georgeff [1983]; McArthur et al. [1982]). The bidding mechanism discussed in this paper for FMS scheduling can be viewed as a mechanism for regulating information exchanges and coordinating activities among agents.

3. Distributed Intelligence in an FMS

An FMS consists of asynchronous flexible cells communicating through a local area network. An important consideration for distributed scheduling in an FMS is the coordination among the asynchronous cells, ensuring that every job can be assigned to the most appropriate cell while attaining global efficiency. Shaw and Whinston [1985, 1988a] and Shaw [1987a&b] proposed a bidding mechanism based on the contract-net framework which treats each cell as a specialized, intelligent agent competing with other cells for jobs. The whole mechanism is highly analogous to the way the bidding process is used to allocate resources in human organization or to select the contractor(s) to whom the jobs can be delegated most economically.

The bidding mechanism is carried out based on the job hierarchy, where each job is decomposed into a set of tasks. Each task may correspond to the operations for a part family, thus can be performed in a single set-up. The job hierarchy is shown in Figure 1.

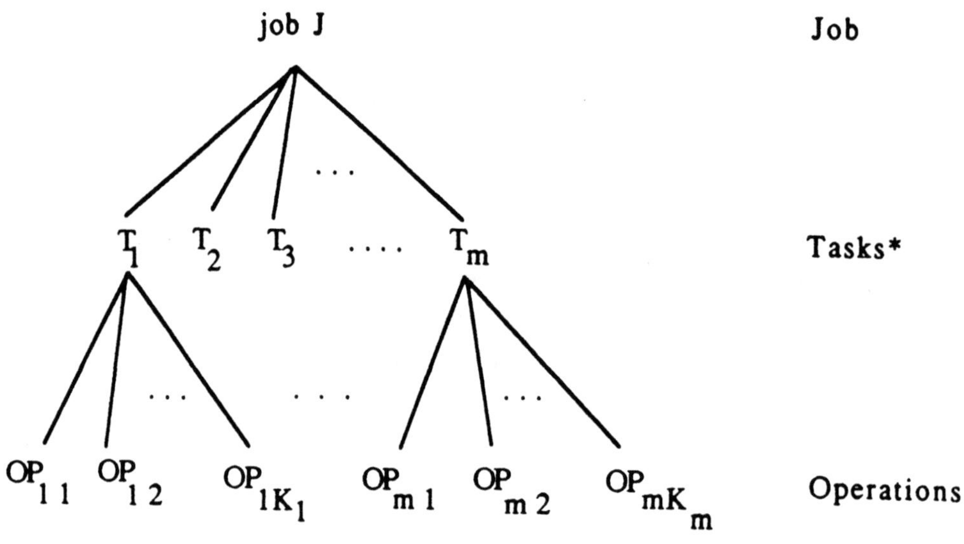

Figure 1. The Job Hierarchy

In this paper we take a step further and treat the FMS, a DAI system, as a type of cognitive organization (Holland [1975]) which can adapt to its environment by monitoring its experience and repeating past actions that appear successful, changing actions that appear unsuccessful.

In essence, the proposed DAI methodology treats an FMS as a market. The knowledge-based controller of each cell is an intelligent agent serving as a middleman in the market to supply jobs to other cells through contracting. That is, each cell can be viewed as an agent who gets paid for performing jobs for others and who would pay to contractors for sharing some of his work. When finished, the workpiece would have been processed by several cells through a series of bidding, thus the credit for completing the job should be properly allocated among the cells. This credit assignment is achieved by viewing the bidding process as a purchase of service. The bid submitted by a cell for a given job is a function of the cell's strength and the matchability between the cell's setup and the job's required operations. As a job is awarded by one cell to the lowest bidder, a hypothetical payment reflecting the cell's ability to perform the job is paid, which in turn will affect the strength of the cells involved. The strength level of a cell reflects its ability to bid for future jobs. The bidding mechanism is described in detail in Section 4.

From the control theory standpoint, the strength of an agent is treated as the state variable, the bid submitted as the control variable, and the bidding process essentially provides the feedback mechanism that, triggered by the output (i.e., the outcome of the bidding process), would change the state value and adjust the subsequent control. Viewing the FMS as a DAI organization with multiple agents, the method uses the bid (control variable) as the coordination between agents and the strength (state variable) as the belief of a cell on its own ability and usefulness. In the effort to improve performance, <u>learning</u> occurs when the agents (cells) modify their beliefs and <u>adaptation</u> occurs when changes are made in the control variable. As a result, the better cells which have been involved in more job completion would increasingly be favored by the bidding process, thus accelerating the subsequent selection decision. The less useful cells on the other hand are forced to adapt and change their configuration by mimicking some of the features of the successful cells in order to attract more jobs. Accordingly, two learning mechanisms are needed: one to determine the apportionment of credit to indicate each cell's past usefulness, the other to identify troubled cells and modify their configurations. The tenet of this learning scheme is that the competition caused by bidding will force the cells to learn, adapt, and evolve.

4. The Bidding Process

The bidding process introduces an element of competition in the DAI system since it is used to determine which of the agents will be assigned the job announced. In an FMS, the qualified cells would submit a bid to earn the right of performing the job--the lower the bid made by a cell, the more likely it is to be selected to get the job. Three parameters jointly determine the size of the bid made by a cell c at time t:

1. Strength (S(c,t)): an indication of past usefulness of c--higher S translates into higher productivity;
2. Specialization of the cell for the given job J (m(c,J)): a parameter indicating how specialized c is to process J, depending on whether the set-up is ready and if any tooling change is required; and
3. Readiness of the cell c in processing J (r(c,J)): a parameter reflecting c's state of readiness for doing J, as influenced by its queue, operating conditions, and the availability of supporting equipments.

The strength of each cell is used as the capital which can be traded for service. Whenever a cell successfully wins a job contract through bidding, its level of strength would increase by the amount of the bid. As a result, the subsequent bid made by the same cell will further increase its appeal because of the higher strength. Thus, the profitability of a cell depends on its ability to successfully perform as many jobs as possible. The amount of the bid is determined by the three parameter: $B(c,J,t) = W \times S(c,J) \times m(c,t) \times r(c,J)$, where m(c,J) and r(c,J) are normalized to the range of [0,1]. W is a constant with usually a small value, such as 1/8 or 1/16.

The bidding process is carried out by the announce-bid-award cycle. Let k be the total number of operations possible in the FMS. The operation requirements of a job J is specified by a message of k-tuples as follows: $\langle O_1, O_2, \ldots, O_k \rangle$, $O_j \in \{1,0,\#\}$, where $O_j = 1$ when jth operation is required, $O_j = 0$ when job operation is not required, and $O_j = \#$ indicates that the capability of doing operation O_j is optional.

The task-announcement message is represented by a similar type of k-tuples: $< t_1, \ldots, t_k >$, $t_j \in \{1,0\}$, where $t_j = 1$ indicates that the jth operation is required; and $t_j = 0$ indicates that the jth operation is <u>not</u> required. The k-tuple used for representing the capabilities of each cell is used as the <u>genetic characteristics</u> of the cell, which will be subject to reconfiguration in the evolution of the system.

When a winning cell accepts the job, its strength is increased by the amount of the bid, that is, $S(c,t+1) = S(c,t) + |B(c,J,t)|$. At the same time, the cell c' which announced the job would reduce its strength by the same amount, $S(c',t+1) = S(c',t) - |B(c,J,t)|$. This exchange between cells c and c' can be interpreted as c' paid c for its service. For the system, the purpose of strengthening c is to reward the cell for being able to provide needed operations more efficiently than others. The more bids a cell wins, the greater its strength is. This scheme essentially encourages the cells to be utilized as much as possible. Marinescu et al. [1987] describes a Petri Net model explicitly considering the strength as the feedback information affecting the future outcomes.

5. An Application Example

We shall use an example in this section to illustrate the bidding procedure for real-time scheduling. Suppose an FMS consists of five flexible cells, with operation loading shown in Figure 2. For the sake of simplicity, let us assume that the travel time between any two cells is a constant T_M. The bidding duration is T_b. The communication delay for transmitting messages is assumed to be insignificant comparing to T_M, T_b and the processing times. T_a is time taken for awarding a task.

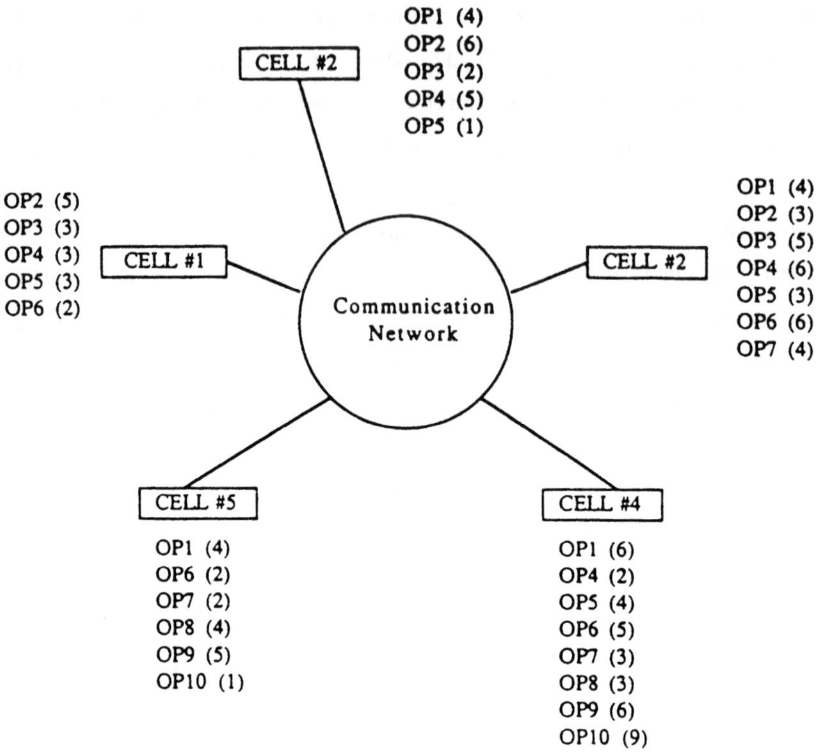

Figure 2. The example system and the operation loading of each Cell (the number in each parenthesis represents the corresponding processing time).

Suppose that a job, Job #003, arrives in the system at time 0. Job #3 has a job hierarchy as shown in Figure 3(a). Figure 3(b) shows that Task 3 cannot begin until both Task 1 and Task 2 are finished. This may happen because Task 1 and Task 2 are performed on two separate components; they are assembled together when finished and then Task 3 is performed. Thus, Tasks 1 and 2 can be announced simultaneously. Initially, an idle cell-host is selected randomly as the dispatcher of the newly arrived job. A manager cell is designated later after the first round of bidding (Cell #2 is the manager cell in this case since it is the awardee with longer processing time).

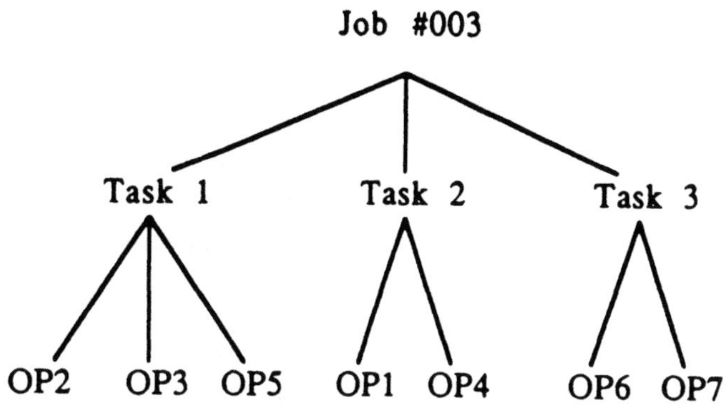

(a)

(b)

Figure 3. (a). The job hierarchy in the example.
(b). The precedence relations between the tasks.

The information flows passed through the communication network to carry out the bidding procedure is shown in Figure 4. Note that the bidding function, f_i, used by Cell i in example is based on the shortest-processing-time (SPT) rule; this rule is used here because of the assumption that the time for transporting parts between any pair of cells is a constant. Shaw [1987a] showed that in general situations, the earliestfinishing-time (EFT) rule performs better than the SPT rule. The EFT rule is calculated by the sum of travel time, processing time and expected queueing time.

Time	Agent	Information Type
0	Dispatcher	Task Announcement (Task 1)
0	Dispatcher	Task Announcement (Task 2)
T_b	CELL #1	Bid (f_1 (Task 1) = 11)
T_b	CELL #3	Bid (f_2 (Task 1) = 9)
T_b	CELL #3	Bid (f_3 (Task 1) = 12)
T_b	CELL #2	Bid (f_2 (Task 2) = 9)
T_b	CELL #3	Bid (f_3 (Task 2) = 10)
T_b	CELL #4	Bid (f_4 (Task 2) = 8)
T_b	Dispatcher	Award (Task 1 --> CELL #2)
T_b	Dispatcher	Award (Task 2 --> CELL #4)
$T_b + T_a$	CELL #2 (Manager)	Accept Task 1
$T_b + T_a$	CELL #4	Accept Task 2
$T_b + T_a + T_M$	CELL #2	(Task 1 Arrives)
$T_b + T_a + T_M$	CELL #4	(Task 2 Arrives)
$T_b + T_a + T_M + 8$	CELL #4	(Finish Task 2)
$T_b + T_a + T_M + 9$	CELL #2	(Finish Task 1)
$T_b + T_a + T_M + 9$	CELL #2	Task Announcement (Task 3)
$T_b + T_a + T_M + 9 + T_b$	CELL #3	Bid (f_3 (Task 3) = 9)
$T_b + T_a + T_M + 9 + T_b$	CELL #4	Bid (f_4 (Task 3) = 8)
$T_b + T_a + T_M + 9 + T_b$	CELL #5	Bid (f_5 (Task 3) = 4)
$T_b + T_a + T_M + 9 + T_b$	CELL #2	Award (Task 3 --> CELL #5)
$2 \times (T_b + T_a) + T_M + 9$	CELL #5	Accept Task 3
$2 \times (T_b + T_a + T_M) + 9$	CELL #5	(Task 3 Arrives)
$2 \times (T_b + T_a + T_M) + 13$	CELL #5	(Finish Task 3)
		Job #003 exists the system

Figure 4. The information flows for task negotiation in the example problem

6. Adaptation Through Genetic Transformation

We view an FMS as a cognitive system capable of adapting to the environment and improving its performance. This process of adaptation can be modeled after that of natural evolution. In evolution, the problem each species faces is one of searching for beneficial adaptations to a complicated and changing environment. The knowledge that each species has learned is embodied in the makeup of the chromosomes of its members. In the general DAI systems, this learning process will change the knowledge distribution among the agents. For the cellular FMS system, this learning process amounts to a sequence of reconfigurations to increase the productivity of the system. This process can be achieved by Holland's Genetic Algorithm.

The Genetic Algorithm was originally inspired by biological genetics, but was adapted by Holland to be a general problem solving technique. Here we use the Genetic Algorithm to improve the performance of a DAI system by searching for the most appropriate configuration. Specifically, the ability of each agent (i.e., a manufacturing cell) is represented by a chromosome. For a given FMS configuration, the bidding mechanism is used to coordinate the solution process among the agents; the strength of each agent is kept up to date to indicate its competitiveness. Genetic operators are then used to "transform" the badly performed cells by incorporating good "genes" in them. This adaptation will continue until the FMS as a whole can perform efficiently. Genetic reconfiguration results from randomly pairing up the selected parent agents and applying genetic operators, such as crossover, mutation, and inversion, to create the new offspring.

For example, the machining capabilities embedded in an FMS cell can be viewed as its genetic structure, using the same k-tuple representation described in the preceding section for representing the cell's capability list and the messages transferred. Periodically, the FMS faces the need of reconfiguration due to the unbalanced utilization of the high-performance cells and the weaker cells.

The process of reconfiguring new cells hinges on the ability to discover and exploit good building blocks (i.e., good genes) based on experience. The building blocks are components of the cells that can be decomposed and recombined easily. The genetic algorithm uses the strengths as the indication of fitnesses to discover good building blocks. The knowledge representation of such a building block is a subset of the K-tuple called the schema. Reconfiguration then is the process of discovering schemata and combine them to form a new one, which in turn provides the new configuration of the cell. The genetic algorithm proves to be an efficient technique for searching without exhaustive enumeration.

7. Conclusions

In this paper we have shown an adaptive organization for FMSs. The methodology presented has these features: (1) it treats an FMS as a market where each cell is analogous to a middleman; (2) the cells use strength as the hypothetical capital for granting contracts or charging services; (3) the bidding process serves as a mechanism to determine task allocations and, at the same time, to update the strength of the cells involved; (4) the bidding mechanism introduces an element of competition, since the strength would affect a cell's ability to bid; and (5) a genetic transformation scheme is used to recombine the genetic characteristics of well-performing cells to produce new cells in place of weak cells. We have shown the application of this method to the scheduling problem of flexible manufacturing systems.

References

Cammarata, S., McArthur, D., and Steeb, R., "Strategies of Cooperation in Distributed Problem Solving," in Proc. 8th Int. Joint Conf. Artificial Intelligence, pp. 767-770, Aug. 1983.

Davis, R. and Smith, R., "Negotiation as a Metaphor for Distributed Problem Solving," Artificial Intelligence, Vol. 20, pp. 63-109, 1983.

Durfee, E. H., Lesser, V. R., and Corkill, D. D., "Coherent Cooperation Among Communicating Problem Solvers," in Proc. 1985 Distributed Artificial Intelligence Workshop, pp. 231-276, Dec. 1985.

Genesereth, M. R., Ginsberg, M. L., and Rosenschein, J. S., "Cooperation Without Communication," in Proc. 5th Nat. Conf. Artificial Intelligence, Philadelphia, PA, Aug. 1986, pp. 51-57.

Georgeff, M., "Communication and Interaction in Multi-Agent Planning," Proceedings AAAI-83, 1983.

Holland, J. H., 1975, Adaptation in Natural and Artificial Systems, the University of Michigan Press, Ann Arbor, MI.

__________, 1986, Induction: Processes of Inference, Learning and Discovery, The MIT Press, Cambridge, MA.

Huhns, M. N., 1987, Distributed Artificial Intelligence, Morgan Kaufmann Publishers, Inc., Los Altos, CA.

Marinescu, D., Whinston, A., and Chaudhury, A., "Nets with Feedback," CSD-TR-731, Computer Science Department, Purdue University, 1987.

McArthur, D., Steeb, R., and Cammarata, S., "A Framework for Distributed Problem Solving," in Proc. 2nd Nat. Conf. Artificial Intelligence, Pittsburgh, PA, Aug. 1982, pp. 181-184.

Michalski, Carbonell, and Mitchell, 1986, Machine learning (II), Morgan Kaufmann, Los Altos, CA.

Parunak, H., "Manufacturing Experience with the Contract Net," in Distributed Artificial Intelligence (M. Huhn, Ed.), Pitman Publishing, London, 1987.

Rosenschein, J. S., "Synchronization of Multi-Agent Plans," in Proc. 2nd Nat. Conf. Artificial Intelligence, Pittsburgh, PA, Aug. 1982, pp. 115-119.

Rosenschein, J. S. and Genesereth, M. R., "Communication and Cooperaation," Stanford Heuristic Programming Project, Stanford, CA, Tech. Rep. HPP-84-5, 1984.

_____________, "Deals Among Rational Agents," in Proc. 9th Int. Joint Conf. Artificial Intelligence, August 1985, pp. 91-99.

Shaw, M., 1987a, "Distributed Scheduling in Computer Integrated Manufacturing: The Use of Local Area Network," International Journal of Production Research.

_____________, 1987b, "FMS Scheduling as Cooperative Problem Solving," Working Paper # 1326, Department of Business Administration, University of Illinois, Champaign, IL.

_____________, and Whinston, A. B., 1985, "Task Bidding and Distributed Planning for Flexible Manufacturing," Proceedings IEEE Conference on Artificial Intelligence Applications, Miami, FL.

_____________, 1988a, "A Distributed Knowledge-Based Approach to Flexible Automation: The Contract-Net Framework," International Journal of Flexible Manufacturing Systems (forthcoming).

_____________, 1988b, "Learning and Adaptation in a Distributed AI System," Proceedings of the AAAI Workshop on Distributed Artificial Intelligence, Lake Arrowhead, CA.

Smith, R., "The Contract-Net Protocol: High-Level Communication and Control in A Distributed Problem Solver," IEEE Transactions on Computers, C:29-12, pp. 1104-1113, 1980.

Combining Predicate Transition Nets and Rule Based Systems for Planning, Programming and Exception Handling in Manufacturing Process Control

R.Dillmann, U. Negretto
Institut für Prozessrechentechnik und Robotik (IPR)
University of Karlsruhe, Germany

Abstract:
This paper presents a method which allows to specify, to program and to simulate the control behavior of flexible assembly systems. A hierarchical top down approach for planning and programming of a robot based assembly cell is discussed. The cell which consists of multiple units like stationary robots, mobile robots and peripherals is used for product assembly operations. Synchronization and exception handling for each of the assembly processes has to be considered. For specification of the cell program predicate transition nets are used. The predicate transition net of the cell can be broken down into subnets and into device specific local nets which define the operation of each specific device. The work is supported by an ESPRIT-CIM project.

1. Introduction

Increasing the flexibility of assembly cells implies the application of an integrated information and control system. Flexibility means that the cell is not restricted to only one product or product family, but enables also the assembly of various different products. To execute and monitor the operations of the multiple cell components the control system needs a symbolic control model for the overall process of the cell, for the processes of the cell components and their subsystems, for the synchronization of them, for conflict resolution and for the parallelism of the processes in the system. In the past various applications of Petri nets and their extensions have been studied. They have been extended to include time, parallelism and synchronization into the process control representation. Systems like EPOS, ESPNET etc. which are mostly used for manufacturing planning are able to generate automatically Petri nets. In process control they are used to specify the discrete behavior of manufacturing cells. Problems arise when exceptions and their resolution have to be specified. Each exception handling type has to be defined separately which quickly yields to very complex nets. It is cumbersome to handle such complex nets. One solution is the use of predicate transition nets (PTs), which can easily be implemented and structured in Prolog. Further advantage of this method is that with Prolog the specification of exceptions and their resolution by rules can be facilitated. Starting from the global functional view of a manufacturing cell a top down approach to each individual functional unit of the cell can be supported. The smallest functional units of the cell are devices which perform elementary operations (EOs). The global cell functionality is represented by global nets, the functionality of local devices is represented by local nets. This paper presents a method based on predicate transition nets and on rule based systems to specify the static and dynamic properties of FMS (Flexible

NATO ASI Series, Vol. F53
Advanced Information Technologies for
Industrial Material Flow Systems
Edited by Shimon Y. Nof, Colin L. Moodie

Manufacturing Systems). The purpose of this work is a task independent representation of the functionality of real manufacturing cells and the simulation of the processes of the flexible cells based on the specified predicate transition nets. The paper is structured as follows:

- In the second chapter a typical robot based assembly cell is introduced. Referencing this example the requirements to methods for modeling such assembly cell types are discussed. Based on this requirements the selected predicate transition nets are introduced.
- In the third chapter the PT-net method is applied to the robot based assembly cell as example. Different specification levels according to the degrees of process abstraction and the considered process control level are presented, following a strict top down approach.
- In chapter four the combination of the PT-Net representation and the rule based language Prolog is discussed in regard to the affinity between both. The net is transformed and implemented into a Prolog notation. The simulation of the process is shown and the achieved results are discussed.
- Concluding, requirements to AI tools for supporting robot based assembly cell planning and programming using graph oriented representations are summarized.

2. System configuration and requirements to modeling of manufacturing cells

2.1 The assembly cell

Planning and programming of robot based assembly cells with interacting components is one of the main research areas of the IPR of the University of Karlsruhe. Since several years the interaction of stationary robots with mobile robots within manufacturing cells is under study. A prototype manufacturing cell for product assemblies is under construction with the purpose of integrating its operational control into a CIM environment. The most important device within the examined assembly cell is the Karlsruhe Autonomous Mobile RObot (KAMRO), see fig. 1. All other components of the cell have a stationary layout such that the KAMRO is able to reach them and to perform assembly operations and interactions with them.
Each operation of the mobile robot is preceded by a definite travelling and docking operation to the desired device to achieve a defined reference position and to start manipulation. The components of the cell are as follows:

- A conveyor is used as transportation device between the cell and its environment. It´s function is to feed the cell with unassembled single parts and to transport the completed assemblies out of the cell.
- A stationary robot (Puma 600) is used to pickup the parts from the conveyor and to put them into a magazine with defined position and orientation.
- The magazine is used as cell material storage device. It serves as material buffer for the incoming parts.
- An assembly station (AS) which consists of an assembly table with fixtures for pallets or parts is the location of the assembly process. It is

planned to extend the cell with multiple assembly stations to allow the assembly of different products in parallel.

- The autonomous mobile robot performs a variety of tasks. First it has to perform transportation tasks within the cell, i.e. it transports parts from the magazine to the assembly station and assemblies from the AS to the conveyor. Second it has to perform the workpiece handling operations to assemble the parts. It grasps the parts from the magazine, transports them to the AS and performs there the assembly task. The KAMRO consists of a mobile omnidirectional platform and of two robots (Puma 260) in portal configuration hanging upside down on gauges fixed on the platform. The platform is also equipped with a part buffer device for transportation of several parts. The capacity can be modified. Travelling, docking and manipulation is supported by tactile sensors, infrared, sonar and vision systems.

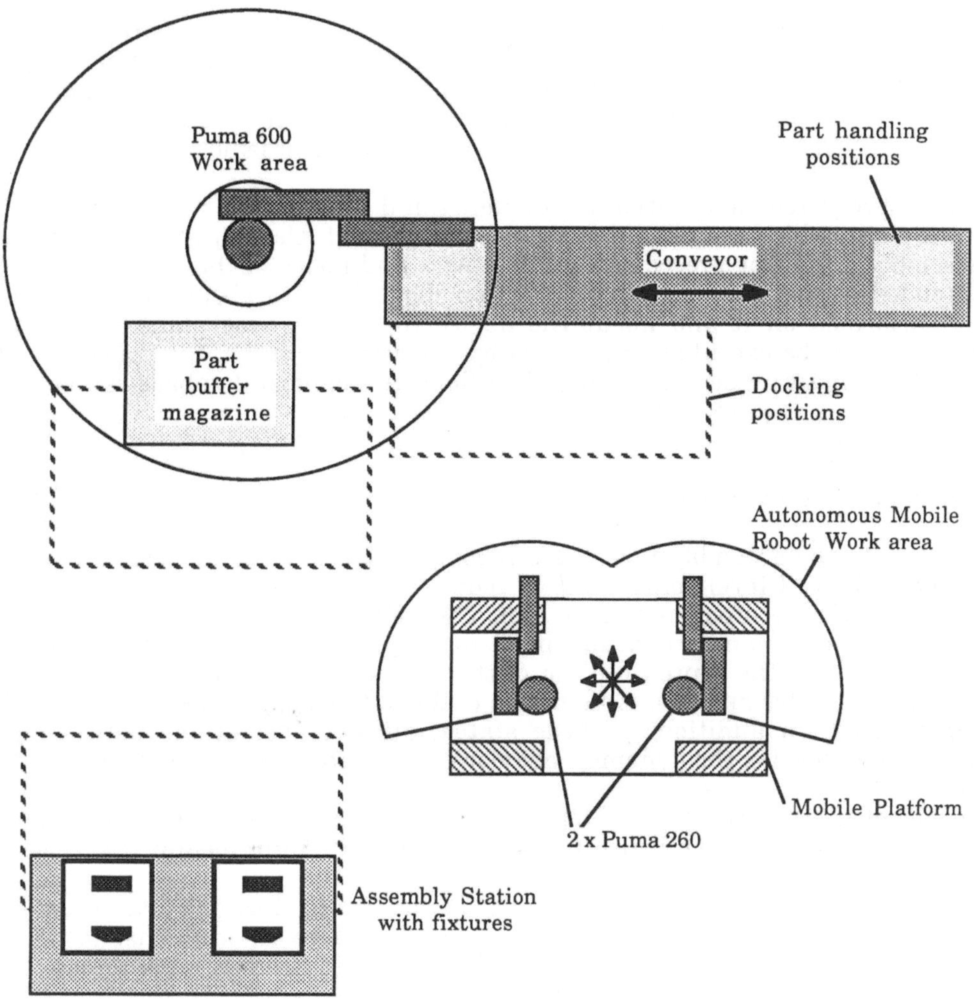

Fig. 1 : The flexible assembly cell

The cell control system is considered to be hierarchical. There is a high cell level control tier which controls the overall cell function. Under its control the cell components are operating each one under a local independent controller. The robot systems and especially the complex KAMRO system is further subdivided into two further control tiers, the high level interpreter (HLI) and the low level interpreter (LLI). The HLI support the autonomy of the mobile robot system i. e. local planning, monitoring, decision making and executing. The LLIs control the execution of elementary operation of the mobile robot, like grasp, insert, release, dock, undock,move etc. Thus, the EOs are considered to represent the smallest functions to be executed in the cell. With this hierarchical control concept simple functional units and very complex functional units may be coordinated within a cell.

For specifying and planning of the cell functionality, the cell program,the control structure and its simulation, the following basic data is required:

- Geometric data (layout,topology, size of cell devices,working area)
- Kinematic data (degrees of freedom, joint types, relation between joints, kinematic chains, kinematics of fixtures, conveyors, etc.)
- Technologic data (storage capacity, material, load capacity, required accuracy, dynamic properties of assembly processes).

Further required information for planning and programming are part specific work plans, which are related to the desired product. Usually they are based on assembly graphs or precedence graphs. Arcs and nodes of these graphs can be extended, so that they contain information about the part, about the operation to be performed on it and about the device to do it. These plans specify the operations to be executed on one part by one component, i.e. the trajectories to handle the part by device <name, identifier>. A technical database can be used to access this required information.

2.2 Model requirements

To execute the assembly task the cell components have to cooperate. For this purpose the cell level control tier is required to synchronize and monitor the processes of all components in the cell. This cell controller needs in addition to the given work plans a task independent representation of all components in the system, i.e. the representation of the real work cell. For this purpose the local device states and as a combination of them, the global cell states have to be represented. In addition to those states, allowed transitions between the different states, i.e. the dynamic behavior of the system, has to be defined in the model.
To model the functionality of flexible assembly cells various methods can be taken into consideration which fulfill the following requirements:

- Operating with various independent components, flexible systems are able to execute time parallel operations. Therefore the model has to support the representation of parallel processes running on different devices and the synchronization between them.
- The control of such complex systems is facilitated through the subdivision of the system into a hierarchy of subsystems. In this case the specification method has to allow hierarchical structuring of the system into

subcomponents. Further the specification of the relations between the subcomponents has to be supported.

- In dependence of the local capabilities of the single device control, i.e. the degree and complexity of autonomously executable operations, the level of detail represented in the model has to vary in accordance. The method can fulfil this requirement if it supports the specification of the components functionality on various level of abstraction. Thus, the model considers the required level of detail for each control level in the hierarchy, see fig.2.

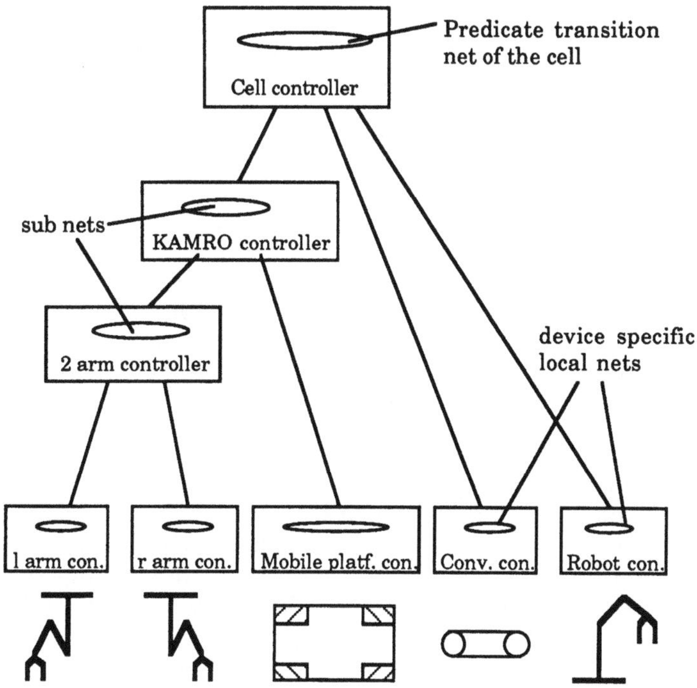

Fig. 2: Control levels in the cell

- At each moment the actual cell state has to be retrievable, because checking of all possible states would require a large search space. In addition it would be reasonable to know the next reachable states to reduce the amount of conditions to be monitored before the next step.
- The processes in flexible manufacturing systems have definite start and end points and are discrete. The simulation of the process in the model must be therefore event driven. In addition to the simulation, which is essential for the verification of the operation flow in the cell, the model has to ensure the applicability of the programs and of the control structure for the real plant.
- During simulation of FMS various problems can arise due to unforeseen conflict situations and deadlocks. The model should be accessible to mathematical analysis and verification methods for deadlock and conflict detection in an early stage.

- In this context always occurring exceptions, which happen in real processes have to be considered. The exceptions have first to be detected, classified and in the following recovery operations have to be executed. The selected method has either to predict autonomous recovery functions or to access dedicated exception handling systems (i.e. expert systems). Further it has to allow a restart from the interrupted system state without the need of restarting the cell from the initial state.
- Finally the specification method should be applicable in each project phase. The information representation has to be the communication link between the involved experts in the different cell control design phases.

To meet the above mentioned requirements we decided to use predicate transition nets, which are Petri nets with extensions proposed in recent publications. In the following a brief introduction is given into the applied method.

2.3 Predicate - transition nets

Petri nets with their various extensions have already been used to model the dynamic properties of discrete systems, like distributed systems, distributed data bases, FMS, communication protocols and others.
To reduce the size of the representation and thus, to provide a more compact and efficient model the concept of predicate - transition nets (PTs), which is an extension of Petri nets, is introduced. PTs are directed graphs with two different types of nodes: predicates and transitions. The relations between the nodes are specified through arcs. An arc always connects a predicate with a transition or vice versa. To each transition and each arc in the net an inscription is associated specifying either variables (to arcs) or logic equations (to transitions). A marking, i.e. the system state, is defined through the individual-token tuples in each predicate at a definite time, also called partial states. A capacity relation is associated to each predicate defining the maximal amount of tokens for each individual type. A firing rule is associated to each transition defining the change of partial states between two or more predicates. Because PTs are abstract models it is possible to give an interpretation to the elements of the net assigning those to elements of the physical system. This assignment is more or less arbitrary and thus very important in the aim of having an exact representation of the systems behavior.
For example a robot could be associated to a token and the execution of a task represented through the partial state "a token <identifier> is on the predicate <identifier> ". Transitions could then be interpreted as events and represent the state changes of the real system. The interpretation defines also the firing rules of the transitions. A transition which is associated to an event fires if and only if it is enabled and if a predefined value specified in the inscription is reached. This value could be a sensor threshold which is exceeded. A transition is enabled if the tokens marking the input predicates are members of the individual token set associated to the arc inscription. When a transition fires the tokens associated to the input arcs are removed from the input predicates. According to the logic equations specified in the transitions inscriptions the information related to the tokens may change. The sets of the output predicates are checked and the predicates are marked with the appropriate tokens.

3. Application example

The chosen interpretation follows the assignment rules stated before. According to this we associate the parts of the desired product assembly to the individual tokens. The token parameters specify precedence values, pointers which refer to geometry data, assembly positions and others. Each possible state of the components is associated to a predicate and accordingly the events defining the changes in the states of the components were associated to transitions.
The states of the components on the highest level of abstraction are:

KAMRO	-	is docked on magazine, robots are inactive
	-	is docked on assembly station, robots are inactive
	-	is docked on conveyor, robots are inactive
	-	is docked on <comp_name>, robots perform operation <op_name, parameters>
	-	is moving from Pos_p_1 to Pos_p_2
Conveyor	-	carries <parts> into the cell
	-	carries <parts> out of the cell
	-	is inactive
Robot	-	performs operation <op_name, parameters>
	-	is inactive
Magazine	-	is filled with <(part a,pos mag_a),(part b, pos mag_b),....>
Assembly station	-	is filled with <(part a,pos ass_a),(part b, pos ass_b),....>

For each component a subnet is modeled which reflects its physical states and the allowed changes between the states through transitions. The transitions associated to the events do not only represent the state changes of components but are also used to synchronize between components.
For example the transition which starts the procedure: "The robot gets part <name, pos> of the conveyor <part a, pos con a>" synchronizes the robot and the conveyor because it triggers the execution of the operation only if the conveyor is inactive and has a part available. The real movement of the robot and the actual transport operation of the conveyor have to be further detailed on adequate levels.
The five subnets, one for each component, build up the cell net. The cell net is a task-independent model of a cell. Simulation and then control of processes in FMS with such nets requires the fulfillment of two additional conditions:

1. The high informal representation has to be detailed to a formal level, where the model can be directly associated to processes and signals.
2. The execution of cell tasks requires the representation of the task work plan on the cell model (cell programming).

To 1.: As mentioned before the PTs formalism provides methods to refine or coarse nodes of the net. In this approach a top-down method to gradually refine structures is followed. The inscriptions become more formal with each step. The nodes can be refined to linear predicate transition sequences or be expanded to subnets.
To 2.: The introduced individual tokens and the inscriptions with variable parameters allow to use the cell model for the task-dependent process simulation. The variables in the inscriptions are associated to task-specific

constants, which direct the flow of individual token in the net. The task description, here for an assembly task, has to be given in a form which specifies the priorities, i.e. the assembly sequence of the individual parts of the product. The use of assembly graphs is an adequate representation, see fig. 3.

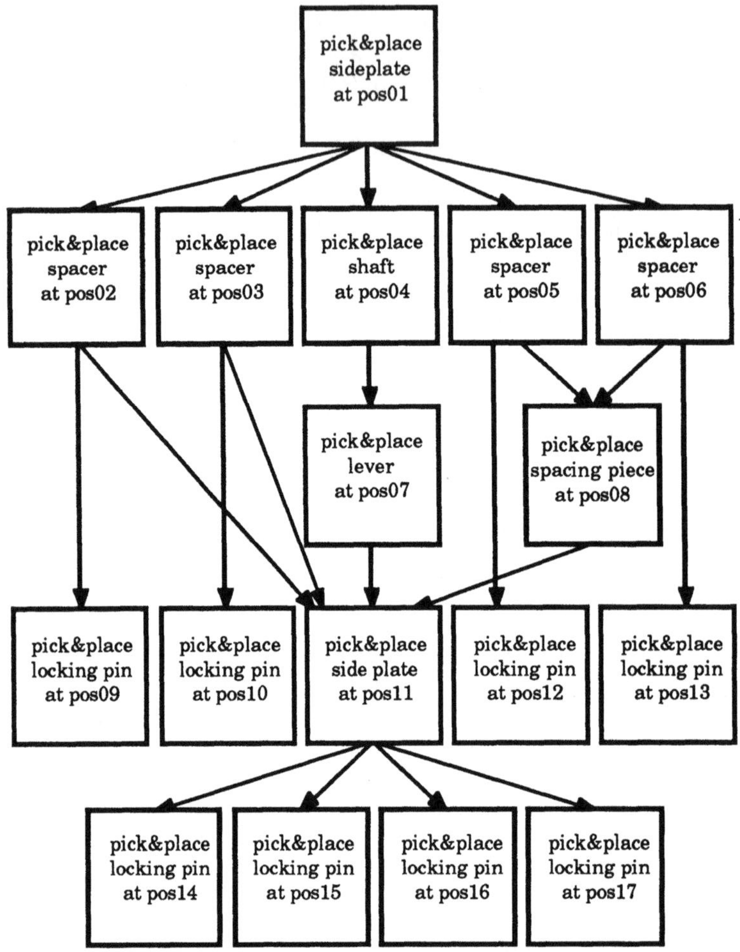

Fig. 3: The assembly graph of the Cranfield benchmark

With this description and the cell model the cell controller is able to monitor and to execute the process. Details will be discussed later. Because of the event driven strategy of the method the consideration of time is not explicitly necessary. The transitions are assumed to fire immediately if they are enabled and the conditions hold. For other applications the introduction of time is quite useful.

To model the concept of time, an additional predicate and two transitions are used. The first transition starts the process and the time (clock, watchdog), the predicate represents the active process and the second transition specifies the end of the process and of the time.

Separating both cell modeling and cell programming the application can be analyzed for different properties, first cell specific properties and then task dependent properties. In the first case, known algorithms are to be applied to

detect conflicts in the net, where strategies have to be selected to resolve these conflicts. In the simplest case a random selection algorithm can be used or else some priorities may be specified to arcs to solve the conflicts. The analysis of the net in dependence of the task can be performed to detect deadlocks due to specified system parameters, for example a buffer with too low capacity.
The performance of the components can be evaluated and optimized through simulation techniques. After detection of inefficiencies changes in the scheduling strategy can be performed. The simulation of the process is executed by a "player"-program, which has a general, task independent form. This program monitors the token flow in the net and executes cyclewise operations specified in the inscriptions. A cycle consists of the following steps:

- Determine enabled transitions through actual marking
- Check the preconditions of enabled transitions if individual tokens of the in_predicates match the parameters of the arc inscriptions
- Firing of the enabled transitions in dependence of the fulfilled conditions. Change of the actual marking.
- Execute the operations connected with the new marked predicates

Concluding, this method provides an efficient model for flexible assembly cells which is used to simulate the processes within the assembly system. There are although lacks in this representation due to the fixed structure of the net and to the fast growing model size for complex systems, despite to the concise representation form. For example modeling of all possible states of a FMS would require also the predictive determination of all states after occurred errors and of the recovery operations to be executed, in order to achieve a state from which the nominal operation can continue. To overcome this weakness the following extension was taken into the approach.

4. Combining PTs and the rule based language Prolog

The combination of PTs and Prolog is reasonable because of the affinity between both methods. This affinity is given by the correspondence between different elements of PTs and elements of Prolog:

- Transitions can be represented by rules, if the input marking and the preconditions of the transitions are associated to the conditional part of the rule and if the operational part of the rule causes the new marking, which would result after firing
- Facts can be associated to markings or to instantiated variables
- The inference engine of Prolog corresponds to the player program of the PTs formalism because it checks the rules if they succeed and performs the related operations. These procedures create new facts or delete old ones and generate the new fact base.

A general description of PTs was implemented in Prolog and adapted to the needs of the system. In the transformation the net properties were saved. Therefore also rules could be defined to detect deadlocks or to check the reachability of a marking. Like in chapter 3, on the general representation of PTs in Prolog the interpretation of the elements of the real system is done in analogy to Prolog rules and facts.

To simulate the overall task process the assembly graph was also transformed in a Prolog notation. To each part of the assembly set the part specific task was associated and stepwise refined to subtasks and elementary operations.
A module generates from this description component specific EO sequences for each active component, i.e. KAMRO, conveyor and robot. In this sequences the priority information of the assembly graph is still kept. This information is necessary for the later application of rescheduling strategies in case of exceptions in the process. It is not needed to program explicitly the parallel procedures and the synchronization points. The cell model and the component specific EO sequences are combined and the process is simulated driven by the inference engine of Prolog. The combination is effected with the instantiation of the variables of the cell model. The part specific parameters are matched with the variables during the simulation. In Fig. 4 The two separated representations and the combination of them are shown.

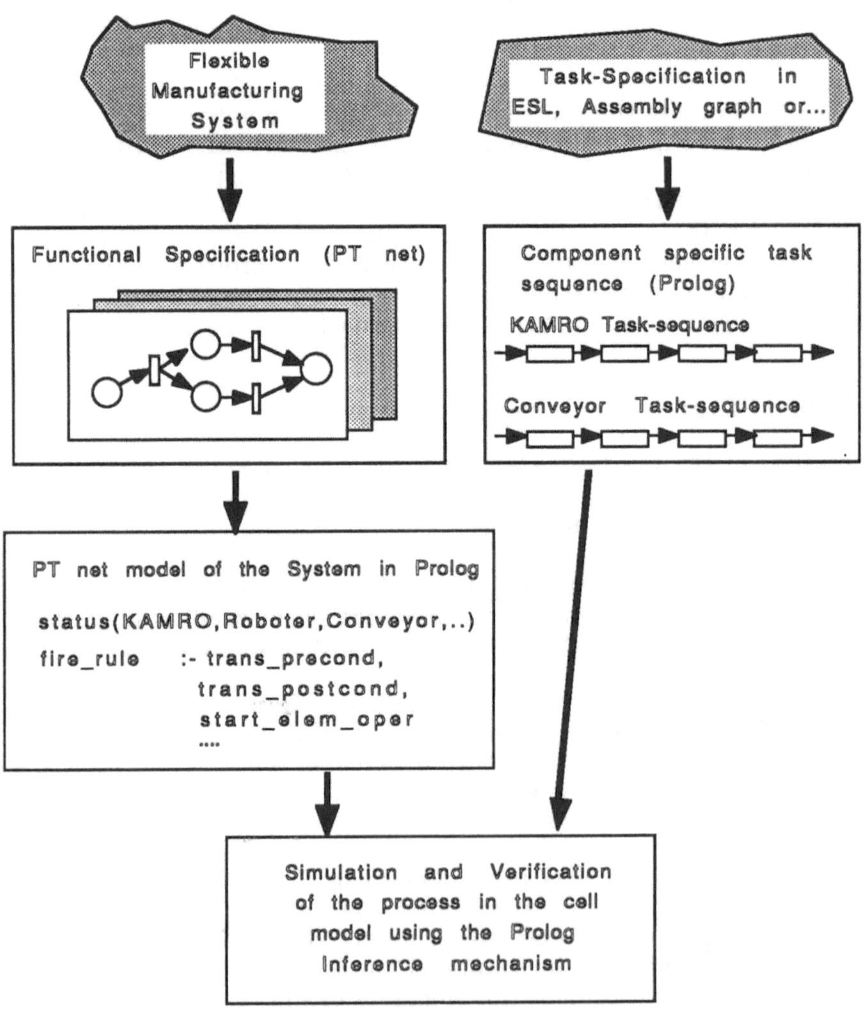

Fig. 4: Representation levels and combination of both models

One of the problems which arise transforming PTs in Prolog is the loss of the parallelism specified in the net. In general Prolog follows a sequential top down processing. The use of existing parallel Prolog versions like Parlog was not taken into consideration because of the problems of these languages. Also the process time of the used sequential Prolog (IFprolog) needed to complete the cycle "matching - conflict resolution - action", was smaller for some order of magnitude as the operation time. Therefore, a quasi parallel behavior of the cell process simulation could be reached. Quasi parallelism means that, although sequential processing, for the modeled cell no difference to real parallelism was measured. This can become a problem in case of much larger and more complex systems.
To gain a first evaluation of the modeled system the operation time of each operation was defined. The quasi parallel behaviour is outlined in fig.5, where the result of a simulation run with predefined operation times is shown. In the next step the integration of sensors is foreseen, so that the predefinition of operation times won´t be necessary.

```
Time_0 : Conveyor begins in_transport of
              {Task (Part_1, Priority(1), parameter(pos_x1)) }
              {Task (Part_2, Priority(2), parameter(pos_x2)) }
              {Task (Part_3, Priority(3), parameter(pos_x3)) },

         kamro begins to move from in_free_space to to_magazine.

Time_1 .
Time_2 .
Time_3 .
Time_4 .

Time_5 : conveyor ended in_transport of
              {Task (Part_1, Priority(1), parameter(pos_x1)) }
              {Task (Part_2, Priority(2), parameter(pos_x2)) }
              {Task (Part_3, Priority(3), parameter(pos_x3)) }

Time_6 : roboter begins pick_and_place of
              {Task (Part_1, Priority(1), parameter(pos_x1)) }
              from on_conveyor to in_magazine,

         kamro ended move from in_free_space to to_magazine.

Time_7 .

Time_8 : roboter ended pick_und_place of
              {Task (Part_1, Priority(1), parameter(pos_x1)) }.

Time_9 : Roboter begins pick_and_place of
              {Task (Part_2, Priority(2), parameter(pos_x2)) }
              from on_conveyor to in_magazine.

Time_10 .
  .
  .
```

Fig. 5: Example of a process simulation run with predefined operation times

In the example first the two asynchronous concurrent operations ´conveyor transport´ and ´KAMRO move´ were started. During the 5 consequent steps the cell states are unchanged. Then, the conveyor stops transporting and changes its state into inactive. In the following the transition to activate the robot is fired and the pick and place operations can start. The following steps are vice versa. The declarative definition of objects and their relations in Prolog facilitates modeling of the system consistently. In advantage of the strict structure of the PTs Prolog offers methods to change the model during execution of the process. This can be done with the Prolog build in predicates "assert" and "retract". This allows to formulate general recovery rules, which after occurred exceptions enable to reach states from which the normal process can continue. The analysis of the execution and the replanning is based on the application of the given rules on the status knowledge before and after the exception and on the planned goal state. The difference of this AI method to the explicit declaration of exception handling mechanisms in PTs relies in the out of the application of the rules implicitly generated conclusions. They result out of different combinations of rules to which a probability factor can be associated. These criteria are used to choose the adequate recovery actions. This replanning can require too much time to recover on-line, but because of the concurrency representation in the model the operation continuation of the components not involved in the error is guaranteed.

5. Conclusion

As shown in this paper the specification of a manufacturing task for a robot based cell and its programming can be done using the concept of PTs. Following a top down approach, the operations of the robot based manufacturing cell can be successively refined till the EO level of each cell component. The basic data and knowledge required is of different type, see fig.6. The program information is represented by PTs, in general by graph structures. An integrated development environment for robot assembly cell programming as requested for CIM systems has to fulfill the following basic requirements:

- Management of the basic data and task knowledge
- Assurance of consistency of the basic data and task knowledge
- Support of the generation, modification and retrieval of the basic data from an object oriented user and user program point of view.

Classical data bases and available CAD systems don´t allow a feasible extension of their information representation capabilities. The effective extension and integration of such systems can be done with the aid of a frame engine. This allows the extension of of the basic data while maintaining the desired functionalities. A frame engine which satisfies the above requirements including handling of predicate transition nets can be realized on the base of Knowledge Craft. In addition to the frame representation this AI tool kit supports Prolog, OPS-5 and Common Lisp.

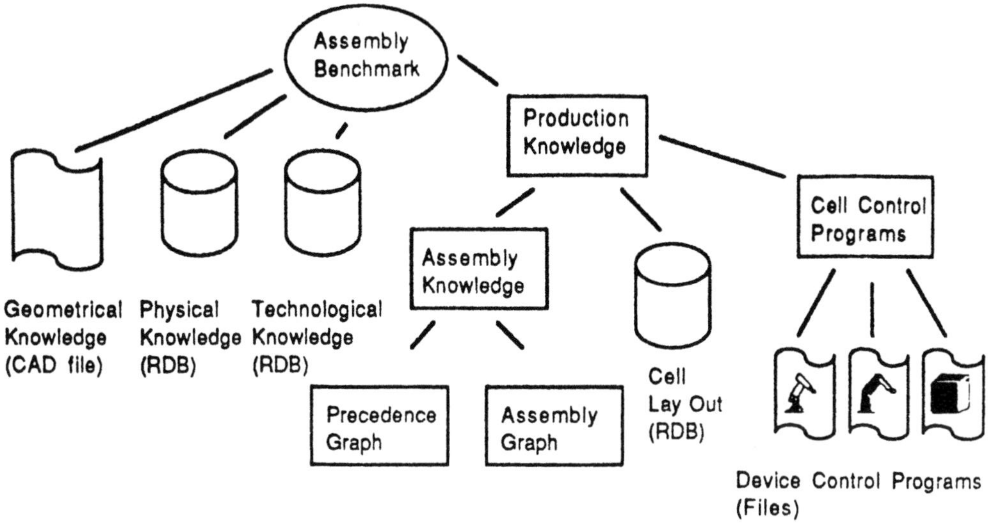

Fig. 6: Different types of knowledge required for programming the assembly benchmark

6. References

Alanche,P. et al : "PSI: a Petri net based simulator for FMS", 5th European workshop on applications and theory of Petri nets, AARHUS, 1984

Valette R. : "Control of flexible Production Systems and Petri Nets", 3rd European Workshop on Theory and Application of Petri Nets, Varenna, Italy, 1982.

Rillo M. : "Petri Nets and Flexible Manufacturing Systems", Internal Report, Sao Paulo University, 1984.

Peterson,J.L.: "Petri Nets Theory and modelling of Systems", Prentice Hall Inc., 1981.

Reisig W. : "Petrinetze - Eine Einführung", Springer Verlag, 1986.

Brauer W., Reisig W. and Rozenberg G. : "Petri Nets: Central Models and Their Properties", Part I and II, Proceedings of an Advanced Course, Bad Honnef, September 1986.

Jensen K. : "Colored Petri Nets and the invariant Methode", Theoretical Computer Science, n. 14, 1981, pg. 317-336.

Genrich H.J. and Lauterbach K. : "System modeling with High Level Petri Nets", Theoretical Computer Science, n. 13,1981, pg. 109 - 136.

Jensen K. : "High Level Petri Nets" in Applications and Theory of Petri Nets, Pagoni A. and Rozenberg G. eds., Informatik Fachberichte, Vol. 66, Springer Verlag, 1983.

Zisman M.D. : "Use of Production Systems for Modeling Asynchronous Concurrent Processes", in Pattern-Directed Inference Systems, Waterman D.A. and Hayes-Roth F. eds., Academic Press, 1978, pg 53-68.

Valette R.: "Nets in Production System", GMD Advanced Course on Petri Nets, Bad Honnef, 1986.

Sahraoui A., Atabakhche M., Courvoisier M. and Valette R.: "Joining Petri Nets and Knowledge Based Systems for Monitoring Purposes", Proceedings of 1987 IEEE International Conference on Robotics and Automation, Raleigh, USA, 1987.

Negretto,U., Rillo,M.: ´´Erweiterte Petri Netze in flexiblen Fertigungs- und Montagesystemen ´´, Robotersysteme,Band 4, Heft 1, 1988, p. 34 - 42

Acknowledgement:

This work is based on a CEC sponsored ESPRIT project (ESPRIT project No.623 : Operational Control of Robot System Integration into CIM). The work on PT-Nets has been performed at the Institute for Real-Time Computer Systems and Robotics (Prof. Dr.-Ing. U.Rembold, Prof. Dr.-Ing. R.Dillmann) at the University of Karlsruhe.

A KNOWLEDGE-BASED SYSTEM FOR PRODUCTION EQUIPMENT SELECTION AND LAYOUT

Andrew Kusiak and Sunderesh S. Heragu
Department of Mechanical and Industrial Engineering
The University of Manitoba
Winnipeg, Manitoba R3T 2N2
Canada

ABSTRACT

In this paper, a knowledge-based system (KBSMD) for selection of machines and material handling equipment and determining a layout is discussed. The approach presented involves solving a model for the equipment selection problem. This results in a selection of the type and number of machines and material handling equipment. Next, the system selects a model for machine layout and solves it using an algorithm. It then uses the knowledge stored in the knowledge-base to incorporate qualitative aspects not considered in the equipment selection and layout models and modifies the solutions accordingly. Thus the system utilizes the optimization and knowledge-based approaches to solve the manufacturing system design problem.

1. INTRODUCTION

The manufacturing industry has witnessed significant developments in recent years. These developments can be measured by the increase in the number of automated systems in the manufacturing industry. The key to success of these systems is effective use of available resources such as machines, tools, fixtures and material handling equipment. Proper use of these resources has increased productivity. At the same time, the design problems related to modern manufacturing systems have become more complex. Designers and users of automated manufacturing systems have attempted to develop new tools to cope with these complexities. Knowledge-based systems represent a class of modern tools that have been applied to improve the design and management functions in automated manufacturing systems.

NATO ASI Series, Vol. F53
Advanced Information Technologies for
Industrial Material Flow Systems
Edited by Shimon Y. Nof, Colin L. Moodie

The design of a manufacturing system involves the:

i) selection of the number and type of machines such as lathe, milling machine, drilling machine, etc.

ii) selection of the number and type of material handling equipment, for example, robot, automated guided vehicle (AGV), etc.

iii) layout of machines on the factory floor. While determining the layout factors such as material handling equipment path, clearance between machines, etc., must be considered.

To solve each of the above problems, optimization techniques have traditionally been used. However, they are not always easy to apply because:

- the data required by the optimization models may not be easily available;

- their scope of applicability may be limited;

- they require the interaction of human experts; and

- the algorithms are often not able to provide optimal solutions for industrial problems because of their complexity.

On the other hand, knowledge-based systems have typically been used to solve problems that are either too complex for mathematical formulation or too difficult to solve using the optimization approach. Some of the difficulties encountered when using optimization techniques may be offset by combining them with knowledge-based systems. A knowledge-based system can be thought of as a computer program which consists of knowledge in a problem domain and therefore can use the knowledge stored to take decisions like a human expert.

Traditionally, human experts have solved problems using the optimization approach; if knowledge-based systems are to be used successfully, they should be used in such a way that they replace human experts rather than the optimization approach.

In this paper, a knowledge-based system named KBSMD for solving the manufacturing system design problem, i.e., production equipment selection and machine layout, is presented. The problem itself is discussed in the next section. Models which can be used to formulate the machines and material handling equipment selection problem and layout problem are provided in section 3. The problem solving approach of the system is described in section 4. In section 5, the components of KBSMD are presented. A numerical example is provided in section 6. Conclusions are drawn in the last section.

2. MANUFACTURING SYSTEM DESIGN

Manufacturing system design is a complex activity because it requires solving a number of problems arranged in a hierarchy (see figure 1). Each problem in the hierarchy may have to be solved more than once. In the remainder of this section, the four-level hierarchical approach to manufacturing system design is presented.

The four-level hierarchical approach involves the following problems: production equipment selection; machine cell formation; machine layout and cell layout (figure 1). Note that group technology (GT) provides the basis for the four-level hierarchical approach to design of automated manufacturing systems.

Production equipment selection: There are three important issues regarding the production equipment selection problem which are discussed below:

i) timing of machine selection;

ii) timing of material handling equipment selection; and

iii) budget consideration.

Timing of machine selection: The type and number of machines required depends on the parts to be manufactured and the manufacturing processes to be used. To select the required machines it is not necessary to know all the

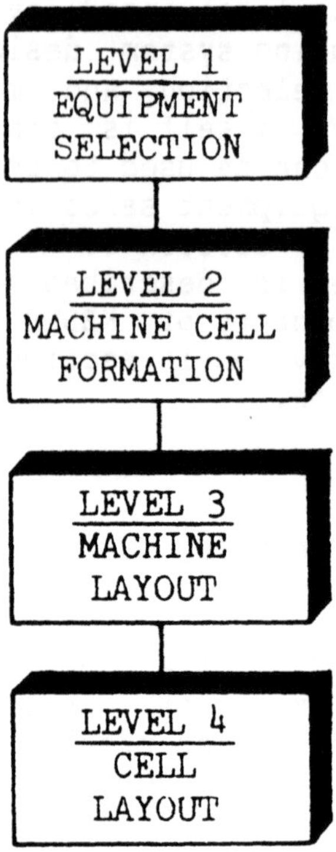

Figure 1. Four-level hierarchical approach to design of automated manufacturing systems

details regarding the parts to be manufactured or manufacturing processes to be used. It is sufficient to know the data concerning representative operations (parts). A representative operation (part) can be defined as an operation (part) which represents a number of other operations (parts) that have the same requirements and characteristics as the representative operation (part). For example, if part P_r is representative of n parts $\{P_1,\ldots,P_n\}$ and if part P_r can be processed using machine M_1 and material handling system H_1, then any of the n parts $\{P_1,\ldots,P_n\}$ can also be processed using the equipment M_1 and H_1.

The required production equipment has to be selected in the initial stages of the design of a manufacturing system. In many instances such a selection may only be a preliminary decision and may have to be modified at a later stage. This aspect is discussed in the subsection on budget consideration.

Timing of material handling equipment selection: The representative operations (parts) may indicate the type of equipment that is required for material handling. However, this is not the only factor which has an impact on the selection of material handling equipment. The layout of machines and cells also determine the type and number of material handling equipment to be selected. For example, using a robot as a handling device imposes the arrangement of machines in a circular pattern (figure 2). If it has already been decided to select an AGV to tend these machines, it may be necessary to revise the decision since a robot handles material more efficiently than an AGV when the layout has a circular pattern.

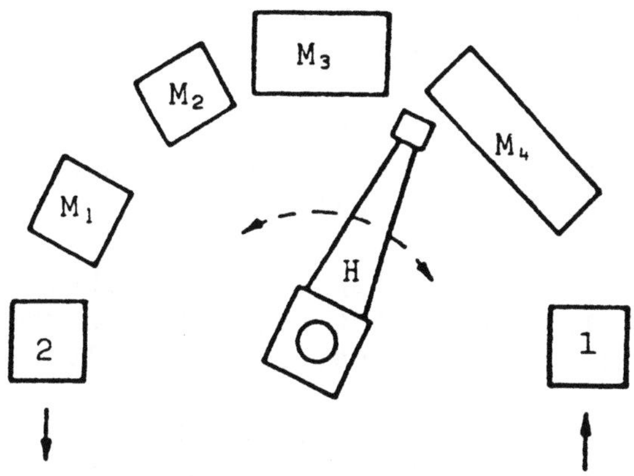

M_1, M_2, M_3, M_4 - machines
1 - pallet with incoming parts
2 - pallet with outgoing parts
H - handling robot

Figure 2. Circular layout of machines

Budget consideration: If budget was not a constraint in the design of a manufacturing system, then the selection of material handling equipment could have been postponed until the machine layout is determined. But in practice, budget frequently imposes a severe constraint on the equipment to be selected and hence it is necessary to at least:

a) determine the type of material handling equipment to be used while solving the equipment selection problem; or

b) allocate a percentage of the budget for material handling equipment which will be selected after the machine layout has been determined.

If procedure (a) is adopted, then, as mentioned before, the equipment selection problem may have to be solved again at the time the machine layout is determined, as the type of material handling equipment selected initially may be found unsuitable for the determined layout. If procedure (b) is adopted, then it calls for a proper allocation of the budget for selection of material handling equipment and involves stochastic data.

Machine cell formation: In some manufacturing systems, there is a need to group machines into cells using the group technology (GT) concept, where parts are grouped into part families and machines into cells based on the similarity of process plans. The implementation of GT provides:

i) better planning and control of the manufacturing process: It is easier to schedule a family of parts than each individual part;

ii) reduction in setup cost: Scheduling parts with identical machining requirements reduces setup time and hence setup cost considerably;

iii) better space utilization: GT implementation results in cells, each being a compact group of machines. As a result, space utilization improves; and

iv) higher rate of equipment utilization: When compared to a functional manufacturing facility, a given part in a cellular manufacturing facility spends much less time in travel. As a result, parts are available for processing which in turn reduces the equipment idle time.

Machine layout: A frequently used objective for arrangement of machines in a cell minimizes material handling cost. The layout of machines in a cell is referred to as the machine layout problem. While solving the layout problem, factors such as space constraints, safety, type of material handling equipment, etc., must be considered.

Cell layout: The cell layout problem involves the location of each machine cell on its corresponding site in a way that minimizes plant-wide material handling costs.

3. MODELING THE EQUIPMENT SELECTION AND LAYOUT PROBLEMS

The machine and material handling selection problems have been solved in the past using a number of approaches. Miller and Davis (1977) have surveyed, classified and compared a number of them. The dynamic resource allocation model developed by Miller and Davis (1978), the machine procurement model in Murty (1983), the aggregate production planning and machine requirements planning model in Behnezhad and Khoshnevis (1988) are examples of the models which have been used to formulate the machine selection problem. Webster and Reed (1971) and Hassan et al. (1985) have developed models for the selection of material handling equipment. Kusiak (1987a) has developed a model that can be used to formulate the production equipment selection problem. Below, an integer programming formulation (Model M1) of the machine and material handling equipment selection problem is presented. To formulate the model the following notation is introduced:

c_{ij} cost of performing operation o_i on machine M_j

h_{ij} cost of handling part P_i using material handling equipment H_j

r_{ij} time required to perform operation o_i on machine M_j

s_{ij} time required to transport part P_i using material handling equipment H_j

T_j time available on machine M_j

S_j time available on material handling equipment H_j

N_i number of operations o_i to be performed

M_i number of parts P_i to be manufactured

U_j cost of machine M_j

V_j cost of material handling equipment H_j

B total budget available

x_{ij} number of operations o_i to be performed on production equipment M_j

y_{ij} number of parts P_i to be transported on material handling equipment H_j

u_j number of machines M_j selected

v_j number of units of material handling equipment H_j selected

Model M1

The objective function of model M1 minimizes the procurement cost of machines and material handling equipment as well as the operating and handling cost of the parts manufactured.

$$\min \sum_{i=1}^{m} \sum_{j=1}^{n} c_{ij} x_{ij} + \sum_{i=1}^{p} \sum_{j=1}^{q} h_{ij} y_{ij} + \sum_{j=1}^{n} U_j u_j + \sum_{j=1}^{q} V_j v_j \qquad (1)$$

$$\text{s.t.} \quad \sum_{j=1}^{n} x_{ij} \geq N_i \qquad i=1,\ldots,m \qquad (2)$$

$$\sum_{i=1}^{m} r_{ij} x_{ij} \leq T_j u_j \qquad j=1,\ldots,n \qquad (3)$$

$$\sum_{j=1}^{q} y_{ij} \geq M_i \qquad i=1,\ldots,p \qquad (4)$$

$$\sum_{i=1}^{p} s_{ij} y_{ij} \leq S_j v_j \qquad j=1,\ldots,q \qquad (5)$$

$$\sum_{j=1}^{n} U_j u_j + \sum_{j=1}^{q} V_j v_j \leq B \qquad (6)$$

$$x_{ij} \geq 0, \text{ integer} \qquad i=1,\ldots,m;\ j=1,\ldots,n \qquad (7)$$

$$y_{ij} \geq 0, \text{ integer} \qquad i=1,\ldots,p;\ j=1,\ldots,q \qquad (8)$$

$$u_j \geq 0, \text{ integer} \qquad j=1,\ldots,n \qquad (9)$$

$$v_j \geq 0, \text{ integer} \qquad j=1,\ldots,q \qquad (10)$$

Constraints (2) and (4) ensure that the required number of operations are performed and the required number of parts are transported, respectively. Constraints (3) and (5) impose that the time available on each machine and material handling equipment is not exceeded. Constraint (6) ensures that the available budget is not exceeded. Constraints (7)-(10) impose nonnegativity and integrality.

The machine layout problem can be divided into two classes:

- single-row; and
- multi-row.

The single-row layout problem can be modeled as a linear program with absolute values in the objective function and constraints while the multi-row layout problem can be modeled as a non-linear program with absolute values in the objective function and constraints (Heragu and Kusiak 1987). It is assumed that the production equipment are square or rectangular in shape. Also, the physical orientation of the machines are assumed to be known. In order to formulate the single-row layout problem (Model M2) the following notation is used:

f_{ij} number of trips to be made between machines i and j

c_{ij} cost per trip between machines i and j

d_{ij} minimum distance by which machines i and j are to be separated horizontally

x_i distance between center of machine i and vertical reference line vrl

l_i length of machine i

The parameters, decision variables and reference lines vrl relevant to model M2 are illustrated in figure 3.

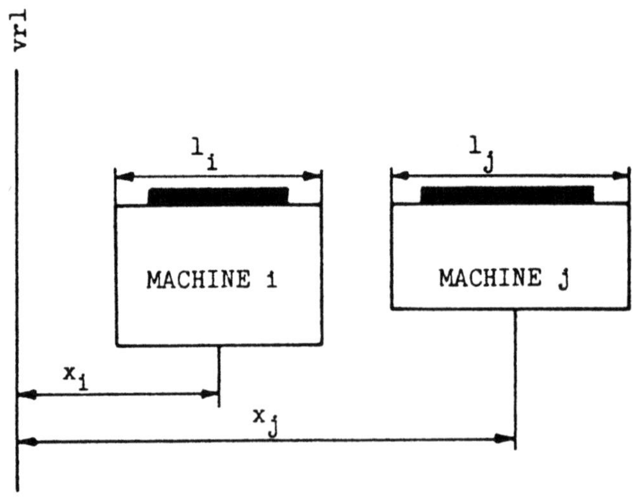

Figure 3. Illustration of decision variables and parameters for the single-row machine layout problem

Model M2

The objective function of model M2 minimizes the total cost involved in making the required trips between the machines.

$$\min \sum_{i=1}^{n-1} \sum_{j=i+1}^{n} c_{ij} f_{ij} (|x_i - x_j|) \qquad (11)$$

$$\text{s.t.} \quad |x_i - x_j| \geq 1/2(l_i + l_j) + d_{ij} \qquad i=1,\ldots,n-1 \quad j=i+1,\ldots,n \qquad (12)$$

Constraint (12) ensures that no two machines in the layout overlap.

To model the multi-row layout problem, the following variable and parameters in addition to those defined for model M2 are introduced.

y_i distance between center of machine i and horizontal reference line hrl

d'_{ij} minimum distance by which machines i and j are to be separated vertically

b_i breadth of machine i

The parameters, decision variables and reference lines vrl relevant to model M3 are illustrated in figure 4.

Model M3

The objective function of model M3 minimizes the total cost involved in making the required trips between the machines.

$$\min \sum_{i=1}^{n-1} \sum_{j=i+1}^{n} c_{ij} f_{ij} (|x_i - x_j| + |y_i - y_j|) \qquad (13)$$

$$\text{s.t.}\quad |x_i - x_j| + Mz_{ij} \geq 1/2(l_i + l_j) + d_{ij} \qquad i=1,\ldots,n-1 \quad j=i+1,\ldots,n \qquad (14)$$

$$|y_i - y_j| + M(1 - z_{ij}) \geq 1/2(b_i + b_j) + d'_{ij} \qquad i=1,\ldots,n-1 \quad j=i+1,\ldots,n \qquad (15)$$

$$z_{ij}(1 - z_{ij}) = 0 \qquad i=1,\ldots,n-1 \quad j=i+1,\ldots,n \qquad (16)$$

Constraints (14) and (15) ensure that no two machines in the layout overlap. Constraint (16) ensures that only one of the two constraints (14) and (15) holds.

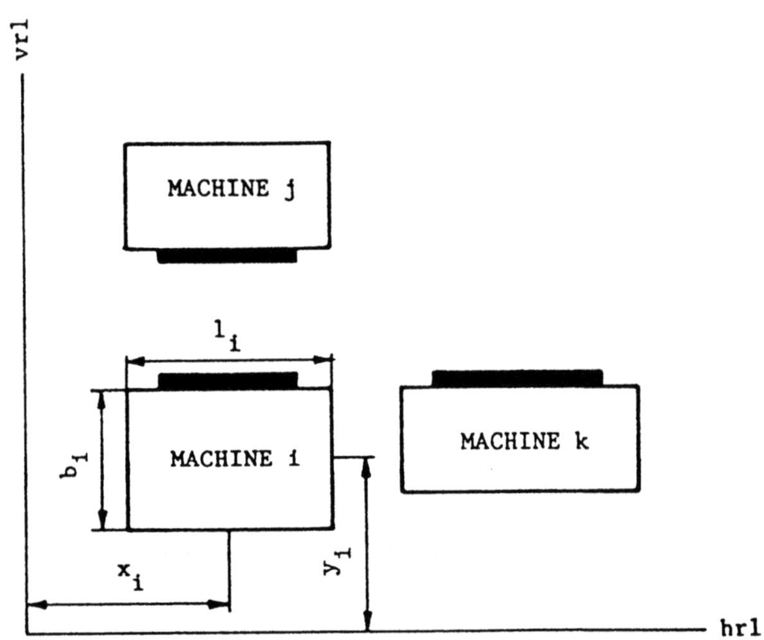

Figure 4. Illustration of decision variables and parameters for the multi-row machine layout problem

Models M2 and M3 can be transformed into an equivalent linear mixed integer programming model and non-linear model. As an example, the linear mixed integer formulation of model M2 is presented below. The following notation is introduced.

$$x_{ij}^{+} = \begin{cases} (x_i - x_j) & \text{if } (x_i - x_j) > 0 \\ 0 & \text{if } (x_i - x_j) \le 0 \end{cases} \tag{17}$$

$$x_{ij}^{-} = \begin{cases} -(x_i - x_j) & \text{if } (x_i - x_j) \le 0 \\ 0 & \text{if } (x_i - x_j) > 0 \end{cases} \tag{18}$$

From the above, it can be easily verified that:

$$|x_i - x_j| = x_{ij}^{+} + x_{ij}^{-} \tag{19}$$

$$(x_i - x_j) = x_{ij}^{+} - x_{ij}^{-} \tag{20}$$

Model M2a

$$\min \sum_{i=1}^{n-1} \sum_{j=i+1}^{n} c_{ij} f_{ij} (x_{ij}^{+} + x_{ij}^{-}) \tag{21}$$

s.t. $x_i - x_j + Mz_{ij} \ge 1/2(l_i + l_j) + d_{ij}$ $\quad i=1,\ldots,n-1;\ j=i+1,\ldots,n$ (22)

$-(x_i - x_j)+M(1-z_{ij}) \ge 1/2(l_i +l_j)+d_{ij}$ $\quad i=1,\ldots,n-1;\ j=i+1,\ldots,n$ (23)

$x_{ij}^{+} - x_{ij}^{-} = x_i - x_j$ $\quad i=1,\ldots,n-1;\ j=i+1,\ldots,n$ (24)

$x_{ij}^{+}, x_{ij}^{-} \ge 0$ $\quad i=1,\ldots,n-1;\ j=i+1,\ldots,n$ (25)

$z_{ij} = 0 \text{ or } 1$ $\quad i=1,\ldots,n-1;\ j=i+1,\ldots,n$ (26)

Constraints (22)-(23) ensure that no two facilities in the layout overlap. Since z_{ij} is a 0 or 1 variable, only one of the constraints (22)-(23) holds. Constraints (24) is identical to expression (20). Constraint (25) ensures nonnegativity while constraint (26) imposes integrality.

If the dimensions of the floor plan are known, one may add suitable constraints and ensure that the equipment is located within the boundaries of the floor plan. For more details refer to Heragu and Kusiak (1987).

4. PROBLEM SOLVING APPROACH FOR MANUFACTURING SYSTEM DESIGN

Knowledge-based systems have been applied to solve the equipment selection and layout problems. For example, Fisher and Maimon (1988) and McGlennon et al. (1988) have developed systems for selecting material handling equipment; Fisher and Nof (1984) and Kumara et al. (1985) have developed systems for facility layout. In this section, the problem solving approach of KBSMD - a system developed for solving the equipment selection and layout problems is presented.

The manufacturing system design problem is a complex problem. It was shown in section 2 that some of its subproblems may have to be solved more than once. In this section a two-phase approach for solving the system design problem is presented. It involves the following nine steps.

<u>Phase 1</u>

Step 1: Collect data from the user.

Step 2: Represent declarative knowledge concerning parts, operations and machines, as frames.

Step 3: Eliminate appropriate machines (material handling equipment) from consideration, so that the number of integer variables in model M1 is reduced.

Step 4: Generate the following matrices as necessary:
(i) operation-machine cost matrix
(ii) operation-machine time matrix
(iii) part-material handling equipment cost matrix

(iv) part-material handling equipment time matrix.

Generate the following vectors:
- (i) time available on each machine
- (ii) time available on each material handling equipment
- (iii) number of times each operation is to be performed
- (iv) number of units of each part to be manufactured
- (v) cost of each machine
- (vi) cost of each material handling equipment.

Step 5: Solve model M1 for the data generated in step 4, using the algorithm stored in the model and algorithm base.

Phase 2

Step 6: Generate the flow matrix data.

Step 7: Select a model (M2 or M3) for the layout problem.

Step 8: Solve the selected model using a suitable algorithm in the model and algorithm base.

Step 9: Evaluate the solution produced by the algorithms, incorporate qualitative factors not considered by the equipment selection and layout models, and modify the solution, if necessary.

In phase 1, the required machines and material handling equipment are selected. The system obtains the required data from the user and stores it in the data base. Note that the system obtains data required for solving the equipment selection as well as layout problems in step 1. Next, the system attempts to eliminate some machines (material handling equipment) from consideration, in order to reduce the number of integer variables in model M1. The system considers procurement cost, machine (material handling equipment) capabilities, etc., to determine whether an equipment is to be eliminated from consideration. A sample rule which is used to eliminate machines from consideration is shown below.

IF machine M_1 can perform operation o_1

AND machine M_2 can perform operation o_2

AND machine M_3 can perform operations o_1 and o_2

AND the cost of machines M_1 and M_2 is greater than that of machine M_3

THEN eliminate machines M_1 and M_2 from consideration.

The system applies knowledge to the data stored in the data base, in order to generate a set of data suitable for model M1. The data generated are in the form of matrices and vectors. Using the algorithm available in the model and algorithm base, the model is solved for the data generated by the system.

In phase 2, the machine layout problem is solved. Based on the machines selected, the system generates the flow data, i.e., frequency of trips between each pair of machines. Then, a model (M2 or M3) is selected. The algorithm presented in Heragu and Kusiak (1987) which is available in the model and algorithm base is used to solve model M2. The solution produced by the algorithms are evaluated, qualitative factors not considered in the models are incorporated in the solution and the revised solution is provided to the user.

5. STRUCTURE OF THE KNOWLEDGE-BASED SYSTEM KBSMD

KBSMD consists of the following five components (figure 5):

- data base;
- knowledge base;
- model and algorithm base;
- explanation module; and
- inference engine.

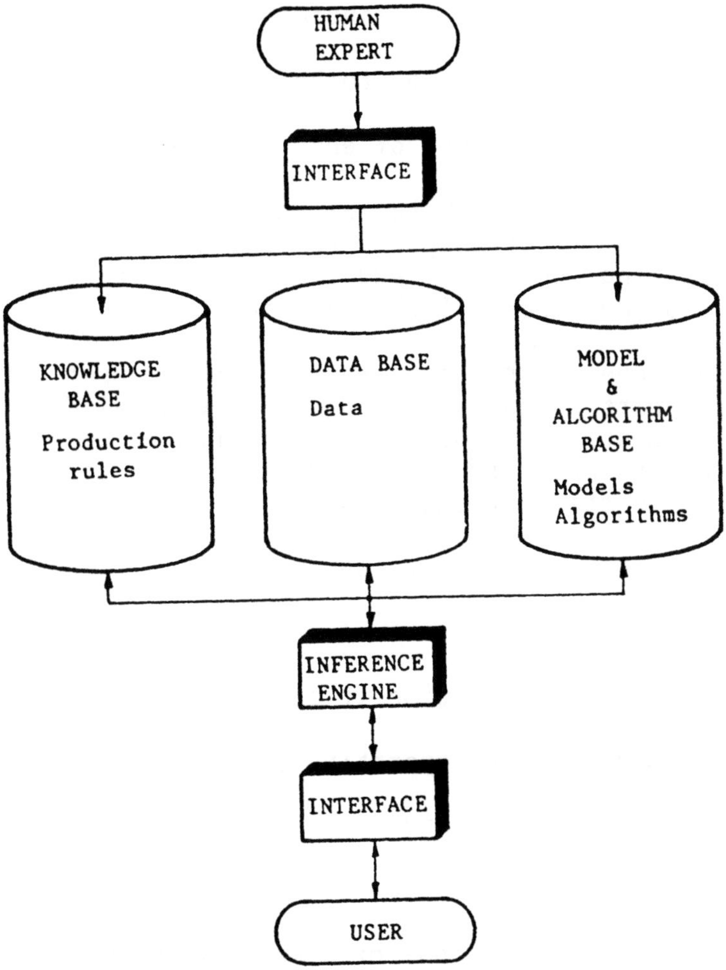

Figure 5. Basic components of the system KBSMD

Data base: The data base consists of two categories of data - data for the equipment selection problem and data for the layout problem. The data are obtained from the user in an interactive mode.

For the first category of data, the user is required to enter only the information concerning parts and machines. For each part, the system requests the following data:

- code;

- number of parts to be manufactured;
- code of each operation to be performed;
- number of operations of each type required; and
- type of available material handling equipment that can be used to transport the part, the corresponding operating cost(s) and time(s) involved, and the procurement cost of the material handling equipment.

If the code of the part entered by the user matches with that of a part already existing in the knowledge base, the system requests only the number of parts to be manufactured; the remaining data are not requested since they are already available in the knowledge base.

For each machine, the following data is requested:

- code;
- code of each operation it can perform;
- the corresponding operating cost and operating time;
- procurement cost; and
- dimensions.

As before, if the code of a machine matches with that of another already existing in the knowledge base, the system does not request the user to enter the operating cost, operating time, procurement cost and dimensions.

The second category of data (pertaining to the layout problem) includes information on the following:

i) clearance required between machines

ii) relationship indicators between each pair of machines

iii) location restriction (if any) for each machine

iv) dimensions of the floor plan.

Details regarding the above data required for solving the layout problem are provided below.

Clearance matrix: Elements of the clearance matrix indicate the minimum distance by which each pair of machines are to be separated if they are located adjacently in the layout.

Relationship indicator matrix: KBSMD uses three relationship indicators, namely: A,O and X, which indicate the adjacency requirements that are to be satisfied while placing machines in the layout. The system uses the relationship indicator matrix not to determine the closeness desired between the machines but to determine whether a pair of machines must: 1) be located in adjacent sites; 2) not be located in adjacent sites; 3) located as suggested by the algorithm which solves the layout problem.

Location restrictions: It may sometimes be desirable to restrict the location of a particular machine to a particular site. Such information may be easily recorded in KBSMD.

Dimensions of the floor plan: This information is required so that KBSMD can determine whether the arrangement of machines violates space constraints. It is assumed that the floor plan is rectangular in shape and the user is required to input the length and breadth of the floor plan. As mentioned previously, the system generates the flow matrix data after the machines have been selected. The entire data belonging to the second category are entered in the form of lists.

Knowledge base: The knowledge base consists of, declarative knowledge and procedural knowledge. Frames are used to represent the declarative knowledge concerning machines and parts.

The data in the data base is used to create declarative knowledge. There are three main types of declarative knowledge in KBSMD, namely knowledge concerning:

- parts;
- operations; and
- machines.

A sample frame belonging to each type of declarative knowledge is shown in the figures below.

```
(part (part_name) (oper_reqd  (oper_1 . . .   oper_N))
                  (no_of_oper (n_1    . . .   n_n))
                  (mhs_avail   (mhs_1 . . .   mhs_n))
                  (mhs_oper_cost (moc_1 . . .  moc_n))
                  (mhs_oper_time (mot_1 . . .  mot_N)))
```

Figure 6. A sample frame representing a part

```
(operation (oper_name) (M_reqd  (M_1    . . .  M_n))
                  (oper_cost    (cost_1 . . .  cost_n))
                  (oper_time    (time_1 . . .  time_n)))
```

Figure 7. A sample frame representing an operation

```
(machine M_name) (oper       (oper_1  . . .  oper_n))
                    (oper_cost (cost_1  . . .  cost_n))
                    (oper_time (time_1  . . .  time_n)))
```

Figure 8. A sample frame representing a machine

Lists are used to represent declarative knowledge pertaining to the machine layout problem.

The entire procedural knowledge is stored in the form of production rules and is divided into seven classes: The following four are used for solving the equipment selection problem:

- Class 1 rules for collecting data from the user;
- Class 2 rules for generating required data in the form of matrices and vectors;
- Class 3 rules for selecting the algorithm required to solve the model; and
- Class 4 rules for incorporating qualitative factors (not considered in the model) into the solution produced by the algorithm.

A sample rule belonging to each class is shown below.

Class 1 Rule:

IF **code of a machine (provided by user) does not match with that of another machine existing in the knowledge base**

THEN request user to provide: code of machine, code of each operation it can perform, corresponding operating cost, time, purchasing cost of machine and its dimensions.

Class 2 Rule:

IF **time available on each machine is provided by user**

THEN generate a time vector which indicates time available on each machine.

Class 3 Rule:

IF **the number of integer variables in model M1 > 25**

THEN apply algorithm A1

ELSE apply algorithm A2.

Note that algorithm A1 refers to the branch-and-bound algorithm (see for example, Schrage 1984). Algorithm A2 is a heuristic algorithm for solving model M1.

Class 4 Rule:

IF material handling equipment H_1 is not suitable to tend machine M_1

AND material handling equipment H_1 is selected to tend machine M_1

THEN select material handling equipment H_2 that can perform the same functions as equipment H_1 and that is suitable for machine M_1.

In addition to the above, KBSMD consists of three classes of rules for solving the layout problem. They are:

i) Class 5 rules for selecting an appropriate model and algorithm for the layout problem;

ii) Class 6 rules for making initial assignments based on input data; and

iii) Class 7 rules for checking whether the layout is implementable and incorporating qualitative factors not considered in the equipment selection and layout models.

As a result of solving the equipment selection problem in phase 1, the required machines and material handling equipment are known. Then, Class 5 rules rules which are capable of selecting an appropriate model and algorithm for solving the layout problem are invoked. Note that the layout problem can be modeled as a linear or a non-linear program (model M2 or M3) depending on whether the layout problem is single-row or multi-row. A sample Class 5 rule which selects the model and algorithm for a given problem is provided below.

Class 5 rule:

IF layout problem is single-row

THEN select model M2

ELSE select model M3.

Class 6 rules are used to make initial assignments. The initial assignments may be specified by the user or decided by the system. For example, if an AGV is used as the material handling equipment, then a battery charging station may be required. It is advantageous to assign machines with maximum flow value between them to adjacent sites near the battery charging station. As a result, the AGV spends less time in travel to the battery charging station (see the sample Class 6 rule presented below).

Class 6 rule:

IF type of material handling equipment used is AGV

AND the assignment of machines i,j with maximum flow value between them is not restricted to any particular site

THEN locate battery charging station near one end of the layout and assign machines i and j to sites which are adjacent to the battery charging station.

User desired assignments have priority over the assignments done by KBSMD. Thus, if the user desires to assign machines with maximum flow value between them to sites which are not near the battery charging station, the system does not attempt to relocate the machines near the battery charging station. Class 6 consists of 12 rules. Next, the layout obtained is evaluated for implementability by Class 7 rules. A layout is implementable if:

i) adjacency requirements (between pairs of machines) specified by the user are met;

ii) location restrictions for each machine specified by the user are satisfied; and

iii) space constraints are not violated.

Class 7 rules check whether the layout is implementable. If a layout is implementable, its solution cost is computed and provided to the user. If not, the solution is modified to make it implementable. Class 7 rules are also capable of incorporating qualitative factors not considered in models M1, M2 or M3.

Model and algorithm base: The branch-and-bound algorithm and the heuristic algorithm used to solve model M1 are stored in the model and algorithm base. In addition, the model and algorithm base includes the heuristic algorithm presented in Heragu and Kusiak (1987) to solve models M2 and M3. Of course, the models M1, M2 and M3 are also stored in the model and algorithm base. The models used by KBSMD are stored in the form of a frame. The frame representation for model M1 is shown in figure 9.

```
(model_M1 (obj_func) ((constraint_1) (constraint_2)
                      (constraint_3) (constraint_4)
                      (constraint_5) (constraint_6)
                      (constraint_7) (constraint_8)
                                     (constraint_9)))
```

Figure 9. Frame representation of model M1

Inference engine: The inference engine employs forward-chaining strategy. It sequentially applies the seven classes of rules beginning from class 1 rules. Its features are similar to the inference engine described in Heragu and Kusiak (1988) and is not discussed in this paper.

7. ILLUSTRATIVE EXAMPLE

In this section, an example problem is solved using KBSMD. The problem is to determine the machines and material handling equipment to be purchased and a single-row machine layout for the following data:

- 20 parts P_1, 15 parts P_2, 18 parts P_3, 10 parts P_4, and 12 parts P_5 to be manufactured;
- available machines: M_1, M_2, M_3, M_4;
- available material handling equipment: H_1, H_2;
- available time on each machine and material handling equipment: 5000;

- cost of machines M_1, M_2, M_3, M_4: \$130,000; \$190,000; \$80,000; and \$255,000 respectively;

- cost of material handling equipment H_1, H_2: \$130,000 and \$200,000 respectively;

- available budget is \$800,000;

- clearance between each pair of machines is 1 unit;

- relationship indicator between each pair of machines is 0;

- if selected, machine 4 is to be located at the right extreme site of the floor plan;

- dimensions of machines M_1, M_2, M_3, and M_4: 10x10, 15x20, 20x20, and 10x20, respectively; and

- dimensions of floor plan: 30x75.

As mentioned before, KBSMD acquires the above data in an interactive mode. In addition, it also obtains information on:

i) machine capability, i.e., operations each machine can perform, cost and time required to perform each operation; and

ii) part requirements, i.e., operations to be performed on each part, alternative material handling equipment that can be used and the corresponding handling cost and time.

The information on machine capability and part requirement is obtained for each machine and each part. This knowledge is then represented using frames. For the above problem, the frames for representing parts, operations and machines is provided below:

```
(part (P1) (oper_reqd (o1 o2 o3))
           (no_of_oper (1 1 1))
           (mhs_avail (H1 H2))
           (mhs_oper_cost (50 60))
           (mhs_oper_time (50 60)))
```

```
(part (P2) (oper_reqd (o2 o3 o5 o6))
           (no_of_oper (1 1 1 1))
           (mhs_avail (H1 H2))
           (mhs_oper_cost (70 70))
           (mhs_oper_time (70 70)))

(part (P3) (oper_reqd (o4 o7 o8))
           (no_of_oper (1 1 1))
           (mhs_avail (H1))
           (mhs_oper_cost (70))
           (mhs_oper_time (70)))

(part (P4) (oper_reqd (o1 o6 o8))
           (no_of_oper (1 1 1))
           (mhs_avail (H1 H2))
           (mhs_oper_cost (10 60))
           (mhs_oper_time (60)))

(part (P5) (oper_reqd (o2 o4 o5))
           (no_of_oper (1 1 1))
           (mhs_avail (H1 H2))
           (mhs_oper_cost (50 50))
           (mhs_oper_time (50 50)))

(operation (o1) (M_reqd (M1 M2 M3))
           (oper_cost (20 15 40))
           (oper_time (20 15 40)))

(operation (o2) (M_reqd (M1))
           (oper_cost (20))
           (oper_time (20)))

(operation (o3) (M_reqd (M1 M3))
           (oper_cost (10 30))
           (oper_time (10 30)))

(operation (o4) (M_reqd (M1 M3))
           (oper_cost (8 20))
           (oper_time (8 20)))

(operation (o5) (M_reqd (M3 M4))
           (oper_cost (10 8))
           (oper_time (10 8)))

(operation (o6) (M_reqd (M2 M4))
           (oper_cost (8 5))
           (oper_time (8 5)))
```

(operation (o_7) (M_reqd (M_2))
(oper_cost (12))
(oper_time (12)))

(operation (o_8) (M_reqd (M_2 M_4))
(oper_cost (15 8))
(oper_time (15 8)))

(machine (M_1) (oper (o_1 o_2 o_3 o_4))
(oper_cost (20 20 10 8))
(oper_time (20 20 10 8)))

(machine (M_2) (oper (o_1 o_6 o_7 o_8))
(oper_cost (15 8 12 15))
(oper_time (15 8 12 15)))

(machine (M_3) (oper (o_1 o_3 o_4 o_5))
(oper_cost (40 30 20 10))
(oper_time (40 30 20 10)))

(machine (M_4) (oper (o_5 o_6 o_8))
(oper_cost (8 5 8))
(oper_time (8 5 8)))

The data provided by the user indicates that the cost of performing operation o_i on machine M_j is equal to the time required to perform operation o_i on machine M_j. Similarly, the cost handling part P_i on material handling equipment H_j is equal to the time required to transport part P_i on material handling equipment H_j. For the above problem the system does not attempt to eliminate any machines from consideration, because, the number of integer variables in model M1 is small. Then it generates the following data:

$[c_{ij}] = [r_{ij}] =$

operation \ machine	M_1	M_2	M_3	M_4
o_1	20	15	40	-
o_2	20	-	-	-
o_3	10	-	30	-
o_4	8	-	20	-
o_5	-	-	10	8
o_6	-	8	-	5
o_7	-	12	-	-
o_8	-	15	-	8

$$[h_{ij}] = [s_{ij}] =$$

part	material handling equipment H_1	H_2
P_1	50	60
P_2	70	70
P_3	70	-
P_4	10	60
P_5	50	50

a vector of operations to be performed $= [N_i] = [30\ 47\ 35\ 30\ 27\ 25\ 18\ 28]^T$

In addition to the above, the system generates vectors which indicate the time available on each machine and material handling equipment, their corresponding procurement costs, and the number of units of each part to be manufactured. Since the number of integer variables is > 25, algorithm A2 is invoked to solve model M1. The resulting solution is provided below:

$$[x_{ij}] =$$

operation	machine M_1	M_2	M_3	M_4
o_1	0	30	0	-
o_2	47	-	-	-
o_3	35	-	0	-
o_4	30	-	0	-
o_5	-	-	27	0
o_6	-	25	-	0
o_7	-	18	-	-
o_8	-	28	-	0

$$[y_{ij}] =$$

part	material handling equipment H_1	H_2
P_1	20	0
P_2	15	0
P_3	18	-
P_4	10	0
P_5	12	0

Machines M_1, M_2, M_3 and material handling equipment H_1, at a cost of $530,000 are selected. The operating and handling costs are $3086 and $4110, respectively. The number of operations of each type to be performed on the machines is provided in matrix $[x_{ij}]$. The number of parts to be transported on the selected material handling equipment is provided in matrix $[y_{ij}]$.

The system then generates the frequency of trips between machines M_1, M_2, and M_3. Since the machines have to be arranged in a single-row, the system selects model M2 and solves it using a suitable algorithm available in the model and algorithm base. The resulting solution which satisfies all the constraints imposed by the user is provided in figure 10.

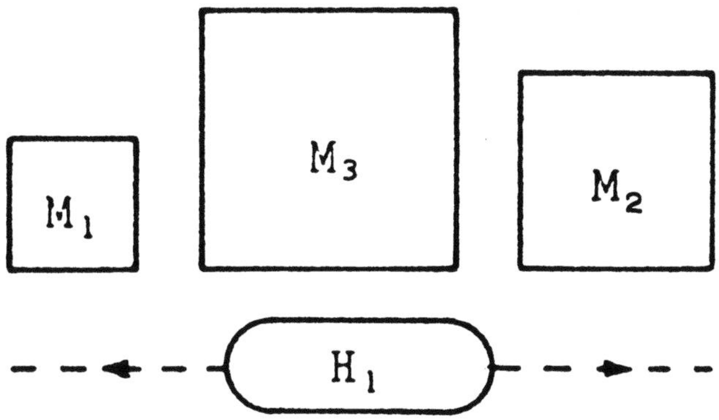

Figure 10. Solution for the single-row layout problem

8. CONCLUSION

In this paper, a new approach for solving the problem of selecting machines and material handling equipment and determining a layout is presented. It involves combining the optimization and knowledge-based system approaches. Such approaches have been attempted in the past and found to be successful (see for example, Dolk and Konsynski (1984), and Kusiak 1987). The system presented in this paper attempts to solve the manufacturing system design problem using the optimization and knowledge-based approaches. It is capable of selecting the machines and material handling equipment required for an automated manufacturing system. It also determines a layout for the machines. The models for the selection of manufacturing equipment available in the literature, are not adequate as they do not consider the integration of machines and material handling equipment. Further research on this aspect may be necessary. The approach presented allows one to incorporate other models and algorithms.

ACKNOWLEDGEMENT

Research presented in this paper has been supported in part by grants from the Natural Sciences and Engineering Research Council of Canada and the University of Manitoba.

REFERENCES

1. Behnezhad. A.R. and Khoshnevis. B.: The effects of manufacturing progress function on machine requirements and aggregate planning. International Journal of Production Research. 26. 309-326 (1988)

2. Dolk. D.R. and Konsynski. B.R.: Knowledge representation for model management systems. IEEE Transactions on Software Engineering. SE-10, 619-628. (1984)

3. Fisher. E.L. and Maimon. O.Z.: Specification and selection of robots. in A. Kusiak (Ed.) Artificial Intelligence: Implications for CIM. IFS Publications. Kempston. UK and Springer-Verlag. New York. NY. (1988)

4. Fisher. E.L. and Nof. S.Y.: FADES: knowledge-based facility design. Annual International Industrial Engineering Conference Proceedings. Chicago, Il. May 6-10. 74-82. (1984)

5. Hassan. M.M.D. Hogg. G.L. and Smith. D.R.: A construction algorithm for the selection and assignment of material handling equipment. International Journal of Production Research. 23. 381-392 (1985)

6. Heragu. S.S. and Kusiak. A.: Efficient models for the facility layout problem. Working Paper #11/87. Department of Mechanical and Industrial Engineering. University of Manitoba. Winnipeg. Canada R3T2N2. (1987)

7. Heragu. S.S. and Kusiak. A.: A knowledge-based system for machine layout (KBML). in A.Kusiak (Ed.) Expert Systems in Manufacturing Design. Society of Manufacturing Engineers. Dearborn. MI. (1988)

8. Kumara. S.R.T. Kashyap. R.L. and Moodie. C.L.: Application of AI techniques to facility layout. Conference on Intelligent Systems and Machines. Department of Computer Science. Oakland University. California. (1985)

9. Kusiak. A.: Artificial intelligence and operations research in flexible manufacturing systems. Information Processing and Operations Research. 25. 2-12. (1987)

10. Kusiak. A.: The production equipment requirements problem. International Journal of Production Research. 25. 319-325. (1987a)

11. McGlennon. J.M. Cassidy. G. and Browne. J.: ROBOSPEC: a prototype expert system for robot selection. in A. Kusiak, (Ed.) Artificial Intelligence: Implications for CIM. IFS Publications. Kempston. UK and Springer-Verlag. New York. NY. (1988)

12. Miller, D.M. and Davis. R.P.: A dynamic resource allocation model for a machine requirements problem. IIE Transactions. 10. 237-243. (1978)

13. Miller. D.M. and Davis. R.P.: The machine requirements problem. International Journal of Production Research. 15. 219-231. (1977)

14. Murty. K.G.: Linear Programming. Wiley. New York. NY. (1983)

15. Schrage. L.E.: Linear, Integer, and Quadratic Programming with LINDO. Scientific Press. Palo Alto. Ca. (1984)

16. Webster. D.B. and Reed. R.: A material handling system selection model. AIIE Transactions. 3. 13-21. (1971)

SPECIFICATIONS FOR KNOWLEDGE REPRESENTATION IN INTELLIGENT SYSTEMS LAYOUT DESIGN ENVIRONMENTS

Benoit Montreuil, Ph.D., P.E.
Departement Operations et Systemes de Decision
Universite Laval
Quebec, Canada

Working in layout design may sometimes be a frustrating experience since it shares a basic attribute with riding a bicycle. Most six years old kids know how to ride a bike. The same way, most employees in the company, be they machine operator, intern student, or president, are able to draw a layout of the plant. They simply have to cut themselves some representative templates of the various cells in the shop, then lay them out over a grid paper, in a similar way as you do a 50-piece puzzle, being careful not to grossly violate the major flow patterns perceived in the system. However, companies have never, and especially not today, been embarked in a sunday afternoon country ride. They are engaged in a Tour-de-France like competition with the best in the business from all over the world.

The 50-piece layout puzzle translates into the spatial configuration of the manufacturing system, within which all subsequent operations are to be performed and subject to which all subsequent production planning and control decisions have to be taken. Layout design has hence high impact on the long term profitability of the firm. The potential marginal value of a few percent improvement in efficiency may well justify requiring Tour-de-France performance from the layout designer, and providing him with the best computer support possible.

The actual research frontier in computer support for systems layout design lies in what we term *intelligent design environments*. Such design environments aim at integrating leading edge artificial intelligence techniques, optimization modeling, simulation modelling, interactive graphics, etc., in order to support the layout design throughout the entire design process.

Pioneering work has been accomplished by Fisher & Nof [1984], Heragu & Kusiak [1988], Kumara et al. [1987], and Montreuil & Banerjee [1988]. All these papers emphasize integration of novel artificial intelligence concepts, languages, and techniques. Included are abstract objects, encapsulation, expert systems, frame, heuristic search, inference engine, inheritance, integrated expert system and optimization, knowledge base, LISP, message passing, meta-knowledge, model-view-

NATO ASI Series, Vol. F53
Advanced Information Technologies for
Industrial Material Flow Systems
Edited by Shimon Y. Nof, Colin L. Moodie

controller paradigm, multi-agent concept, object oriented paradigm, open system, predicate logic, PROLOG, rule programming, SMALLTALK, syntactic pattern recognition, etc.

In order for intelligent systems layout design environments to mature, significant research efforts have to be undertaken in several directions. These include domain knowledge representation, overall environment architecture, internal knowledge representation, heuristic search strategies, optimization modelling, simulation modelling, interenvironment interfacing, etc.

This paper concentrates on specification of domain knowledge representation requirements. We purposefully stay away from technical AI terminology in order for the focus to be explicitly on systems layout design knowledge. The paper begins by exposing the classical layout design problem formulation, then justifies a need for more elaborate knowledge representation. The core of the paper segregates knowledge representation requirements into ten sections: space, cell, building, flow, relationships, life cycle dynamics, hierarchical design, logical design, operational dynamics, and multi-agent concept.

CLASSICAL LAYOUT DESIGN PROBLEM FORMULATION

Given a set of cells and a building, a layout is generally defined as follows for design purposes. Each cell is spatially defined as the discrete combination of a prespecified number of unit squares. A certain number of constraints are imposed on feasible combinations, examples are maximal overall length-to-width ratios, and fixed rectangular shape. Similarly, a building is spatially defined through a grid matrix of unit squares. Specific configuration constraints are imposed by forbidding some specified unit grid squares. Furthermore, some cells may be prefixed to specific unit grid squares. A layout is then defined as a feasible discrete assignment of unit cell squares to unit building squares, as illustrated in Figure 1.

The layout design objective is generally stated as finding a feasible layout which *optimizes proximity* between cells depending upon their degree of *interaction*, as these terms are defined below. The interaction between every pair of cells is represented by a prespecified single value. Depending upon the school of thought, this value is the intercell flow (Cameron [1952]), the intercell relationship (Muther, [1961]), or a combination of these two. Basically, this value can take any number from $-\infty$ to $+\infty$. A negative value represents a repulsive interaction while a positive value represents an attractive interaction. The amplitude of the absolute value indicates the relative importance of the intercell interaction.

Proximity between two cells in a layout is either measured through the distance between their centroids or as whether the cells are adjacent or not. When intercentroid distance is selected, the euclidean metric is used when it is judged that straight line travel is approximately possible (e.g. crane travel), otherwise, and in most cases, the rectilinear metric is used to approximate restricted aisle travel.

As indication of the actual widespread use of the above definition of the layout design problem, papers by Golany & Rosenblatt [1987], Johnson [1982], Picone & Wilhelm [1984], and Scriabin & Vergin [1985], all attempt to find the discrete assignment of cell squares to building squares minimizing the sum, over all pairs of cells, of the product of their intercell flow times their intercentroid rectilinear distance. Similarly, Foulds [1983] and Montreuil et al. [1987] attempt to maximize the sum, overall pairs of cells, of the product of their intercell adjacency times their intercell flow (or relationship).

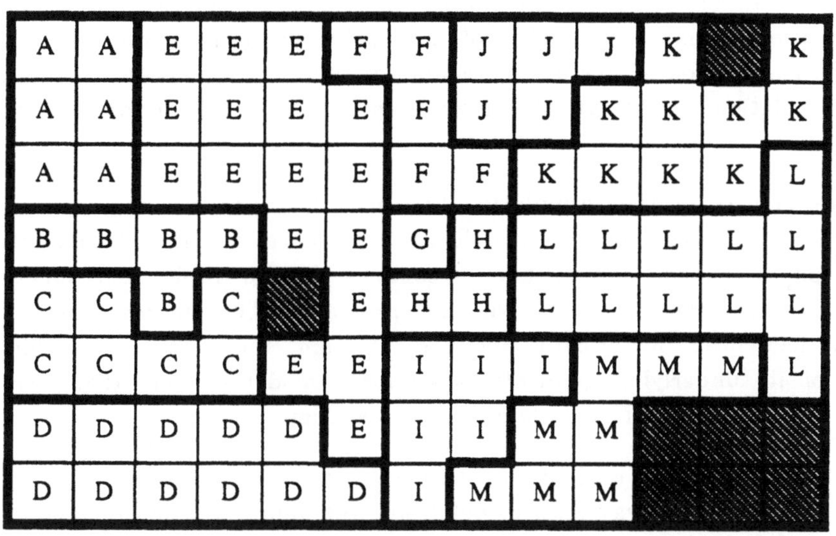

Figure 1. Discrete assignment of cells to locations satisfying a maximal 3:1 length-to-width for each cell.

AGGREGATION AND ABSTRACTION OF REALITY.

It is astounding to contrast the simplicity of the aggregate classical layout design problem formulation with descriptions of the layout design problem presented in classical textbooks such as Apple [1963], Francis & White [1974], Moore [1962], Muther [1961], Reed [1961], and Tompkins and White [1984]. These include exhaustive coverage of a multiplicity of considerations including, but not limited to, material handling, spatial utilization, manpower, safety, security, technology, process, organization, storage, expansion, flexibility, modularity, adaptability, aisle, and services.

Visual displays of existing layouts include much more details than in the classical formulation, including specific manufacturing equipments; laboratories; offices; walls; doors; aisles; storage zones and equipments; material handling equipment and routes; electrical, plumbing, heating and air conditioning, computer and communications services; building floor, columns and height specifications, etc.

Most researchers recognize that the classical formulation intrinsically assumes a decomposition of the real layout design into two successive phases, being the aggregate and detailed layout design phases (e.g., Montreuil et al., 1987). The more subtle assumption is that most of the potential gains are derived from the aggregate design phase. Realization of this latter assumption raises the question as whether the classical formulation works at the most appropriate level of aggregation and abstraction.

The discreteness of the formulation has been historically recognized essential through the combination of, first, difficult representation and manipulation of continuous-space defined layouts and, second, with the rapid explosion of the continuous model (Drezner [1980]) of the problem. The aggregateness of the formulation has been historically justified through the extreme combinatorial complexity of its underlying quadratic assignment model Lawler [1963], rendering untractable the treatment of more detailed considerations.

Three major reasons lead us to sustain that an intelligent layout design environment should in no way restrict definition and representation of layout design to the classical formulation, and should support an array of layout considerations whose minimal set is defined in this paper.

First, several major recent research achievements have either been derived by explicitly replacing or enriching the classical formulation, or do not need the classical assumptions to remain tractable. Some of these will be explicited in latter sections.

Second, most relevant computer-oriented layout design research lies in the field of heuristic problem solving and heuristic optimization. The underlying skeleton of the heuristic methodologies proposed is rarely strictly dependent upon the classical assumptions in order to remain valid. Furthermore, most heuristic methodologies are inherently adaptable to support more elaborate problem definition.

Third, and probably the most important, the concept of moulding the problem definition to fit the capabilities of the model or problem-solving strategy is not valid for a design environment, and in fact for any artificial intelligence oriented problem solving environment. Such an environment should support problem definition at any level of precision, aggregation and abstraction whose representation is tractable and evaluation of solution is amenable to computer treatment. The environment should then be capable of translating the problem into a definition fitting embedded model(s) and/or problem-solving strategy(ies) as well as translating their "solution" back to the generic representation supported by the design environment.

Following our rejection of the classical layout design problem formulation as the target problem representation within an intelligent design environment, the following sections concentrate on providing a sound framework for knowledge representation in an intelligent systems layout design environment.

SPACE REPRESENTATION

Discrete grid representation of space should be replaced by the much richer continuous space representation (e.g. Driscoll & Sangi [1986], Ketcham & Malstrom [1984], and Montreuil & Banerjee [1988]).

The actual norm in layout design softwares is to represent space through a grid matrix. This representation is very simple to implement in a software. It is particularly effective in checking for avoidance of spatial conflicts between cells (over-imposition), for cell area counting, and for scanning a layout. It is however our claim that most softwares could be elegantly recoded using a continuous space representation given actually available standard graphic routines.

A good example is provided by the classical CRAFT software (Buffa et al. [1964]). CRAFT only uses the grid to store the incumbent layout, to find adjacent cells, to compute the location of a cell centroid, to verify satisfaction of cell shape constraints (length-to-width ratio, splitting avoidance), to switch location between two cells with the same area, and to switch location between two adjacent cells.

Using a very simple continuous spatial representation, each cell can be spatially defined using two records. The first stores the world coordinates of a reference point on the cell boundary. The second record stores the cell shape through the relative extreme coordinates of the closed circuit of horizontal and vertical segments defining the cell contour. The building is defined similarly.

A layout can easily be redrawn from the set of defining records for each cell. Adjacent cells share a common boundary segment, which is straight-forward to verify by inspecting the cell records. There exist standard routines for computing the centroid of a cell as defined above, based on basic algebraic geometry. Verification of cell shape constraints is much easier to perform using a continuous boundary representation. Discretely switching location between two cells with the same area simply requires to exchange their sets of spatial records. Switching location of two adjacent cells is performed using a myopic feasibility-ensuring scanning procedure in CRAFT, no investigation for the "best" way to interchange them being performed. In a continuous boundary representation, the closed spatial envelope for the union of the two cells can easily be defined. Then a myopic feasibility-ensuring procedure for envelope splitting can be efficiently used to switch the two cells. More elegant, yet efficient, procedures are easily definable.

This type of exhaustive comparison will not be repeated in the remainder of the paper. However, we felt necessary to perform at least one to clearly state the very down-to-earth feasibility of our statements. In this CRAFT example we have shown how all grid related functions can be efficiently performed in a continuous space representation, while unaffecting in any way all other parts of the CRAFT software. Therefore the discrete grid representation is absolutely not necessary to the CRAFT heuristic.

Furthermore, recent models and methods such as the Cut Tree design skeleton approach (Montreuil and Ratliff [1988]), the DISCON approach (Drezner [1980]), the Matching Approach (Montreuil and Ratliff [1987]), and the Planar Graph Approach (Carrie et al. [1978]), lend themselves inherently to a continuous space representation. As an illustration, Figure 2 presents a design generated using the Cut Tree Layout Approach, under a continuous space representation.

From a design perspective, continuous space representation is easily capable of sustaining notions such as restriction to grid location and modular standard shapes. However it becomes a designer's explicit decision, and not an imposition. Most actual manufacturing system layouts can be represented using horizontal and vertical boundary segments. However, certain new designs have efficiently taken advantage of diagonal aisles. Cells whose internal handling is performed using a robot lend themselves nicely to circular configurations; these have been investigated by Chandru and Samuel [1988], Drezner [1980] and Heragu & Kusiak [1988].

Manufacturing systems layout research has historically focused on bidimensional layout, even in multistory applications. However practitioners are beginning to consider tridimensional space as a resource by strategically locating cells on different floors to permit direct handling, by using mezzanines to permit more compactness of critical zones and saving floor space, and using Automated Storage and Retrieval Systems (AS/RS) to handle work-in-process movement between cells located at various tridimensional coordinates adjacent to the AS/RS, etc. Intelligent layout design environments would gain by supporting tridimensional space representations. Weinzapel and Handel [1975] provide a basic pionering framework oriented toward architectural design.

Once a design environment supports continuous space representation, it should permit acquisition of spatial information through transfer from computer-aided drafting systems (e.g. Collier [1983]), and potentially from solid modelling systems. This rapidly permits to capture detailed spatial knowledge from existing drawings and libraries, which can then be aggregated to a level pertinent for design purposes.

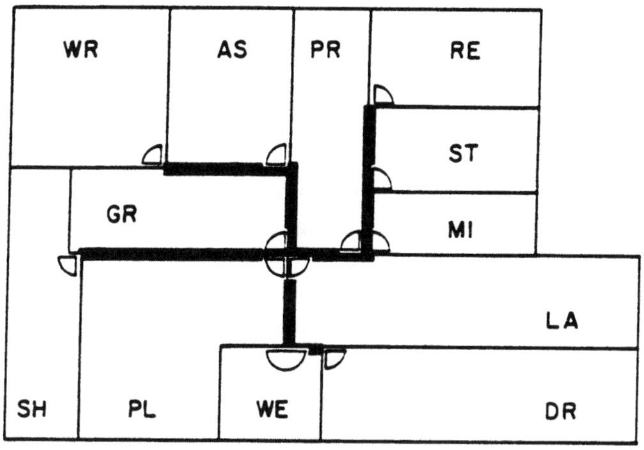

Figure 2. Cut tree based layout design generted using a continuous space representation for the cells, I/O stations, flow network, and fixed building.

CELL REPRESENTATION

Throughout this paper we use the term cell to represent any entity to be laid out, be it an equipment, a workstation, an office, a group technology cell, a flexible manufacturing cell, a storage zone, a department, a laboratory, an organizational unit, etc. This unifies and simplifies terminology.

Concerning cell representation, three major aspects need enrichment: representation of Input/Output stations and representation of spatial configuration requirements, and cell attributes representation.

The classical formulation does not support the notion of Input/Output stations (Pickup/Delivery points) in a cell. A cell is conceived as a homogeneous block, therefore all interactions between cells are assumed to be associated with the cell centroid. In fact, the centroid travel assumption makes sense when the designer has no idea nor control on how the cell is to be internally laid out and interfaced to the rest of the system; or when the cell consists of parallel processors to and from which all materials are handled directly, with no dominant movement patterns. Even in these cases, one can define the cell centroid as its global Input/Output station. In general, however, cells have definite Input/Output stations to handle incoming and/or outgoing parts, tools, personnel, etc. As Montreuil [1987], Montreuil and Ratliff [1988a] and [1988b], O'Brien and Abdel Barr [1980], and Warnecke and Dangelmaier [1983] demonstrate, incorporating Input/Output stations is essential for valid flow representation. It is interesting to note that most layout design softwares can easily be altered to take advantage of the concept of I/O station.

Generally, each layout design software has imposed its own cell spatial configuration capabilities to the layout designer. Examples are one fixed unit square configuration, one (or a few) designer predefined rectangular configuration(s), and any discrete unit square configuration totalling the cell area and within a designer prespecified overall length-to-width ratio. An intelligent design environment should deal explicitly with flexibility in cell configuration. At an extreme case a cell may be known to require a given shape (most often from technological considerations). At the other extreme, maximum flexibility in cell shape may be allowed within some extreme boundaries (e.g. length-to-width ratio). Cells may require a rectangular or circular shape, may have one of a set of standard modular shapes, etc.

Limiting spatial cell configuration to specification of its external shape does not remain valid when including I/O stations and when one deals with open plan cells. The relative position of the I/O stations may well influence spatial requirements. As for open plan cells, a typical example is a cell englobing a single processing equipment. The spatial requirements include physical equipment location, employee ingress/outgress, work-in-progress, I/O station, and equipment clearance. Some of these areas may be shared with other cells depending upon the gross relative positioning of cells, and hence reduce the total space requirements.

In cases where a cell contains a number of significant physical objects, the only way to validate a given shape may be to verify the existence of a feasible internal layout. Therefore the design environment should permit representation and manipulation of physical objects and zones internal to a cell (e.g., Jacobs [1987], Ketcham & Malstrom [1984], O'Brien & Abdel Barr [1980]). This concept may be extrapolated to treating hierarchies of cells, as discussed in a latter section. There also exist cases where the spatial requirements for a cell are proportional to the intensity of flow associated with the cell, as examplified by a storage cell. An intelligent design environment should support the fact that configuration requirements are rarely subject to 0-1 feasibility, but that certain feasible configurations are more internally efficient than others.

A design environment should recognize the critical distinction between the gross and net spatial requirements for a cell. When aisles are not laid out concurrently to the cells, the layout-without-aisle is in fact a gross layout in which aisle space should be accounted for. The classical approach is to boost all cell area requirements by a fraction accounting for the expected overall ratio of aisle space versus cell space in the final net layout including instanciation of the aisles. The designer generally refers to the net cell when specifying the spatial requirements, these must be then translated adequately into gross requirements.

A layout design environment should permit the designer to input and manipulate a multiplicity of attributes for each cell (e.g. Sato & Owen [1980]). These may include, but are not restricted to, organizational categories, environmental requirements, energy and services requirements, technological specifications, and production related information.

BUILDING REPRESENTATION

The treatment of buildings in layout design literature is generally much to simplistic, for both the cases of a green-grass construction and of an existing building.

An existing building normally includes fixed number of stories and mezzanines; fixed cells and equipments; fixed services; fixed floor zone capacities (e.g. weight, height); fixed walls, partitions and columns; fixed windows; fixed docks; fixed parking and access; fixed handling equipment (e.g. crane in a given zone, elevator), etc. All these fixed items may have significant impact on layout design and should therefore be represented explicitly.

Furthermore all these *fixed* items may be *unfixed*, but with potentially significant cost, resource, and time commitments. These alteration considerations should also be represented explicitly. Some research has been done on layout modification (e.g. ref. on CRAFT-M by Hicks & Cowan [1976]). The general assumption is that each cell can be moved away from its actual location, incurring a fixed cost for moving it plus a cost proportional to the moving distance. In many instances, this approximation is inadequate; the total modification cost for a relayout being related to the overall set of modifications required.

When laying out a new construction with no prespecified building, one needs to understand that a building will have to be specified and built before the layout is to be implemented. All aspects mentioned above for an existing building must now be specified. There exist many relations between the designs of the layout, of the building, of the services, and of the site; in terms of overall efficiency, cost and aesthetics. Interaction with the architect and construction engineering firms is critical (Shaw and Hicks [1986]).

An intelligent design environment should permit adequate representation of building/services knowledge, support evaluation of critical aspects and interaction with other building/services design environments.

FLOW REPRESENTATION

Flow knowledge can generaly be divided in two categories: flow requirement and flow travel. Flow requirement specifies *what* and *how much* must be flowed through the layout, while flow travel specifies *how* and *where* it is to be flowed through the layout. In the classical formulation, flow requirement is defined for each pair of cells as how many equivalent-trips/period are estimated between the cells, while flow travel is defined as free rectilinear (or euclidean) flow between centroids of any two cells.

First consider a single flow. The first elements of representation are the two *entities* (cell or I/O station) involved, and the direction of flow. A flow may be stated simply as "between cell A and cell B". It may alternatively be defined as "from output station #2 of cell A to input station #4 of cell B", or "from the output station of cell A to the input station of the nearest cell among cells B, C, and D". The final elements of representation are the item to be flowed, the unit handling load, and information on number of trips per period. These are typical information imbedded in a material handling planning sheet (Tompkins & White [1984]). The item and the unit load have a strong impact on what are the potential handling systems that will activate these flows. Raw materials, parts, finished products, people, energy, utilities, and information may require significantly different flow systems. As a strict minimum, there must be support for relatively weighing these various flows; more sophisticated *costing* schemes are possible. The quantity information may be a deterministic number of trips, an absolute number, or a fraction of an overall production volume. Stochastic information may also be available (e.g. Shore & Tompkins [1980]). Examples are a flow with stochastic distribution function having estimated parameters (e.g. μ, σ), and a specific flow for each of a set of scenarios (with probability of scenario occurence).

Representation of hierarchical groupings of flow is critical in layout design. A basic grouping is what is known as a *routing*. Such a routing for a given item may be a linear sequence of flows. It may also be a much more elaborate *precedence diagram*, with optional and parallel flows (Tompkins & White [1984]). The most aggregate flow grouping sums up all flows from each (cell, I/O station) to each (cell, I/O station). Special groupings may include all flows related to a family of items, all flows addressable (or expected to be addressed) by a given material handling system (e.g. crane, AGVS, fork lift, etc.) as done in COFAD (Tompkins & Reed [1976]), all flows associated with a subset of (cells, I/O stations), etc. The importance of specific groupings is strongly dependent upon flow travel considerations.

The classical flow travel assumption is that it does not need explicit consideration in layout design and that using one of the two following extreme hypotheses is sufficient for flow travel representativity. The first hypothesis corresponds to free flow, meaning that each flow may be assumed to be travelling in straight rectilinear or euclidean path from its source to its destination. This assumes that the layout realization has no impact on travel means. The second hypothesis corresponds to proximity measure through intercell adjacency. It implicitly assumes that flow travel is very efficient between adjacent cells, and deteriorates completely when flowing between nonadjacent cells, intercell distance then becoming unsignificant is compared to intercell adjacency.

Researchers have recently demonstrated that explicit consideration of flow travel in layout design may permit to achieve significantly higher flow efficiency. O'Brien and Abdel Barr [1980] and Warnecke & Dangelmaier [1983] have used explicit aisle travel throughout the layout design process. Warnecke & Dangelmaier (see also Warnecke et al. [1982]) have concurrently laid out net cells and aisles in their layout construction algorithm. O'Brien and Abdel Barr have exchanged net cells given aisle locations and locally altered aisles in their layout improvement algorithm. Their layout output is a net layout including the cell layout, the I/O station locations, and the aisle layout.

Montreuil [1987] has introduced a methodology integrating the designs of the (gross) cell layout, the I/O station locations, and the flow network. A flow network is a logical network linking the I/O stations and along which flow is restricted to travel. A flow network can represent aisle travel, an AGVS path set, a conveyor network, etc. Figure 3 provides an illustration of a design integrating cell layout, multiple I/O stations, and flow network. Montreuil and Venkatadri [1988b] have developed an efficient model to then translate such a design into a net layout including aisles and net cells. A manufacturing system may in general have several flow networks, coupled with free flow within certain floor zones. Figure 4 illustrates a net layout combining an aisle layout and an AGVS directed flow network. Evaluation of travel through flow networks can generally be in terms of distance, time, and/or cost. Consideration of flow travel will eventually lead to address the notion of integrated design of layout and material handling *system*, which is recognized as an aimed ideal (Tompkins & White [1984]).

RELATIONSHIPS REPRESENTATION

The concept of intercell relationships intrinsically depicts the multiobjective nature of systems layout design. However it must be significantly enriched within an intelligent design environment.

The classical Muther's AEIOUX relationships classification is generally used to state whether intercell proximity is extremely important (A) down to not important (U), or whether intercell non-proximity is extremely important (X). Each intercell relationship has to be prespecified by the designer based on his qualitative assessment of multiple factors such as flow, ease of supervision, common personnel, convenience, safety, etc. In most layout softwares, the designer has also to input a weight for each relationship class (e.g. A:64, E:16, I:4, O:1, U:0, X:-64), or this weight is preset in the software. These weights are necessary to weights sustain quantitative additive evaluations.

In an intelligent design environment, Muther's classification should be alterable to fit either simpler or more complex relationships specifications needs. In very rough-cut studies, the designer may desire to use only two classes (e.g. Important, Not Important); while in some other studies, the six-class scheme may be blocking adequate relationships knowledge acquisition.

It is also critical to recognize that Muther's classification intermixes two different elements of information: proximity relations and importance relations. In fact, an

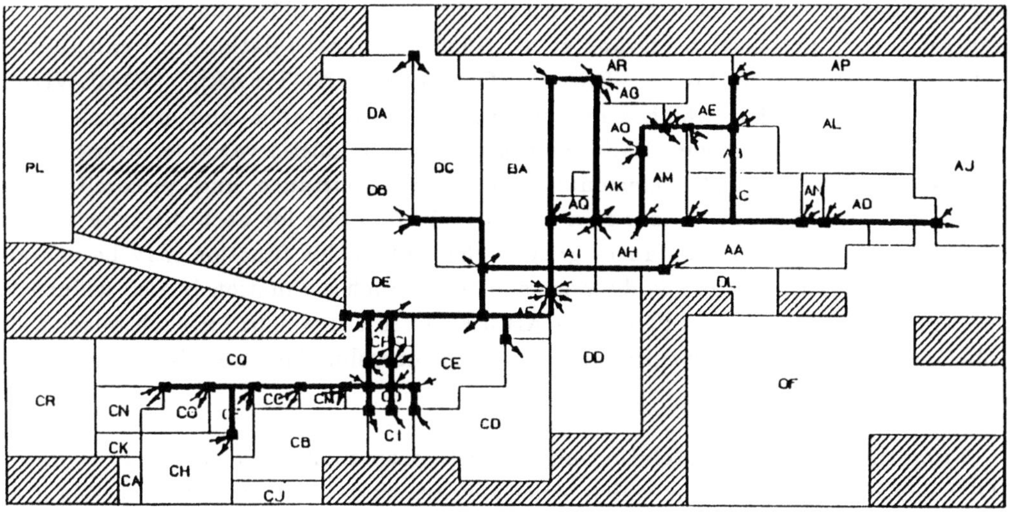

Figure 3. Illustration of design integrating cell layout, multiple I/O stations, and flow network.

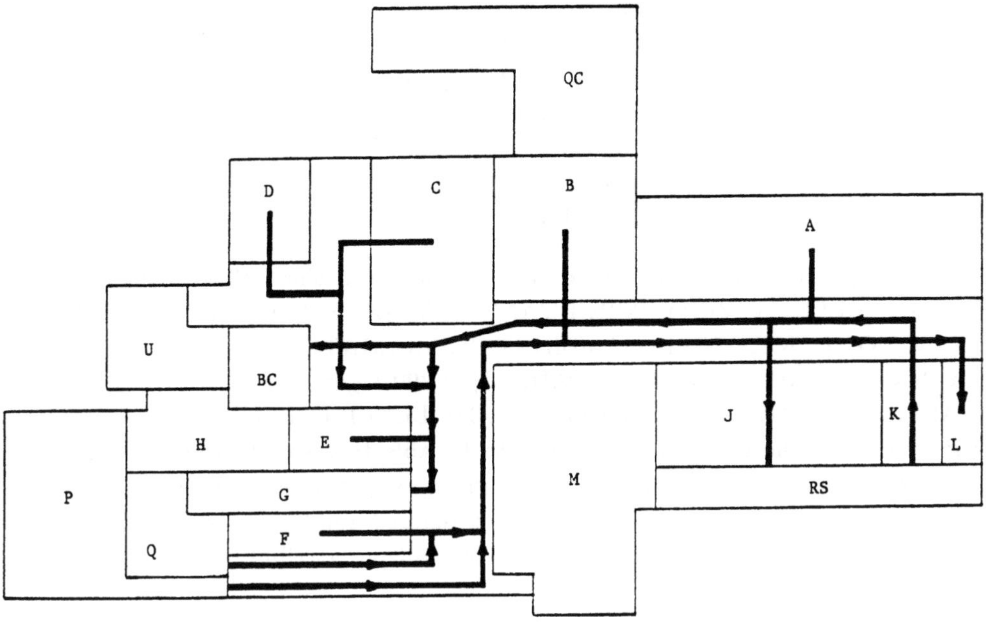

Figure 4. Overimposing an AGVS directed flow network on a net layout with instanciated aisles.

intelligent design environment should sustain the richer classification scheme based on fuzzy set theory proposed by Evans et al. [1986]. Typical relationships can then be [CRITICAL] [VERY-NEAR], [VERY-IMPORTANT] [NOT-FAR], [SOMEWHAT-IMPORTANT] [NEAR], which combine explicitly importance and proximity in a quasi-natural language mode. Each relationship is a linguistic variable, whose weight depends upon its linguistic elements concerning importance and proximity. For example, the designer has to specify a membership function for each distance and the [VERY FAR] element; for distances greater than 200m this function may be one and then reduce toward zero as the distance shortens. Such a function is also defined for importance elements such as [UNDESIRABLE]. Then the weight for each relationship is obtained through the cross products of the membership functions (see Evans et al. [1986] for details).

From this scheme, the Relationships concept may easily be enlarged to englobe a [CONSTRAINT] relationships which strictly states that a layout is not feasible if the relationship is violated. Note also that explicit distances can be used, an example could state [CELLS A,B] [CONSTRAINT] [DISTANCE $>$ 20m].

Relationships representation should also be enriched to sustain relations between a cell and some specific element of the building, such as an entrance, the outside, a window, an elevator, a floor zone, etc. An example relationship could state [CELL A] [CRITICAL] [NOT IN] [FLOOR ZONE z]. Similarly relationships should not be limited to pairs of entities. For example, a relationships could state [CELLS A,B,C] [CRITICAL] [ADJACENT]. **AND** and **OR** statements should be supported. An example would state [CELL A] [IMPORTANT] [NEAR] [CELLS B,C, OR D].

A critical element of relationship representation is to sustain input of relationships justification. This means that the designer may elect to input [CELL A,B] [X rel.] [due to vibration from cell B].

Relationships representation could also sustain function-defined relationships. Examples are [CELL A] [VERY IMPORTANT] [AMBIENT NOISE LEVEL] [$\leq$90Db], [CELLS A TO J] [CRITICAL] [AMBIENT TEMPERATURE] [$\leq 25\,^{\circ}$C], [FLOOR ZONE z] [CONSTRAINT] [WEIGH LOADING] [$\leq$200 kg/m^2] and [CELLS A AND B] [DESIRABLE] [SURROUNDING TRAFFIC] [LOW LEVEL ($\leq$10 trips/period)]. These respectively require the design environment to be able to estimate the ambient noise level, temperature, and surrounding traffic given a proposed (partial) layout. A preliminary requirement to sustaining function-defined relationships is thus for the design environment to support representation of multiple attributes for each cell and building element.

Relationships weighs are classicially perceived as gross relative approximations. However, in information rich organizations, these weighs may be conveying financial information. For example, Liggett and Mitchell [1981] incorporate relationships equivalent to [CELL A] [FLOOR ZONE z] [IN] [$500] when starting a fixed cost to assign cell A into zone z. In general, function defined relationships lend themselves to more rigorous weighing.

Sustaining function-defined relationships leads naturally to generalization toward autogeneration of subsets of relationships. The first level of autogeneration sustains generation of specific relationships from high-level statements and attributes entered by the designer. A simple example is provided by translation of flow requirement knowledge into intercell relationships (e.g. Montreuil [1984]). Another example is provided by [CELL A] [DESIRABLE] [QUIET ENVIRONMENT]. The environment would interpret this statement, transform it into specific requirements for noise, traffic, etc., and then create specific relationships. The higher level of autogeneration requires the environment to *recognize* cell types from their description and attribute set, state high-level requirements based on this recognition and mapping with generic requirements for a cell type, and then generate specific relationships. Recognition capability requires preestablished and adaptative cell (type) libraries scanned through some inference engine. Some basic work on autogeneration of relationships is presented in Eastman [1975], Fisher and Nof [1984], Kumara et al. [1987], Rosenman and Gero [1985], and Sato and Owen [1980].

LIFE CYCLE DYNAMICS REPRESENTATION

The classical layout design formulation depicts layout design as a static process. At a given point in time, a layout must be designed based on estimated requirements for the future. The suggested overall design methodology requires the designer to dynamically monitor the evolution of estimated requirement sets. Then when there is significant expected departure from the original requirement set, the designer should proceed through a new static layout design study. Recent pioneering work by Rosenblatt [1986] and Montreuil & Venkatadri [1988a] have clearly demonstrated the significant potential offered by treating layout design as a dynamic process. This means that the output of a layout study should consist of a set of multiple layouts, as illustrated in Figure 5. The first layout is the one to be implemented right away. The other layouts are temporally phased layouts whose actual realizations depend on evolution of expected future requirements throughout the system's life cycle, as compared to prespecified prospective scenarios (see Montreuil and Nof [1988]).

The historical justification for not treating life cycle dynamics (or macrodynamics) in layout design has been that the static case is already so complex

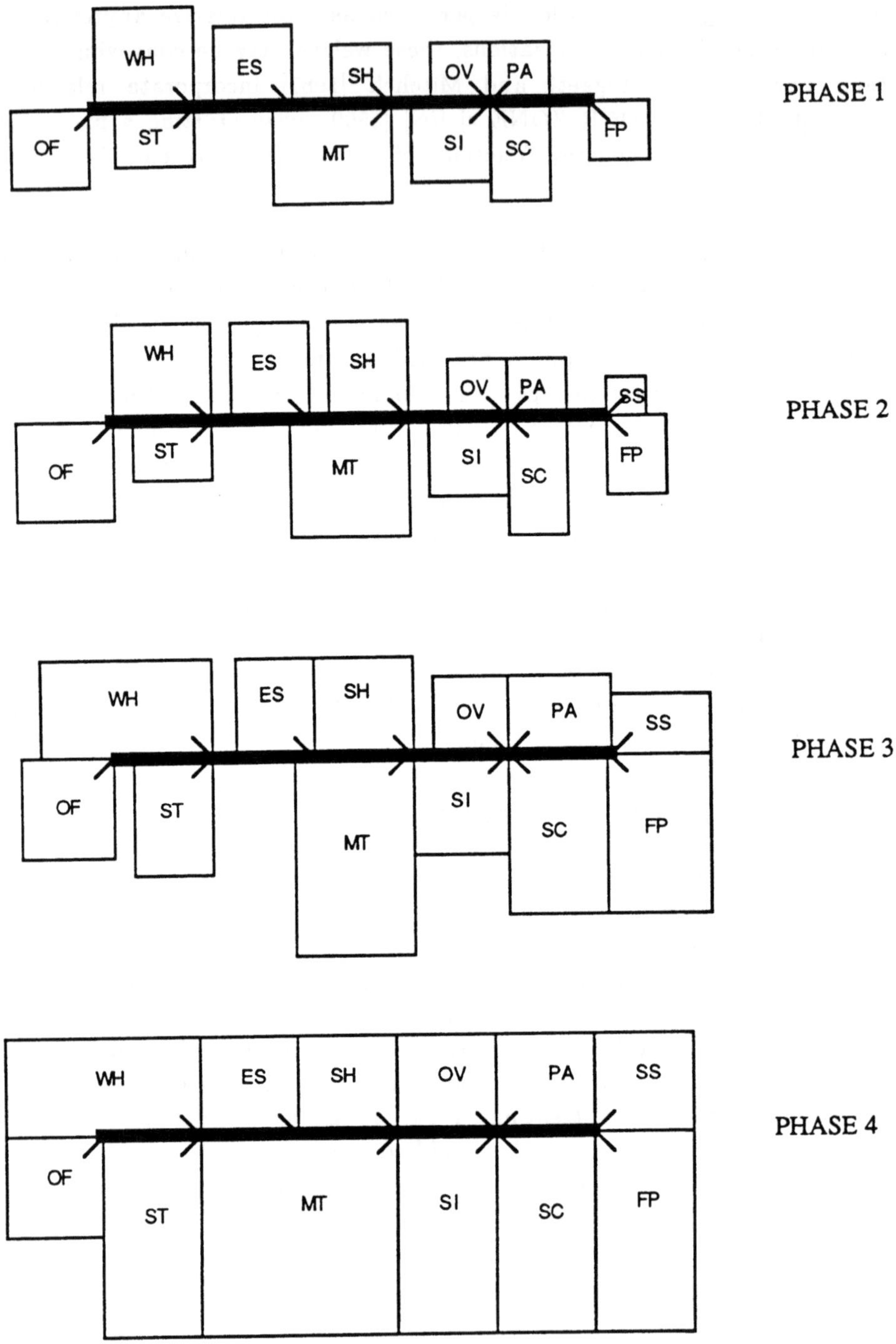

Figure 5. Temporally phased layout designs for a manufacturing system under planned expansion.

that adding dynamic considerations would render the overall problem intractable for any practical purposes. An interesting counterargument comes from the fact that both dynamic models proposed recently are solvable in polynomial time. Rosenblatt [1986] presented a multiphase layout selection model. Given a set of pregenerated layouts for each phase, the model uses the shortest path algorithm to determine the layout to be selected for each phase in order to minimize the overall sum of transition and operations costs. The operations cost for each layout in each phase as well as the transition cost from a layout in phase p to another specific layout in phase (p+1) are prespecified by the designer. The model proposed by Montreuil and Venkatadri [1988a] is specialized for either the expansion phase or the decline phase of manufacturing systems with rigid facilities, and for sustaining a proactive strategic interpolative design philosophy. Their linear programming model efficiently generates optimal layouts for each of multiple phases; given the cell sets, flow sets, and relationships sets for each phase, a goal layout for the final expansion phase (or initial layout of declining phase), and various designer specified constraints, for an array of alternative objectives.

From a knowledge representation perspective, the treatment of life cycle dynamics requires supporting multiple time-phased cell sets, building(s), flow sets, relationship sets, and layouts. It also requires support of the notion of probabilistic future requirements and of prospective scenarios with associated probability tree. The environment should be able to aggregate requirements for specific time windows.

When layout dynamics is considered, the notion of layout transition becomes highly significant. Layout transition (or changeover, turnover) refers to the set of activities required to transform an existing layout L1 into a proposed layout L2. Lilly and Driscoll [1985] have suggested that instant transition and phased transition plans be developed using an interactive methodology combining interactive graphics, heuristic move scheduling, and a sophisticated cost evaluation to generate the activities to be executed in each period until relayout is completed. The first representation issue is for the environment to be able to recognize the set of spatial interference conflicts between two layouts. The second issue deals with supporting representation of a schedule for cell moves and building alteration, including specific cell move times. The third issue relates with representation of various transition costs (see Lilly & Driscoll [1985] for details).

HIERARCHICAL DESIGN REPRESENTATION

Large scale layout design cases may require the designer, or design team, to use a multistage hierarchical design strategy as opposed to a globalist strategy (Liggett [1985], Montreuil & Ramudhin [1986]).

Approaching layout design through a globalist strategy simply consists in working at the most precise level of cell definition for which the information is available through the organization and which is pertinent for layout design. The envisioned advantages are that it offers the designer maximum level of freedom and full control on the layout of each and every cell, given the set of design constraints. The disadvantages are that the sheer size of the globalist design task may impact strongly on time and complexity associated with generating the design, and that the task being treated as one single activity, it must be accomplished in a homogeneous sequential mode.

According to a hierarchical strategy, the designer voluntarily decomposes the global layout task into a hierarchy of subcases. He then solves all subcases from top to bottom of the hierarchy, the higher level layouts being considered fixed when laying out a lower level stage. The designer obtains the global layout by integrating the layouts designed for each hierarchical stage. The main advantages are to offer a way to harness large scale cases by decomposing them into a hierarchy of more easily tractable stages, and that the multistage decomposition permits parallel processing of stages, which may lead to improved resource and schedule management. The disadvantages are that the designer loses degrees of freedom and may suffer from local suboptimization, and that the knowledge representation and manipulation requirements are more sophisticated.

There are basically two types of hierarchical decomposition for layout design: cell hierarchy and spatial hierarchy. A cell based hierarchical strategy (Montreuil and Ramudhin [1986]) capitalizes on the fact that most manufacturing systems lend themselves to forming a logical hierarchy of cells. For example, in the cell hierarchy of a functional manufacturing system depicted in Figure 6, the Turret Lathe Cell is an element of the Turning Cell, which itself is an element of the Removal Processes Cell, which is an element of the Machining Cell, which joins with other high-level cells to form the manufacturing system. The first design subtask is to lay out the high level cells within the manufacturing system. Then, at the second level, for each high level cell such as the Machining Cell, its internal cells must be laid out. This process is repeated until the last level is laid out; here including layout of the specific Lathe and Screw Machine Cells within the Turning Cell.

A spatial hierarchical strategy (Liggett [1985]) decomposes the available space into a hierarchy of spatial zones such as depicted in Figure 7, which has various buildings within the overall site, various floors for each building, various defined floor zones within each floor, then cells laid out within each floor zone. The first design subtask is to *assign* cells to buildings, then the second-level subtasks consists to *assign* cells to floors within each building, the third level *assigns* cells to a floor zone, and the final levels *lay out* cells within a floor zone.

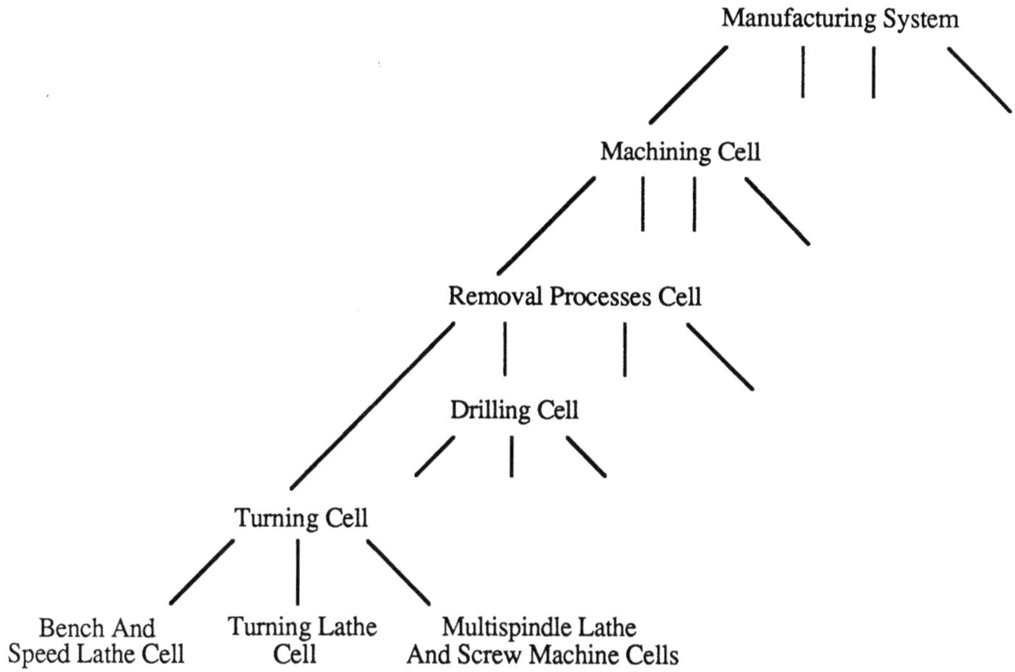

Figure 6. Typical cell hierarchy for a functional manufacturing system

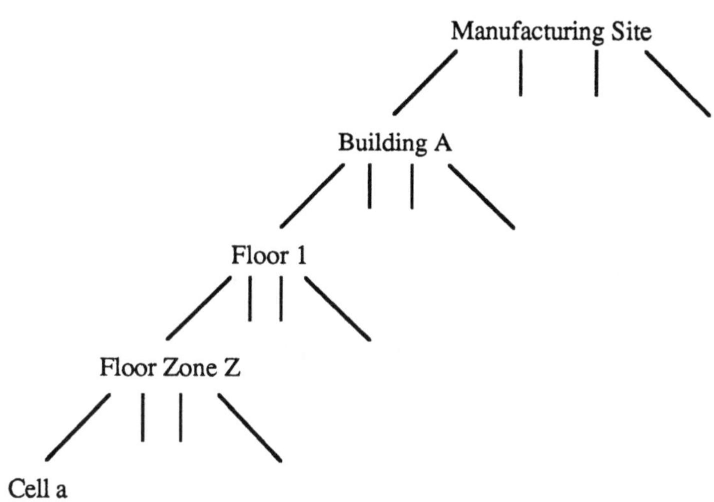

Figure 7. Typical spatial hierarchy within a site with multifloor buildings

An intelligent layout design environment should support both strategies as well as any combination of these strategies. From a representation perspective, this flexibility can be achieved by first conceptualizing a zone as a spatial equivalent of a cell, with the same representation capabilities. The environment should further be able to create and modify concurrent hierarchies of cells (zones). This can be achieved by including in a parent cell representation the list of its child-cells, supporting aggregation of flow requirements between child-cells and cells not in the parent cell (e.g. [A1 to B: 50], [A2 to B: 60] $\Rightarrow$ [A to B: 110]), and similarly supporting aggregation of relationships (including cell-to-zone relationships). Flow networks should be definable to the interzone (or inter-parent-cell) level. A critical element is that the environment should be able to support multiple concurrent layouts, to associate these to specific cells or zones, and to merge selected children layouts in order to disaggregate layouts for parent cells. Finally, the environment should permit using various models to assign cell to zones (often referred to a block stacking), to assign group cells into cell hierarchies (e.g. production flow analysis and group technology (Burbidge [1971]), and to design layouts at any level of a hierarchy.

LOGICAL DESIGN REPRESENTATION

Recognition that manufacturing systems layout design is in fact an organic element of the overall manufacturing systems design process leads to proposing that interaction with other design elements and support of high-level system considerations should be required from an intelligent layout design environment. A natural framework for satisfying this requirement is the concept of logical vs. physical design of production facilities proposed by Montreuil and Nof [1988], enriching upon previous research by Nof [1982].

Given requirement sets for product mix, product quantity demands, equipment types and capacity, space and capital availability, and other factors usually uncontrollable by the designer, then the concepts of logical and physical design can be defined as follows. Logical design involves that part of facility design that seeks to satisfy the requirement set in terms of the facility logic. Physical design physically instanciates facilities that satisfy the requirement sets given a logical design. The logical decisions most pertinent to systems layout design concern equipment grouping decisions, process logic, flow logic, storage logic, production control logic.

Equipment grouping logic is best visualized through the traditional decision between line flow production, cellular (GT) production, and functional (job shop) production organizations. Most actual manufacturing systems have hybrid combinations. The impact of these on layout design is highly appreciated and documented. Figure 8 illustrates the structural impact of equipment grouping logic on layout design. Process logic selects specific routings (flow sets) for specific parts, and therefore impacts the level of flow flexibility imbedded in the system and directly influences the efficiency of flow networks and layouts. Variations in parts-to-cell assignments are suggested through many layout studies.

The impact of storage logic is best examplified through the specification of work-in-progress (WIP) storage policies and zone locations. Centralized versus distributed WIP storage, buffer size in each cell, etc., all have significant impact on flow requirements as well as space requirements.

Flow logic impact is examplified by decisions such as design of logical flow network, specification of material handling equipment topology, specification of the number and type of Input/Output stations for each cell, and specification of unit loads for parts. These decisions impact directly upon layout design.

Production control logic, as examplified through Just-In-Time versus Lot-Inventory control logics, specification of lot sizes and transfer sizes, etc., all influence flow requirements and the criticality of short travel distances and times.

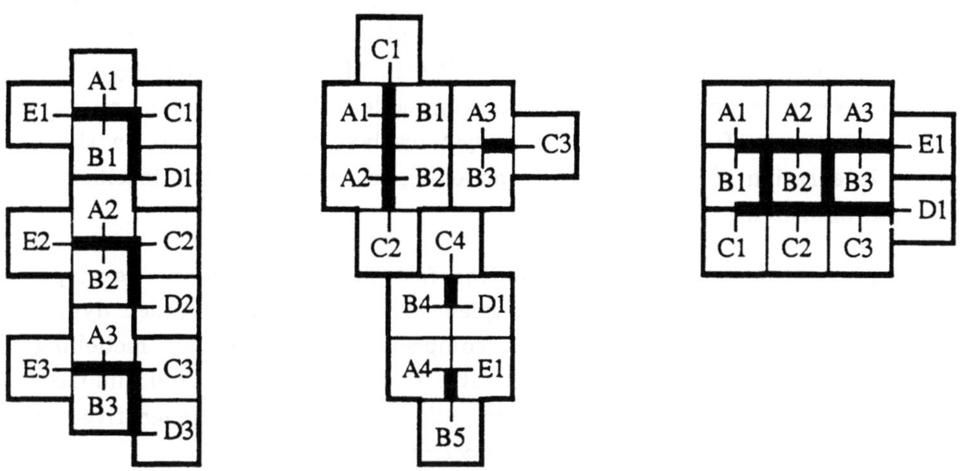

Figure 8. Proposed layout designs for each of three logical designs based on multifamily line, single family group technology, and job shop production logics; given the same requirement sets.

Some of these decisions are uncontrollable by the layout designer, however, he may have significant impact on several of them. Therefore, an intelligent layout design environment should conceptually recognize logical design considerations and support their efficient translation into layout requirements. For example, if a designer desires to evaluate the impact on layout of partial decentralization of the WIP storage, the environment should require minimal designer input and interaction in order to create the desired storage cells and alter the flow requirements. The layout design environment should further be capable of acquiring and transmitting knowledge from/to product design environments, process design environments, handling systems design environments, production planning and control environments, etc. Due to the very limited sustaining body of research actually performed in this domain, we do not feel appropriate at this time to elaborate in more detail about specific knowledge representation requirements.

OPERATIONAL DYNAMICS REPRESENTATION

As a last major topical section on domain knowledge representations, we address the controversial notion of operational dynamics in layout design. Operational dynamics (or microdynamics) refers to the minute-to-minute, day-to-day set of processing activities, movements, equipment failures, unexpected events, etc., dynamically occuring within any layout once implemented. The impact of these on the quality of a layout is actually best analyzed and evaluated through simulation and animation models (e.g. Musselman & Duket [1984], Mackulak & Glenney [1984]), and in certain cases through approximate queueing models. The controversy resides on the relative importance of microdynamics in systems layout design, and on the relative degree of integration of microdynamics modelling to be imbedded in intelligent layout design environments.

Our proposition is that in view of the fast paced stand-alone development of simulation and animation modelling, coupled with the heavily detailed dynamic and stochastic knowledge requirements for comprehensive simulation modelling, then layout design environments should concentrate on specific localized simulation and animation modelling capabilities, and on efficient, partly autogenerative, interface with *intelligent systems simulation and animation environments.*

Specific localized simulation and animation modelling is best examplified when attempting to estimate spatial requirements for centralized Work-In-Progress storage, traffic congestion on flow network links and intersections, as well as specific impact of travel times. An intelligent layout design environment should have the capability to sustain such simulation-based and queueing-based estimations.

Full blown intelligent systems simulation and animation environments should be envisioned as independent agents providing sophisticated dynamic and stochastic evaluation of a layout design. The challenge for layout design environments is to sustain a critical level of microdynamic knowledge representation so that it can express pertinent evaluation requests to the simulation environment, efficiently transfer relevant knowledge to the simulation environment in order to efficiently generate a pertinent simulation model, and comprehend evaluation results provided by the simulation environment. This domain of interaction between layout design and simulation environments is however evidently in its early infancy and will require significant research and development.

MULTI-AGENT REPRESENTATION

Classically, layout design softwares have been conceived through a single algorithm/single criterion philosophy. This type of single-agent philosophy is rapidly moving toward obsolescence. Basically, design environments should move away from model competition and toward collaboration, based on the premise that some model or approach behaves better in some cases than others and that by combining the output of different models, one can achieve better designs.

The actual leading edge in such multi-agent collaboration is best illustrated through the CADFLO system (Montreuil [1988]). CADFLO offers the designer an array of agents capable of sustaining knowledge preanalysis (Montreuil and Stump [1988]); basic approaches such as SLP's Relationships Diagramming (Muther [1961] and Travel Charting (Smith [1955]); interactive multicriteria cell/station selection (for placement) models; interactive location mapping for placement of a cell (Liggett and Mitchell [1981]); DISCON-type approach (Drezner [1980]) for approximate cell centroid placement; Planar Graph methods (Foulds [1983]) to propose intercell adjacency graph skeletons; Cut Tree Design Approach (Montreuil and Ratliff [1988]) to generate layouts from design skeletons based on a cut tree material flow network; a model optimizing I/O station location (Montreuil & Ratliff [1988]); flow network design models (Montreuil [1982]); multiple flow and relationships based evaluation models with free flow travel, adjacency and flow network travel, as well as evaluation models for space utilization and ease of expansion; interfacing with CAD software such as AUTOCAD for detailed drafting, with spreadsheets for transfer of design information and evaluation in order to sustain case specific analyses and financial evaluations, and with word processing softwares for project documentation. CADFLO provides the designer with full control on access, parameterization and execution of these agents.

Intelligent design environment should not be considered synonymous to autonomous design environment since we do not anticipate the human design to be replaced, but that more appropriately the human designer and the design environment should converge toward an efficient symbiosis.

The first implication is that, since the human designer is to remain a critical agent, then graphical representation and sensorial interface with the designer are to be of upmost criticality. The actual recognized trend toward powerful interactive graphics is to be pursued with improved graphics primitives creation and manipulation; better usage of colors; multiple abstraction levels of cell display; intensive usage of overlays (levels), zoom and pan functions, enriched and standardized drawing object libraries; etc. Coexistence of menu driven interface and natural language dialog is to be supported. The design environment should support multiwindow interfacing. Each window may support interface with a specific agent; display a specific zoom and/or overlay combination, an alternative layout for a given phase of life cycle, or the layout of an existing related manufacturing system; etc.

Based on the framework proposed by Montreuil & Banerjee [1988], the other virtual agents can be typified for illustrative purposes as evaluators, analyzers, proposers, optimizers, supporters, tutors, and monitors. Each of these may imbed functional algorithm(s), inference engine(s) for heuristic search, expert system(s), independent knowledge base, etc. The design environment should sustain efficient sharing of knowledge and coordination between these virtual agents themselves, as well as with the human designer agent.

The design environment should not be conceptually limited to centralized sequential processing. It should lend itself naturally to technologically upcoming distributed and parallel processing. Finally, knowledge learning capabilities (Kumara et al. [1987]) should in the future be incorporated into all aspects of knowledge discussed in previous sections as well as in the multiple agents.

CONCLUSIONS

Throughout this paper we have attempted to emphasize the importance of the domain knowledge representation capabilities in sustaining systems layout design. We have hence gone in much depth into layout design related domain knowledge, in fact further than any single software can sustain at the present time. This is consistent with our stated aim toward next generation intelligent design environments.

It is clear that implementing such environments will require extensive research into overall environment architecture, internal knowledge representation, etc.

However, what we consider the most critical challenge does not reside in purely technical artificial intelligence. It consists in ensuring that such a design environment is conceived so that is can be harnessed naturally by designers in the field without extensive training. In other words, such an intelligent design environment should be so refined internally that it becomes externally simple and powerful.

A specific but critical consideration in this regard is that it should support as well the designer performing a rough-cut design study with very limited design information and design time available, as the designer involved in the design of layouts for leading edge intelligent manufacturing systems. In fact, it should be like the ideal bicycle, adaptable to weekday city runs, sunday afternoon country rides, mountain sporting events, and Tour-de-France performances.

ACKNOWLEDGEMENTS

This work has been partially completed while the author was a faculty member in the School of Industrial Engineering at Purdue University (W. Lafayette, IN, USA) and in Purdue's Engineering Research Center for Intelligent Manufacturing Systems. The research has been supported in part by the U.S. National Science Foundation under Grant Number CDR-8500022. Finally, the author wants to thank Prashant Banerjee, Dilip Chhajed, Jocelyn Drolet, Daniel Stump, Uday Venkatadri, and Detlef Weber, who are graduate students at Purdue University, for insightful comments and discussions that have influenced this research.

REFERENCES

1. Apple, J.M., 1963, *Plant Layout and Material Handling,* 2nd ed., Ronald Press Co., N.Y., U.S.A.
2. Buffa, E.S., G.C. Armour, and T.E. Vollman, 1964, "Allocating Facilities with CRAFT", *Harvard Business Review,* Vol. 42, No. 2, March, pp. 136-159.
3. Burbidge, J.L., 1971, "Production Flow Analysis", *The Production Engineer,* April, pp. 139-152.
4. Cameron, D.C., 1952, "Travel Charts - A Tool for Analyzing Material Movement Problems", *Modern Materials Handling,* Jan., pp. 37-40.
5. Carrie, A.S., J.M. Moore, R. Roczniak, and J.J. Seppanen, 1978, "Graph Theory and Computer Aided Facilities Design", *OMEGA, the Int. Jl. of Mgmt. Sci.,* Vol. 6, No. 4, pp. 353-361.
6. Chandru, V. and R. Samuel, 1988, "Optimal Algorithms for Computing Minimal Circumscribing Gleichdicke, *Research Report, IE School, Purdue University,* W. Lafayette, IN, USA.
7. Collier, L.M., 1983, "Use of the Computer in Facilities Planning: YES", *IE,* March, 6p.

8. Drezner, Z., 1980, "DISCON: A New Method for the Layout Problem", *Oper. Res.,* Vol. 28, No. 6, Nov., pp. 1375-1384.
9. Driscoll, J., and N.A. Sangi, *The Development of Computer Aided Facilities Layout (CAFL) Systems: International Survey 1985-86, Survey Report and Results,* Univ. of Liverpool, Liverpool, England, 119p.
10. Eastman, C.M., ed., 1975, *Spatial Synthesis in Computer-Aided Building Design,* Wiley, NY, USA.
11. Fisher, E.L., and S.Y. Nof, 1984, "FADES: Knowledge-Based Facility Design", *Proceedings of 1984 IIE Conference,* Chicago, pp. 74-82.
12. Francis, R.L. and J.A. White, 1974, *Facility Layout and Location: An Analytical Approach,* Prentice-Hall, Englewood Cliffs, USA.
13. Foulds, L.R., 1983, "Techniques for Facility Layout: Deciding which Pairs of Activities Should be Adjacent", *Mgmt. Sci.,* Vol. 29, No. 12, Dec., pp. 1414-1426.
14. Golany, B. and M.J. Rosenblatt, 1984, "A Heuristic Algorithm for the Quadratic Assignment Formulation to the Plant Layout Problem", *Proceedings of the IXth ICPR,* Cincinnati, USA, pp. 2224-2229.
15. Heragu, S.S. and A. Kusiak, 1988, "Knowledge Based System for Machine Layout (KBML)", *Proceedings of Orlando IIE Conference, pp. 159-164.*
16. Hicks, P.E., and T.E. Cowan, 1976, "CRAFT-M for Layout Rearrangement", *IE,* May, pp. 30-35.
17. Jacobs, F.R., 1987, "A Layout Planning System with Multiple Criteria and a Variable Domain Representation", *Mgmt. Sci.,* Vol. 33, No. 8, pp. 1020-1034.
18. Johnson, R.V., 1982, "SPACECRAFT for Multi-Floor Layout Planning", *Mgmt. Sci.,* Vol. 28, No. 4, April, pp. 407-417.
19. Ketcham, R.L., and E.M. Malstrom, 1984, "A Computer Assisted Facilities Layout Algorithm Using Graphics", *Fall IIE Conference Procedings,* 8p.
20. Kumara, S.R.T., R.L. Kashyap and C.L. Moodie, 1987, "Expert System for Industrial Facilities Layout Planning and Analysis", *Computers and Ind. Engrg,* Vol. 12, No. 2, pp. 143-152.
21. Lawler, E.L., 1963, "The Quadratic Assignment Problem", *Mgmt. Sci.,* Vol. 9, No. 4, pp. 586-599.
22. Liggett, R.S., 1985, "Optimal Spatial Arrangement as a Quadratic Assignment Problem", *Design Optimization,* Ed. by J.S. Gero, Academic Press, Orlando, USA, pp. 1-40.
23. Liggett, R.S., and W.J. Mitchell, 1981, "Interactive Graphic Floor Plan Layout Method", *CAD,* Vol. 13, No. 5, pp. 289-298.
24. Mackulak, G.T., and N. Glenney, 1984, "Simulation of Facility Designs on a Micro-Computer", *Proceedings of 1984 Winter Simulation Conference,* 8p.
25. Montreuil, B., 1988, *CADFLO 3.1 User's Manual,* Systemes Espace Temps, Quebec, Canada.
26. Montreuil, B., 1987, "Integrated Design of Cell Layout, Input/Output Station Configuration, and Flow Network of Manufacturing Systems", *Integrated and Intelligent Manufacturing: Analysis and Synthesis,* ASME Pub. XXXX, pp. XXX
27. Montreuil, B., 1985, "Domesticating CRAFT, CORELAP, PLANET et al.", *Computerized Facilities Planning, selected readings,* ed. by H.L. Hales, IE and

Mgmt Press, Atlanta, USA, pp. 42-44.
28. Montreuil, B., 1982, *Interactive Optimization Based Facilities Layout,* Ph.D. Dissertation, Georgia Institute of Technology, Altanta, USA.
29. Montreuil, B., and P. Banerjee, 1988, "Object Knowledge Environment for Manufacturing Systems Layout Design", *International Journal of Intelligent Systems,* to appear.
30. Montreuil, B. and S.Y. Nof, 1988, "Approaches for Logical vs. Physical Design of Intelligent Production Facilities", *Recent Developments in Production Research,* Ed. A. Mital, Elsevier Sciences Publ.
31. Montreuil, B. and A. Ramudhin, 1986, "Globalist vs. Hierarchical Strategies for Large Scale Facilities Layout", *ORSA/TIMS Conference,* Los Angeles, USA; available as *IE Research Memorandum 86-13, Purdue University,* W. Lafayette, IN., USA.
32. Montreuil, B., and H.D. Ratliff, 1988a, "Optimizing the Location of Input/Output Stations within Facilities Layout", to appear in *Engineering Costs and Production Economics.*
33. Montreuil, B., and H.D. Ratliff, 1988b, "Utilizing Cut Trees as Design Skeletons for Facility Layout", to appear in *IIE Transactions.*
34. Montreuil, B., H.D. Ratliff, and M. Goetschalckx, 1987, "Matching Based Interactive Facility Layout", *IIE Trans.,* Vol. 19, No. 3, Sept., pp. 271-279.
35. Montreuil, B., and D. Stump, 1988, "Knowledge Preanalysis for Manufacturing Systems Layout Design", *Proceedings of Orlando IIE Conference, pp. XX*
36. Montreuil, B., and U. Venkatadri, 1988a, "Strategic Interpolative Design of Dynamic Manufacturing Systems Layouts", submitted for journal publication, available as *IE Research Memorandum No. 88-7, Purdue University,* W. Lafayette, IN., USA.
37. Montreuil, B., and U. Venkatadri, 1988b, "On Layout Netting: An Efficient Model for Instanciating Aisles into an Aggregate Layout Design", submitted for journal publication, available as *IE Research Memorandum 88-11, Purdue University,* W. Lafayette, IN., USA.
38. Moore, J.M., 1962, *Plant Layout and Design,* MacMillan, NY., USA.
39. Musselman, K.J., and S.D. Duket, 1984, "Simulation in Computer Integrated Manufacturing", *Proceedings of 1984 Fall IE Conference,* 7p.
40. Muther, R., 1961, *Systematic Layout Planning,* Ind. Educ. Inst., Boston, Mass., USA.
41. Nof, S.Y., 1982, "On the Structure and Logic of Typical Material Flow Systems", *Int. J. Prod. Res.,* Vol. 20, No. 5, pp. 575-590.
42. O'Brien, C. and S.E.Z. Abdel Barr, 1980, "An Interactive Approach to Computer Aided Facility Layout", *IJPR,* Vol. 18, No. 2, pp. 201-211.
43. Picone, C.J. and W.E. Wilhelm, 1984, "A Perturbation Scheme to Improve Hillier's Solution to the Facilities Layout Problem", *Mgmt. Sci.,* Vol. 30, No. 10, pp. 1238-1249.
44. Reed, R., 1961, *Plant Layout: Factors, Principles, and Techniques,* Irwin, Homewood, Ill., USA.

45. Rosenblatt, M.J., 1986, "The Dynamics of Plant Layout", *Mgmt. Sci.,* Vol. 32, No. 1, Jan., pp. 76-86.
46. Rosenmann, M.A. and J.S. Gero, 1985, "A System for Integrated Optimal Design", in *Design Optimization,* ed. by J.S. Gero, Academic Press, Orlando, USA, pp. 259-294.
47. Sato, K., and C.L. Owen, 1980, "A Prestructuring Model for System Arrangement Problems", *Proceedings of 1980 Design Automation Conference,* ACM, 11p.
48. Scriabin, M. and R.C. Vergin, 1984, "A Cluster-Analytic Approach to Facility Layout", *Mgmt. Sci.,* Vol. 31, No. 1, Jan., pp. 33-49.
49. Shaw, L.M. and P. Hicks, 1986, "Facility Design: The IE/Architect Team Approach", *IIE Seminar Document,* Boston, Mass., USA.
50. Shore, R.H., and J.A. Tompkins, 1980, "Flexible Facilities Design", *IIE Transactions,* Vol. 12, No. 2, June, pp. 200-205.
51. Smith, W.P., 1955, "Travel Charting-First Aid for Plant Layout", *Journal of IE,* Vol. 6, No. 1, Jan., pp. 13-25, 26.
52. Tompkins, J.A., and R. Reed Jr., 1975, "COFAD-A New Approach to Computerized Layout", *Modern Materials Handling,* Vol. 30, No. 5, April, pp. 40-43.
53. Tompkins, J.A. and J.A. White, 1984, *Facilities Planning,* Wiley, NY., USA.
54. Warnecke, H.J. and W. Dangelmaier, 1983, "Progress in Computer Aided Plant Layout", *Technical Report, Institute for Manufacturing Engineering and Automation-Fraunhofer,* Stuttgart, Fed. Rep. of Germany.
55. Warnecke, H.J., W. Dangelmaier, and H. Kuhnle, 1982, "Computer-Aided Layout Planning", *Material Flow,* Vol. 1, pp. 35-48.
56. Weinzapfel, G.E., and Handel, S., 1975, "IMAGE: Computer Assistant for Architectural Design", *Spatial Synthesis in Computer-Aided Building Design,* Ed. by C.M. Eastman, Wiley, NY, USA, pp. 61-97.

Robotics and Communications

THE IMPACT OF ADVANCES IN INFORMATION TECHNOLOGY ON INTERACTIVE ROBOTICS

S.Y. Nof
School of Industrial Engineering
Purdue University
W. Lafayette, IN 47907
E-Mail: nof@gb.ecn.purdue.edu

Abstract

Automated knowledge-based machine interaction, or interactive robotics, is critical to the future development of robotic operations. Such interaction should include higher levels of intelligent behavior, as cooperation, substitution, context-driven response, and mutual error-recovery. This article details our recent and on-going research projects of such robotic machine interaction.

KEY-WORDS: Intelligent machine interaction; cooperation models; real-time activity control; knowledge-based control.

1. INTRODUCTION

The objective of this article is to describe the need for robotic machine interaction and discuss our recent and on-going research projects in two main areas: multi-robotic systems, and interactive robotic devices (IRD) for handicapped persons. In both areas we have found that knowledge-based interaction increases significantly the flexibility, reliability, and responsiveness of the system.

Intelligent Machine Interaction Background

Machine interaction is fundamental to flexibility, and can be illustrated in the area of FMS. The computer controlled manufacturing plant can be viewed as having a hierarchical structure [De et al., 1985]. The lower levels of this structure consist of stand-alone manufacturing elements such as CNC-machine tools, automatic welders, robots, etc. The operation of several machine-tools and material handling devices can be incorporated into a single unified system which is usually referred to as a machine cell and constitutes the second level of hierarchy. Cells behave as independent and intelligent units with the cell-host (a computer). The cell-host provides the means of autonomy by having under its control, at least within its local control, the machine-tools and the material handling devices. The operation of many such cells can be integrated by a

Research reported here was funded in part by NSF Grant No. DMC-8719845, "Models of Cooperative Robotic Work", and NSF Grant No. EET-8714698, "Interactive Robotic Devices for Handicapped Children".

NATO ASI Series, Vol. F53
Advanced Information Technologies for Industrial Material Flow Systems
Edited by Shimon Y. Nof, Colin L. Moodie

central computer control system with the aid of an automated material handling system. But the larger the scale of such a system, the worse its performance under a central controller. Therefore, intelligent machine interaction is becoming important.

The computational system of a FMS should have the capability to decide on the following issues, among others:

(i) Allocation and sequencing of jobs through machine-cells;

(ii) Allocation of resources such as tool, fixtures and material-handling devices among the cells; and

(iii) Coordination of the activities of various cells to bring about an efficient utilization of the resources.

A FMS can therefore be viewed as a team of intelligent modules (for instance, with each cell as a module) that share a common goal, and has the following salient characteristics:

(i) A physical distribution of modules (i.e. cells) which are connected together by communication and material-handling links and can interact.

(ii) An autonomous and intelligent nature of each of the modules; that is, when left on its own, it can compute decisions over a wide range of issues that affect its performance without human or higher level intervention. Each module has its own set of goals, and its own computational capability.

(iii) An intermodule communication system for sending messages for instructions, requests for resources, control and record updating, in sum, interaction among modules.

Intermodule communication and distributed control

The distributed information structure of automated facilities is characterized by a lack of global and centralized knowledge. In a distributed computation system with autonomous and heterogeneous nodes, the message protocol helps to integrate and hold the system together. The message protocol is an ideal way [Gray, 1986] to structure the interaction between nodes in a distributed computation system. It has the benefits of encapsulation, abstraction and autonomy.

A distributed system as a collection of autonomous interacting nodes is more naturally suited to 'requestor-server' approach for inter-cell interaction than a 'remote-procedure-call' type input/output approach. The 'requestor-server' approach is based on the idea of coroutines as a set of autonomous programs communicating with other program modules. A description of the approach can be found in Marlin [1980].

Modeling the distributed decision-making structure

The basic decisions which need to be made in a distributed system, for instance, at the FMS level, are task allocation and resource allocation among cells. In order to understand and plan intelligent machine interaction, models are needed. Economic theory has long concerned itself with the central problem of optimal allocation of resources among consumers in order to maximize their utilities. Depending on the nature of interaction between agents, economic theory attempts to outline optimal behavior in several different frameworks, and therefore can be useful for research in this area. Three examples that have been looked at are: market structure, game-theory and team theory.

A variant of market structure is the auction scheme. Auction model has been investigated as a mechanism for task allocation among machine cells in a FMS in [Shaw and Whinston, 1985]. Game theory under the title of 'cooperative games' has studied the interaction of n-agents who with cooperation attempt to improve their individual payoffs, over what they can get without cooperation. Formal analysis of such a situation was first provided by Nash [1953].

Game theory does not concern itself with the acquisition of necessary information for players to play. In team theory [Marschak and Radner, 1972] the central concern is the economic way of providing and allocating information to a team of decision makers who are striving to optimize a common set of goals. An extensive form of solution to this problem has been studied in [Moore and Whinston, 1986].

The common issues that the architectural design of a distributed system of interacting machines must address are:

- The number and type of modules and the nature of intermodule interaction and communication;

- The degree of intelligence and autonomy of the modules. Autonomy is defined as the independence exercised by the modules in choosing their own tasks and the methods of execution; and

- The type of decentralized control allowed in the structure. A decentralized structure is defined as a collection of autonomous modules.

For complex information sharing and distributed decision-making structures there are two extreme examples [Simon, 1981], namely: (1) a market, which is a highly decentralized structure; and (2) a hierarchical organization (like that of an army). Between these two extreme examples, there are a variety of other structures which combine in various degrees the features of autonomy and decentralization found in the market structure and the features of rigid control found in the hierarchical structure. For example, these find a parallel in the world of FMS such as: (1) functional hierarchies, and (2) product hierarchies. The type of architecture has a large influence on the coordinating mechanism. Some studies, such as [Shaw, 1984], have investigated the relationship between coordination schemes and the types of structure.

Concurrency and temporal logic

Concurrency is an essential and pervasive issue in the study of dynamic and distributed systems with interacting machines. Petri nets have been used to model the concurrency among interrelated manufacturing processes (for example, Nof et al. [1980]). Petri nets avoid the use of explicit representation of time [Reisig, 1982] while describing the degree of dependence or concurrency among processes.

However, in most occasions, explicit representation of time is unavoidable for real-time control. For intelligent systems to exercise real-time dynamic control, the system must be able to relate and reason about casual relationships among processes spread over time.

Temporal logic as proposed by Turner [1984] can be of use in modeling FMS structures. Allen [1983] has introduced temporal reasoning in AI application and it was extended to database application in [De et al., 1984]. Such computational approaches should be explored for their possible extension to real-time intelligent control applications.

Activity controllers for cooperative multi-robot work

The objective of machine interaction by cooperation [Nof, 1985; Nof and Hanna, 1987] is to increase the capabilities of industrial robots and interacting machines working together on production tasks. Some control algorithms, heuristics, and strategies have already been developed for cooperative work with shared resources, but more are needed and are being developed for new concepts of robotic cell spatial design. Analytic and simulation models are needed to analyze the operating characteristics of robotic cells based on existing and future technology. Computation is needed for analysis and control of compliance, part insertions, part kitting, bin picking, recovery, rework, recirculation, dynamic task allocation and so on. Intelligent control procedures are needed for robotic assembly and part handling, including the integration of sensory information inputs.

2. MULTI-ROBOT COOPERATION

Background

Robotic workstations employing two or more robots can be advantageous when resource and task sharing lead to reduced costs. But increased levels of sharing among robots require more complex control, and potential conflicts among the robots can decrease effectiveness, e.g., production rate. In previous work [Maimon and Nof, 1983, 1985, 1986] an activity controller was designed and implemented for analysis, optimization and real-time coordination of multi-robot work. Subsequently, Hanna and

Nof [1986, 1987] investigated models and methods to measure and determine the appropriate cooperation and sharing levels for given objectives. An effective method was developed to plan multi-robot workstations, particularly, for measurement of relevant operational characteristics and determination of the desirable levels of sharing [Hanna, 1986].

A multi-robot system can be defined as a group of robots working with machines, material-handling devices, part feeders and fixtures to accomplish certain production objectives under some combined computer coordination control. A number of approaches to plan motions of multiple robot arms in common workspace have been developed and tested [for example, Freund, 1986; Zheng and Luh, 1986; Lee and Lee, 1987]. At the robot machine-level control these path-planning schemes include trajectory planning to avoid collisions with stationary known obstacles, or motions in time-varying environments. The attempt there is to develop a single kinematic chain of the individual robot arms that move together. Lirov [1986] developed a method based on machine-tool control for synchronizing the smooth motions of multiple arms with stepper motors.

The value of using multiple, independent, non-cooperating robot arms was analyzed by Grossman et al. [1985]. They found that under fairly general conditions of spatial contention, there is little merit in having more than two independent, non-cooperating arms in the same workspace. On the other hand, Maimon and Nof [1983, 1985, 1986], and Hanna and Nof [1986, 1987], who focused on systems with multi-robot cooperation (MRC), analyzed multi-robot systems with coordinated, cooperating arms, and found significant advantages relative to independent non-cooperating arms, in wide ranges of industrial applications.

Planning multi-robot activities at the cell level by several researchers has assumed that physical collisions are prevented either by the path-planning techniques mentioned above, or by the aid of vision and proximity sensors. For instance, optimal sequencing procedures for robot activities in machine tending applications were developed by Wilhelm and Sarin [1985], assuming robot-machine collisions are avoided. Chang et al. [1986] applied a time-buffer method to balance multiple arms without cooperation, based on the longest cycle-time. By decoupling the arms operations they simplified or even eliminated synchronization contentions at the cost of less efficiency. A different approach was suggested by Maimon and Nof [1983], who developed a hierarchical activity controller for multiple robot arms with real-time coordination of some task-sharing and resource-sharing functions. This controller is now being extended by knowledge-based functions to enable real-time intelligent interaction among the robots to improve their overall effectiveness.

Classification and measurement of cooperation among interacting robots.

Cooperation among machines is a form of intelligent machine interaction. We divide the modes of cooperation possible among interacting robots to *task-cooperation,* and *resource-sharing. Task-cooperation* itself can be of three types: Mandatory, Optional, or Concurrent.

- **Mandatory task cooperation:**

This is a cooperation mode that is required when certain operations *must* be carried out by two or more robots simultaneously. In general, designing a work-cell with such a cooperation mode should be avoided as much as possible, since the productivity per each individual robot here is diminished.

A typical example of the mandatory task cooperation mode is where a number of arms must lift together a cumbersome structure while maintaining its position and orientation over a period of time.

- **Optional task cooperation:**

This cooperation mode is designed for certain operations that can be carried out by one out of several robots. In general, this cooperation mode may increase the productivity and reliability of the whole cell: Robots can share in performing a larger variety of tasks as soon as they become available; Robots can substitute other robots when the latter malfunction.

- **Concurrent task cooperation:**

This is another mode of cooperation by design, which may increase the cell reliability and productivity. It is found when several robots can carry out simultaneously, in parallel, several operations, e.g., on a number of product facets, thus cutting the total completion time. Unlike the two previous cooperation modes, this one is characterized both by the ability of different robots to cooperate, and by the relative reduction in task time that depends on the number of cooperating robots.

Other types of cooperation modes, relatively more complex, exist too. For instance, complex conditional combinations of mandatory, optional and concurrent cooperation may be found. However, they are not considered in this article.

The complete study has included a number of probabilistic and simulation analyses of the various cooperation modes. Two types of multi-robot cells were considered: for point operations such as drilling or riveting of generally flat parts, and for progressive assembly and inspection of electromechanical products. The main research questions that were studied:

- How much concurrent, optional, and combination of concurrent and optional cooperation should be designed into a specified cell?

• How do different resource sharing and task cooperation combinations influence the cell performance?

The above questions were analyzed extensively for the two types of multi-robot cells both from the production performance point of view, and the cost-effectiveness point of view. Some results are illustrated in Figures 1, 2, 3, 4.

Observations

1. Cooperation through resource sharing and full optional task cooperation by several machines is illustrated in Figures 1, 2, 3. Graph (A) in Figure 1 indicates the value of the optional cooperation: Waiting to be processed decreases when more interacting robots are available. In conflict, graph (B) shows that because of resource sharing, there is more waiting when there are more interacting robots. Similarly, the number of conflicts goes up in an increasing rate (Fig. 2), and as a result, the production rate of the cell reaches the point of decreasing rate of growth (Fig. 3). The optimum number of interacting robots in the cases analyzed was between 3 and 4.

   The effects of varying the level of cooperation among the interacting machine are illustrated in Figure 4. While in Figs. 1-3 it is assumed that full optional cooperation is possible, in 4(a) it is varied: from 0 (no optional cooperation), through four partial levels where only part of the production process has full optional cooperation, to 100%. In 4(b) the level of resource sharing is varied from 0 to 100%, and in 4(c) the level of mandatory cooperation is varied from 0 to 10%.

2. As shown in Fig. 4(a), for $.60 < S_o < .80$ the production rate decreases. This can be explained by the imbalance in allocating operations among the robots in this region of partial cooperation. It is similar to the imbalance in the case of partial levels of resource sharing. As a result, only operations of the task that are optionally shared will be performed faster, but in this case, it does not result in shortening the total task time.

3. Because of the conflict over shared resources, the level of optional cooperation was found to have significant impact on PR in this study for levels of $S_o = 62\%$ or below. But no significant improvement was found between the various higher levels of optional task cooperation, between 62% and 100%. (The case of optional cooperation with no resource sharing is trivial, because it amounts simply to an increased capacity of the cell.)

4. The level of resource sharing was also found to have significant negative impact on PR above a level of $S_r = 82\%$ (Fig. 4(b)), for the case of relatively high part arrival-rate. For the relatively low part arrival-rate, it was found that in the example cell there is no significant decline in PR even at resource sharing level of $S_r = 100\%$.

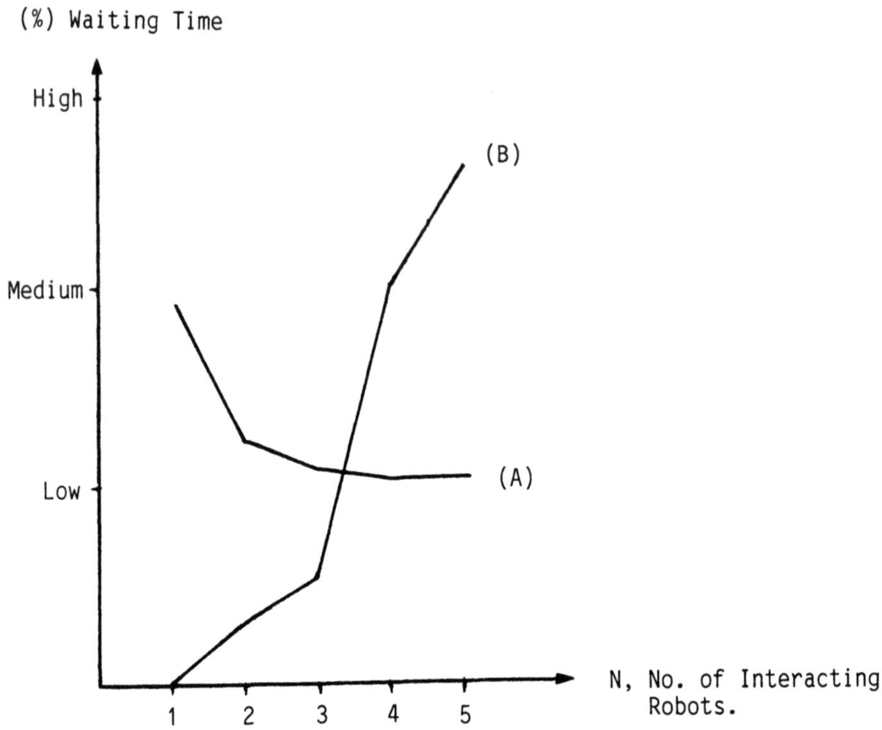

Figure 1. Waiting time in multi-robot cells with full optional cooperation due to conflicts over shared resources. (No mandatory or concurrent cooperation)

A: Percentage of time a part is waiting for a robot.
B: Waiting due to conflicts.

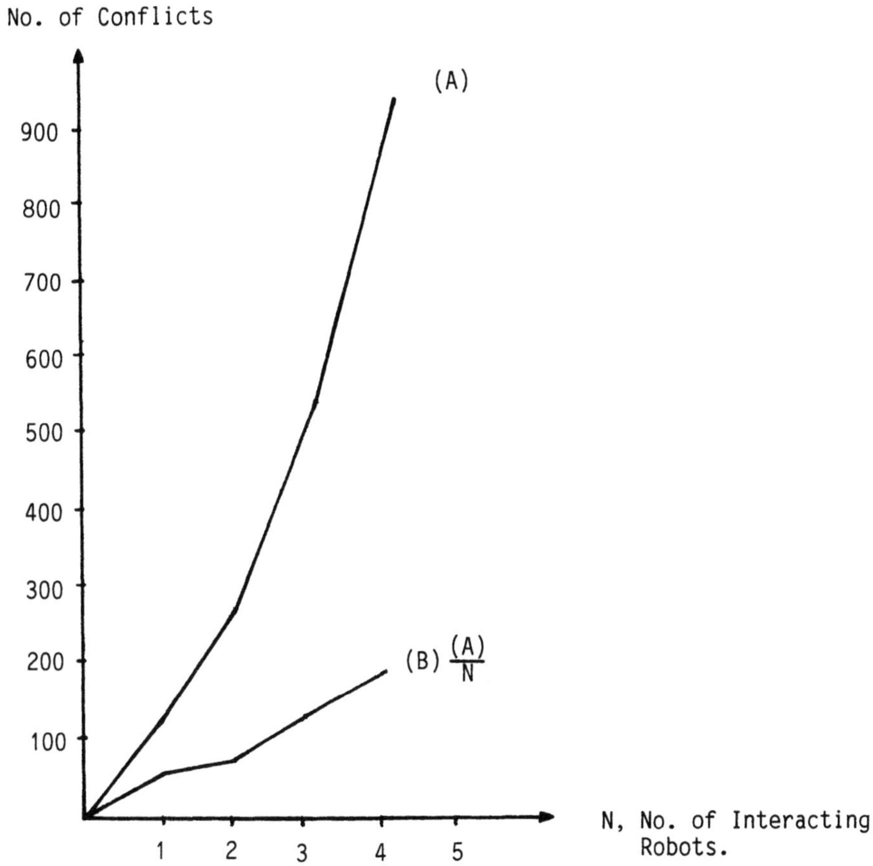

Figure 2. No. of conflicts over shared resources during cell operation. (Same cells as in Figure 1.)

A: Average number of conflicts;
B: Average number of conflicts per robot.

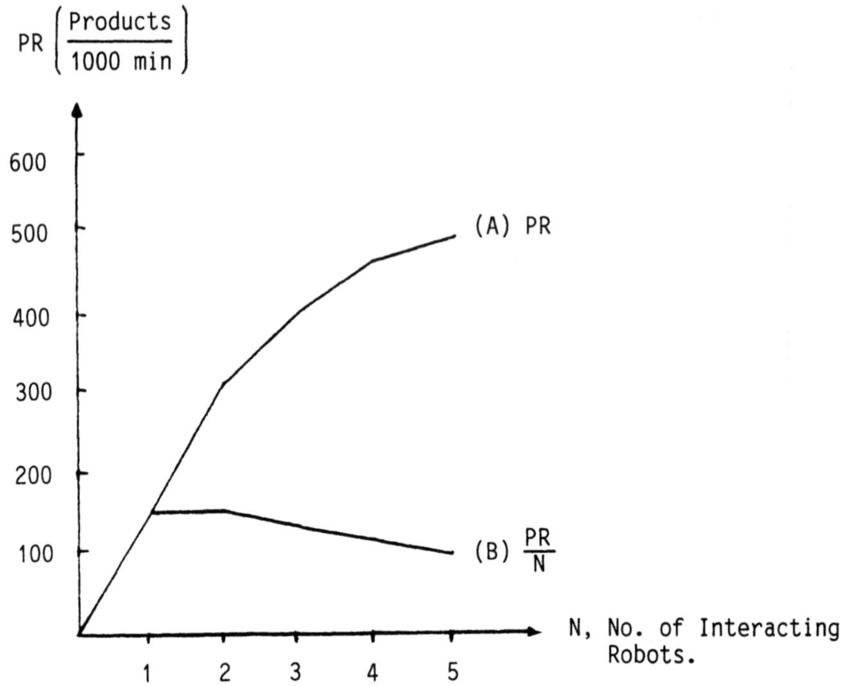

Figure 3. Average production rate (PR) vs. no. of cooperating robots (N).
(Same cells as in Figure 1.)

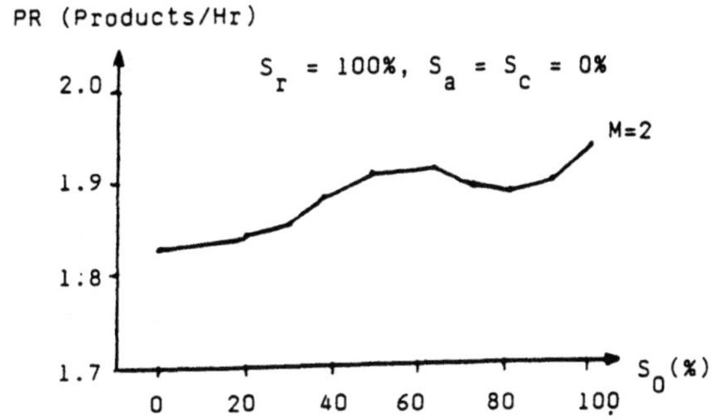

(a): Production rate, PR vs. level of optional task cooperation, S_o.

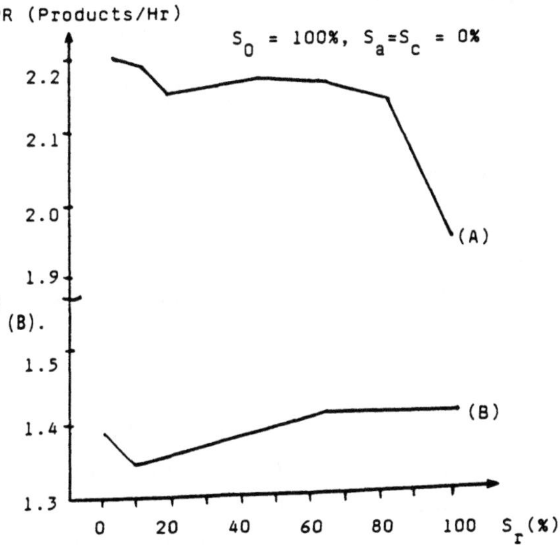

(b): Production rate, PR vs. level of resource sharing, S_r, in a two-robot cell. Part interarrival-time distributed u(4,5) min. (A), or u (14,15) min. (B) (B).

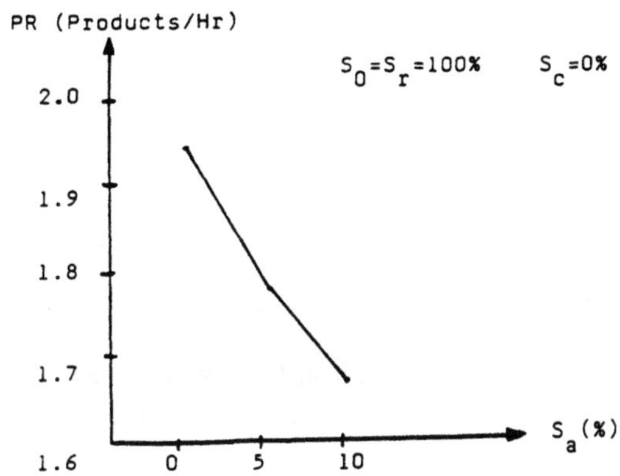

(c): Production rate, PR vs. mandatory cooperation level, S_2.

Figure 4. Production Rate of final products vs. different levels of task cooperation and resource sharing, in a cell with N = 2 robots.

5. Mandatory cooperation significantly reduced performance in terms of PR, as illustrated in Fig. 4(c), since one of the robots is tied up in this case part of the time in serving as a fixture.

Design recommendations for cells with interacting machines can be developed based on these observations. In multi-robot systems it is possible to achieve relatively higher production rates, better resource utilization, higher reliability and overall higher cost-effectives production by task cooperation and resource sharing among the interacting robots. Based on the experimental studies described above and other analyses included in the study, specific indications have been found about the impact of the various levels of task and resource cooperation. We presently evaluate alternative interaction rules and intelligence procedures to further improve the system performance and reliability.

3. DEVELOPMENT OF THE INTERACTIVE ROBOTIC DEVICE

Background

Another form of interactive robotics is the interaction of humans with robots. In most cases of human-robot systems this has been viewed as a hierarcial system in which the human is the supervisor, while the robot serves as slave. Thus, intelligent interaction has been developed mainly to improve the information transfer between robots and humans.

An emerging interest in the future, however, is the development of intelligent interfaces combined with knowledge-based interaction systems. Here the objective is to provide assistive intelligence not only to the interaction process but also to the interacting parties themselves. A good example is our resarch project to develop and investigate the IRD, Interactive Robotic Device, aimed at aiding severely handicapped children.

The components of the IRD system are shown in Fig. 5. Interaction takes place through the interface component, controlled by an extensive IRD interaction control sub-system, and gains intelligence input for the goal activities from the interaction knowledge-base. Thus, any intelligence deficiencies of the user, in this case a handicapped child, can be supplemented dynamically when and as long as they exist.

The same IRD concept is viewed as potentially applicable to other situations associated with severe handicaps. For example, humans in space, workers in hostile environments, and geriatrics.

IRD for Handicapped Children

In order to provide the necessary manipulative abilities that meet the cognitive needs of children with severe physical impairments and intellectual delays a robotic device system was specified. At minimum, such system will enable the child training to occur at two successive levels of development. At the first level, the system must be

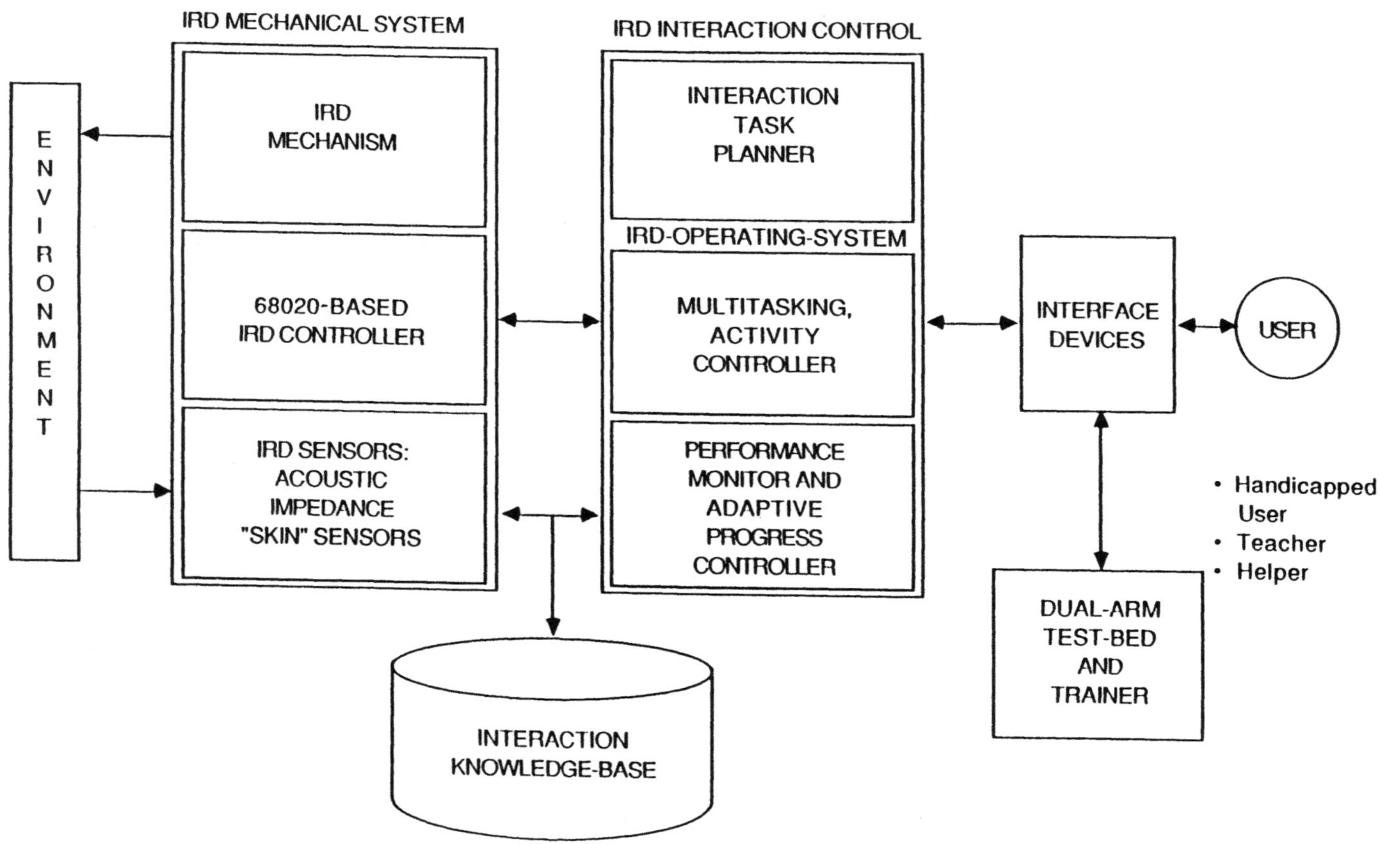

FIGURE 1. COMPONENTS OF THE INTERACTIVE ROBOTIC DEVICE (IRD) SYSTEM

capable of providing the child with a means of using simple actions to repeat or prolong an interesting stimulus, or event produced with an object.

At the second level, the robotic system must allow the child to control component actions and to interact in the process of coordinating these actions into more complex sequences.

Because such training is considered part of the educational program for the child, the system should also permit a teacher to periodically call forth data concerning the child's performance and progress.

The research completed to date with the system, called IRD-1, represents initial activities aimed at examining the feasibility of implementing such a system with commercially available computers, interfaces and small educational robot arms. Now we are developing knowledge-based interaction between child, computer, and robot (see Fig. 5), following the IRD concept discussed above.

Method and Experimentation

The first level of the interactive robotic device (IRD) was divided further into two sublevels [Nof et al., 1987]. At level 1.1, the IRD has been programmed to respond to the closure of a single switch by performing a randomly selected action on a sequence of toys. At level 1.2, the IRD has been programmed to respond to a two-step sequence. First it responds to the closure of one of several switches by the child selecting a specified toy; Then it responds to the closure of one of several switches by the child, in order to perform an action upon the object specified.

The first sublevel demonstrates simple cause-and-effect relationships, while the second sublevel begins to differentiate the properties of objects and to associate classes of actions with certain objects.

Level 1.1 was implemented using an IBM RS-1 robot and two toys capable of producing interesting auditory and visual consequences when manipulated. Programming provided both a brief demonstration and response pause opportunities, in which the student child could initiate or maintain control in order to repeat the object manipulation. Data collected included initial and response pause actuations, and interresponse times and average interresponse time for each child. Though selected for its programming language and real-time monitoring capabilities, the RS-1 served mainly for initial feasibility studies due to its lack of portability, and noise levels arising from its hydraulic components, that mask desirable auditory effects from the toys.

Implementation of Level 1.2 used a five degree-of-freedom Mitsubishi Movemaster robot and controller, a portable computer, a Seraid hardware/software interface (keyboard emulator), and a membrane keyboard. Three toys (tennis ball can filled with small ball bearings and encircled with metal "sleigh" bells, a plastic slinky, a "Nerd Honker" bicycle horn) were modified by adding jigs and fixtures to permit manipulation and placement by the robot on a toy-rack when not in use.

An interaction session begins with the IRD performing a "ready ritual". Then the IRD responds to the child touching one of three large drawings (13 cm x 13.5 cm) of the toys on the membrane keyboard by picking up a toy and moving it to a "home" position and performing a "pause ritual".

If the child touches one of three "action squares" (13 cm x 6 cm) located below the toy drawings, then one of three routines pre-programmed for that particular toy is selected by the controller at random, and performed. If the child does not respond, the toy is returned to the rack and another "pause ritual" is performed.

At the conclusion of each action routine, the IRD performs a pause ritual to indicate readiness to proceed. Data collection programming permits information concerning toy selections, actions selected for each toy, number of actuations per toy selection, and interresponse intervals to be gathered, for each child interacting with the device, for subsequent analysis.

These initial activities, including a preliminary field evaluation with handicapped children, did suggest that future investigation of Interactive Robotic Device (IRD) system is feasible and warranted. One prominent feature currently being addressed is knowledge-based component to carry out on-going performance monitoring, to interpret data, and generate automatically information concerning student progress.

4. CONCLUSIONS

Interaction among machines can have significant advantages provided it is designed well and is controlled by intelligent, knowledge-based controllers. In this article, two different interaction types were described: in multi-robot cells, where robots interact with resource and task sharing; between an IRD and a handicapped child, where interaction involves both intelligent interface and knowledge-based support for lacking human cognitive and manipulation skills.

Impact of Advances in Information Technology

Six major recent trends in information technology can influence dramatically developments in material flow systems [Nof, 1987], three in software and three in hardware.

- Knowledge-based systems
- Data-based technology
- Object-oriented systems

- Workstations

- Supercomputing

- Machine intelligence

In studying the nature of robotic machine interaction, it has become evident that such interaction is strongly associated with:

- Communication

- Coordination

- Cooperation

These, in turn, depend on information exchange, and can be impacted by the major trends identified above. The first two, knowledge-based systems and data-base technology, are components of the IRD, essential for intelligent control. Object oriented programming and systems provide heterarchical computation, that is useful for non-hierarchical interaction architectures. Workstation integration will influencee the development of effective interactive environments. The last two, supercomputing and machine intelligence, including intelligent sensors, are required for intelligent interaction that is dependent on fast, complex calculations and inference, and on reliable acquisition of information and distributed decision making in real-time.

Research work is needed in the above direction. Some specific questions to study are: When is group-based robotic machine interaction needed? Under what conditions is it justified? How to enable it in specific cases? What is the optimal number of interacting machines and level of interaction for specific cases? And finally, how to optimize the interaction?

References

1. Allen, J. (1983). "Maintaining knowledge about temporal intervals", *Communication of ACM,* 26.

2. Chang, C.A., Goldman, J., and Yaralan, S., (1986). "A time buffer method for multi-arm operational planning, *Material Flow,* 3(1-3), 187-197.

3. Freund, E. (1986). "On the design of multi-robot systems", *Proc. of IEEE Intl. Conf. on Robotics and Automation,* 477-490.

4. Gray, N.J. (1986). "an approach to decentralized computer system", *IEEE Transactions on Software Engineering,* 12, July.

5. Grossman, D.D., Evans, R.C., and Summers, P.D. (1985). "The value of multiple independent robot arms", *Robotics and CIM,* 2(2), 135-142.

6. Hanna, D. (1986). "Operational characteristics of multi-robot cells", Unpublished M.Sc. Thesis (in Hebrew), Faculty of Industrial Engineering & Management, Technion, Haifa, Israel

7. Lee, B.H. and Lee, C.S.B. (1987). "Collision-free motion planning of two robots", *IEEE Trans. on Sys. Man and Cyb.*

8. Lirov, Y. (1986). "On synchroized proportional motion of multi-robot systems with stepper motors and shared controller", *Robotica,* 4, 151.154.

9. Maimon, O.Z., and Nof. S.Y. (1983). "Activity controller for a multiple robot assembly cell", in *Control of Mfg. Processes and Robotic Systems,* (D.E. Hardt and W.J. Book, Eds.), ASME Publication, 267-284.

10. Maimon, O.Z., and Nof, S.Y. (1986). "Analysis of multi-robot systems", *IIE Transactions,* 18(3), 226-234.

11. Marlin, C.E., *Coroutines, A Programming Methodology,* Springer Verlag, New York, 1980.

12. Marschak, J., and Radner, R. (1972). *Economic theory of teams* Yale University Press.

13. Moore, J.C. and Whinston, A.B., (1986). "A Model for Decision-Making with Sequential Information-Acquistion", *Decision Support Systems,* Vol. 2, No. 4, Dec., 285-307.

14. Nash, J.F. (1953). "Two-person cooperative games", *Econometrica,* 21.

15. Nof. S.Y. (1985). *Handbook of Industrial Robotics,* John Wiley & Sons, Inc., NY.

16. Nof, S.Y., "The Impact of Advances in Information Technology on Robotic Material Handling", *Proc. of the 1987 Material Handling Research Forum,* Georgia Institute of Technology, Atlanta, GA., September 1987.

17. Nof, S.Y., and Hanna, D. (1987). "A study of multi-robot workstations with process and resource sharing", *Proc. of the IXth Intl. Conf. o Prod. Res.,* Cincinnati, Ohio, August

18. Nof, S.Y. and Maimon, O.Z. (1985). "Activity control in an assembly station with multiple robots", (in Hebrew), *Journal of the Israeli Association of Engineers and Architects,* 44(4), October 1984, 7-10.

19. Nof. S.Y., Whinston, A.B., and Bulles, W.I. (1980). "Control and decision support in automatic manufacturing systems" *AIIE Transactions,* 12, 2.

20. Nof, S.Y., Widmer, N.S., and Karlan, G.R. (1987). "Interactive robotic devices for cognition and manipulation skill development in severely handicapped children", *Research Memorandum, 87-2,* IIES, Purdue University, W. Lafayette, IN

21. Reisig, W. (1982). *Petrinets: an introduction,* Springer Verlag, NY.

22. Shaw, M.J. (1984). "The design of a distributed knowledge-based system for the intelligent manufacturing system", Ph.D. Thesis, Krannert School of Management, Purdue University, West Lafayette, IN.

23. Shaw, M.J. and Whinston, A.B. (1985). "Automatic planning and flexible scheduling – a knowledge based approach", *IEEE* CH 2152, July.

24. Simon, H.A. (1981). *The Sciences of the Artificial,* MIT Press, Cambridge, MA.

25. Turner, R. (1984). *Logic for artificial Intelligence,* Springer Verlag, NY.

26. Wilhelm, W.E., and Sarin, S.C. (1985). "A structure for sequencing robot activities in machine tending applications", *Int. J. of Prod. Res.,* 23(1), 47-64.

27. Zheng, Y.F. and Luh, J.Y.S. (1986). "Joint torques for control of two coordinated moving robots", *Proc. of IEEE Intl. Conf. on Robotics and Automation,* 3, San Francisco, CA, 1375-1380.

The Karlsruhe Autonomous Mobile

Assembly Robot

by

Prof. Dr.-Ing. Ulrich Rembold
Institute for Realtime Computer Systems and Robotics
University of Karlsruhe
7500 Karlsruhe

Federal Republic of Germany

Summary

In this paper the architecture and functions of an autonomous mobile system are described. For the operation of such a system knowledge-based planning, execution and supervision modules are necessary which are supported by a multi-sensor system. The individual functions of such a vehicle are explained with the help of an autonomous mobile assembly robot which is being developed at the University of Karlsruhe. The vehicle contains a navigator, a docking module and an assembly planner. Navigation is done with the help of a road map under the direction of the navigator. The docking maneuver is controlled by sensors and the docking supervisor. The assembly of the two robot arms is prepared by the planner and controlled by a hierarchy of sensors. The robot actions are planned and controlled by several expert systems.

1 Introduction

For several years, various autonomous mobile robots are being developed in Europe, Japan and the United States. Typical areas of application are mining, material movement, work in atomic reactors, inspection of under-water pipelines, work in outer space, leading blind people, transportation of patients in a hospital, etc. The first results of these research endeavors indicate that many basic problems still have to be solved until a real autonomous mobile vehicle can be created; e.g. the development of an integrated sensor system for the robot is a very complex effort. To recognize stationary and moving objects from a driving vehicle is several orders of magnitude more complex than the identification of workpieces by a stationary camera system. In most cases the autonomous system needs various sensors. For processing of multi-sensor signals, science has not found a solution to date. An additional problem imposes the presentation and processing of the knowledge needed for planning and following a route or trajectory which is necessary to execute an assignment. Unexpected obstacles have to be recognized, and if necessary an alternate coarse of action has to be planned.

At the University of Karlsruhe an autonomous mobile robot for the performance of assembly tasks is being developed. The assignment of the system is to retrieve parts from a storage, to bring them to a work table and to assemble them to a product. All assignments have to be done autonomously, according to a defined manufacturing plan which is given to the system.

With autonomous mobile robots it is possible to develop manufacturing plants of great flexibility. Any combination of machine tools may be selected according to a virtual manufacturing concept. E.g., an autonomous assembly system equipped with robot arms is capable of working at various assembly stations. For welding or riveting tasks, the robot can move along a large object, such as the hull of a ship and perform the desired operations. An increase in flexibility can only be obtained by the use of knowledge based planning, execution and supervision modules which are sensor supported. In addition, omnidirectional drive systems have to be conceived, capable of giving the vehicle a three-dimensional flexibility, including turning on a spot.

NATO ASI Series, Vol. F53
Advanced Information Technologies for
Industrial Material Flow Systems
Edited by Shimon Y. Nof, Colin L. Moodie

2 *Components of an Autonomous Mobile System*

An autonomous system must be capable of planning and executing a task according to a given assignment. When a plan is available, its execution can start. A complex sensor system must be activated which leads and supervises the travel of the vehicle. Furthermore, it is necessary to recognize and solve conflicts with the help of a knowledge processing module. The basic components of an autonomous intelligent robot are shown in Fig. 1. To conceive and build these components, expertise of many disciplines such as physics, electronics, computer science, mechanical engineering, etc. is required. It is very important to design good interfaces between the functional components of the system.

The most difficult task is building the software components. This is a universal problem with automation efforts involving computers. Designing software for autonomous vehicles is, however, complicated by the fact that very little is known about their basic concepts.

3 *Mechanics and Drive System*

Many present research efforts on autonomous vehicles use experience gained from guided vehicle experiments. Most of these vehicles are not suitable for autonomous operations. Their mobility is limited because they employ a drive system with 2 degrees of freedom and a rather simple control system.

For autonomous mobile robots drive systems with 3 degrees of mobility are of interest. However, there are no matured control systems available. This was a major consideration when the Karlsruhe autonomous robot was designed. Fig. 2 shows the robot and its drive systems. It has four active Mecanum wheels. Along the circumference of each wheel, passive rollers are fastened in a fixed angular orientation with regard to the main axis. The rollers of the front wheels are arranged in a positive, and those of the rear wheels in a negative herringbone pattern. In case the vehicle drives forward, all active wheels are rotated in the same direction. For a sideward motion the front wheels are rotated in the opposite direction of the rear wheels. Both sets of wheels rotate at the same speed. Thus the vector of the drive force acts perpendicular to the centerline of the vehicle and pushes the vehicle sidewards via the passive rollers. This vehicle can also be moved in any direction or it is able to rotate on a spot under the guidance of a proper control circuit.

4 *The Sensor System*

A sensor system of the Karlsruhe autonomous mobile robot consists of various sensors which are interconnected by a hierarchical control concept. The sensors furnish the planning and supervision modules with information about the status of the robot world. For each of the three major tasks of the vehicle, the navigation, docking and assembly an own sensor system is provided.

The sensoric has the following assignments:

- locating workpieces in storage
- supervising the vehicle navigation
- controlling the docking maneuver
- identifying the workpieces and their location and orientation on the assembly table
- supervising the assembly
- inspecting the completed assembly

For the navigation of the autonomous vehicle, a multisensor system is necessary. A distinction is made between vehicle based internal and external sensor and world based sensors. Internal sensors are increment decoders in the drive wheels, the compass, inclinometer, etc. External sensors are TV cameras, range finders, approach and contact sensors, etc. World based sensor systems use sonic, infrared, laser or radiotelemetry principles. The Karlsruhe robot will be guided by several cameras, sonic sensors and a ranging device. The vehicle must constantly monitor its path with its sensor system and must be able to react to unforeseeable events. Recognition of objects is done by extracting from the sensor data of the scenario specific features and comparing these with a sensor hypotheses obtained from a world model. For scenes with many and complex objects the support of an expert system is needed for the sensor evaluation.

The docking maneuver will be supported by optical and mechanical proximity sensors. Thereby, for coarse positioning a vision system is used and for fine positioning mechanical feelers.

Recognition of parts for assembly will be done with a 2D vision system which also determines the position and orientation of the object. For the supervision of the assembly process a 3D vision system is being developed. Operations such as parts mating, fastening and aligning are monitored by force/torque, approach and touch sensors. Thereby, the vision system is the most important sensor; it is connected with the other sensors to a multisensor module to supervise the complex assembly operation.

5 Planning and Supervision

5.1 Global Planning and Supervision

The task of the global planning and supervision system is the conception of an action plan for the autonomous mobile robot and the supervision of the execution of the plan. With the autonomous mobile robot of the University of Karlsruhe, there are three functions to be performed, Fig. 3. For each function, an individual planning module is required.

A typical task for this system may be the instruction to perform an assembly of a product. All necessary information about the part, obtained from the assembly drawing and the bill of materials, is sent to the action module, Fig. 4. The action module retrieves knowledge from the world model about the inventory, product delivery date, the configuration of the material storage, the assembly workstation and the layout of the plant. The vehicle must be brought into a suitable fixed position for the retrieval of the parts and for assembly. For both of these operations knowledge about the docking maneuver must be available. An additional task is the supervision and synchronization of the navigation, docking and assembly. For these operations the corresponding parameter, transfer procedures, interfaces and synchronization protocols must be provided.

For the assembly of the product the necessary operational steps must be determined and ordered in proper sequence, Fig. 5. The number of products to be manufactured determines the required number of parts. With this knowledge the retrieval of the material, the material flow and the navigation route are determined. In case of conflicts, it may be necessary to perform the planning repeatedly or to divide the material flow into several subactivities.

The majority of the tasks has to be executed sequentially. However, many tasks can be solved in quasi-parallel operations and must be prepared for parallel execution by the task scheduler. The task scheduler is also the executive which initiates the processing of the tasks and which controls asynchronous events. The scheduler employs algorithms for path and time optimization under the boundary conditions that the precedence requirements of the task are obeyed, duedates and priority assignments are met and the load capacity of the robot is not exceeded

The scheduler performs the route planning under global considerations with the help of a very abstract model. The geometric model is presented by a graph, whereby the nodes represent locations and the edges segments of the path. The edges are directed and represent distances. At stochastic intervals a data acqusition system reports the actual state of the world, for example about the traffic density, jams, road blocks, station times, queues, etc. With the help of online information, the scheduler selects the proper workstation in case there are several alternatives. The execution planner reports to the scheduler at periodic intervals the completion of a task or possible problems which had arisen. In case of a problem, it may be necessary to do the task at another station or to abort the whole endeavor and to remove the task from the queue.

The scheduler gives the task to the task planner for execution.There are three task planners, one each for the navigation, docking and assembly.

The output of the planning modules is entered into the execution and supervision modules which are preparing the instruction for the low level interpreters. Characteristic of these modules is the ability to perform their own planning operations. The plans are further refined, brought to execution and the actions are supervised. In case of deviations from the plan, a situation analysis will be made and a new plan will be generated.

If a task is not solvable, the system has to give information about the cause of the problem; e.g. if a sensor has failed and a corrective action has to be initiated. For this purpose, it is required to store knowledge for the recognition and solution of problems. Thus, the system must have an expert system which is able fo responding to any unusual situation.

5.2 Path Planning and Navigation System

With the help of the path planning and navigation module, the vehicle will be capable of determining its path along the plant floor. It will try to plan an optimal route between the start and goal position, and it will attempt to stay on the path, avoiding obstacles and possible collisions. The navigation system will be tailored to the application. Since unexpected obstacles can be met on the path, the navigation becomes very complex. In this case the world has to be monitored constantly and the vehicle must be capable of recognizing dangerous situations and performing corrective actions. For the solution of conflicts, knowledge about new strategies is needed. The planner must have access to a map to know the action space of the vehicle and to interprete the map. For the evaluation of online sensor data, it is necessary to store in a world model sensor hypotheses for every situation to be expected.

Path planning is done according to a hierarchical scheme, whereby different tasks are solved at different levels, Fig. 6.

1. At the planning level the planner draws up a plausible and collision free path by using expert knowledge. However, at first the vehicle has to determine its start position. This can be done by starting at a known position or by sensors. The planner may try to search for an optimal plan in case there are alternative solutions. Optimization criteria may be the shortest path or the route with the least number of obstacles to be encountered.
2. On the navigation level there is a local planning module which has knowledge about the path, its objects and possible obstacles. Fine planning of the navigation is done on the basis of the plan which was drafted on the previous level, thereby all local situations must be considered. It is important to recognize unknown obstacles by the use of sensors and to report any abnormal situation to the planning system. All problems must be considered for further planning.
3. On the pilot level the elementary control functions of the vehicle are performed. For trajectory planning a map of the world is used showing the path to be travelled in detail. For simple operations only a two-dimensional map of the routes, branches and crossings is needed. With the help of the map trajectory segments, like straight lines or curves are calculated. The passage of doorways or underpaths must be investigated with a three-dimensional map. The navigator corrects the map with the aid of realtime sensor data. A wrong way or obstacles entering or learing the path must be recognized by the sensors. The information is processed quantitatively and qualitatively in order to plan a necessary corrective action. Possible alternative moves are checked with the planner at the higher level. If the action is approved detailed move instructions are sent to the pilot level for execution.

5.3 Planning and Supervision of the Docking

As soon as the autonomous mobile robot has reached its goal, it must lock itself into a stable work position to be able to retrieve parts or to perform an assembly. Two docking maneuvers will be investigated:

1. The robot moves into a coarse position and tries to perform fine positioning with the aid of sensors. Thereby, the vehicle aligns its own coordinate system with the reference coordinate system of the workbench.
2. The robot locks itself into a coarse position and establishes, by activating sensors, its exact location with regard to the reference position of the work area. Any actions to be performed in the work area are corrected by coordinate transformation to account for the offset. Docking is overseen by a supervisor and possible problems are corrected by an expert system.

5.4 Planning and Supervision of the Assembly

For assembly the robot will try to recognize the parts, their location and orientation. An assembly plan may be prepared, either online or offline. Fig. 7 shows the overall structure of the assembly planning and supervision module. By means of an interface, the user or the sensor system describes the parts to be assembled. The planner fetches from a CAD databank and from the world model information about the geometry of the workpieces, the robot world and the required fixtures. Planning is done via a hierarchy of tiers until at the lowest level the robot motion commands are generated. The individual planning steps are, Fig. 8:

1. Drafting of a precedence graph to determine the sequence of assembly
2. At the strategic planning tier the assembly motions, fixtures and auxiliary peripherals are determined according to a virtual assembly cell concept. From a set of alternative solutions a strategy is selected which will render the best solution for the described problem.
3. At this level the equipment to be used is selected. The virtual assembly plan is mapped into the real world by defining the robots, grippers and fixtures for the operations. At this level a distinction between coarse and fine planning is made. For coarse planning the start and goal positions and the trajectories are determined. The fine positioning is concerned with parts mating and the required fine motions.
4. A collision investigation is made at this level. No collision may occur between cooperating robots, equipment or workpieces. In case of a possible collision, a new planning cycle is necessary.
5. The instruction code for robot joint motors and the peripherals is generated at this level. For example, the sensor system must obtain the control information to monitor the world and report back the world status. This information is sent to the planning module via the error recovery module, Fig. 7.

The planning module sends the code to the execution module for processing, Fig. 7. Here, every line of code is checked by the monitor to verify that it can be executed. The monitor contains an up-to-date report on the state of the system from the sensors. In case the state of the assembly area has changed, e.g. by the displacement of an object, the code is not consistent with the present situation and will not be transmitted to the robot. A message is sent to the error recovery module to find the cause of the problem. Thereafter, the planning module is activated anew in order to suggest an alternative assembly method. Simultaneously, a message is sent to the user interface to report the problem. Planning, supervision and error recovery are supported by various expert systems.

6 *The World Model*

A world model can be considered in a broad sense as the extension of the CAD databank. The CAD databank contains information about the workpiece, its geometry, physical data and information about the manufacturing process. However, an autonomous robot must know more about its world. The Karlsruhe robot needs information from the world model for the navigation, docking and assembly. Thus, three different types of world models are needed in which the following information is stored.

- geometry of the objects and their geometric relations
- functions, operation parameters and capability of the objects (e.g. of the vehicle, robots and auxiliary equipment)
- interpolation routines for the vehicle motions and the robot trajectories
- motion parameters for all moving parts (speed, acceleration and deceleration)
- sensor hypotheses for planning and supervision of the action sequences
- information about the effect of tolerances for the different operations
- preferred routes and standard motions
- behavior rules for the avoidance and removal of errors and collision

The world model for the assembly robots needs the following additional information:

- assembly parameters such as fits, mating forces and torques, as well as approach and work speeds

Since the world may change during the execution of a task, part of the world model has to be conceived as a dynamic world model. For this reason, the world is constantly monitored by sensors. For example, if a part falls over, the event must be reported for updating of the world model. Thus, information for the following tasks has to be supplied to the dynamic world model:

- evaluation result of the online sensor system

- reports about changes in the world
- updates about the world model

Part of the world model for the navigator is shown in Fig. 9. The modeling method employed divides the work space recursively into cubes until there are only free and occupied cubes left. Thus, the obstacles are presented by an occupied cube structure. This information is stored in the computer. With the aid of the cube structure, the computer is able of calculating the optimal path the vehicle should take in a three dimensional space.

7 *Knowledge Acquisition and World Modeling*

To plan, execute and supervise its actions, an autonomous mobile vehicle must be able to make decisions on several levels. Thus, is must be equipped with a hierarchy of knowledge based modules and submodules. A special problem presents the acquisition and storage of expert knowledge. For some functions, knowledge may be stored in independent databanks, in other cases knowledge has to be shared via a blackboard architecture.

The Karlsruhe autonomous vehicle will have the following knowledge-based modules.

1. The global action planner and supervisor on the level of the manufacturing cell
2. The route planner and supervisor for the navigation
3. The action planner and supervisor for the assembly
4. The sensor data processor for the navigation
5. The sensor data processor for the assembly
6. The error recognition and recovery module for the navigation
7. The error recognition and recovery module for the assembly
8. The expert systems are presently being implemented with the aid of OPS5 and installed in various computers of the robot. Other expert shells are being investigated.

An interactive modeler for the robot and the workspace is used for the system design and the investigation about the robot motions. It is possible to observe the actions of all components of the assembly system and to watch for possible conflicts.

8 *The Computer System*

With the presently available computers two different computer systems are necessary for the planning and control of the autonomous vehicle. Planning is done offline on a powerful scientific computer and the execution and supervision is done online by a vehicle based realtime computer. The main computer contains the global planner and various expert systems for the planning and supervision of the subfunctions of the vehicle. For example, the main computer gives the action program to the vehicle computer for execution. The vehicle computer interpretes stepwise the instructions and executes them. In addition, expert knowledge is given to the vehicle computer to process sensor information and to solve conflicts which may arise during the navigation, docking or assembly. Since the size of the vehicle computer is restricted, it only can solve simple problems. In serious situations the main computer will be notified and it in turn tries to find a solution. It will also prepare and issue a situation report for the operator.

Acknowledgement

The autonomous mobile robot of the University of Karlsruhe is being developed by the Institute of Realtime Computer Systems and Robotics (Prof. Dr.-Ing. U. Rembold and Prof. Dr.-Ing. habil. R. Dillmann). The work is funded by the Deutsche Forschungsgemeinschaft as part of a cooperative research project on artificial intelligence (Sonderforschungsbereich Künstliche Intelligenz).

Literature

1. Rembold, U. and Levi, P.: "Development Trends for Expert Systems for Robots", Der Betriebsleiter, March April 1987
2. Rembold, U. and Dillmann, R.: "Artificial Intelligence in Robotics", Proceedings of the IFAC SYROCO 85 Conference, Oct. 85, Barcelona, Spain
3. Rembold, U., Blume, C., Dillmann, R. and Levi. P.: "Intelligent Robots", a series of 4 articles, VDI Zeitschrift, Vol. 127, No. 19, 20, 21 and 22, 1985
4. Frommherz, B.: "Robot Action Planning", CIM EUROPE 1987 Conference, Cheshire, England, May 1987
5. Soetadji, T.: "Methods for Solving the Rout Planning for a Navigation System for an Autonomous Mobile Robot", Dissertation at the University of Karlsruhe, 1987

1. Mechanics and drive system

2. Sensor system
   internal sensors
   external sensors

3. Planner and navigator
   planner
   navigator
   expert system
   knowledge base
   meta knowledge

4. World model
   static component
   dynamic component

5. Knowledge acquisition and world modeling

6. The computer system

Fig. 1: Components of an autonomous robot

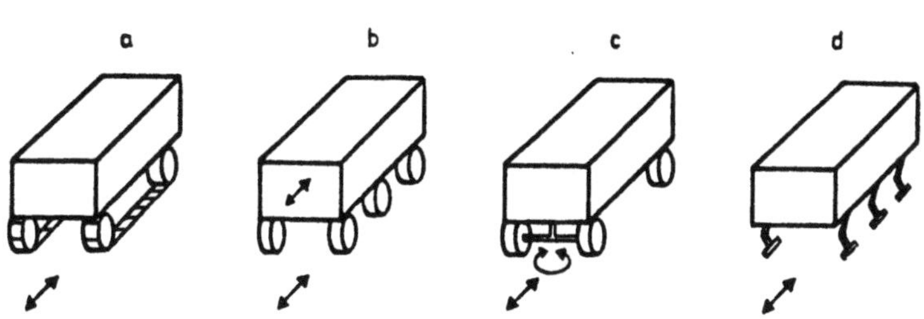

Fig. 2: Various means of propulsion for autonomous vehicles with two degrees of mobility

a. Caterpillar treads
b. Fixed axes
c. Rotational front axis
d. Legs

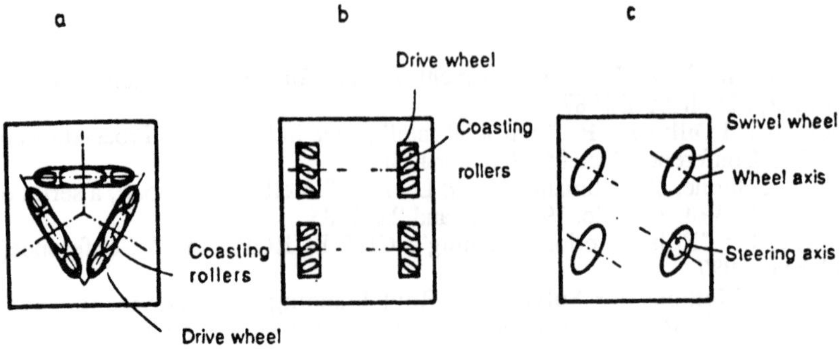

Fig. 3: Various means of propulsion for autonomous vehicles
with three degrees of mobility
a. Three independently powered drive wheels with passive rollers
b. Four independently powered drive wheels with passive rollers
c. Four independently powered and steered swivel wheels

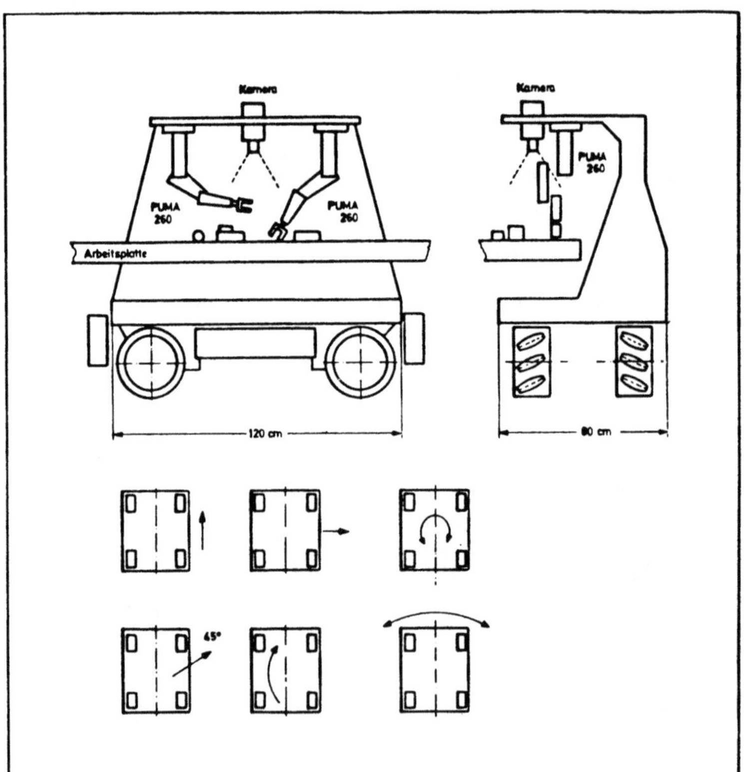

Fig. 4: The Karlsruhe autonomous mobile assembly robot

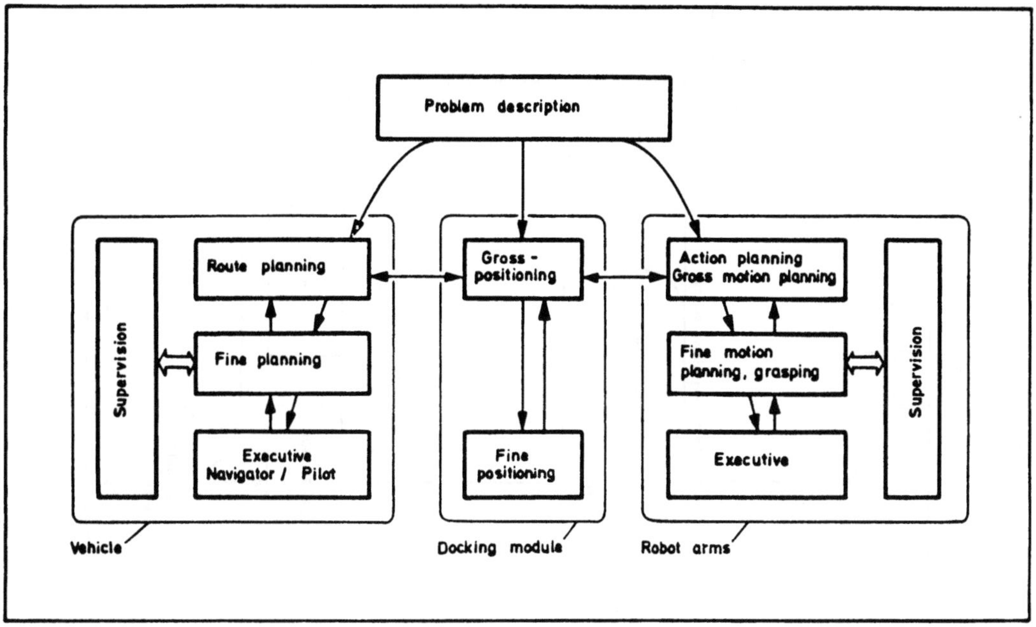

Fig. 5: Global planning system of the Karlsruhe autonomous mobile robot

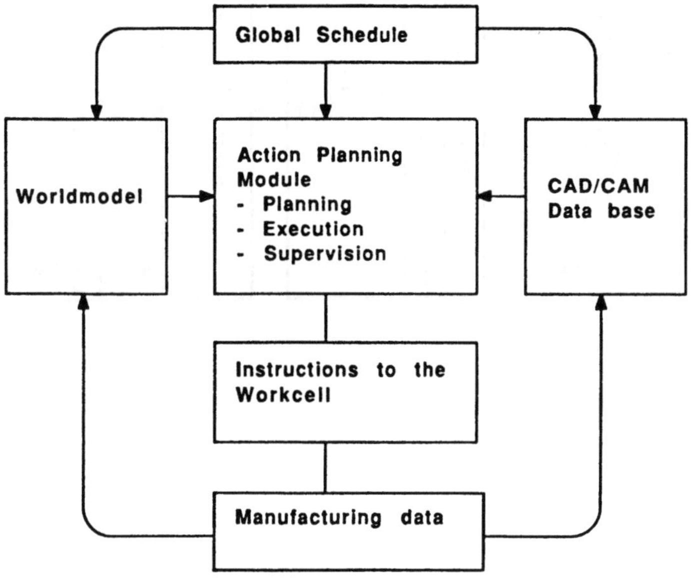

Fig. 6: Global action planning module for an autonomous mobile system

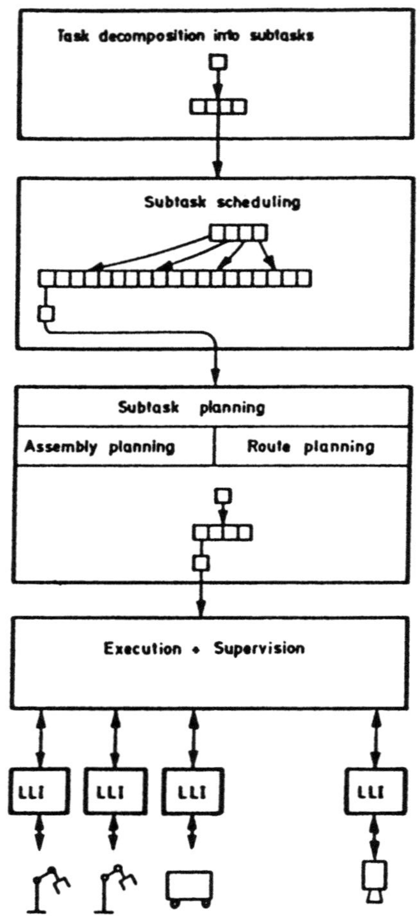

Fig. 7: Schema of planning execution and supervision functions

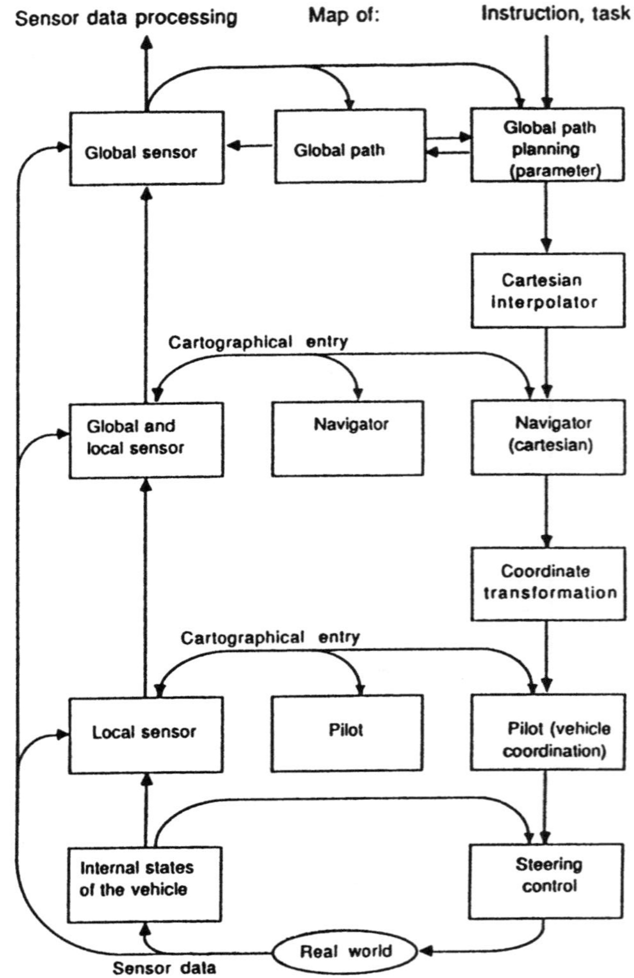

Fig. 8: The hierarchical navigation system

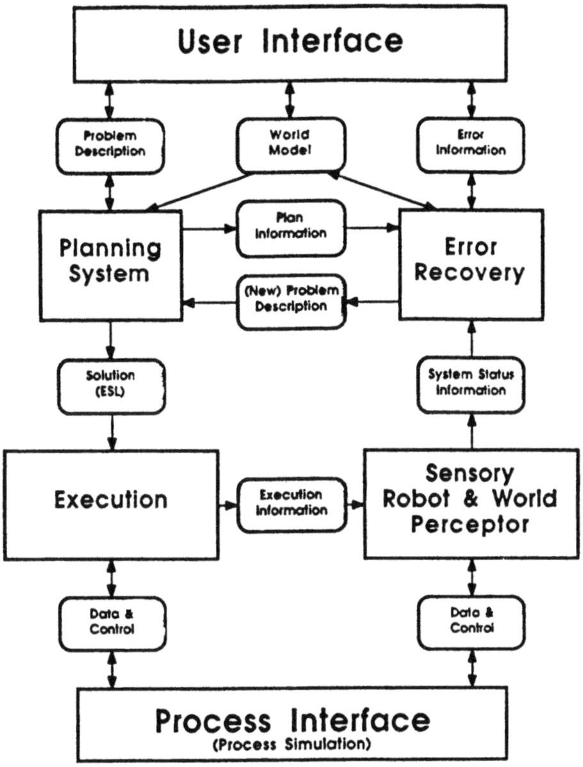

Fig. 9: Schema of the assembly planning and supervision functions

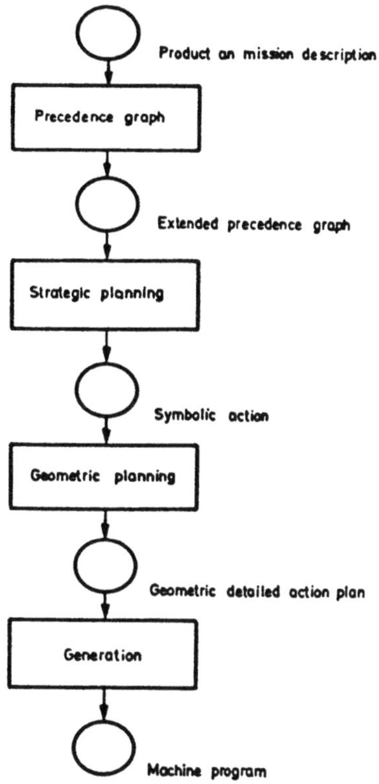

Fig. 10: Different tiers of assembly planning module

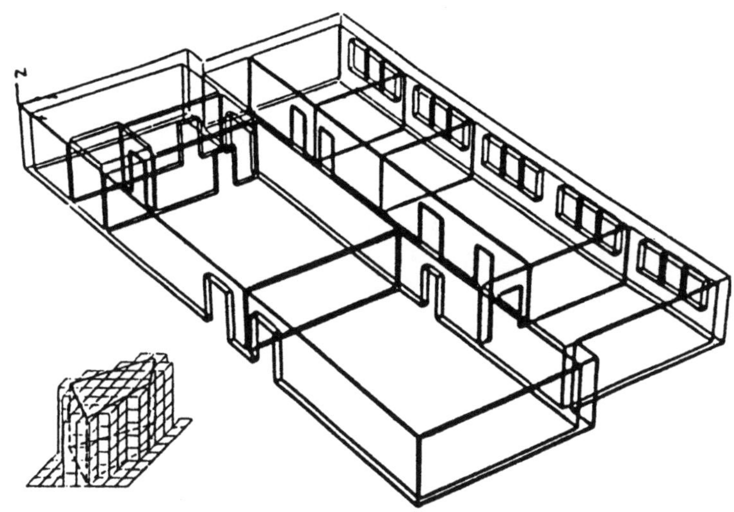

a) Modeling method

b) Model of the robot world

Fig. 11: Modeling of the robot work area using the oct-tree cube space method

Knowledge Representation and Integration of Robot Assembly Applications

S. Schneider
Institut für Prozeßrechentechnik und Robotik (IPR)
University of Karlsruhe, Germany

Abstract:

The knowledge representation and the design of data models are influenced by the various applications which are operating on these representation schemes. For the activities of design, planning, programming and simulation the same robot objects are referenced but different knowledge representations are used.
The requirements of automatic assembly planning, graphic supported offline programming, computer aided design and sensor integration for the representation of knowledge will be discussed in this paper. The results of this work will support the integration of complex systems like the KAMRO (Karlsruhe Autonomous Mobile Robot) into a CIM environment. The relevant objects of this research area are robots, parts, cell layouts, control programs etc. which are represented on various layers of abstraction with different data representations. This different data representations result from the special needs of the invidual activities. The merging of all separate activities into an integrated production systems working on common knowledge bases is necessary. For this purpose different knowledge representations like neutral CAD formats, various databases and AI tools have to be integrated. An object oriented representation of the most important objects is proposed as an approach to an integrated production system, because a homogen representation of all the specific knowledge lead to a inefficient data management effort. The management of the large data objects should be done by conventional means like databases or file management systems.

NATO ASI Series, Vol. F53
Advanced Information Technologies for Industrial Material Flow Systems
Edited by Shimon Y. Nof, Colin L. Moodie

1. The Situation

The research area of automated part assembly by autonomous mobile robots is characterized by various heterogenous activities and their associated knowledge bases. The following typical applications are taken into consideration :

- offline programming and graphic simulation of mobile robots
- navigation and control of automous systems
- part assembly planning
- cell layout planning
- grasp planning

In the following, some examples will explain the abstraction layers and the different knowledge representations used for robotics applications.

Geometric models of assembly cells and assembly parts are the inputs for the navigation and assembly planning algorithmens. CAD formats like IGES, FEMGEN, VDA FS, SET or CAD*I [BEY86] are used for the geometric representation of the robot world. Complex geometric descriptions can be stored using these file oriented formats. A general overall representation of the geometric shape, the volume-, free form shapes, and FEM (finite element model) models in one model has not been done until today and is still a topic of research .

Modern CAD formats like IGES and CAD*I have user defined entities to describe attributes for technical informations like roughness, center of gravity, standards etc. . Technical databases are another tool to integrate technical and geometric knowledge. Interactive query languages allow an efficient access to large databases.

Artificial Intelligence representation methods like rules, Petri nets or frame concepts are used for the representation of the planning knowledge. The knowledge about the objects and the relation between the objects is of technical, physical and geometric nature. The output of the planning algorithms are assembly graphs, precedence graphs and control programs. Modern AI tools use object oriented methods including inheritance, methods, encapsulation and message passing.

The various representation methods are the result of independent research development and applications. Their inclusion into CIM systems leads to integration and data conversion problems.

2. An Systematic Approach for Integrated Information Representation

This section describes a proposal for an object oriented and frame based representation of the central data objects for robots. The following architecture of a central information management and representation module serves as a tool, which allows a quick access to the existing knowledge. Furthermore, this module is a knowledge base for advanced, AI planning, navigation [SOET87], complex sensor and simulation systems etc. .

The structure of the robot knowledge will be discussed by analyzing the different activities and their requirements to data- and knowledge bases. One possible structure of a global knowledge model is the division into submodels with well defined relations between the submodels. The division into submodels may be usefull, because various knowledge representations can be choosen for the submodels. Most of this submodels, Fig. 1, are already existing, and their knowledge representation schemes are defined by the used data access methods of the robotic application. The following submodels can be defined for robot part assembly applications:

- geometric model
- physical model consisting of
    - kinematical model
    - dynamical model
- production model
- process- / cell state model
- application specific models

This submodels use neutral file formats, relational databases [CODD81], rule or object oriented representations etc according to the requirements of their applications. Most robot applications need more than one submodel as input for planning and programming. In addition the output of these activities has to be stored. This is the reason why the submodels can not be handled as independent knowledge bases. This consistency has to be assured. A concrete example is the relation between the production model and the geometric and physical model. Objects which are described in the production model are referencing geometric shapes and physical attributes in the other submodels. This separation into submodels is a useful technique for preventing inconsistencies and redundencies.

This leads to the general question : How to integrate these submodels ? The idea of the integration is to examine objects which can be found in several submodels. Thereby common objects have to be grouped into classes which could be robots, assembly parts, assemblies, cell layout, robot programs and so on. This classes have to be defined for all submodels and the semantic equivalence must be assured.

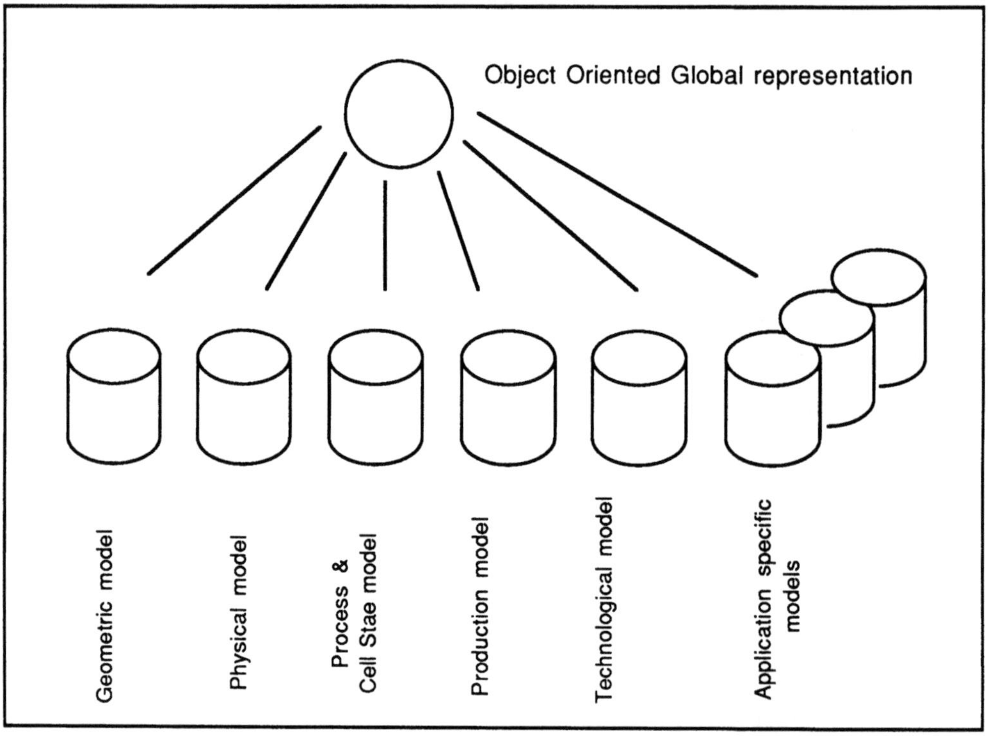

Figure 1 : Robotic knowledge submodels

Equivalent classes and their members have to be defined for simple representations in files and their file management systems. The use of a class dependent suffix for files as example 'PUMA_260.ROBOT' for the member PUMA_260 of the class ROBOT is one possibility. Another possibility is the use of common memory regions (directory structures) for the member files of the same class.

The class and member concept ist directly supported by hierachical database systems such as IMS from IBM.

Relational databases need a simple relation for the management of the classes and their members. Such a relation may have the following form :

CLASS	CLASS_MEMBER
ROBOT	PUMA_260
ROBOT	BOSCH_SR800
CELL	ASSEMBLY_LAY_OUT
CELL	KAMRO_LAY_OUT
CODE	WELDING_CODE
CODE	INSERT_PIN_CODE

Object oriented representations can be integrated directly because they are based on the classification concept. A central object oriented representation can be realized with a class concept for the most important objects, if the submodels have equivalences for the defined classes and their members.

The module for the central representation of the submodels has two main tasks. One important task is the representation of the meta knowledge. This knowledge consists of the representation form of an object, the access method, the information about the location and the structure of the knowledge (datatypes, record formats and so on). This capability will help to generate the query procedures for the application if the structure of the knowledge has to be changed during its implementation. Modules with this function are called 'Common Data Dictionaries' (CDD) in database term analogy.
The second requirement of the central representation module is one which is typical for a modern information management system. Whereby the existing knowledge of an object is presented using graphic support. The manufacturing engineer should be able to control the design and manufacturing progress of a product by using this central representation module. The problems of remote databases and different access methods must be hidden in this representation form of an object. This module can be used to determine whether the requirements for the next manufacturing step are fullfilled or not.

3. An Example of the Information Management of the European Assembly Benchmark

The knowledge representation of the 'European Assembly Benchmark' for its various production phases from the design of the parts to the assembly benchmark will be shown as example in this chapter. A generic approach to extend the knowledge to the product life cycle is also illustrated. The assembly benchmark, Fig. 2, is a mechanical pendulum, consisting of shaft, sideplates, spacers, levers, locking pins etc.

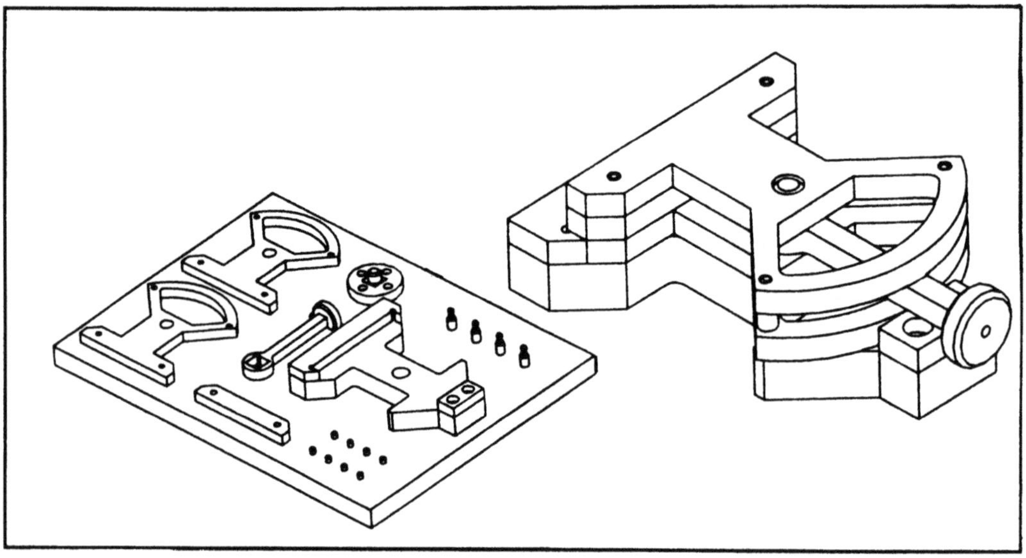

Figure 2: The European Assembly Benchmark

The first step in the life cycle of the product is the conception of the product. This process is performed by the designer. The second step is the geometric and functional description of the product which can be guided by CAD systems. The results of the computer aided design is the description of the geometric, physical and technical knowledge of the assembly and its parts. The geometric knowledge will be represented in a CAD database or a neutral file format. The technical and physical knowledge may be stored in the database of the CAD system or in an independent technical database [EBER84,WEDE81] ,Fig. 3 .

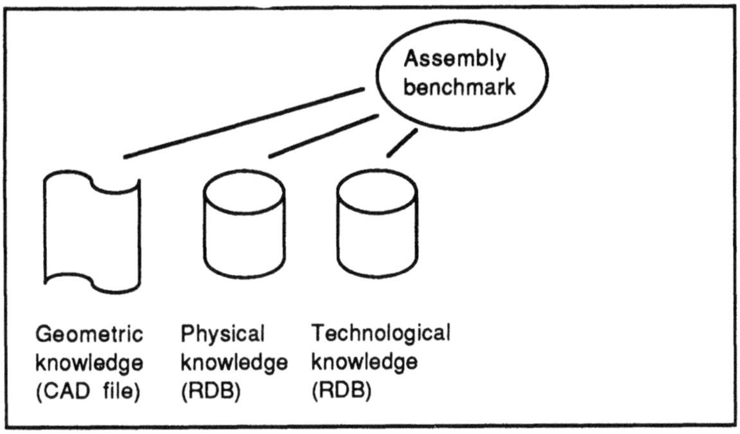

Figure 3 : Presentation of the knowledge after the design activity

The assembly planning system is using the geometric, technical and physical information of the assembly benchmark and its parts to generate work plans, precedence graphs, assembly graphs, Fig. 4 . AI shells such as OPS5, KEE or KNOWLEDGECRAFT are often used to represent this planning knowledge. A transparent access to this planning knowledge should be guaranteed by the proposed central representation module.

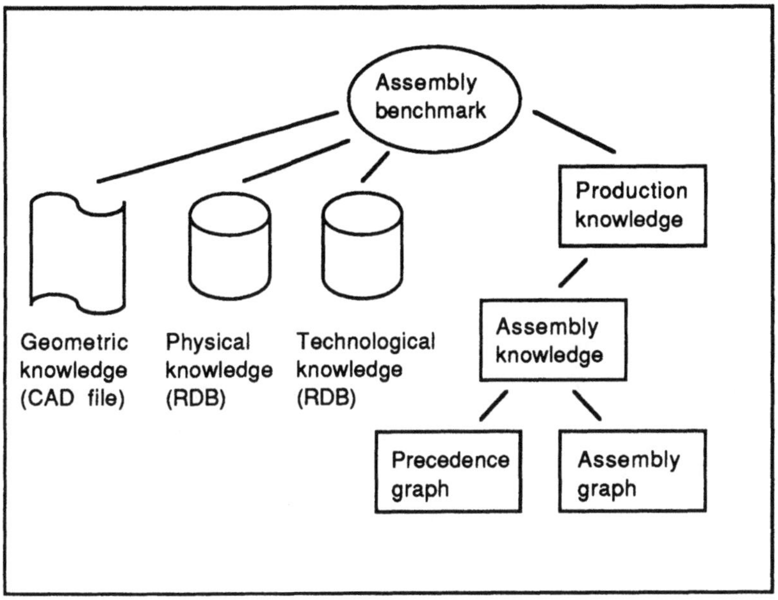

Figure 4 : Presentation of the knowledge from the assembly planning activity

The layout of the flexible production cell can be planned with the help of the assembly knowledge and the information about the number of parts which have to be assembled. The data above the existing resources like conveyors, robots, fixtures,magazins etc. are also required to generate the cell layout. In the following example, a relational database is proposed as representation tool for this knowledge because the explicit offline programming system ROSI [DILL86,HORN86] is using the cell layout for the graphic visualization, Fig.5.

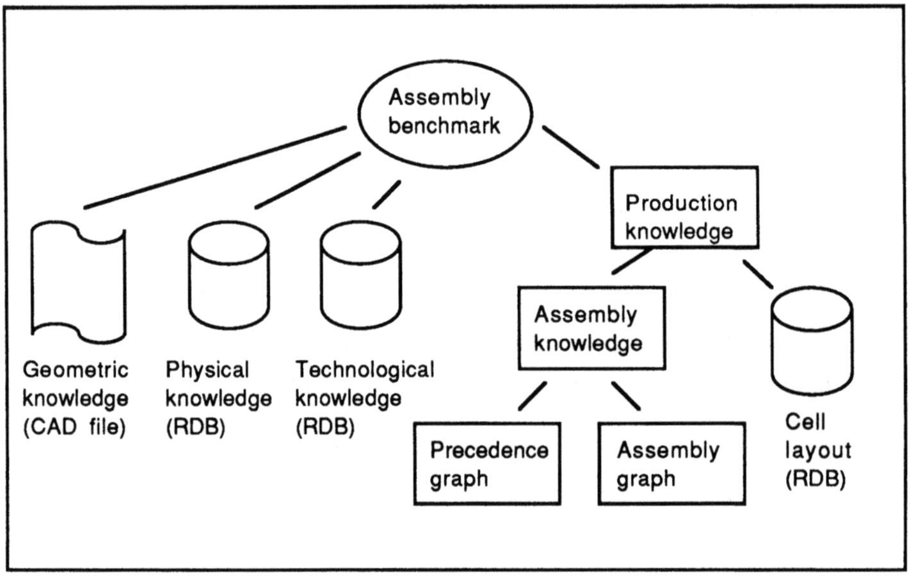

Figure 5 : Presentation of the knowledge from the cell layout activity

Programming of the cell components can be performed if the process requirements, technical restrictions and other planning knowledge are available. The cell layout, the geometric, physical and technical knowledge of the assembly parts and the planning requirements and restrictions will be used to generate the control programs for the cell components. An additional result of this activity may be an optimized cell layout which can be generated by a graphically supported offline programming system. The generated program code for the various devices is represented in device dependent languages. The files with this cell control programs should be known to the central representation module, Fig. 6.

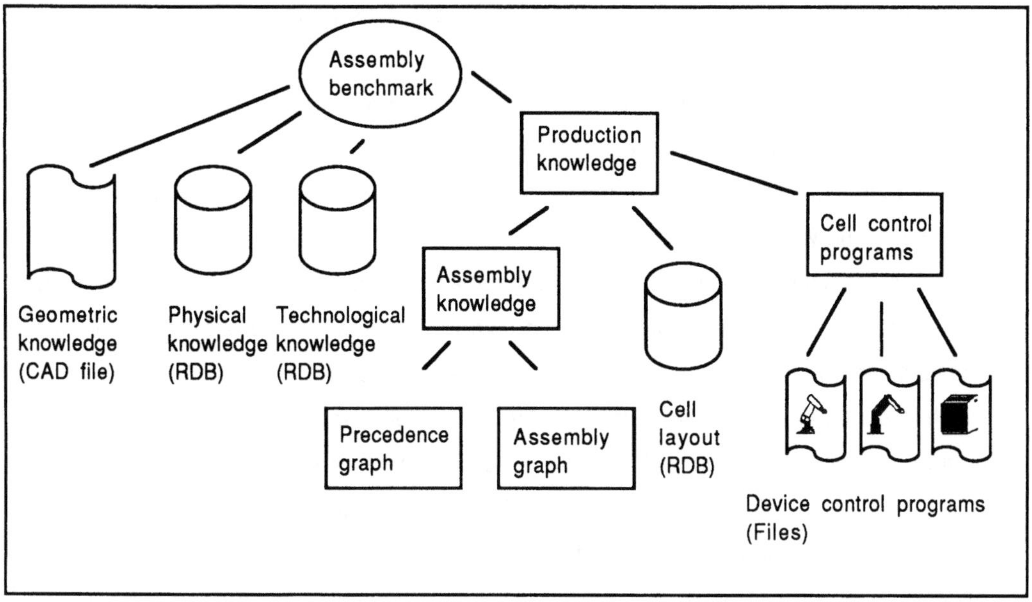

Figure 6 : Knowledge representation after offline programming activity

4. Representation of the Robotic Knowledge with Frames

The proposed way of representing the declarative knowledge is in terms of 'frames' [MINS75, NILS80, WINS77]. The frames contain 'slots' which can be filled with expressions or frames themself [HAYES]. Frames are data structures in which the knowledge of an object in the different submodels is represented together. This representation is useful for modularity accesibility of the robotic knowledge. In addition, a frame structure often allow ways to specify default values for pieces of information about an object when that information is not explicitly given [NAU83]. Modern AI shells like KEE and KNOWLEDGE CRAFT provide useful tools to edit frames and to integrate conventional data access methodes to a frame.

In this paper a frame system is proposed as a representation concept for the global robot knowledge. The different classes of objects (assemblies, parts, production cells, robots etc.) can be realized by different classes of frames. The knowledge of the objects in the different submodels are represented in the slots of the frames.

There are various types of slots to represent knowledge. There are :

- Slots for simple attributes with default values; they are used for the representation of names, masses, partnumber, weight etc.
- slots which contain object frames, Fig. 7; they represent parts of assemblies or production knowledge.
- Slots can consist of access procedures to databases using other representations shemes. The mechanismen can be used to hide various forms of representation, if this is possible. An example of a transformation of a representation scheme is a cell layout which is stored in a relational database. The list of cell components can be generated by a query procedure which will put the result of the query into a frame of the type CELL_LAY_OUT. The operation will only be performed if the cell layout slot is activated. The goal of this technique is to hide other representation forms if only general knowledge of an object is required. The representation form can´t be hidden if the slot has to present very simple structured knowledge as a file with geometric information or a robot program. The only thing which can be done is to open the file and to list it up.

User friendly editors and programming tools can be used if this frame based representation is imbedded in an AI shell such as KNOWLEDGECRAFT. Furthermore, this knowledge can be used by planning and navigation systems which are running under the same AI shell. Expert systems for planning tasks can use this frame based representation as an interface to conventional databases because the data conversion should be hidden by the frame mechanism.

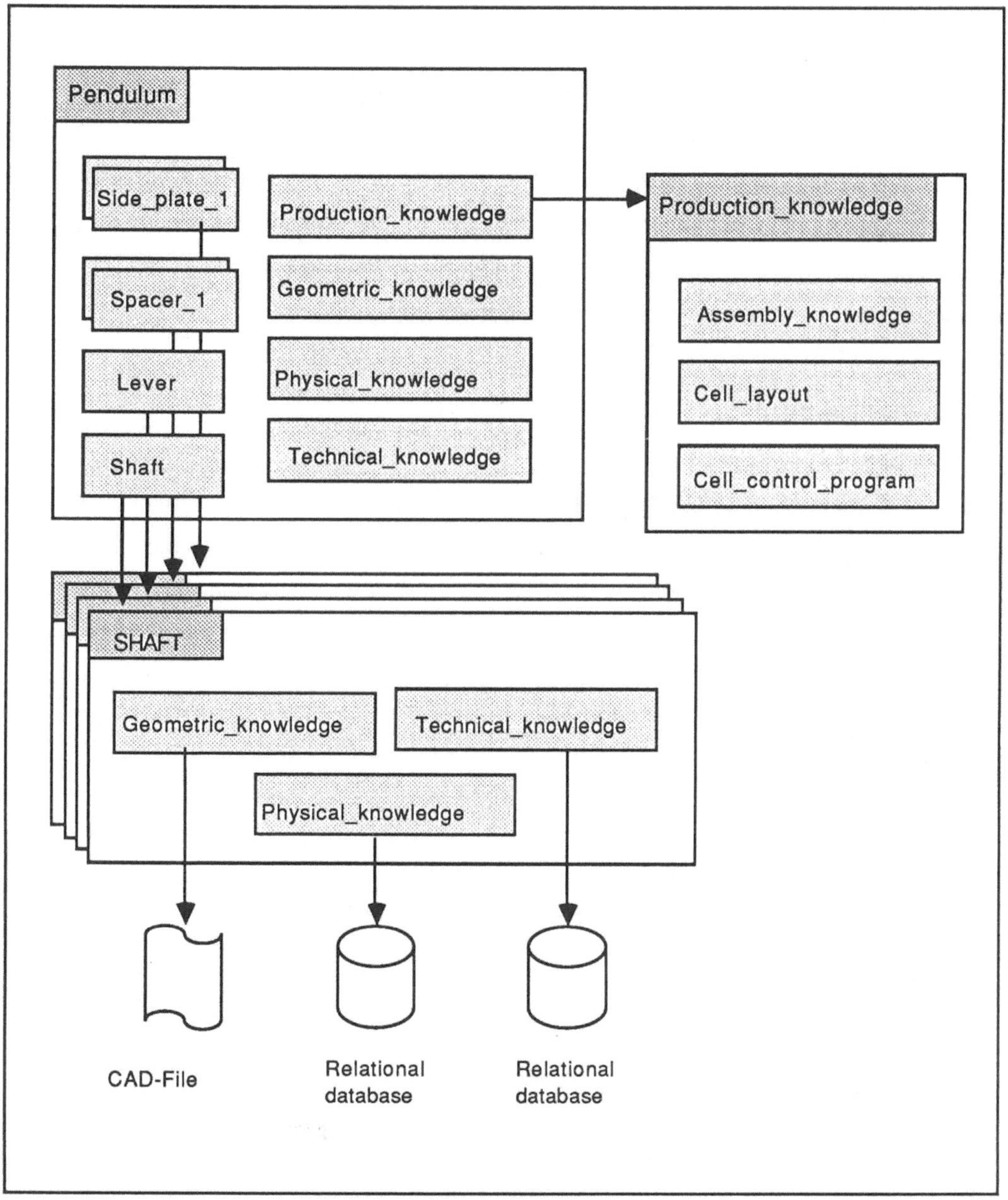

Figure 7 : Frame representation of the European Assembly Benchmark

Another long term goal of this work is the creation of a component which can generate new data access procedures for conventional applications if the knowledge structure has to be changed.

5. Conclusion

A frame based representation is a flexible tool for the organization of heterogenous knowledge of the robots area. Existing knowledge structures can be adapted to changing requirements at the implementation time. The manufacturing engineer can use this central representation module as an information system for the existing knowledge of objects without knowing details of special query languages and data access methods. This will be obtained by the encapsulation of the most important object classes.

Acknowledgement

This work is based on a CAD/CAM research project sponsored by the government of Baden Würtemberg and on a CEC sponsored ESPRIT project (Project No. 623: Operational Control of Robot System Integration into CIM). The work was performed at the Institute of Real-Time Computer Systems and Robotics (Prof.Dr.-Ing. U. Rembold and Prof.Dr.-Ing. R. Dillmann) at the University of Karlsruhe.

6. References

[ANSI78] ANSI/X3/SPARC : The ANSI/X3/SPARC DBMS Framework
Report of the Study Group on Databases Management Systems,
Information Systems, Vol. 3 (1978), pp. 173 191

[BEY86] Bey I.; Leuridan J.; Schlechtendahl E.G.
Specification of a CAD*I Neutral File for Solids
Version 2.1 edited by E.G. Schlechtendahl Springer-Verlag 1986

[CODD81] Codd E.F.;
The Capabilities of Relational Database Management Systems
IBM Research Report RJ3132, San Jose, Cal. (1981)

[DILL86] Dillmann,R.;Hornung,B.;Huck,M.;
Interactive Programming of Robots Using Textual Programming and Simulation Techniques
Proc.16th ISIR, Bruselles, October 1986

[EBER84] Eberlein W.;
CAD-Datenbanksysteme; Springer Verlag 1984

[HAYES] Hayes P.J.;
The Logic of Frames
Readings in Knowledge Representation edited by R.J. Brachman pp.287-295

[HORN86] Hornung,B.;Huck,M.;
Ein benutzerorientiertes Verfahren zur Erstellung von Roboterberwegungsprogrammen mit graphischer Visualisierung
VDI-Bericht 598, Tagung in Langen, May 1986

[MINS75] Minsky M.;
A Framework Representing Knowledge
The Psychology of Computer Vision, P. H. Winston ed. McGraw-Hill
New York, 1975 pp 211-277

[NAU83] Nau Dana S.;
Expert Computer Systems
COMPUTER Feb. 1988 pp.63-85

[NILS80] Nilsson N. J.;
Principles of Artificial Intelligence, Tioga, Palo Alto, Calif. 1980

[SOET87] Soetadji T.;
Fortschrittsberichte VDI
Methode zum Lösen des Routenplannungsproblems für ein Navigationssystem eines autonomen mobilen Roboters
Reihe 10: Informatik/ Kommunikationstechnik Nr. 70 VDI Verlag

[WEDE81] Wedekind H.;
Datenbanksysteme I; Wissensschaftverlag, 2. Auflage (1981)

[WINS77] Winston P.H.;
Artifical Intelligence, Addison-Wesley Reading, Mass. , 1977

Results of the Development of a Manipulator with a very large reach

Martin-Christoph Wanner, Group Manager
Alexander Schließmann
Robotic and Sensor Systems
IPA-Stuttgart

Nobelstraße 12
7000 Stuttgart 80
Federal Republic of Germany

A B S T R A C T

The paper gives an overview of the work performed in a national project by the companies Putzmeister, AEG and Dornier together with the IPA-Stuttgart.
The principal hardware and software is now completed and tested. The first results show that the specifications of the system can be realized.

1. General remarks

Manipulators with very large reach are very important for construction engineering applications /1/. However these machines are quite difficult from the technical point of view.
Conventional industrial robots are in general low payload devices and this has been state of the art since many years. The laws of physics mitigate against the easy implementation of large devices. As machines become larger, the structure and drives natural frequency decrease, limiting the system gain and rapidly degrading the accuracy and speed of operation of the machine to an impractical level.
To compensate, much more sophisticated mechanical hardware, controller strategies on the system and servo level, sensor systems and drives have to be employed. Not surprisingly the first development of larger manipulators worked on the master/slave principle, where the operator's natural abilities permit many of the machines deficiencies to overcome.
This paper gives a short overview of the measures realized.

NATO ASI Series, Vol. F53
Advanced Information Technologies for
Industrial Material Flow Systems
Edited by Shimon Y. Nof, Colin L. Moodie

2. Mechanical and controller hardware

The layout of the realized mobile manipulator is shown in Figure 1. The main features compared with conventional machinery are:

- o Increase of stiffnes of the arm structure by the factor 5
- o Placement of the valves near to the actuator components
- o Use of high resolution resolvers as path measuring systems due to their reliability in difficult environment
- o Calibration of the zero position by means ot tough-triggers, a system which can be operated by workers in the construction site
- o Use of high tensile steel in order to reduce weight
- o For the first time software tools developed at the IPA for the calculation of the forces and moments were used to find the optimum design

The mechnical hardware was designed and built by Putzmeister, Aichtal. Alltogether two units were built, one unit was aquired by KfK on a commercial contract.
For the controller hardware we started with the existing industrial controller Robotronic 500 from AEG. In this context several hardware-modifications were realized, to mention some of them:

- o Power supply 24V for the whole controller system (and hydraulic servo system)
- o Development of new housings for integration on a mobile truck for the specified environment
- o New electronic hardware for the resolvers considering cable length of more than 50m and EMV-problems
- o Use of an intelligent joystick with advanced man-machine features
- o Use of new NEC-processor with high computation power

The descision to start with an already existing industrial standard turned out to be a happy one considering the time to realize a reliable solution.

3. System level of the controller

Various new software items developed at the IPA together with the AEG had been implemented.
The coordinate transformation software turned out to be a very important factor for the béhaviour of our total system. Altogehter three totally different approaches were realized and tested against the following requirements

- short cycle time (less than 100μs on the AEG-Robotronic 500)
- Smooth movements of the individual axes

- collision avoidance has to be realized within the given hardware in real-time for a limited number of obstacles
- the selection of independent and dependent axes should depend on the current situation of the manipulator
- compensation of the deflection should be realized within this context

Results of the selected approach are shown in Figure 2 for a horizontal and in Figure 3 for a vertical movement. The implementations of the software on the AEG-controller showed a real-time behaviour of less than 30 s. In this case the computation of the deflection is running in the background.

4. Servo level of the Controller

Various simulations showed that the possible measures on the servo level are restricted due to the given valves and the computation power of the current controller H/W. The simulation however showed good results regarding the system's accuracy at the TCP. Figure 4 shows the results of the simulation of the hydraulic system for the horizontal movement shown in Figure 2, Figure 5 shows the results of the movements of Figure 3. These results achieved are quite remarkable considering the difficult system with changing loads and masses.
Considering the sensor system from Dornier to relize an analytical model of the environment it is too early to present here first results. The problems here are closely related to the reduction of the complexity of the analytical model measured by the laser.

5. Conclusion

With the exercise of the of the development of a manipulator with very large reach a great step forward was realized in the last year. The project manager of the TP5-project would like to thank Putzmeister-Maschinenfabrik, AEG-Aktiengesellschaft and Dornier for their fine and fruitful cooperation.

/1/ Wanner et al: Design of a manipulator with very large reach for applications in civil engineering.
Forth International Symposium on Robotics and Artificial Intelligence in Building Construction, Haifa, 1987

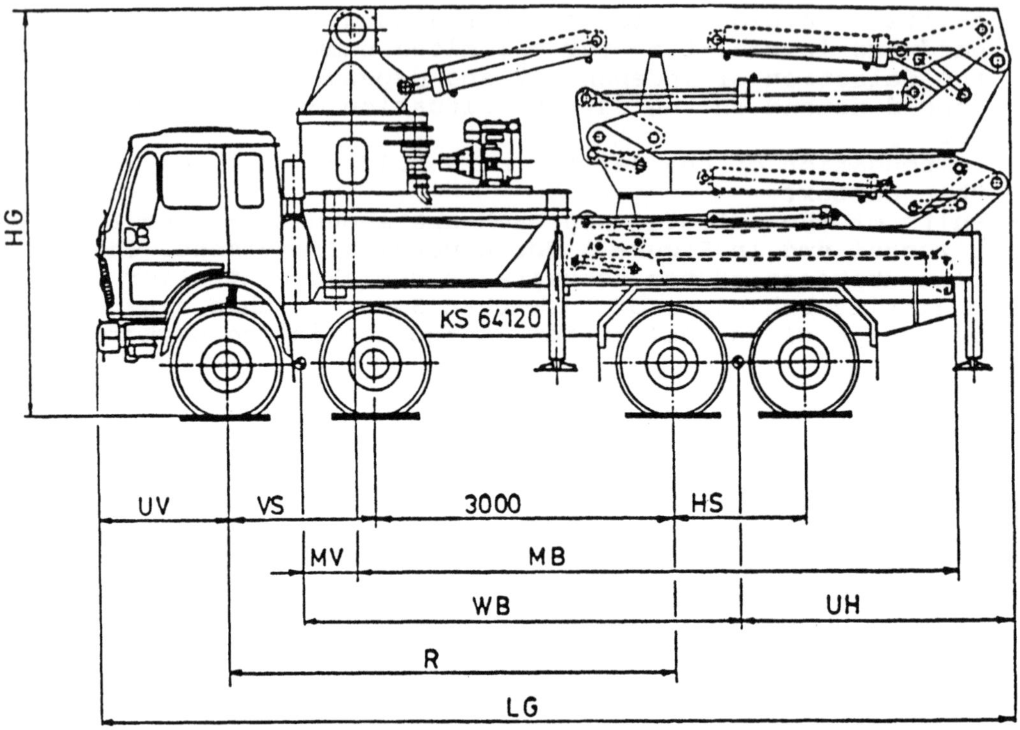

Figure 1: Mobile Manipulator FH22

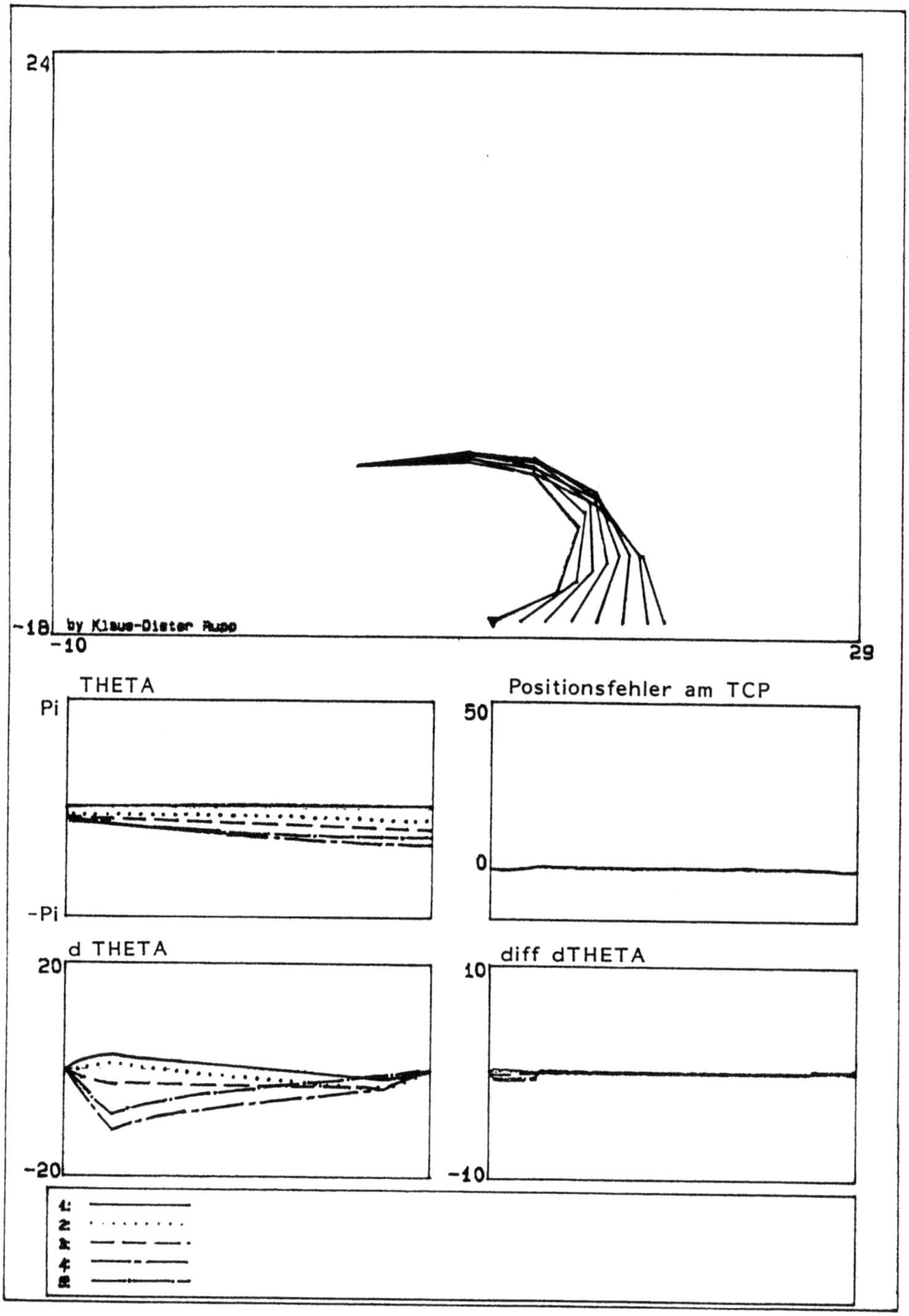

Figure 2: Coordinate transformation - horizontal movement

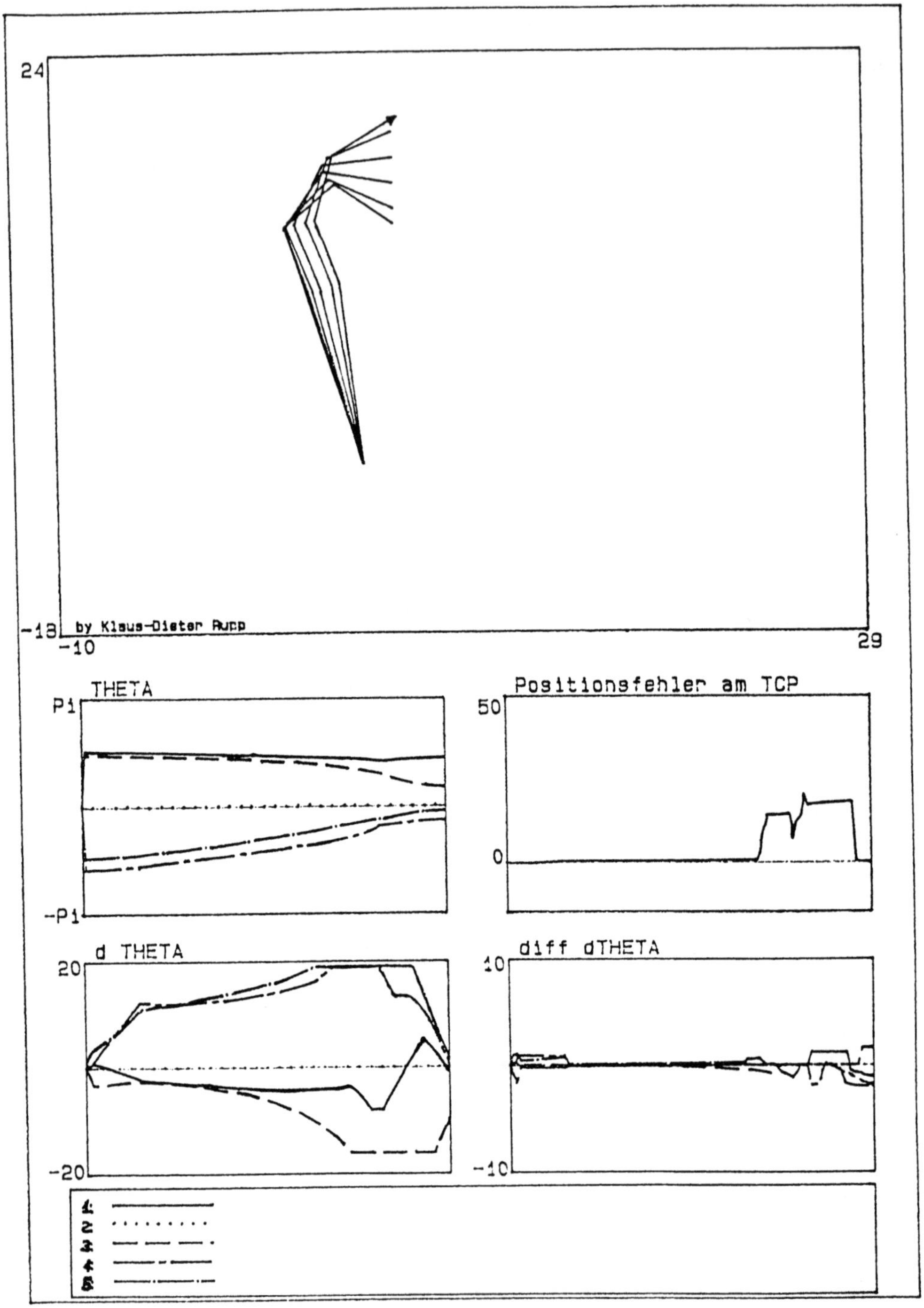

Figure 3: Coordinate transformation - vertical movement

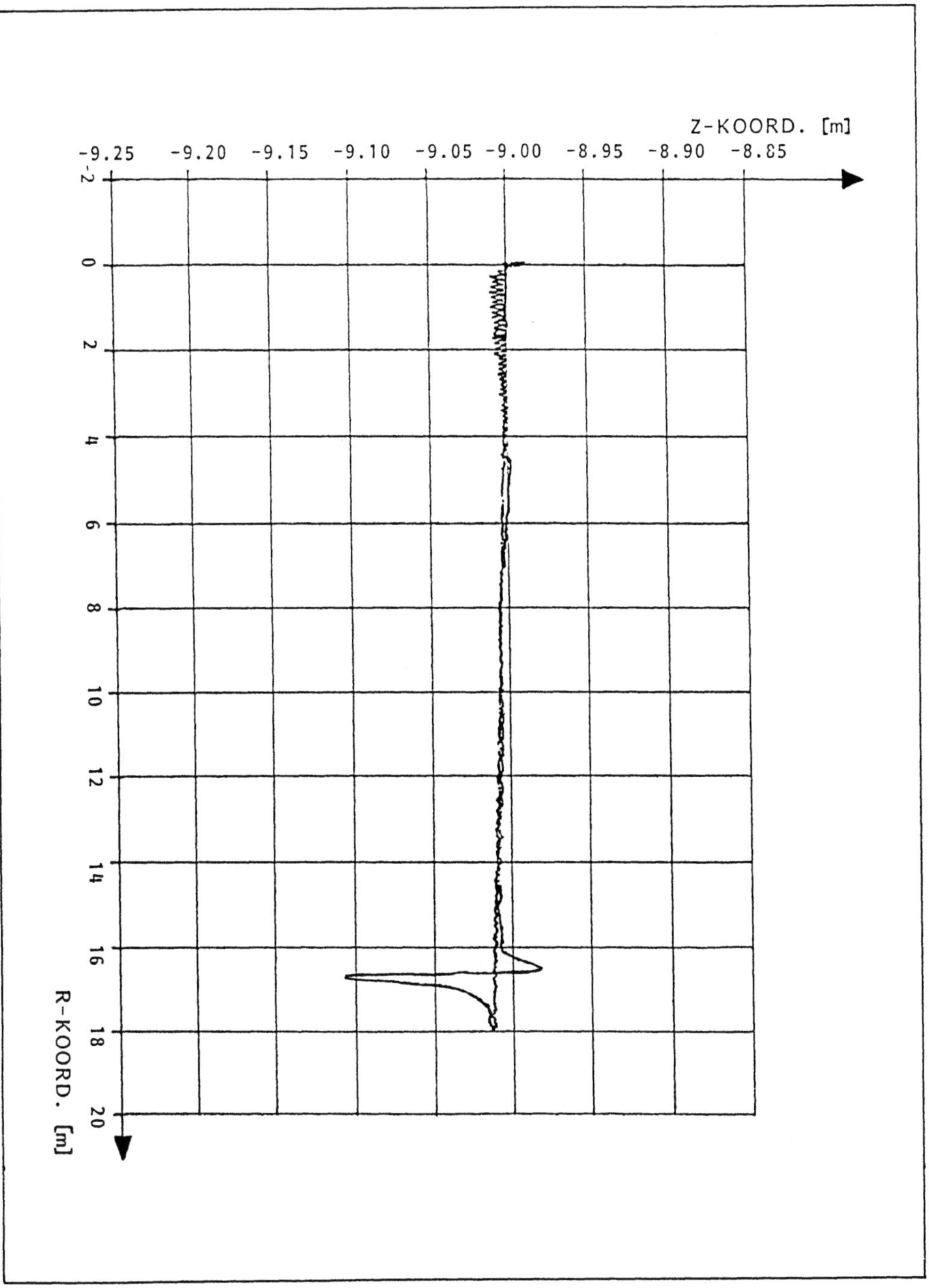

Figure 4: Behaviour at the TCP for the horizontal movement of Figure 2

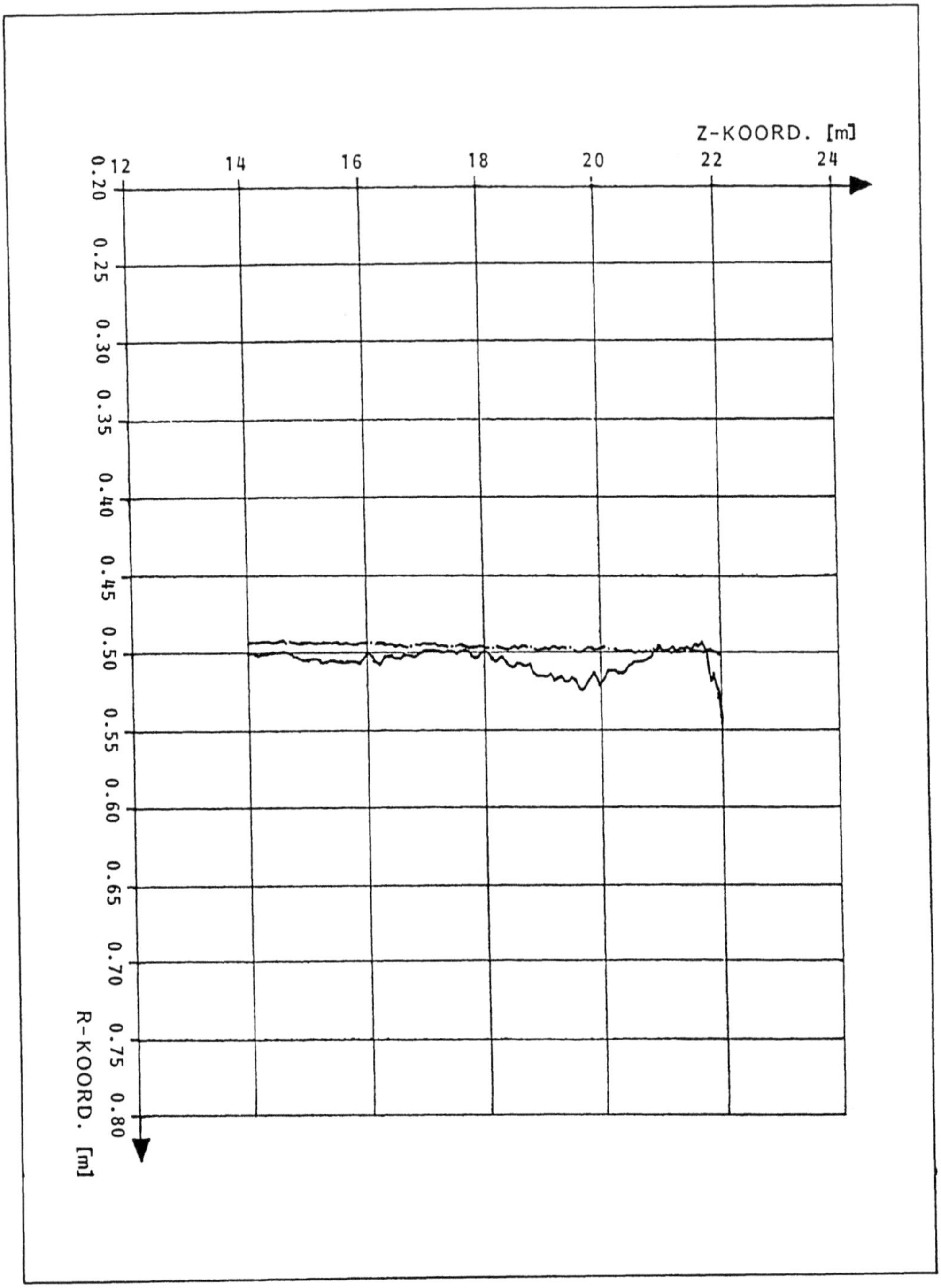

Figure 5: Behaviour at the TCP for the vertical movement of Figure 3

Global Path Planning Approaches for Material Movements in a Worksite

Jean-Claude Latombe

Robotics Laboratory, Computer Science Department
Stanford University, Stanford, CA 94305, USA

Abstract

In various environments (e.g., manufacturing shopfloors, civil engineering construction sites, space) there is an increasing need to efficiently transport objects from locations to other locations. Although most practical material transportation robotic system built so far have been quite primitive, we believe that in many areas one can significantly gain in efficiency, reliability and flexibility by automatically planning the motions of the transportation devices. While process planning provides a high-level 'logical' and possibly 'temporal' specification of material movements, motion planning says how these movements are to be 'physically' carried out. Thus, motion planning is the natural intermediate stage between process planning and task execution. In this paper, we survey techniques for planning mobile robot paths among obstacles, which have been developed over the last few years. We focus on the so-called 'global' techniques. We describe in detail the three most common approaches, which are based on the notions of cell decomposition, free space retraction, and visibility graph, respectively. Within the first two approaches, we survey both the so-called exact and approximate techniques. Although this paper is far from exploring all the facets of motion planning, it gives a fundamental and detailed presentation of issues, which are of general interest to all motion planning problems. These issues are likely to be of prime importance in future material transportation systems.

Key-Words

Material Tranportation, Robotics, Autonomous Robots, Spatial Reasoning, Motion Planning, Path Planning.

1 Introduction

In various environments there is an increasing need to efficiently transport objects from locations to other locations. For instance, in a machining shop-floor blanks and partly cut parts, as well as tools (e.g., fixtures, cutting tools) have to be moved from storage areas or machines to other machines, where the machining process plans proceed further. In a civil engineering construction site, a large variety of materials such as earth, steel bars, concrete beams, pipes, and prefabricated components have to be frequently moved over a large and unstructured working area, in a much un-repetitive fashion. In space, construction of platforms will require moving frame components out from the aircraft to the partially assembled platform structure. In most of these application areas, material

NATO ASI Series, Vol. F53
Advanced Information Technologies for
Industrial Material Flow Systems
Edited by Shimon Y. Nof, Colin L. Moodie

handling and transportation turn out to be a critical issue responsible for an important share of the operating costs, and also for many delays. Mobile robots or similar devices are likely candidates to improve the way these tasks are currently performed.

Robotized material movements in a worksite requires that the transportation robots be provided with advanced capabilities for planning their motions. In fact, motion planning is the natural intermediate stage between process planning (or task planning), which determines and specifies the various operations that have to be executed in order to perform a task and the actual execution of these operations. Indeed, process planning provides a 'logical' and possibly 'temporal' specification of material movements. However, it does not say how these movements are to be 'physically' carried out. This is the purpose of motion planning.

We believe that in many areas one can significantly gain in efficiciency, reliability and flexibility by automating the motion planning stage. We envision an architecture integrating process planning and motion planning in a hierarchical fashion. In many cases, process planning will be performed off-line, possibly in a non-automated fashion, while motion planning will be an on-line capability implemented in the robot controller. In situations where process plans may often change, both process and motion planning capabilities will have to be available on-line; then, motion planning will still occur at a more detailed level and probably within a shorter time perspective, but it will directly return feedback constraints to the process planning stage.

Most practical material transportation robotic systems built so far have been quite primitive, using only low-level control techniques. In these systems, motion planning has been done by hand and expressed in the form of task-specific programs driving the robots through the planned paths. Often, control is simplified further by materializing potential paths using specifically designed environmental features (e.g., wiring guides). Typically, robots controlled in this manner can only follow predetermined routes, limiting their application to carefully engineered environments and, in these environments, restricting their flexibility and versatility. In addition, the burden of programming the robots is put on the users.

On the other hand, most of the early robotic projects based on the use of Artificial Intelligence techniques attempted (rather unsuccessfully) to directly connect a high-level planning and monitoring system to a low-level execution controller, thus skipping the motion planning level. In fact, for a long period of time, motion planning has been considered as a secondary and relatively easy problem. But, experience has shown that efficient techniques are needed at this intermediate level. This is not surprising. A robot operates in a physical world in which geometry plays an important role, and it seems rather natural that a somewhat autonomous robot be equipped with advanced techniques for reasoning about this world. Experience and theoretical studies (e.g., [43,49]) have also shown that motion planning is quite complex, both from the conceptual and computational points of view.

Motion planning requires to make use of spatial knowledge about the workspace in order to plan collision-free paths among fixed and moving obstacles. It must also be able to deal with lack of knowledge about the environment and to generate robust plans allowing to face contingencies, such as unexpected obstacles. It should also be capable of reasoning about uncertainties in object locations, robot control and sensing, and able to generate sensory-based motion plans coping with these uncertainties. Therefore, motion planning has many facets. In this paper, we present techniques for planning paths among obstacles, which have been developed over the last few years. Our discussion is limited in scope and focuses on basic issues along the following lines:

1. We concentrate on the motion planning problem for a *moving rigid object*. This means that we mainly address the case of a mobile robot. The presented techniques are directly applicable to ground-based mobile robots, whose workspace can be realistically approximated as a two-dimensional workspace. The principle of most of the techniques can easily be extended to free-flying robots operating in three-dimensional workspaces, at the expense of increased computations. The problem of planning the motions of an articulated object, such as a manipulator arm, is not addressed in this paper, although several of the presented techniques can be generalized to this case (see [49,33]).

2. We concentrate on the so-called *global* path planning techniques. All these techniques consist of, first constructing a representation of the global topology of the space of collision-free positions and orientations of the robot in the form of a graph, and next searching this graph for a path. Another family of path planning techniques are known as the *local* techniques. These techniques base their decisions only on local considerations. Since they avoid to explicitly represent free space topology, they may occasionally be faster than global techniques. But because they tend to search very large graphs, their worst-case computational complexity is much higher than the worst-case complexity of global methods. The most well-known and successful local path planning technique is the Potential Field method [22].

3. We assume that the robot has a complete and accurate world model prior to planning. This means that we do not consider the case where the workspace is 'discovered' by the robot as it moves, nor the case where there are significant uncertainties in the robot's model. Dealing with incomplete knowledge is often considered an easier problem when local techniques are used rather than global ones. However, we believe that this is basically an architectural issue, in which one has to appropriately interweave motion planning within a limited subset of the workspace (the portion of space for which the robot has a model) and execution (which allows the robot to extend its spatial knowledge of the workspace through sensing). Dealing with uncertainties is more difficult. It requires to produce sensory-based motion plans, not just paths. Other techniques than thosed presented below are needed to that purpose. We refer the reader to [24] for a presentation of the motion planning problem in the presence of uncertainty.

4. We only consider the case of a single robot moving among fixed obstacles. Several of the methods presented below can easily be extended, in principle, to handle the case of multiple robots. One general approach is to consider the set of all the robots as a single multiple-bodied robot with many degrees of freedom. However, motion planning is known to be NP-hard in the number of degree of freedom [43], so that this approach results in a substantial increase of computational complexity. Another approach to multiple-robot path planning is the so-called *prioritarized* approach [18], which consists

of planning the motion of one robot at a time using one of the techniques described in this paper, and considering the robots whose motions have already been planned as obstacles to the other robot. This approach, however, is not guaranteed to succeed whenever collision-free paths exist for all the robots. The case of moving obstacles is easy to handle as long as there are no bounds on the velocity and the acceleration of the robots. Dealing with such bounds in a rigorous fashion makes the problem much harder.

Thus, this paper is far from exploring all the facets of motion planning. However, it gives a fundamental and detailed presentation of issues which are of general interest to the other facets of motion planning. Furthermore, these issues are likely to be of prime importance in future material transportation systems, if we want to increase the flexibility and ease of use of these systems. Path planning techniques also have more specific applications; for instance, moving long bars, pipes and beams through a building under construction is known to raise tough planning problems interfering with the building construction process, whether material handling is robotized or not. Nevertheless, due to the limited scope or the paper and its theoretical nature, the reader should be constantly aware that a practical implementation of any of the techniques described below requires a lot of domain-specific engineering.

There exist two major global approaches to motion planning, namely *decomposition* and *retraction*. These approaches are 'universal' in the sense that they are applicable to path planning in general, not just to specific instances of this problem. Decomposition consists of representing 'free space', i.e. the set of collision-free positions and orientations of the robot as a finite collection of cells. Path planning is then reduced to finding a sequence of free cells such that any two successive cells in the sequence are 'adjacent'. Retraction consists of mapping free space on a network of curves, so that path planning is reduced to searching this network.

Within both approaches, there are two sorts of techniques, the *exact* and the *approximate* ones. The exact techniques are based on rather involved mathematics and are complete in the sense that they can produce a path whenever one exists. The approximate techniques are based on various approximations and/or assumptions, and they are not complete. Implementation of the exact techniques raises difficult issues, such as doing exact computations with rational or algebraic numbers. Implementing the approximate techniques is usually easier.

Exact decomposition techniques are presented in Sections 3 through 5. Approximate decomposition techniques are described in Sections 6 and 7. The concept of exact retraction is introduced in Section 8, and an approximate method based on this concept is detailed in Section 9. Finally, Section 10 describes a third approach to path planning, known as the *visibility graph* approach. Unlike the previous two, this latter approach is not universal, but, due to its conceptual simplicity, it is a popular one.

Our presentation of path planning methods makes intensive use of the concept of *configuration space*. We describe this concept in the following section and we formalize the path planning problem.

2 Configuration Space

The path planning problem may be formulated as the problem of constructing a continuous collision-free sequence of positions and orientations of an object (the robot) moving among obstacles from an initial position/orientation to a final one. Configuration space is a conceptual tool that makes the constraints on the moving object explicit.

Let us consider an object $\mathcal{A}$, which we indifferently call the *moving object* or the *robot*. $\mathcal{A}$ can move in a space, $\mathcal{W}$, called *workspace*, of dimension 2 or 3. Typically, in the case of a ground-based mobile robot, $\mathcal{W}$ is approximated as a space of dimension 2 and is isomorphic to $\Re^2$. In the case of a general manipulator arm or a three-dimensional free-flying object, its dimension is 3 and it is isomorphic to $\Re^3$.

Let us assume that a Cartesian frame $\mathcal{F}_\mathcal{A}$ (resp. $\mathcal{F}_\mathcal{W}$) has been attached to $\mathcal{A}$ (resp. $\mathcal{W}$).

DEFINITION 1: *A* **configuration** $\mathbf{c}$ *of* $\mathcal{A}$ *is a specification of the position and orientation of* $\mathcal{F}_\mathcal{A}$ *with respect to* $\mathcal{F}_\mathcal{W}$*. The* **configuration space** *of* $\mathcal{A}$ *is the space, denoted* $\mathcal{C}$*, of all the possible configurations of* $\mathcal{A}$*. The subset of* $\mathcal{W}$ *occupied by* $\mathcal{A}$ *at configuration* $\mathbf{c}$ *is denoted* $\mathcal{A}(\mathbf{c})$*. In the same fashion, the point* a *on* $\mathcal{A}$ *at configuration* $\mathbf{c}$ *is denoted* $a(\mathbf{c})$ *in* $\mathcal{W}$*.*

The configuration space $\mathcal{C}$ of $\mathcal{A}$ is a manifold of dimension m, which can be embedded in an Euclidean space of dimension N ($N \geq m$) [1,51]. The topology on $\mathcal{C}$ is the subspace topology induced by the Euclidean metric in the N-dimensional space or, equivalently, by the following distance function d on $\mathcal{C}$:

$$\forall \mathbf{c}, \mathbf{c}' \in \mathcal{C} \;\; [d(\mathbf{c}, \mathbf{c}') = \max_{a \in \mathcal{A}} distance(a(\mathbf{c}), a(\mathbf{c}'))]$$

where *distance* denotes the Euclidean distance in $\mathcal{W}$.

For instance:

- If $\mathcal{A}$ is only allowed to translate in $\mathcal{W} = \Re^k$ ($k = 2$ or 3), then $\mathcal{A}$'s configuration can be defined as the coordinates of the origin of $\mathcal{F}_\mathcal{A}$ in $\mathcal{F}_\mathcal{W}$. Thus, $\mathcal{C} = \Re^k$ (more precisely, $\mathcal{C}$ is isomorphic to $\Re^k$) and $m = N = k$.

- If $\mathcal{A}$ translates and rotates in $\Re^2$, $\mathcal{A}$'s configuration can be represented as a list of three parameters (x, y, θ), the coordinates x and y of the origin of $\mathcal{F}_\mathcal{A}$ and the angle θ between the x-axes of $\mathcal{F}_\mathcal{A}$ and $\mathcal{F}_\mathcal{W}$. This parametrization, however, does no make explicit the fact that in this case $\mathcal{C}$ is a cylinder, i.e. that the orientation is defined by an angle modulo 2π. In fact, $\mathcal{C} = \Re^2 \times S^1$, where S^1 is the unit circle in $\Re^2$. Thus, $\mathcal{C}$ is a manifold of dimension 3 (a 3-cylinder) that can be embedded in an Euclidean space of dimension 4 [51].

- If $\mathcal{A}$ translates and rotates in $\Re^3$, $\mathcal{A}$'s configuration can be represented as a list of six parameters, say the three coordinates of the origin of $\mathcal{F}_\mathcal{A}$ and the three Euler angles [6] specifying the orientations of $\mathcal{F}_\mathcal{A}$'s axes with respect to $\mathcal{F}_\mathcal{W}$. Again this parametrization does not reflect the topology of $\mathcal{C}$. In this case, $\mathcal{C} = \Re^3 \times SO(3)$, where SO(3) is the

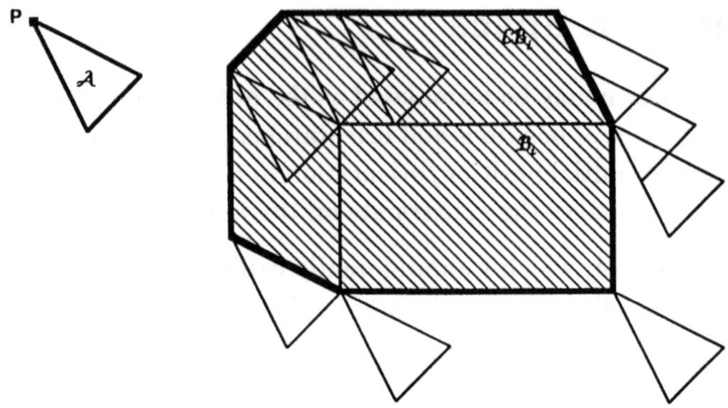

Figure 1: **C-obstacle in the Convex Polygonal Case**

special orthogonal group made of 3×3 matrices with orthonormal rows (and columns) and determinant 1. The space of 3×3 matrices with real coefficients is a vector space of dimension 9. Othonormality of the rows generates 6 independent polynomial equations specifying two three-dimensional subsets of this space. The constraint that the determinant should be +1 selects one of these subsets. The set $\Re^3 \times SO(3)$ is a manifold of dimension 6, which can be embedded in the Euclidean space of dimension 12. The constraints on the matrices in $SO(3)$ determines the manifold in $\Re^{12}$.

One of the reasons why path planning is difficult is that, except in the translational case, $\mathcal{C}$ is a multiply-connected manifold. This can easily be illustrated in the case where $\mathcal{C} = \Re^2 \times S^1$. Any two paths which make, respectively, i and j $(i \neq j)$ net turns around S^1 are not homotopic (i.e. one cannot be continuously deformed into the other) [21].

Let us assume now that the workspace $\mathcal{W}$ contains obstacles $\mathcal{B}_i$ $(i = 1, 2, ...)$. We model all objects, both $\mathcal{A}$ and the $\mathcal{B}_i$s, as closed bounded (i.e., compact) subsets of $\mathcal{W}$. Each $\mathcal{B}_i$ is mapped in $\mathcal{C}$ as another region, $\mathcal{CB}_i$, called a C-obstacle and defined as follows:

DEFINITION 2: *The obstacle region $\mathcal{B}_i$ in $\mathcal{W}$ maps into $\mathcal{C}$ as the region* $\mathcal{CB}_i = \{\mathbf{c} \in \mathcal{C} \ / \ \mathcal{A}(\mathbf{c}) \cap \mathcal{B}_i \neq \emptyset\}$. *$\mathcal{CB}_i$ is called a* **C-obstacle**.

Figure 1 illustrates the case where both $\mathcal{A}$ and $\mathcal{B}_i$ are convex polygonal regions in $\Re^2$. If $\mathcal{A}$ keeps a fixed orientation, then $\mathcal{C} = \Re^2$, and the C-obstacle corresponding to $\mathcal{B}_i$ is shown in Figure 1. It is easy to see that $\mathcal{CB}_i$ is another convex polygonal region. If $\mathcal{A}$ can both rotate and translate, then $\mathcal{C} = \Re^2 \times S^1$, and the representation of $\mathcal{CB}_i$ in a (x, y, θ) Cartesian frame, where $\theta \in [0, 2\pi)$ represents the orientation of $\mathcal{A}$ (modulo 2π) is a three-dimensional volume bounded by patches of ruled surfaces. Every cross-section through this volume is a polygonal region representing the C-obstacle in $\Re^2$ for a fixed orientation of $\mathcal{A}$ [32].

There exist rather straighforward techniques for computing the representation of $\mathcal{CB}_i$

from the representation of $\mathcal{A}$ and $\mathcal{B}_i$, when both objects are convex polygons or polyhedra [32,16]. The computed representation of $\mathcal{CB}_i$ when both $\mathcal{A}$ and $\mathcal{B}_i$ are convex polygons is of the form:

$$(x, y, \theta) \in \mathcal{CB}_i \Leftrightarrow \bigwedge_k [\theta \in \delta_{ik} \Rightarrow (a_{ik}(\theta)x + b_{ik}(\theta)y + c_{ik}(\theta) \leq 0)]$$

where δ_{ik} is a subinterval of $[0, 2\pi)$ and $a_{ik}(\theta)$, $b_{ik}(\theta)$, and $c_{ik}(\theta)$ are continuous function of θ. The inequality $a_{ik}(\theta)x + b_{ik}(\theta)y + c_{ik}(\theta) \leq 0$ is called a **C-constraint** (the equality represents a ruled surface). The interval δ_{ik} defines the range of orientations of $\mathcal{A}$ in which the C-constraint applies. In the case where $\mathcal{A}$ has a fixed orientation (i.e., it can only translate), the representation simplifies to an expression of the form:

$$(x, y, \theta) \in \mathcal{CB}_i \Leftrightarrow \bigwedge_{k'} (a_{ik'}x + b_{ik'}y + c_{ik'} \leq 0).$$

If, more generally, both $\mathcal{A}$ and $\mathcal{B}_i$ are represented as semi-algebraic sets, then $\mathcal{CB}_i$ is also a semi-algebraic set, whose representation is computable (see Section 5).

C-obstacles explicitly describe the constraints on the motions of $\mathcal{A}$. A second reason why path planning is difficult is that, whenever the moving object $\mathcal{A}$ is allowed to rotate, C-obstacles may be complex regions bounded by curved surfaces.

The set of collision-free positions and orientations of the robot $\mathcal{A}$ is defined as follows:

DEFINITION 3: Free space *is* $\mathcal{C}_{free} = \{\mathbf{c} \in \mathcal{C} \ / \ \mathcal{A}(\mathbf{c}) \cap (\bigcup_i \mathcal{B}_i) = \emptyset\}$.

In the same way, we can define valid space as the set of configurations of $\mathcal{A}$ where the interior of $\mathcal{A}$ and the interiors of the $\mathcal{B}_i$s have null intersection:

DEFINITION 4: Valid Space *is* $\mathcal{C}_{valid} = \{\mathbf{c} \in \mathcal{C} \ / \ int(\mathcal{A}(\mathbf{c})) \cap (\bigcup_i int(\mathcal{B}_i)) = \emptyset\}$[1].

At a valid configuration, $\mathcal{A}$ may touch an obstacle.

A collision-free (more simply, a free) path of $\mathcal{A}$ in $\mathcal{C}$ is formalized as follows:

DEFINITION 4: *A* **free path** *of* $\mathcal{A}$ *between an initial configuration* $\mathbf{c}_1$ *and a final configuration* $\mathbf{c}_2$ *is a continuous map* $\tau : [0, 1] \to \mathcal{C}_{free}$, *with* $\tau(0) = \mathbf{c}_1$ *and* $\tau(1) = \mathbf{c}_2$.

Thus, given a configuration space $\mathcal{C}$, C-obstacles $\mathcal{CB}_i$ in this space, and an initial and final configurations $\mathbf{c}_1$ and $\mathbf{c}_2$, collision-free path planning consists of constructing a mapping τ as defined above. We now describe various techniques applicable to this problem.

3 Exact Decomposition: The Translational Case

The principle of an exact decomposition method is to decompose free space into non-overlapping cells whose union is equal to free space. Any such decomposition is not

[1]Throughout this paper, $int(S)$, $cl(S)$, and $\partial(S)$ respectively denote the interior, the closure, and the boundary of the subset $S \subset E$, where E is a topoligical space.

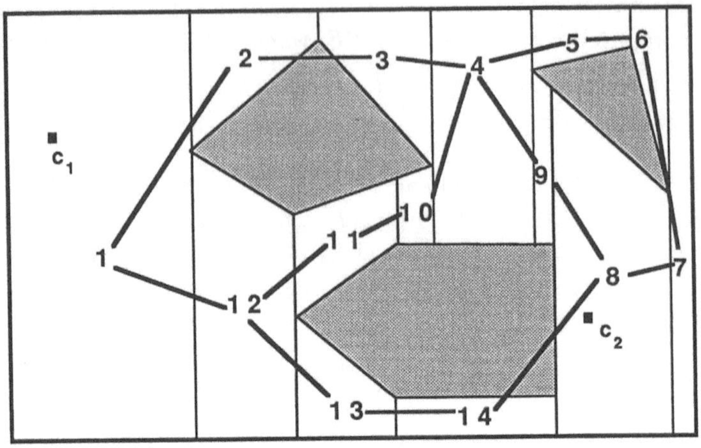

Figure 2: **Vertical Trapezoidal Decomposition of Free Space**

appropriate, however. Cells should have the following two characteristics. First, they should be simple enough so that we can find a path between any two points in a cell. Second, it should be possible to test adjacency between cells and to find a path crossing the common portion of boundary between two adjacent cells.

In this section, we introduce the exact cell decomposition approach in the simple case where the moving object $\mathcal{A}$ is a polygon only allowed to translate among fixed polygonal obstacles. By decomposing the moving object and the obstacles into convex polygons, we can easily construct possibly overlapping convex polygonal C-obstacles $\mathcal{CB}_i$, $i = 1$ to q (see Section 2). Let n be the total number of vertices of these C-obstacles.

Let us assume that $\mathcal{C}_{free}$ is bounded by a rectangle R whose edges are parallel to the x and y-axes embedded in $\mathcal{C}$. Thus $\mathcal{C}_{free} = int(R) - \bigcup_i \mathcal{CB}_i$. The edges of R are equivalent to mechanical stops on the physical translations of $\mathcal{A}$ along the x and y-axes.

DEFINITION 5: *A* **open convex decomposition** *of $\mathcal{C}_{free}$ is a finite collection of open convex polygons, called* **cells**, *such that any two cells do not intersect and the closed union of all the cells is equal to $cl(\mathcal{C}_{free})$. Two cells c and c′ are* **adjacent** *if only if $cl(c) \cap cl(c')$ is a line segment of positive length (i.e., a set of non-null measure).*

Figure 2 shows a possible convex decomposition of the free space and the *connectivity graph* representing the adjacency relation between the cells. Cells are labelled by numbers and the non-directed arcs of the connectivity graph are displayed as bold lines connecting these numbers.

Consider an initial configuration $\mathbf{c}_1$ and a final configuration $\mathbf{c}_2$, which are both in free space. We want to generate a free path from $\mathbf{c}_1$ to $\mathbf{c}_2$. The exact decomposition approach in this specific case proceeds as follows [12]:

1. Generate a convex decomposition of $\mathcal{C}_{free}$.

2. Construct the connectivity graph CG representing the adjacency relation between cells.

3. Search CG for a path linking the cell containing $\mathbf{c}_1$ to the cell containing $\mathbf{c}_2$.

4. If the search terminates successfully return the corresponding sequence of cells; otherwise, return failure.

The output is a sequence $c_1, ..., c_p$ of cells such that $\mathbf{c}_1 \in c_1$, $\mathbf{c}_2 \in c_p$ and $\forall j \in [1, p-1]$, c_j and c_{j+1} are adjacent. Let β_j designate the common edge of c_j and c_{j+1}, i.e. $\beta_j = cl(c_j) \cap cl(c_{j+1})$. One way to transform the sequence of cells into a path is to consider the midpoints Q_j of every edge β_j, and to connect $\mathbf{c}_1$ to Q_1, Q_j to Q_{j+1} ($j = 1$ to $p-2$), and Q_{p-1} to $\mathbf{c}_2$ by straight line segments.

The generation of a convex decomposition of a polygon is a classical problem in Computational Geometry. The 'Optimal Convex Decomposition' problem is solvable in time polynomial in the number n of edges [13] when there is no hole. But, as shown by Lingas [29], the presence of holes makes this problem NP-hard. Nevertheless, a non-optimal decomposition can be generated rather efficiently as shown below.

A simple technique consists of sweeping a line L parallel to the y-axis, inside R, from left to right ('line-sweep' paradigm [42]). Whenever the line encounters a new vertex X, if X is not inside a C-obstacle, one vertical line segment is created connecting X to the edges immediately above and below it (see Figure 2). The C-obstacles edges, the boundary of R and the vertical line segments determine a *vertical trapezoidal decomposition* of $\mathcal{C}_{free}$. Two cells of this decomposition are adjacent if their closures share a portion of a vertical segment. Before line sweeping, the vertices of the C-obstacles can be sorted along the x-axis in $O(n \log n)$. Then, as the vertical line L is swept from left to right the segments of the intersections $L \cap \mathcal{CB}_i$ (for all $i \in [1, q]$) are kept dynamically in a balanced tree. During line sweeping, using this structure, it is possible to concurrently create the vertical line segments emanating from vertices, generate the connectivity graph CG between the trapezoidal cells, and identify the cells containing the initial and final configurations $\mathbf{c}_1$ and $\mathbf{c}_2$. This whole sweep-line technique produces the decomposition of $\mathcal{C}_{free}$ in $O(n \log n)$ time. The number of generated trapezoidal cells is $O(n)$. The number of arcs in CG is also $O(n)$.

Graph searching can be done in various ways. A systematic 'breadth-first' exploration of the graph takes $O(n)$ time. Indeed, each node is explored at most once and each arc is traversed at most once. Average time can be improved by using heuristics to guide the search (see [36]).

Therefore, the total time complexity of the method described above is $O(n \log n)$.

The method can be extended to the three-dimensional case. But the decomposition phase is more time consumming. In particular, the optimal decomposition of a polyhedron without hole is known to be NP-hard [29]. However, non-optimal decompositions of polyhedra with holes are still possible in reasonable time.

4 Exact Decomposition: Case of a Segment

We now illustrate the exact decomposition approach to path planning in a more difficult, but still specific case, where C-obstacles are bounded by curved surfaces. The robot $\mathcal{A}$ is a line segment (sometimes called a *ladder*) moving in a two-dimensional workspace $\mathcal{W}$ among non-intersecting compact (possibly non-convex) polygonal obstacles $\mathcal{B}_i$. $\mathcal{A}$'s end-points are denoted P and Q. The configuration space $\mathcal{C}$ of $\mathcal{A}$ is $\Re^2 \times S^1$. We represent each configuration by a triplet (x, y, θ), where x and y are the coordinates of P in $\mathcal{F}_\mathcal{W}$ and $\theta \in [0, 2\pi)$ is the angle (modulo 2π) between the x-axis of $\mathcal{F}_\mathcal{W}$ and the segment PQ.

The method presented below is due to Schwartz and Sharir [47]. The basic idea is to decompose the set of $\mathcal{A}$'s positions into *regular regions*, to 'lift' these regions into *cells*, and to represent the adjacency relation among the cells as a *connectivity graph*. The regular regions are such that the C-obstacles maintain a constant 'structure' in the cylinders above these regions, and the cells projecting on a regular region make this structure explicit. The boundaries of the regular regions are called *critical curves*.

The time complexity of the technique described below is $O(n^5)$, where n is the number of vertices (or edges) of the obstacles. Assymptotically more efficient techniques have been recently proposed (e.g. [27]). However, the method described below is a good illustration of the exact decomposition approach and its principle is rather simple.

Throughout this section, we assume that the boundary of every obstacle in the workspace is partitioned into *open* edges and vertices.

4.1 Critical and Regular Positions of the Segment

DEFINITION 6: *A position (x, y) of $\mathcal{A}$ is* **admissible** *if there exists at least one orientation θ such that $(x, y, \theta) \in \mathcal{C}_{free}$.*

For each admissible position (x, y) of $\mathcal{A}$, the set $F(x, y)$ of all free orientations of $\mathcal{A}$ – i.e., $F(x, y) = \{\theta \ / \ (x, y, \theta) \in \mathcal{C}_{free}\}$ – is a finite collection of open intervals. Each interval endpoint is a contact orientation of $\mathcal{A}$, i.e. $\mathcal{A}$ touches the boundary of at least one obstacle. In a similar fashion, we denote $V(x, y) = \{\theta \ / \ (x, y, \theta) \in \mathcal{C}_{valid}\}$ the set of all valid orientations of $\mathcal{A}$ at position (x, y). It is made of the closure of the open intervals in $F(x, y)$ and eventual isolated orientations. The set $L(x, y) = V(x, y) - F(x, y)$ contains a finite number of orientations, which are called **limit orientations** of $\mathcal{A}$ at (x, y).

DEFINITION 7: *Let θ be a limit orientation of $\mathcal{A}$ at (x, y) and s an obstacle's edge or vertex touched by $\mathcal{A}(x, y, \theta)$. If for all arbitrarily small clockwise (resp. counterclockwise) rotations of $\mathcal{A}$ about P, s (if s is an edge) or at least one edge abutting at s (if s is a vertex) intersects PQ, then s is called a* **clockwise** *(resp.* **counterclockwise***)* **stop** *at (x, y).*

DEFINITION 8: *A limit orientation θ at (x, y) is* **exceptional** *if and only if*

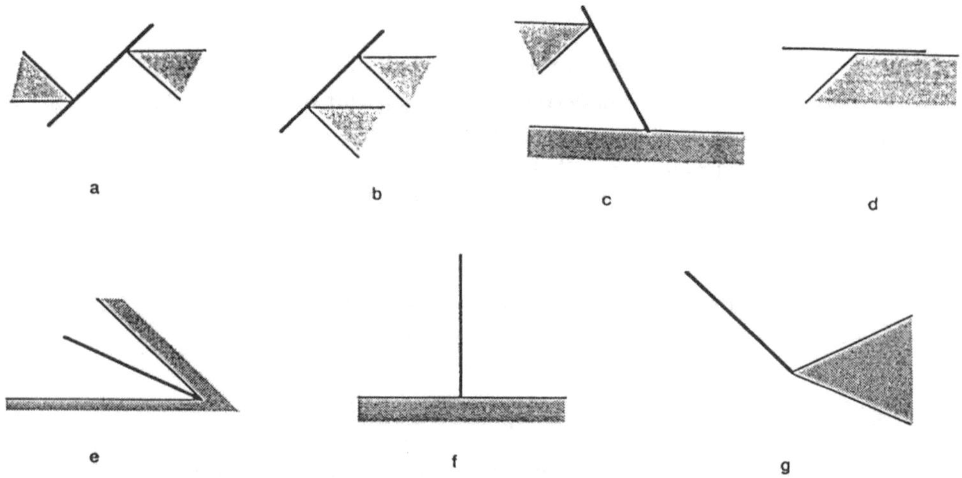

Figure 3: **Exceptional Orientations of the Moving Objects**

$\mathcal{A}(x, y, \theta)$ does not touch a unique stop. A position (x, y) is **regular** *if it has no exceptional orientation; otherwise it is* **critical**.

Thus, a limit orientation θ is exceptional at (x, y) if $\mathcal{A}(x, y, \theta)$ is touching more than one stop, or if it is touching no stop. $\mathcal{A}(x, y, \theta)$ touches more than one stop if it touches two vertices (Figure 3 *a* and *b*), or a vertex and an edge (Figure 3 *c* and *d*), or a concave vertex (Figure 3 *e*). Notice that in the latter case, the same vertex is counted as two stops (a clockwise one and a counterclockwise one). $\mathcal{A}(x, y, \theta)$ touches no stop if the only contact of $\mathcal{A}$ with an obstacle is at the Q extremity in one of the following two ways: either $\mathcal{A}(x, y, \theta)$ is perpendicular to an edge with Q touching that edge (Figure 3 *f*), or Q coincides with a convex vertex X and, in a sufficiently small neighborhood of X, the line perpendicular to PQ at Q intersects obstacles only at X (Figure 3 *g*).

Let (x, y) be a regular position of $\mathcal{A}$. The set $V(x, y)$ consists of finitely many disjoint closed intervals, none being an isolated point. For each such interval $[\theta_1, \theta_2]$, the unique stop touched by $\mathcal{A}(x, y, \theta_1)$ (resp. $\mathcal{A}(x, y, \theta_2)$) is a clockwise (resp. counterclockwise) stop. If all orientations of $\mathcal{A}$ are free, then $F(x, y) = [0, 2\pi)$, where $[0, 2\pi)$ is the 'circular' interval with no extremity.

If θ is a limit orientation at the regular position (x, y), $s(x, y, \theta)$ designates the unique stop touched by $\mathcal{A}(x, y, \theta)$. We denote $\sigma(x, y)$ the set of all the pairs $[s(x, y, \theta), s(x, y, \theta')]$ such that $s(x, y, \theta)$ (resp. $s(x, y, \theta')$) is a clockwise (resp. counterclockwise) stop at (x, y) and the interval $(\theta, \theta') \subseteq F(x, y)$. If $F(x, y) = [0, 2\pi)$, then we write $\sigma(x, y) = \{[\Omega, \Omega]\}$, where Ω designates a 'nonexistent' stop. Given a pair $[s_1, s_2] \neq [\Omega, \Omega]$ in $\sigma(x, y, \theta)$, we denote $\lambda_1(x, y, s_1)$ the unique orientation such that $\mathcal{A}$ positioned at (x, y) touches s_1 in such a way that s_1 is clockwise stop. Similarly, we denote $\lambda_2(x, y, s_2)$ the unique orientation such that $\mathcal{A}$ positioned at (x, y) touches s_2 in such a way that s_2 is a counterclockwise stop. By convention: $\lambda_1(x, y, \Omega) = 0$ and $\lambda_2(x, y, \Omega) = 2\pi$.

One can show the following:

LEMMA 1: *Let (x, y) be a regular position of $\mathcal{A}$. There exists an open neighborhood U of this position that only consists of regular positions and such that, $\forall(x', y') \in U$, we have $\sigma(x', y') = \sigma(x, y)$ (i.e., there is a 1-1 onto correspondance between the angular intervals of $F(x', y')$ and those of $F(x, y)$). Moreover, if $[s_1, s_2]$ belongs to $\sigma(x', y') = \sigma(x, y)$, then $\lambda_1(x', y', s_1)$ and $\lambda_2(x', y', s_2)$ are continuous functions of (x', y') for $(x', y') \in U$.*

Thus the set of regular positions of $\mathcal{A}$ is an open subset of $\Re^2$. We now analyze how this set is separated into regions by critical positions.

4.2 Critical Curves and Regular Regions

The locus of all critical positions of $\mathcal{A}$ is the union of a finite collection of curves, called **critical curves**. The intuition behind these curves is that $\sigma(x, y)$, i.e. the structure of C-obstacles above the position (x, y), changes when a critical curve is crossed.

Let d be the length of the segment $\mathcal{A}$. Critical curves are obtained[2], by considering the exceptional orientations shown in Figure 3:

(1) For each obstacle edge E, the line segment at distance d from E is a **Type I** critical curve (see Figure 4 *a*).

(2) For each obstacle vertex X, the circular arc of radius d about X and bounded by the two half-lines containing the edges adjacent to X is a **Type II** critical curve (see Figure 4 *b* and *c*).

(3) For each pair of obstacle vertices X_1 and X_2, the line segment traced by P as $\mathcal{A}$ translates while touching both X_1 and X_2 is a **Type III** critical curve (see Figure 4 *d*).

(4) For each obstacle edge E and convex vertex X at the extremity of E, the line segment traced by P as $\mathcal{A}$ slides along E is a **Type IV** critical curve (see Figure 4 *e*).

(5) For each obstacle edge E and convex vertex X not at the extremity of E, the curve traced by P as $\mathcal{A}$ moves touching both E and X is a **Type V** critical curve (see Figure 4 *f*).

The Type V critical curve is a conchoid of Nicomedes, which is described by an algebraic equation of degree 4.

The collection of critical curves is finite in a workspace with finitely many obstacle edges. Each curve extremity is located either on an obstacle boundary, or at the extremity of another critical curve. In addition, every critical curve is a smooth algebraic curve of degree 1 (Types I, III, and IV), 2 (Type II), or 4 (Type V). Thus, every two critical curves can have only finitely many intersections, except if they coincide. A coincidence happens, for example, when two parallel edges are separated by the distance $2d$. In

[2]This classification differs slightly from that of Schwartz and Sharir [47]. The differences, however, are only superficial.

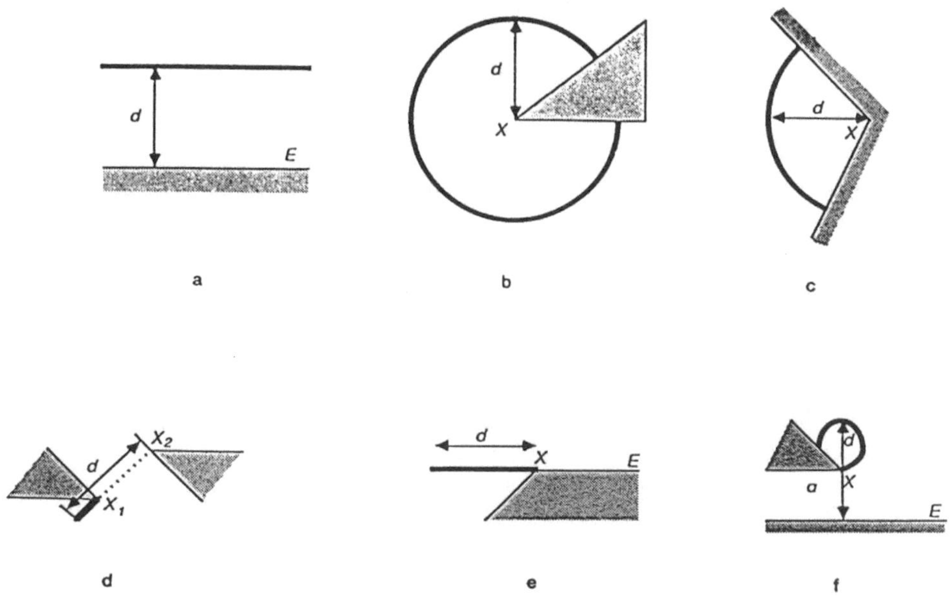

Figure 4: **Critical Curves**

theory, all coincidences can be eliminated by shifting some obstacles by very small amounts. Thus, we will assume no curve coincidence.

Notice here that several critical curves constructed as above are likely to be supersets of critical positions of $\mathcal{A}$. Indeed, each curve is built by considering one or two features (edge and/or vertex). Hence, the exceptional orientations for some sections of these curves, which make these curve critical, may correspond to non-valid configurations. As we will see, such a 'redundant' section of a critical curve needs not be explicitely removed.

It follows from Lemma 1 and the above discussion that:

LEMMA 2: *The set of admissible positions of $\mathcal{A}$ is divided by critical curves into a finite collection of open connected regions R, called* **regular region**. *The set $\sigma(x,y)$ is constant in each such region. In addition, for any such region R, any $(x,y) \in R$ and any pair $[s_1, s_2] \in \sigma(x,y)$, the region*

$$c(R, s_1, s_2) = \{(x, y, \theta) \;/\; (x,y) \in R \text{ and } \theta \in (\lambda_1(x,y,s_1), \lambda_2(x,y,s_2))\}$$

is an open connected region of C_{free}. It is called a **cell**[3].

Each cell $C = c(R, s_1, s_2)$ is homeomorphic to $\Re^3$, hence has no hole.

[3]Throughout this paper, the word *cell* is given several formal definitions. Within each section, however, the word keeps the same definition.

For any cell $C = c(R, s_1, s_2)$, $\phi_C(x, y)$ and $\psi_C(x, y)$ designate the limit clockwise and counterclockwise orientations at (x, y) such that:

$$(x, y, \theta) \in C \quad \Leftrightarrow \quad (x, y) \in R \text{ and } \theta \in (\phi_C(x, y), \psi_C(x, y)).$$

In addition, we write $\sigma(R) = \sigma(x, y)$, where (x, y) is an arbitrary point in R.

The boundary of any regular region consists of closed sections of critical curves and closed sections of obstacles edges.

4.3 Connectivity Graph

At this point we have decomposed free space into open connected cells separated by 'vertical' surfaces projecting on the (x, y)-plane along the critical curves. Let us now examine how these surfaces can be crossed in order to connect adjacent cells. We will analyze the crossing rules in more detail in the next subsection.

The following lemma drawn from Differential Geometry allows us to discard some transitions between cells which would be quite cumbersome to analyze.

LEMMA 3: *If there is a free path τ' between two free configurations projecting on the (x, y)-plane at regular positions, then there is another free path $\tau : t \in [0, 1] \mapsto (x(t), y(t), \theta(t)) \in C_{free}$ between the same two configurations that satisfies:*
- The curve $(x(t), y(t))$ has a well-defined tangent everywhere along its length.
- The curve $(x(t), y(t))$ does not pass through any point common to two critical curves.
- At each intersection of $(x(t), y(t))$ with a critical curve β, there is a neighborhood U of this point in which $\theta(t)$ is constant and the tangent vector $(\dot{x}(t), \dot{y}(t))$ is constant and lies transversal to β.

Let $C = c(R, s_1, s_2)$ be a cell. We extend by continuity the domain of ϕ_C and ψ_C to the closure $cl(R)$ of R.

DEFINITION 9: *Two cells $C = c(R, s_1, s_2)$ and $C' = c(R', s'_1, s'_2)$ are* **adjacent** *if and only if R and R' share an open section β of a critical curve and, for all $(x, y) \in \beta$, $(\phi_C(x, y), \psi_C(x, y)) \cap (\phi_{C'}(x, y), \psi_{C'}(x, y)) \neq \emptyset$.*

It immediately follows from Lemma 3 that if two cells C and C' are adjacent, any configuration in C can be connected to any configuration in C' by a differentiable path whose projection on the (x, y)-plane crosses β transversally, with a constant orientation in some neighborhood of the crossing point.

DEFINITION 10: *The* **connectivity graph** *CG is the graph whose nodes are all cells $c(R, s_1, s_2)$, where R is a maximal regular region in the (x, y)-plane and $[s_1, s_2] \in \sigma(R)$. An undirected arc connects any two nodes labeled by adjacent cells.*

It follows from the above that:

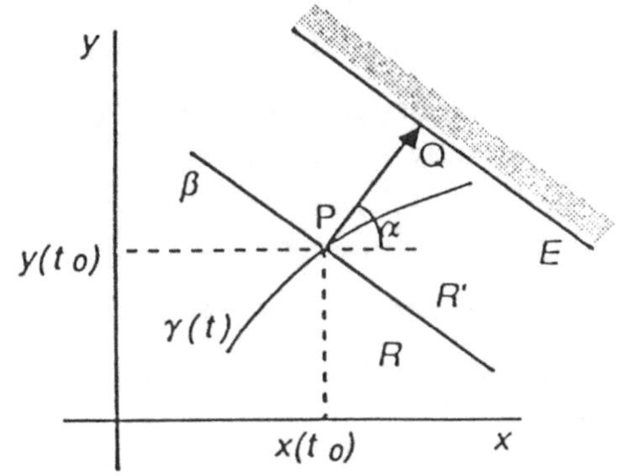

Figure 5: **Crossing a Type I Curve**

THEOREM 1: *Given two configurations $\mathbf{c}_1 = (x_1, y_1, \theta_1)$ and $\mathbf{c}_2 = (x_2, y_2, \theta_2)$ in C_{free}, such that neither (x_1, y_1), nor (x_2, y_2) lies on a critical curve, there exists a free path between them if and only if the two cells containing $\mathbf{c}_1$ and $\mathbf{c}_2$ are connected by a path in the connectivity graph CG.*

This theorem is the basis for the algorithm sketched in Subsection 4.5. Before presenting the algorithm, let us analyze the crossing of a critical curve in more detail.

4.4 Crossing Rules

Let β designate a critical curve section not intersected by any other critical curve. C and C' designate two cells $c(R, s_1, s_2)$ and $c(R', s'_1, s'_2)$, such that $[s_1, s_2] \in \sigma(R)$, $[s'_1, s'_2] \in \sigma(R')$, and β forms part of the boundaries of R and R'. $\tau : t \in [0,1] \mapsto (x(t), y(t), \theta(t))$ is a path such that the projected curve $\gamma(t) = (x(t), y(t))$ crosses β transversally at $t = t_0$, with $\theta(t)$ constant and equal to θ_0 in some interval $(t_0 - \varepsilon, t_0 + \varepsilon)$.

When $t \to t_0$ with $t < t_0$:

$$(\phi_C(x(t), y(t)), \psi_C(x(t), y(t))) \to (\phi_C(x(t_0), y(t_0)), \psi_C(x(t_0), y(t_0))).$$

When $t \to t_0$ with $t > t_0$:

$$(\phi_{C'}(x(t), y(t)), \psi_{C'}(x(t), y(t))) \to (\phi_{C'}(x(t_0), y(t_0)), \psi_{C'}(x(t_0), y(t_0))).$$

The **crossing rule** for a certain type of curve section β is the description of all the connections between cells C and C'. Let us consider the case of a Type I curve as an example, and analyze it in detail.

If β is a Type I curve section, it is a line segment parallel to an obstacle edge E at distance d. Let R' be the region between β and E and α the orientation of the segment $\mathcal{A}$ when P is on β and Q on E (see Figure 5).

If $\alpha \notin (\phi_C(x(t_0), y(t_0)), \psi_C(x(t_0), y(t_0)))$, then the two cells C and C' are adjacent if and only if $(\phi_C(x(t_0), y(t_0)), \psi_C(x(t_0), y(t_0))) = (\phi_{C'}(x(t_0), y(t_0)), \psi_{C'}(x(t_0), y(t_0)))$.

If $\alpha \in (\phi_C(x(t_0), y(t_0)), \psi_C(x(t_0), y(t_0)))$, then when $\gamma(t)$ crosses β the interval of free orientations breaks into exactly two subintervals. The two cells C and C' are adjacent if and only if $(\phi_{C'}(x(t_0), y(t_0)), \psi_{C'}(x(t_0), y(t_0))) = (\phi_C(x(t_0), y(t_0)), \alpha)$ or $(\phi_{C'}(x(t_0), y(t_0)), \psi_{C'}(x(t_0), y(t_0))) = (\alpha, \psi_C(x(t_0), y(t_0)))$.

Thus, the crossing rule for a Type I critical curve section is:

The two cells $c(R, s_1, s_2)$ and $c(R', s_1', s_2')$ are adjacent if and only if $[s_1', s_2'] = [s_1, s_2]$, or $[s_1', s_2'] = [s_1, E]$, or $[s_1', s_2'] = [E, s_2]$.

It is straightforward to analyze in the same fashion all the other cases. Such an analysis shows that whenever $\gamma(t)$ crosses β the set $\sigma(x, y)$ is changing in one of the following ways:

- One pair in $\sigma(x, y)$ disappears, or one new pair appears.

- One element in one pair in $\sigma(x, y)$ changes.

The crossing rules for the different types of critical curves can be summarized by:

Connect $c(R, s_1, s_2)$ to $c(R', s_1, s_2)$ for each $[s_1, s_2] \in \sigma(R) \cap \sigma(R')$, and connect each $c(R, s_1, s_2)$, $[s_1, s_2] \in \sigma(R) - \sigma(R')$, if any, to each $c(R', s_1', s_2')$, $[s_1', s_2'] \in \sigma(R') - \sigma(R)$, if any.

This generic rule handles redundant sections of critical curves, since in that case $\sigma(R) = \sigma(R')$.

4.5 Sketch of the Algorithm

Given a set of compact polygonal obstacles, the length d of the segment $\mathcal{A}$, and the initial and final free configurations $\mathbf{c}_1 = (x_1, y_1, \theta_1)$ and $\mathbf{c}_2 = (x_2, y_2, \theta_2)$ of $\mathcal{A}$, such that neither configuration projects on the (x, y)-plane on a critical curve, the path finding algorithm sketched below generates a free path from $\mathbf{c}_1$ to $\mathbf{c}_2$ whenever one exists, and returns failure otherwise.

This algorithm does not explicitly build the regular regions by collecting the curve sections bounding them together. Instead, for every curve section β, it gives an orientation to β and we designate the two regions separated by β by $right(\beta)$ and $left(\beta)$. As a consequence, the algorithm builds a connectivity graph which is slightly different from the one defined above. In this new graph, each cell appears as many times as there are curve sections forming the boundary of the regular region on which the cell projects.

Since each regular region may be bounded by both portions of critical curves and portions of obstacles edges, it is convenient in the algorithm to treat edges as a special kind of critical curve which cannot be crossed. In addition, since each curve section β may be followed in two directions, it is also convenient to implicitly treat it as a pair of oppositely oriented sections (β^+, β^-), such that $right(\beta^+)$ and $left(\beta^-)$ (resp. $left(\beta^+)$ and $right(\beta^-)$) are identified.

At several points in the algorithm, we have to compute $\sigma(x, y)$ and the limit orientations corresponding to the clockwise and counterclockwise stops appearing in $\sigma(x, y)$. This can be done by rotating a half-line emanating from (x, y) and using a kind of line-sweep algorithm.

With these preliminary remarks, the algorithm is the following:

1. **Compute the boundary of the regular regions.** Compute the set of all critical curves (including eventual redundant sections). Add all obstacle edges to this set. Find intersection points of the closures of these curves with each other. For each curve, sort these intersection points according to their order on this curve (this is easily done by defining a suitable parametrization for each possible type of curve). Decompose each curve into sections, each lying between two consecutive intersection points.

2. **Compute the adjacency relation between cells.** For every curve section β, which is part of a critical curve, compute $\sigma(right(\beta))$ and $\sigma(left(\beta))$, use the general crossing rule, and build a map $ADJACENT$ which maps every cell projecting on $right(\beta)$ into the adjacent cells projecting on $left(\beta)$. If $\sigma(right(\beta)) = \sigma(left(\beta))$, β is redundant and may be removed from further consideration. For every curve section β which is part of an obstacle edge, set the map $ADJACENT$ to $\emptyset$.

3. **Form clusters of intersecting curve sections.** At every intersection point Z of two or more curve sections, form the circular list of all curve sections emerging from Z sorted in clockwise order (this list is called a *cluster*). To this end, compute and sort the outgoing tangential directions of the curve sections. Whenever two such directions are equal, compute and sort higher-order derivative directions.

4. **Locate initial and final cells.** Draw the line segment between (x_1, y_1) and (x_2, y_2). Compute the intersections of this segment with all curve sections. Find the curve sections β_1 and β_2 whose intersections with the line segment are nearest from (x_1, y_1) and (x_2, y_2), respectively. Identify the side of β_1 (resp. β_2) that contains (x_1, y_1) (resp. (x_2, y_2)). Find the pair $[s_{11}, s_{21}]$ (resp. $[s_{12}, s_{22}]$) such that $\theta_1 \in (\lambda_1(x_1, y_1, s_{11}), \lambda_2(x_1, y_1, s_{21}))$ (resp. $\theta_2 \in (\lambda_1(x_2, y_2, s_{12}), \lambda_2(x_2, y_2, s_{22}))$). Let C_i, $i = 1,2$, be the cell containing (x_i, y_i, θ_i).

5. **Search the connectivity graph.** Build a connectivity graph defined as follows. Each node is a cell $c(R, s_1, s_2)$ where R is identified as one side of a curve section β, i.e. $right(\beta)$ or $left(\beta)$. Two nodes $c(R, s_1, s_2)$ and $c(R', s_1', s_2')$ are linked by an undirected arc if one of the following two conditions is satisfied:

   - $R = left(\beta)$, $R' = right(\beta')$, β' follows β in a cluster, and $[s_1, s_2] = [s_1', s_2']$.
   - $R = right(\beta)$, $R' = left(\beta)$, and $c(R', s_1', s_2') \in ADJACENT(c(R, s_1, s_2))$.

   Initialize the graph with node C_1. Then, expand each node that has not been expanded yet. Stop either when the node C_2 has been generated, or when there is no more nodes to expand. In the first case, output the sequence of cells on the path connecting C_1 to C_2 in the graph. In the second case, indicate failure.

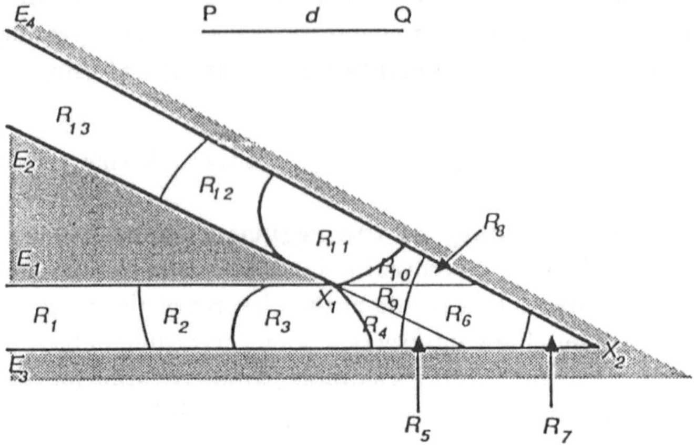

Figure 6: **The Corner Example**

The union of the set of cells output by the algorithm and the set of common boundaries between every two successive cells is a region of $\mathcal{C}_{free}$ homeomorphic to $\Re^3$. It is 'easy' to transform the sequence into a free path, by following and crossing the specified curve sections appropriately and maintaining the orientation in the appropriate open interval.

Notice that in order to implement the algorithm, one would have to pay attention to details not considered above. One delicate issue is related to the sorting of intersection points. With limited precision arithmetic, the identification of two separately calculated intersection points as being the same point is indeed problematical.

A detailed analysis of the algorithm shows that its time complexity is $O(n^5)$.

4.6 Example

Consider the 'corner example' shown at Figure 6. Only the non-redundant critical curves sections are shown in the figure (i.e., we assume that the redundant ones are removed at Step 2 of the algorithm). There are 13 regular regions denoted R_1 through R_{13}. We leave the identification of each critical curve section as an exercise to the reader.

We have:

$$\begin{aligned}
\sigma(R_1) &= \{[E_1, E_3], [E_3, E_1]\}\\
\sigma(R_2) &= \{[E_1, E_3], [E_3, X_1]\}\\
\sigma(R_3) &= \{[E_1, E_3], [E_3, E_4]\}\\
\sigma(R_4) &= \{[E_1, E_3], [E_3, E_4], [E_4, X_1]\}\\
\sigma(R_5) &= \{[E_1, E_3], [E_4, X_1]\}\\
\sigma(R_6) &= \{[E_1, E_3], [E_4, E_2]\}\\
\sigma(R_7) &= \{[E_4, E_3]\}\\
\sigma(R_8) &= \{[E_4, E_2], [X_1, E_3]\}\\
\sigma(R_9) &= \{[E_4, E_2], [E_1, E_3], [E_3, E_4]\}\\
\sigma(R_{10}) &= \{[E_4, E_2], [X_1, E_3], [E_3, E_4]\}\\
\sigma(R_{11}) &= \{[E_4, E_2], [E_1, E_4]\}
\end{aligned}$$

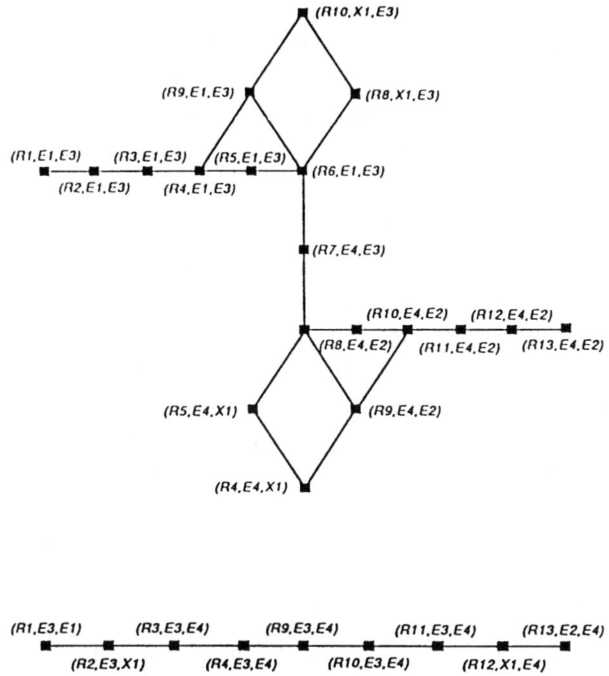

Figure 7: **Connectivity Graph for the Corner Example**

$$\sigma(R_{12}) = \{[E_4, E_2], [X_1, E_4]\}$$
$$\sigma(R_{13}) = \{[E_4, E_2], [E_2, E_4]\}$$

Figure 7 shows the resulting connectivity graph. This graph contains two connected components, which means that $\mathcal{C}_{free}$ has also two connected components. Practically, this implies that the segement $\mathcal{A}$ cannot rotate around the corner. But it can move from one end of the corner to the other with a 'forward - backward' or 'backward - forward' motion.

5 Exact Decomposition: General Method

In the previous two sections we illustrated the exact cell decomposition approach to path planning in two specific cases. Below, we conclude our presentation of the approach by the description of a general method which put no limitation[4] on the dimension k of the workspace ($k = 2$ or 3) and on the possible motions of the robot $\mathcal{A}$. The only constraint imposed by the method is that both the moving object $\mathcal{A}$ and the obstacles $\mathcal{B}_i$ be described as semi-algebraic sets (see Subsection 5.1).

This general method works according to the same principle as the methods described in the previous section. It consists of exactly partitioning $\mathcal{C}_{free}$ into a finite collection of semi-algebraic connected cells, computing the adjacency relation between the cells and representing it as a graph, searching this graph, and transforming the obtained sequence

[4]This general method is also applicable to articulated objects, i.e. manipulator arms.

of cells into a path. As in Section 4, cells in $\mathcal{C}_{free}$ are obtained by first computing their projection in a lower-dimensional space and then lifting them in $\mathcal{C}$. Since the dimension m of $\mathcal{C}$ may be greater than 3, the projection/lift mechanism is recursive, that is, it produces projections of cells on spaces of dimensions $m-1, m-2, ..., 1$, before lifting them back through the same sequence of spaces. Due to this recursive computation, this path finding method is sometimes called 'projection method'.

The method is due to Schwartz and Sharir [48], who considered the more general case of an articulated robot (e.g., a manipulator arm) with many degrees of freedom. It makes use of a well-known result by Collins [15] for deciding the satisfiability of Tarski sentences. We expose this result first. The general path finding algorithm based on this result is sketched next. Our presentation, which is inspired from [54], concentrates on the main ideas of the method. The actual algorithm requires to pay attention to many details, including exact computation with algebraic numbers, which are not covered here. We refer the reader to the original paper for more detail.

5.1 Semi-Algebraic Representation of C-Obstacles

Let $Q[X_1; ...; X_n]$ be the ring of polynomials in n real variables with rational coefficients.

Let us assume that every object in $\mathcal{W}$ is represented as a conjunction of inequations of the form $P(x) \leq 0$, where $x \in \Re^k$ and $P \in Q[X_1; ...; X_k]$. A point $x_0 \in \Re^k$ belongs to the object if and only if the all the inequalities are verified at this point.

Let us denote $\Lambda_{\mathcal{A}}(a)$ and $\Lambda_{\mathcal{B}_i}(b)$, with a and $b \in \Re^k$, the representation of $\mathcal{A}$ and an obstacle $\mathcal{B}_i$, respectively. Without loss of generality, we assume that $\Lambda_{\mathcal{A}}$ represents $\mathcal{A}$ at configuration $\mathbf{0}$, i.e. when $\mathcal{F}_{\mathcal{A}}$ and $\mathcal{F}_{\mathcal{W}}$ coincide. As mentioned previously, the configuration space $\mathcal{C}$ can always be embedded in an Euclidean space of some dimension N. Any configuration $\mathbf{c}$ of $\mathcal{A}$ can be represented as a vector of N real numbers related by $N-m$ independent polynomial equations with rational coefficients expressing the fact that $\mathbf{c}$ belongs to a manifold of dimension m.

Let us denote $\mathbf{x}$ the representation of $\mathbf{c}$ as a N-dimensional vector and $\Gamma(\mathbf{x})$ the conjunction of the $N-m$ polynomial equations constraining this representation. Let $\Upsilon(\mathbf{x}, a, b)$, with $a \in \Re^k$ and $b \in \Re^k$, be the conjunction of polynomial equations expressing that a point, a, fixed with respect to $\mathcal{F}_{\mathcal{A}}$ is such that $a(\mathbf{x})$ coincides with a point, b, fixed with respect to $\mathcal{F}_{\mathcal{W}}$.

The C-obstacle corresponding to $\mathcal{B}_i$ can be represented in $\Re^N$ by the following expression:

$$\mathcal{CB}_i = \{\mathbf{x} \in \Re^N \;/\; \exists a \in \Re^k \exists b \in \Re^k : \Lambda_{\mathcal{A}}(a) \wedge \Lambda_{\mathcal{B}_i}(b) \wedge \Upsilon(\mathbf{x}, a, b) \wedge \Gamma(\mathbf{x})\}.$$

Such an expression is an instance of a Tarski sentence over $\Re^N$.

DEFINITION 11: *An* **atomic polynomial expression** *over $\Re^j$ is one of the form:*

$$P(x) \bowtie 0$$

where $x \in \Re^j$ and $P \in Q[X_1; ...; X_j]$ and $\bowtie$ can be any comparator in $\{=, \neq, >, <, \geq, \leq\}$. A **polynomial expression** *over $\Re^j$ is any finite boolean combination of atomic polyno-*

mial expressions over $\Re^j$. A subset of $\Re^j$ whose points verify a polynomial expression is called a **semi-algebraic set**. *A* **Tarski sentence** *over $\Re^j$ is any polynomial expression over $\Re^j$ prefixed by a finite number of $\exists$ and $\forall$ quantifiers.*

Tarski showed that any Tarski sentence has an equivalent quantifier-free polynomial expression [52]. Collins [15] gave a constructive proof of this result (see Subsection 5.2). Thus:

THEOREM 2: *If both the moving object $\mathcal{A}$ and the obstacles $\mathcal{B}_i$ are represented as semi-algebraic sets in $\Re^k$, then all C-obstacles, hence C_{free}, are semi-algebraic subsets of $\Re^N$.*

5.2 Cylindrical Algebraic Decomposition of $\Re^N$

DEFINITION 12: *A* **decomposition** *of $\Re^N$ is a finite collection Δ of disjoint connected subsets of $\Re^N$ whose union is $\Re^N$. Each element c of Δ is called a* **cell**. *Δ is an* **algebraic decomposition** *if each cell in Δ is a semi-algebraic set homeomorphic to $\Re^j$, with $j = 0, 1, ..., N$ (if $j = 0$, then the cell consists of a single algebraic point).*

The fact that each cell in an algebraic decomposition is homeomorphic to some $\Re^j$ means that it contains no hole.

Two cells c_1 and c_2 of a decomposition Δ are **adjacent**[5] if and only $(cl(c_1) \cap c_2) \cup (c_1 \cap cl(c_2)) \neq \emptyset$.

A **sample** of an algrebraic decomposition Δ of $\Re^N$ is a function $\sigma : \Delta \rightarrow \Re^N$ such that for all $c \in \Delta$, $\sigma(c) \in c$ and each coordinate of $\sigma(c)$ is an algebraic number[6].

The **projection** of a point $\mathbf{x}_N = (x_1, ..., x_N) \in \Re^N$ is $\mathbf{x}_{N-1} = (x_1, ..., x_{N-1}) \in \Re^{N-1}$, obtained by omitting the Nth component of the point. The projection of a subset of $\Re^N$ is the set of projections of its members.

DEFINITION 13: *A* **cylindrical algebraic decomposition** *Δ_N of $\Re^N$ ($N > 0$) is an algebraic decomposition recursively defined as follows:*
- For $N > 1$, there exists a cylindrical algebraic decomposition Δ_{N-1} of $\Re^{N-1}$ such that for each cell c in Δ_N there is a cell c' in Δ_{N-1} that is the projection of c. Δ_{N-1} is called the **base decomposition** *of Δ_N, and c is said to be* **based** *on c'.*
- For $N = 1$, Δ_1 is just a partitioning of $\Re$ into a finite set of algebraic numbers and into the finite and infinite open intervals bounded by these numbers.

The collection of all the cells c in Δ_N which have the same projection c' in Δ_{N-1} is called the **cylinder** over c'.

[5]This definition of adjacency is consistent with the definitions given in the previous two sections. The difference of formulation comes from the fact that here we consider cells of different dimensions.

[6]The set of *algebraic numbers* is the set of real roots of polynomials in $Q[X]$.

A cylindrical sample of a cylindrical algebraic decomposition of $\Re^N$, Δ_N, is a sample $\sigma : \Delta_N \rightarrow \Re^N$, such that:
- if c_1 and c_2 are cells of Δ_N having the same base then $\sigma(c_1)$ and $\sigma(c_2)$ have the same projection in $\Re^{N-1}$
- if $N > 1$, σ recursively induces a cylindrical sample of the base decomposition Δ_{N-1} of Δ_N.

DEFINITION 14: *Let signum(a) be -1, 0, or $+1$ whenever $a < 0$, $a = 0$, or $a > 0$, respectively. Let $\mathcal{F}$ be a set of functions of N real variables. A decomposition Δ of $\Re^N$ is said to be* **$\mathcal{F}$-invariant** *if, for each cell c in Δ and each function f in $\mathcal{F}$, signum$(f(\mathbf{x}))$ is constant as $\mathbf{x} = (x_1, ..., x_N)$ ranges over c.*

THEOREM 3 (Collins): *Given any set $\mathcal{P}$ of polynomials in $Q[X_1; ...; X_N]$ there is an effective algorithm which constructs a $\mathcal{P}$-invariant cylindrical algebraic decomposition Δ_N of $\Re^N$ together with a cylindrical sample of Δ. The time complexity of the algorithm is double-exponential in n.*

The constructive proof of this theorem is the Collins algorithm constructing the $\mathcal{P}$-invariant decomposition Δ_N (we do not describe it here). Each of the cells produced by the algorithm is described by a quantifier-free polynomial expression over $\Re^N$. The algorithm also produces the base decomposition Δ_{N-1} of Δ_N, the base decomposition Δ_{N-2} of Δ_{N-1}, etc. The Collins algorithm takes time polynomial in the number of polynomials in $\mathcal{P}$ (*geometric* complexity) and in their maximum degree (*algebraic* complexity)[7], with double exponential dependence on N. The number of cells generated by the algorithm has the same order of magnitude[8].

5.3 Application to the First-Order Theory of Reals

An immediate application of the Collins algorithm is the construction of a procedure capable of deciding the satisfiability of input Tarski sentences. Two more specific applications are of special interest here: truth decision and quantifier elimination. We illustrate these two applications below on simple examples. It is easy to generalize these examples into general procedures. In the subsequent subsection we will use quantification elimination to construct a cylindrical algebraic decomposition of $\mathcal{C}_{free}$ and truth decision to test adjacency between cells of this decomposition.

- **Truth decision:** The problem is to determine the truth value of a Tarski sentence that includes no free variable. For instance, let us consider the sentence $(\exists x)(\forall y)[\phi(x, y)]$, where both x and y are variables in $\Re$. Let P be the set of polynomials appearing in expression ϕ. Using Collins algorithm, we can build a P-invariant cylindrical algebraic decomposition Δ of $\Re^2$, a base decomposition Δ' of Δ, and a cylindrical sample σ : $\Delta \rightarrow \Re^2$ of Δ. The above given sentence is true if and only if: $\forall c' \in \Delta'$, there exists a cell $c \in \Delta$ contained in the cylinder over c' such that $\sigma(c) = (x_c, y_c)$ and $\phi(x_c, y_c)$ holds.

---

[7] The algorithm is also polynomial in the magnitude of coefficients in $\mathcal{P}$.
[8] The number of cells has no dependence on the magnitude of coefficients.

- **Quantifier elimination:** The problem is to transform a given Tarski sentence with quantifiers into a quantifier-free polynomial expression. For instance consider the sentence $(\exists y)\phi(x,y)$, where ϕ is a polynomial expression over $\Re^2$. Variable x is free, so that the sentence defines a semi-algebraic subset of $\Re$. We wish to define this subset by a quantifier-free polynomial expression over $\Re$. Let Δ, Δ', and σ be as above. We denote ψ_c the quantifier-free polynomial expression defining cell c, for any cell c in Δ or Δ'. Let $\Psi(x)$ be the disjunction of all the $\psi_{c'}$, where c' is the base cell of at least one $c \in \Delta$ such that $\sigma(c) = (x_c, y_c)$ and $\phi(x_c, y_c)$ holds. The two Tarski sentences $(\exists y)\phi(x,y)$ and $\Psi(x)$ define the same semi-algebraic subset of $\Re$.

5.4 General Approach for Path Planning

We now show how the Collins result can be applied to path planning. We assume that the moving object $\mathcal{A}$ and the obstacles $\mathcal{B}_i$ are represented as semi-algebraic subsets of $\Re^k$. Let the defining formula of $\mathcal{A}$ and $\mathcal{B}_i$ be $\Lambda_{\mathcal{A}}(a)$ and $\Lambda_{\mathcal{B}_i}(b)$, respectively, where a and $b \in \Re^k$. As noted in Subsection 5.1, the C-obstacle $\mathcal{CB}_i$ corresponding to the actual obstacle $\mathcal{B}_i$ in the workspace can be represented by a Tarski sentence, and:

$$\bigcup_i \mathcal{CB}_i = \{\mathbf{x} \in \Re^N \ / \ \exists a \in \Re^k \ \exists b \in \Re^k : \ \Lambda_{\mathcal{A}}(a) \wedge (\bigvee_i \Lambda_{\mathcal{B}_i}(b)) \wedge \Upsilon(\mathbf{x}, a, b) \wedge \Gamma(\mathbf{x})\}.$$

We can use the Collins algorithm to generate a quantifier-free polynomial expression over $\Re^N$, $\Psi(\mathbf{x}) = \bigvee \psi_c(\mathbf{x})$, defining $\mathcal{C}_{free}$, together with a $\mathcal{P}$-invariant cylindrical algebraic decomposition Δ of $\Re^N$ ($\mathcal{P}$ is the set of polynomials in Ψ), the successive base decompositions of Δ, and a cylindrical sample σ of Δ. This computation can be done in doubly exponential time in the dimension N. Then, for each cell $c \in \Delta$, we can check if it is part of $\mathcal{C}_{free}$ by testing if $\sigma(c)$ satisfies Ψ.

In order to construct a path from an initial configuration $\mathbf{c}_1$ represented by $\mathbf{x}_1$ to a final configuration $\mathbf{c}_2$ represented by $\mathbf{x}_2$, we have to find a sequence $\{c_1, c_2, ..., c_p\}$ of cells of Δ, such that $\mathbf{x}_1 \in c_1$, $\mathbf{x}_2 \in c_p$, and c_i is adjacent to c_{i+1} for all $i \in [1, p-1]$. By evaluating the polynomial expression defining each cell in $\mathcal{C}_{free}$, with $\mathbf{x} = \mathbf{x}_1$ and $\mathbf{x} = \mathbf{x}_2$, we can identify the first and last cells of the sequence. Constructing the rest of the sequence requires to build and search the connectivity graph CG representing the adjacency relation on the set of cells partitioning $\mathcal{C}_{free}$. The adjacency condition for two cells c_1 and c_2 is $(cl(c_1) \cap c_2) \cup (c_1 \cap cl(c_2)) \neq \emptyset$, which is equivalent to $(cl(c_1) \cap c_2 \neq \emptyset) \vee (c_1 \cap cl(c_2) \neq \emptyset)$. Any of the two conditions in the later disjunction can be expressed as a Tarski sentence. For instance, $c_1 \cap cl(c_2) \neq \emptyset$ holds if and only if the following Tarski sentence is satisfied [2]:

$$(\exists \mathbf{x})(\forall \varepsilon)(\exists \mathbf{y})[(\varepsilon > 0) \wedge \phi_{c_1}(\mathbf{x}) \Rightarrow (distance(\mathbf{x}, \mathbf{y}) < \varepsilon) \wedge \phi_{c_2}(\mathbf{y})].$$

(Read: It exists a point $\mathbf{x}$ in c_1 such that for every $\varepsilon > 0$, there exists a point $\mathbf{y}$ in c_2 distant of $\mathbf{x}$ by less than ε.)

Thus, testing adjacency between two cells and consequently building the connectivity graph CG are decidable problems that can be solved by applying the Collins algorithm to the above Tarski sentence.

As mentioned in Subsection 5.2 the time complexity for generating the cell decomposition of $\mathcal{C}_{free}$ is polynomial in the number of polynomials in $\mathcal{P}$ and in their maximum degree, with a double exponential dependence on N. For a rigid object N is fixed. Therefore, the decomposition of $\mathcal{C}_{free}$ is generated in polynomial time. The number of produced cells is of the same order, so that the connectivity graph CG can be built in polynomial time if adjacency of two cells can be checked in polynomial time. The number of polynomials in the expression ψ_c defining each cell in the decomposition of $\mathcal{C}_{free}$ and their maximum degree are polynomial since the algorithm that constructs these expressions is itself polynomial. Therefore, since the number of variables appearing in the Tarski sentences representing adjacency between two cells is $2N+1$, hence fixed, the construction of the connectivity graph CG is polynomial. However, it has a multiple exponential dependence on N.

A more efficient test of cell adjacency than the above one is proposed in [48]. We will only sketch it here. The improved test requires that $\mathcal{C}$ be represented[9] as a Euclidean space $\Re^m$, where m is the dimension of $\mathcal{C}$, and that the cylindrical decomposition Δ of $\Re^m$ be 'well-based' (see [48]), a concept which recursively applies to the base decomposition of Δ. Since a well-based $\mathcal{P}$-invariant cylindrical algebraic decomposition of an Euclidean space may not exist, without rotating the coordinate axes, this leads to some additional technical complications, which we will not present here. Well-basedness implies that the closure of any cell $c \in \Delta$ is a union of cells in Δ. In addition, we can assume that the initial and final configurations of the robot are contained in cells of the same dimension as $\mathcal{C}$, i.e. m, since the other cells form a measure zero subset of $\mathcal{C}$ (if the assumption is not satisfied at one configuration, an arbitrarily small shift of the configuration will achieve it). Then it can be proven that, without losing completeness, path finding can be confined to those cells in Δ that have dimension m or dimension $m-1$ (i.e., codimension 1), where m is the dimension of $\mathcal{C}$. Thus, adjacency has only to be tested between a cell of dimension m and a cell of dimension $m-1$, which can be done rather efficiently. If two such cells are adjacent, the second is contained in the closure of the first. Using this test, Schwartz and Sharir describe a general path-planning algorithm whose time complexity is polynomial in the number of polynomials in $\mathcal{P}$ and in their maximal degree, with a double dependence on m.

Consider that CG has been constructed and that a sequence of adjacent cells $\{c_1, ..., c_p\}$ has been found by searching it. We now want to generate a path τ from the initial to the goal configuration, which completely lies in the sequence of cells (recall that each cell is a connected set). We assume that $\mathcal{C}$ has been represented as a Euclidean space $\Re^m$ (see the above paragraph) and that the decomposition Δ is well-based, so that $c_1, ..., c_p$ are alternatively m-dimensional and $(m-1)$-dimensional cells (odd indices correspond to m-dimensional cells and even indices correspond to $(m-1)$-dimensional cells). Without loss of completeness, we can construct τ as the concatenation of $(p+1)/2$ sub-paths connecting $\mathbf{x}_1$ to $\sigma(c_2)$, $\sigma(c_2)$ to $\sigma(c_4)$, ..., $\sigma(c_{p-3})$ to $\sigma(c_{p-1})$, and $\sigma(c_{p-1})$ to $\mathbf{x}_2$. Let us consider that the projection τ' of τ has already been generated within $c'_1, ..., c'_p$,

[9] If $\mathcal{C}$ is of the form $\Re^k \times S^1$, this requirement leads to consider two distinct copies of $\Re^m$ $(m = k+1)$. If $\mathcal{C}$ is of the form $\Re^3 \times SO(3)$ $(m = 6)$, it leads to remove one singular orientation, but since the corresponding set of points in $\mathcal{C}$ has a dimension less or equal to $m-2$, i.e. a codimension at least 2, omission of these points does not affect the connectivity of $\mathcal{C}_{free}$.

the base cells of $c_1, \ldots, c_p$. Each sub-path in τ' can be lifted in $c_1, \ldots, c_p$ as follows. Assume i to be odd. The x_m coordinate along the sub-path of τ in any cell c_i can be obtained by linearly interpolating between the values of x_m at the beginning and the end of the sub-path, using the proportion of distance between the boundaries of c_i in the $+x_m$ and $-x_m$ directions and by extending the interpolation to the closure of c_i, which includes c_{i+1}. Two cases have to be distinguished in order to construct the interpolation equation: (1) the two cells c_i and c_{i+1} have the same base cell (i.e., $c'_i = c'_{i+1}$), and (2) the two cells c_i and c_{i+1} have adjacent base cells ($c'_i \neq c'_{i+1}$). In both cases, if c_i is a semi-infinite cell, an artificial boundary has to be generated by adding (resp. substracting) a fixed appropriate amount to the cell boundary below (resp. above) the sub-path in order to make the interpolation possible. The projected path τ' can be generated in the same way by lifting in Δ' a path τ'' generated in the base decomposition of Δ'. Thus, recursively, the problem is reduced to the generation of a path in a base decomposition in $\Re$, which is trivial. The constructed free path is piecewise of class C^2.

Therefore:

THEOREM 4 (Schwartz and Sharir): *The problem of planning a free path of a semi-algebraic rigid object $\mathcal{A}$ among semi-algebraic fixed rigid obstacles $\mathcal{B}_i$ can be solved in time polynomial in the number of polynomials defining the objects in the workspace and in the maximal degree of these polynomials.*

So, if both $\mathcal{A}$ and the obstacles are polygonal (resp. polyhedral) objects in a two-dimensional (resp. three-dimensional) workspace, the time complexity of the above method is polynomial in the number of edges (resp. faces) of the objects.

The above general approach to path finding requires performing exact computations with algebraic numbers. This is a delicate issue not treated here. The paper by Schwartz and Sharir [47] provides many details on exact computation with algebraic numbers. It also analyzes the effect of the growth in size of the integers involved in the computations on the total time complexity of the algorithm. It turns out that the algorithm still requires time polynomial in the number of polynomials defining the object and in the maximal degree of these polynomials. However, it is also polynomial in the size of the coefficients in the polynomials.

The Collins decomposition generates many cells allowing its application to the general problem of deciding Tarski sentences. It is probably too general for path finding. The more specific (and more elaborate) method presented in Section 4, although based on the same general idea, generates fewer cells in the three-dimensional configuration space of the segment robot. Each of these cells turns out to be a union of the Collins cells which would have been produced by the above method[10]. This suggests that decompositions coarser than the Collins one may still be appropriate for path planning and that general algorithms of less time complexity than the algorithm presented above exist. Such an approach has been explored by Canny [10], resulting in an algorithm which is simply exponential in the number of degrees of freedom.

[10]The trapezoidal decomposition presented in Section 3 also produces cells that are unions of Collins cells.

6 Approximate Decomposition: Translational Case

In this section and the next, we describe another type of path finding decomposition methods based on approximate decomposition of configuration space. We will only consider the cases of polygonal and polyhedral workspaces.

An approximate decomposition method typically consists of decomposing $\mathcal{C}$ into a collection of disjoint open rectangloids. Two cells are adjacent if their closures share a portion of an edge. Each cell is compared to the C-constraints defining the C-obstacles and is labelled **EMPTY**, **FULL**, or **MIXED**, depending on whether it lies completely inside $\mathcal{C}_{free}$, completely outside, or partly inside and outside. The problem of finding a free path is thus transformed into the problem of finding a sequence of adjacent **EMPTY** cells. If no such sequence is found, **MIXED** cells may be decomposed further if there is some chance that a free path traverse them. This leads to an interesting hierarchical decomposition of configuration space, which allows to adapt the size of the cells to the occupancy of $\mathcal{C}$ by C-obstacles.

Unlike exact decomposition methods, the methods described below are not based on a preliminary analysis of critical surves or surfaces. Hence, the boundaries of the cells which they produce have no 'physical' meaning. Instead, the rationale of approximate decomposition methods is the simplicity of the shape of the cells in order to allow repetition of elementary computations.

As with exact decomposition, we begin our presentation of approximate decomposition methods with the simple case where the moving object $\mathcal{A}$ is a polygon only allowed to translate among fixed polygonal obstacles. Each C-obstacle $\mathcal{CB}_i$, $i = 1$ to q, is a convex polygon that can be represented as a conjunction of C-constraints (see Section 2):

$$(x, y) \in \mathcal{CB}_i \Leftrightarrow \bigwedge_k (a_{ik}x + b_{ik}y + c_{ik} \leq 0).$$

As in Section 3, we assume that $\mathcal{C}_{free} = int(R) - \bigcup_i \mathcal{CB}_i$, where R is a given rectangle whose faces are oriented according to the x and y-axes embedded in $\mathcal{C}$. The method described below is based on an approach first introduced by Brooks and Lozano-Pérez [31,7,8].

Consider an initial configuration $\mathbf{c}_1$ and a final configuration $\mathbf{c}_2$, both in free space. We want to generate a free path from $\mathbf{c}_1$ to $\mathbf{c}_2$. The overall path finding algorithm proceeds in a hierarchical fashion as follows:

1. Generate an initial decomposition of R into disjoint open rectangular cells whose closed union is equal to $cl(R)$, and label every cell by **EMPTY**, **FULL**, or **MIXED**.

2. For $i = 1, 2$ do: while the cell c containing $\mathbf{c}_i$ is **MIXED** and larger than a preset size, decompose c into smaller cells and label the new cells; at the end of the iteration, if c is still not **EMPTY**, return failure; otherwise, if $i = 1$, c is called the *initial cell* and if $i = 2$, it is called the *final cell.*

3. Build the connectivity graph $CG_{\mathbf{EMPTY}}$ representing the adjacency relation between **EMPTY** cells and search it for a path from the initial cell to the final cell.

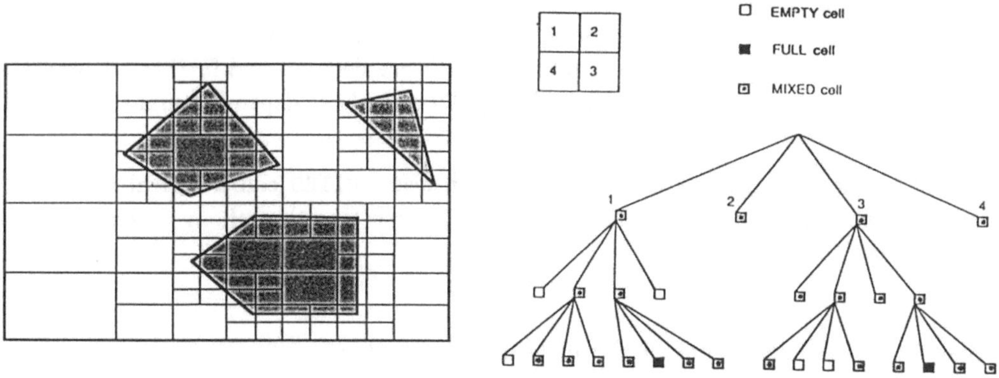

Figure 8: **Example of a Quadtree**

4. If the search of Step 3 terminates successfully, return the corresponding sequence of cells.

5. If the search of Step 3 terminates unsuccessfully, consider all **EMPTY** cells and those **MIXED** cells which are larger than a preset size. Build the connectivity graph $CG_{\texttt{MIXED}}$ representing the adjacency relation between these cells, and search it for a path from the initial cell to the final cell.

6. If the search of Step 5 terminates successfully, decompose the **MIXED** cells along the path into smaller cells, label these new cells, and go back to Step 3.

7. If the search of Step 5 terminates unsuccessfully, return failure.

If successful, this algorithm returns a sequence of rectangular cells. A free path can easily be produced by linking $\mathbf{c}_1$ to $\mathbf{c}_2$ by a succession of straight segments passing through the mid-points of the edges separating consecutive cells.

The hierarchical decomposition of R can be done by recursively dividing R and the successively obtained **MIXED** cells into four identical rectangles. Such a decomposition produces a quadtree [46,20,34,4].

DEFINITION 15: *A* **quadtree** *is a tree of degree 4. Each node represents a rectangle, and is labelled* **EMPTY**, **FULL**, *or* **MIXED**. *A node may have successors only if it is labelled* **MIXED**. *The four successors of a* **MIXED** *node representing a rectangle r represent four rectangles of the same size obtained by decomposing r.*

Figure 8 shows the quadtree decomposition of a simple configuration space and a portion of the corresponding quadtree graph. The depth of a node in the quadtree determines the size of the corresponding cell. The height of the quadtree determines the resolution of the decomposition. During the construction of the quadtree, it is easy to maintain a representation of the adjacency relation between the leaves of the quadtree.

In the above algorithm we must specify the resolution σ_{init} of the initial decomposition. All **MIXED** cells whose depth is less or equal to σ_{init} are decomposed. Later on only those **MIXED** cells which are located on a potential path are decomposed further. A maximal resolution σ_{max} must also be specified. No **MIXED** cell whose depth is equal to σ_{max} gets decomposed further.

In order to label cells, one can associate with every **MIXED** cell a logical expression, called **C-sentence**, specifying the C-obstacles intersecting the cell. A C-sentence is of the form [8]:

$$\bigvee_i \bigwedge_k e_{ik}$$

where e_{ik} is a C-constraint of the form $a_{ik}x + b_{ik}y + c_{ik} \leq 0$. Each conjunct $\bigwedge_k e_{ik}$ corresponds to a particular C-obstacle $\mathcal{CB}_i$. When a cell c is decomposed into smaller cells the C-sentence associated with c is used to label each new **MIXED** cell. During the labelling process (described below), it may get simplified and the simplified C-sentence is associated to the new cell. The rectangle R may be regarded as the unique successor of a fictitious cell whose C-sentence is the disjunction of the expressions describing the C-obstacles.

The algorithm for labelling a new cell and simplifying the C-sentence is shown below [8]. Its input parameters are c and S. c designates the new cell. S designates a C-sentence, whose initial value is the C-sentence associated with the parent of c. The cell c is said to be **inside** a C-constraint e_{ik} if all points (x, y) in c satisfy e_{ik}; it is **outside** e_{ik} if all points (x, y) in c contradict e_{ik}; it is **cut** by the C-constraint if it is neither inside, nor outside e_{ik}. The computation for testing whether a cell c is inside a C-constraint e_{ik}, or outside, or neither is quite simple:
- c is inside e_{ik} if its four vertices verify: $a_{ik}x + b_{ik}y + c_{ik} \leq 0$.
- c is outside e_{ik} if its four vertices verify: $a_{ik}x + b_{ik}y + c_{ik} \geq 0$.
- c is cut by e_{ik} otherwise.

```
procedure LABEL(c,S)
for each conjunct Λ in S do
      for each C-constraint e in Λ do
            if c is inside e
                  then remove e from Λ;
                  else if c is outside e
                        then  remove Λ from S;
                              exit from loop;
                  endif;
            endif;
      enddo;
      if Λ is empty
            then  label c with FULL;
                  exit from procedure;
      endif;
enddo;
if S is empty
```

then label c with **EMPTY**;
else label c with **MIXED**;
associate S with c;
endif;
endprocedure;

Attachment of simplified C-sentences to **MIXED** cells may considerably accelerate the labelling process, since the number of C-constraints to be considered usually decreases rapidly with the size of the cell.

Notice, however, that the procedure LABEL is somewhat conservative. Indeed, it treats each line $a_{ik}x + b_{ik}y + c_{ik} = 0$, as an infinite line, independently of the other lines, and therefore some cells may get labelled **MIXED**, while they do not actually intersect any C-obstacle. However, such an 'error', which never leads to generate incorrect paths, is ultimately corrected, when the cell gets decomposed further.

Unlike exact decomposition methods, the above method is not complete since it cannot be guaranteed to find a free path, whenever such a path exists. This is due to the fact that it approximates free space as a collection of cells of predefined shape. Thus, at any resolution, the collection of **EMPTY** cells is a conservative approximation of $\mathcal{C}_{free}$, i.e. the closed union of the **EMPTY** cells is included in $\mathcal{C}_{free}$. However, at the eventual expense of generating a very large number of cells, the representation can be as close as we want from the exact free space by setting σ_{max} appropriately. The above method guarantees that every **MIXED** cell which may contain a portion of a free path and whose depth is less than σ_{max} will be decomposed into smaller cells if a free path is not found before. Thus, the overall path finding method is said to be **resolution-complete**.

The above approximate decomposition method can easily be extended to the three-dimensional case. The translating object $\mathcal{A}$ and the obstacles are polyhedra. The C-obstacles $\mathcal{CB}_i$ are convex polyhedra. We assume that $\mathcal{C}_{free} = int(R) - \bigcup_i \mathcal{CB}_i$, where R is a parallelepiped whose faces are oriented according to the x, y, and z-axes embedded in $\mathcal{C}$. The path finding method remains basically the same. Only the representation of the decomposition of R is changed into a tree of degree 8, called an **octree** [20,34,4]. An octree is obtained by recursively decomposing a parallelepiped into 8 identical parallelepipeds. Comparing a cell c of an octree to a C-constraint, now of the form $a_{ik}x + b_{ik}y + c_{ik}z + d_{ik} \leq 0$, requires checking the 8 vertices against the constraint.

Notice that, although the extension to the three-dimensional case is conceptually straightforward, the number of cells in an octree augments much quicker with the height of the tree than the number of cells in a quadtree.

7 Approximate Decomposition with Rotation

Extending the previous method to the case of a moving polygonal/polyhedral object $\mathcal{A}$ allowed to both translate and rotate among polygonal/polyhedral obstacles is not difficult, conceptually. However, C-constraints are more complex than in the translational case. Indeed, each of them is applicable only for some orientations of $\mathcal{A}$. Moreover, C-constraints determine curved C-surfaces.

In Subsection 7.1, we directly extend the hierarchical cell decomposition method described above to the rotational case. In Subsections 7.2 and 7.3, we present variants of the approximate decomposition approach to path finding.

We only treat the case of a two-dimensional workspace. Extension to the three-dimensional case is usually not difficult, but may result into impractical algorithms. Indeed, the number of cells to be generated is roughly exponential in the dimension of $\mathcal{C}$. One practical way to deal with a three-dimensional object allowed to both translate and rotate is to limit the number degrees of freedom to be considered at every position. For example, one could allow the object to rotate only in some intermediate regions. There is no general method, however, to determine these intermediate regions.

Hence, throughout this section $\mathcal{C} = \Re^2 \times S^1$. For practical reasons, we will represent it as $\Re^2 \times [0, 2\pi]$. We will also assume that the range of free positions of $\mathcal{A}$ is bounded by a rectangle $R \subset \Re^2$. Thus, the representation of $\mathcal{C}_{free}$ is bounded by the $R \times [0, 2\pi]$ parallelepiped.

7.1 Hierarchical Cell Decomposition

The approximate cell decomposition of the previous section can readily be extended to the case of an object allowed to both translate and rotate in a two-dimensional workspace. Indeed, we can represent the three-dimensional parallellepiped $R \times [0, 2\pi]$ by an octree and apply the same path finding and cell labelling algorithms. However, there are some differences related to how cells are to be compared to C-constraints.

Each C-obstacle $\mathcal{CB}_i$ now determines in $R \times [0, 2\pi]$ a volume bounded by patches of ruled surfaces. We have:

$$(x, y, \theta) \in \mathcal{CB}_i \Leftrightarrow \bigwedge_k [\theta \in \delta_{ik} \Rightarrow (a_{ik}(\theta)x + b_{ik}(\theta)y + c_{ik}(\theta) \leq 0)]$$

where δ_{ik} is the angular interval in which the C-constraint in the right-hand side applies (see Section 2). Thus, the C-sentence associated with each **MIXED** cell still has the form:

$$\bigvee_i \bigwedge_k e_{ik}$$

but e_{ik} now denotes a conditional C-constraint of the form $\theta \in \delta_{ik} \Rightarrow (a_{ik}(\theta)x + b_{ik}(\theta)y + c_{ik}(\theta) \leq 0)$.

Let $c = (x_c, x'_c) \times (y_c, y'_c) \times (\theta_c, \theta'_c)$ be a parallelepipedic cell. c is **inside** e_{ik} if and only if:
- $(\theta_c, \theta'_c) \subset \delta_{ik}$ (i.e., the constraint applies in the whole cell),
- and $\forall (x, y, \theta) \in c$: $a_{ik}(\theta)x + b_{ik}(\theta)y + c_{ik}(\theta) \leq 0$.
It is **outside** e_{ik} if and only if:
- $(\theta_c, \theta'_c) \cap \delta_{ik} = \emptyset$ (i.e., the constraint applies nowhere in the cell),
- or $\forall (x, y, \theta) \in (x_c, x'_c) \times (y_c, y'_c) \times [(\theta_c, \theta'_c) \cap \delta_{ik}]$: $a_{ik}(\theta)x + b_{ik}(\theta)y + c_{ik}(\theta) > 0$.

This means that we have to compare a parallelepiped $(x_c, x'_c) \times (y_c, y'_c) \times (\theta_1, \theta_2) \subseteq c$, where $(\theta_1, \theta_2) = (\theta_c, \theta'_c) \cap \delta_{ik}$, to a constraint $a_{ik}(\theta)x + b_{ik}(\theta)y + c_{ik}(\theta) \leq 0$.

Consider the projection on the (x, y)-plane of the portion of the C-surface $a_{ik}(\theta)x + b_{ik}(\theta)y + c_{ik}(\theta) = 0$, which is comprised within the interval (θ_1, θ_2). Let (x, y) be a point outside this projection; the segment connecting (x, y, θ_1) to (x, y, θ_2) is completely inside or completely outside the constraints $a_{ik}(\theta)x + b_{ik}(\theta) + c_{ik}(\theta) \leq 0$. Inversely, let (x, y) be a point inside the projection of the surface; the segment connecting (x, y, θ_1) to (x, y, θ_2) crosses the C-surface, so that part of it lies inside the constraint and part of it lies outside. Thus, we can compare a cell to a constraint by first projecting the (θ_1, θ_2) slice of the C-surface on the (x, y)-plane, and then by comparing a rectangle to this projection. The computations necessary for this comparaison are rather straightforward. We refer the reader to [8] for a detailed description.

Extending the above method to a polyhedron allowed to both translate and rotate in three-dimensional workspace requires to approximate a six-dimensional configuration space by a 2^6-tree (a quadtree is a 2^2-tree, an octree is a 2^3-tree). This is very likely to be impractical. Indeed, in such a tree, the number of cells at depth 1 is 64; at depth 2, it is 4096, at depth 3, it is 262144, etc. And at depth 3, each dimension is divided in only 8 intevals; for the angular axis, these intervals have an amplitude of $\pi/4$. Except in very favorable situations, such a coarse resolution prevents the method from finding a free path.

7.2 Orientation Slicing

This method, due to Lozano-Pérez [31,32], is based on a decomposition of the range $[0, 2\pi]$ of possible orientations of $\mathcal{A}$ into a finite number of intervals. The basic idea is to consider the areas swept out by $\mathcal{A}$ in all these intervals (see Figure 9) and to treat these areas as new objects only allowed to translate. A path is generated as a sequence of sub-paths, each for one of these objects, such that any two consecutive sub-paths corresponds to adjacent angular intervals.

The overall algorithm is the following:

1. Decompose the range of orientations of $\mathcal{A}$, $[0, 2\pi]$, into a finite numbe of consecutive closed intervals $[\theta_k, \theta'_k]$, $k = 1$ to p, such that $\theta_1 = 0$, $\theta'_k = \theta_{k+1}$ (for all $k \in [1, p-1]$), and $\theta'_p = 2\pi$ (θ_k is allowed to be equal to θ'_k).

2. For every $k \in [1, p]$, approximate the area swept out by $\mathcal{A}$ when its position (x, y) is fixed and when its orientation θ varies from θ_k to θ'_k by a bounding convex polygon denoted $\mathcal{SA}_k$. Attach a frame $\mathcal{F}_{\mathcal{SA}_k}$ to $\mathcal{SA}_k$, which coincides with $\mathcal{F}_{\mathcal{A}}$ when $\mathcal{A}$'s orientation is θ.

3. For every $k \in [1, p]$, treat the polygon $\mathcal{SA}_k$ as a moving object only allowed to translate. The configuration space of $\mathcal{SA}_k$ is $\mathcal{C}^k = \Re^2$. Compute the C-obstacles $\mathcal{CB}_{ki}$ (convex polygons) in $\mathcal{C}^k$.

4. For every $k \in [1, p]$, decompose $\mathcal{C}^k_{free}$ into convex polygons, called **cells**, and construct the connectivity graph CG_k representing the adjacency relation between these cells. If θ_1 (resp. θ_2) $\in [\theta_k, \theta'_k]$, then label the cell containing (x_1, y_1) (resp. (x_2, y_2)) as the *initial* (resp. *final*) cell.

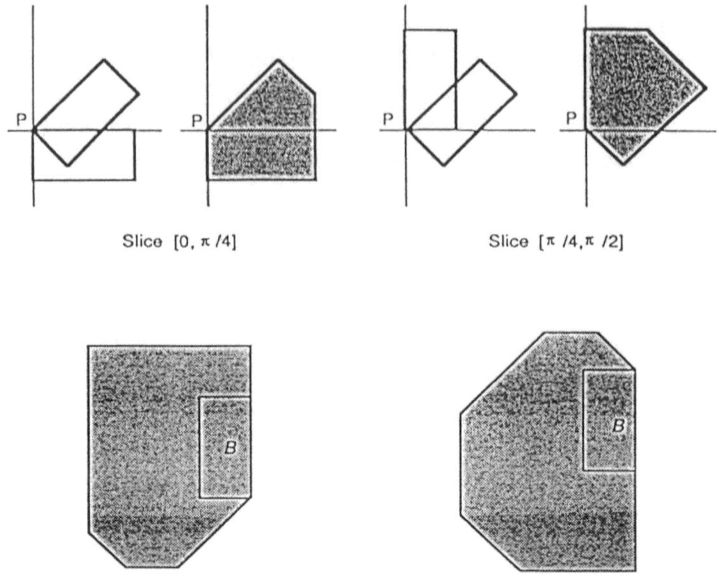

Figure 9: **Illustration of the Orientation Slicing Method**

5. Combine the graphs CG_k into a new connectivity graph CG by linking any two nodes $X_1 \in CG_{k_1}$ and $X_2 \in CG_{k_2}$ $(k_1 \neq k_2)$, whenever these two nodes represent overlapping cells when $\mathcal{C}^{k_1}$ and $\mathcal{C}^{k_2}$ are superposed, and $\theta'_{k_1} = \theta_{k_2}$ (modulo 2π).

6. Search CG for a path linking the initial cell to the final cell. If the search terminates successfully, then return the sequence of cells along the path; otherwise, return failure.

We can regard $\mathcal{CB}_{k,i}$ as a bounding polygonal approximation of the projection of the slice through $\mathcal{CB}_i$ comprised between θ_k and θ'_k. Thus $\mathcal{CB}_{k,i} \times [\theta_k, \theta'_k]$ is a prism bounding $\mathcal{CB}_i$ within the $[\theta_k, \theta'_k]$ slice. Any two cells in CG_k can be adjacent only by their vertical faces. Two cells in CG, which are not contained in the same slice, can be adjacent if they belong to two adjacent slices and if the intersection of the top face of one with the bottom face of the other is a two-dimensional region.

The above method may fail to find a free path if the intervals $[\theta_k, \theta'_k]$ are too large. In case of failure, it is possible to refine these intervals and to run the method again. However, unlike with the method of Subsection 7.1, there is no obvious information that can be used to select which intervals has to be refined. In addition, there is no simple way to capitalize on the computations performed for one interval in order to simplify future computations when this interval gets decomposed. Therefore, refining all intervals may result in excessive computing time.

Extension of the above method to a polyhedral object moving in a three-dimensional workspace is simple in principle, but impractical in most cases.

7.3 Position Partitioning

This method, whose principle is due to Germain [19], has several similarities with the exact decomposition method described in Section 4. However, it does not make use of the critical curves defined there.

Assume, without loss of generality, that the origin of $\mathcal{F}_\mathcal{A}$ is in $\mathcal{A}$. Then, the set of possibly admissible position of $\mathcal{A}$ (i.e., the set of positions which admit free orientations of $\mathcal{A}$) is the empty subset of the workspace, $\mathcal{E} = int(R) - \bigcup \mathcal{B}_i$. The position partitioning method consists of decomposing this set into a finite collection of open convex polygons R_k, $k = 1$ to p, such that $\bigcup_k cl(R_k) \subseteq cl(\mathcal{E})$ and $\forall k \neq k'$ $R_k \cap R_{k'} = \emptyset$. For each polygon R_k, we compute the range of orientations of $\mathcal{A}$ which are free for all positions of $\mathcal{A}$ in R_k. Let us denote $F(R_k)$ this range. Then, we approximate $\mathcal{C}_{free}$ by prismatic cells of the form $R_k \times I_{kl}$, where $I_{kl} = (\theta_{kl}, \theta'_{kl})$ is a maximal connected interval in $F(R_k)$. Finally, we construct the connectivity graph among these cells and we search the graph. All these computations are described in more detail below.

The decomposition of $\mathcal{E}$ into convex polygons may be an exact decomposition – then, $\bigcup_k cl(R_k) = cl(\mathcal{E})$ – or it may be an approximate decomposition – then, $\bigcup_k cl(R_k) \subset cl(\mathcal{E})$. Approximate decomposition makes it possible using only very simple polygons, e.g. rectangles.

The computation of $F(R_k)$ is the core of the position partitioning method. Below, we give two lemmas which reduce this computation to the computation of the set of valid orientations of $\mathcal{A}$ at a single position among transformed obstacles.

The set of valid orientations of $\mathcal{A}$ at (x, y) is $V(x, y) = \{\theta \;/\; (x, y, \theta) \in \mathcal{C}_{valid}\}$. For any $R_k \subset \mathcal{E}$, we define:

$$V(R_k) = \bigcap_{(x,y)\in R_k} V(x, y).$$

The first lemma states that the set of free orientations of $\mathcal{A}$ over the open region R_k is equal to the set of valid orientations of $\mathcal{A}$ over the closure of R_k. For more clarity, we denote $\overline{R_k}$ the closure of R_k. Thus: $\overline{R_k} = cl(R_k)$.

LEMMA 4: $F(R_k) = V(\overline{R_k})$.

Let $V_\mathcal{B}(x, y) = \{\theta \;/\; int(\mathcal{A}(x, y, \theta)) \cap \mathcal{B} = \emptyset\}$, where $\mathcal{B}$ is a closed region of $\Re^2$. Thus, the set of valid orientations of $\mathcal{A}$ at position (x, y) among the obstacles $\mathcal{B}_i$ is $V(x, y) = V_{\bigcup_i \mathcal{B}_i}(x, y)$. The following lemma reduces the computation of $V(\overline{R_k})$ to the computation of $V_{\mathcal{B}'}(x_k, y_k)$ at a single point (x_k, y_k) defined with respect to R_k, for a certain region $\mathcal{B}'$ obtained by 'growing' $\mathcal{B} = \bigcup_i \mathcal{B}_i$ by $\ominus \overline{R_k}$. This is illustrated by Figure 10 in the typical case where (x_k, y_k) is a vertex of R_k.

LEMMA 5: *Let* $\mathcal{B} = \bigcup_i \mathcal{B}_i$. *We have*

$$V(\overline{R_k}) = V_{\mathcal{B} \ominus \overline{R_k}(0,0)}(x_k, y_k)$$

where (x_k, y_k) *is a point fixed relatively to* R_k, $\overline{R_k}(0,0)$ *designates the polygon* $\overline{R_k}$ *obtained by translating the point* (x_k, y_k) *at the origin of the* (x, y)*-plane, and* $\ominus$ *is the*

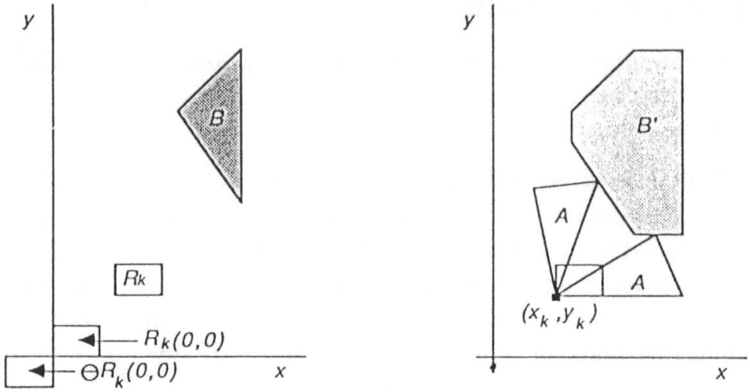

Figure 10: **Growing an Obstacle by a Cell**

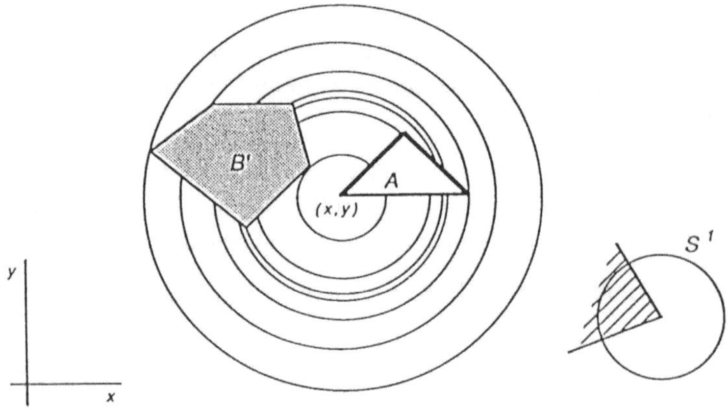

Figure 11: **Computation of $V_{\mathcal{B}'}(x,y)$**

Minkowski operator for set difference.

The region $\mathcal{B} \ominus \overline{R_k}(0,0)$ is a closed polygonal region if $\Re^2$. Therefore, $V_{\mathcal{B}\ominus\overline{R_k}(0,0)}(x_k, y_k)$ (hence $F(R_k)$) is a finite union of closed angular intervals. Some of these intervals may consist of a single orientation. It is easy to compute $V_{\mathcal{B}'}(x,y)$ using a technique inspired from the line-sweep paradigm, after having determined the intersections of the circles centered at (x,y) and passing through the vertices of $\mathcal{A}$ (resp. $\mathcal{B}'$), with the edges of $\mathcal{B}'$ (resp. $\mathcal{A}$) [33], as Figure 11 illustrates.

At this point, we know how to approximate free space as a collection of parallelepipedic cells of the form $R_k \times I_{kl}$. Two cells $R_k \times I_{kl}$ and $R_{k'} \times I_{k'l'}$ are adjacent if and only if:
- The boundaries of R_k and $R_{k'}$ share an open portion of an edge.
- $I_{kl} \cap I_{k'l'}$ is an interval of non-null measure.
It is easy to verify that two adjacent cells can be connected by a free path.

Path planning consists of searching the connectivity graph CG representing the adjacency relation between cells.

This method may fail to generate a free path while one exists. In case of failure, one may attempt to refine the decomposition of the empty space $\mathcal{E}$ and run again the method. However, like with the orientation slicing method, no simple information is available to determine which part of the decomposition to refine. Clearly, there is a tradeoff between the resolution of the decomposition and the computing cost.

Extending the position partitioning method to the case of a polyhedron moving in a three-dimensional workspace is delicate. Indeed, the set $V_{\mathcal{B}\ominus\overline{R_k}(0,0)}(x_k, y_k, z_k)$ would be more difficult to characterize and of course to compute.

8 Retraction Method (Principle)

An alternative to the decomposition approach to path planning is the retraction approach. The basic idea is to approximate free space as a net of curves contained in free space and describing the topology of free space. A mapping (retraction) associates a unique point on these curves to any free configuration. Once the net of curves has been generated, path planning is reduced to the combinatorial search problem of finding a path in the net from the retraction of the initial configuration to the retraction of the final configuration.

In this section, we present the theoretical principle of retraction in the simple case of a polygonal object $\mathcal{A}$ translating among polygonal obstacles; in this case, the net of curves is a Voronoi diagram, a well-known construct in Computational geometry [17,23,28]. In the next section, we will present an ad-hoc method based on the retraction idea, applicable to the case of a two-dimensional object translating and rotating among polygonal obstacles. Additional information on the retraction approach can be found in the literature (e.g. [37,38,39,40,10]).

The configuration space of $\mathcal{A}$ is $\mathcal{C} = \Re^2$, and the C-obstacles are (possibly overlapping) convex polygons $\mathcal{CB}_i$. Assume that the range of possible positions of $\mathcal{A}$ is bounded by a polygon R. Thus: $\mathcal{C}_{free} = R - \bigcup_i \mathcal{CB}_i$. The boundary of $\mathcal{C}_{free}$ is represented as a set of vertices and open edges.

DEFINITION 16: *Let $\beta = \partial(\mathcal{C}_{free})$. For any $\mathbf{c} \in \mathcal{C}_{free}$, let:*

$$clearance(\mathbf{c}) = \min_{(x,y)\in\beta} distance(\mathbf{c}, (x, y))$$

and

$$near(\mathbf{c}) = \{(x, y) \;/\; distance(\mathbf{c}, (x, y)) = clearance(\mathbf{c})\}$$

where 'distance' denotes the Euclidean distance in $\Re^2$. The **Voronoi diagram** *of $\mathcal{C}_{free}$ is the set $\{\mathbf{c} \in \mathcal{C}_{free} \;/\; card(near(\mathbf{c})) > 1\}$, where $card(E)$ is the cardinal of set E. It is denoted $Vor(\mathcal{C}_{free})$.*

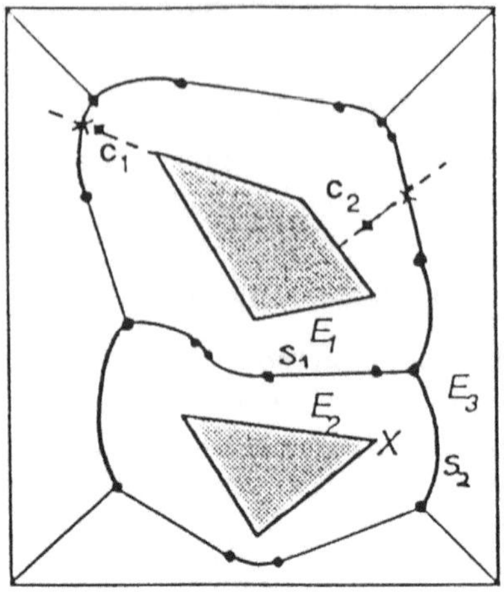

Figure 12: **Example of Voronoi Diagram**

Figure 12 shows a simple Voronoi diagram in a configuration space containing two C-obstacles. One can see that $Vor(\mathcal{C}_{free})$ consists of a finite collection of straight and parabolic curve segments. A straight segment in $Vor(\mathcal{C}_{free})$ is the set of configurations that are closest from the same pair of edges. A parabolic segment is the set of configurations that are closest from the same pair consisting of one edge and a vertex. The pair (edge,edge) or (edge,vertex) of closest elements from any configuration on a straight or parabolic segment determines the equation of the curve supporting the segment. Each straight or parabolic segment is called an **arc** of the Voronoi diagram. Each endpoint of a segment is called an **edge** of the diagram.

Let us now consider a free configuration $\mathbf{c} \notin Vor(\mathcal{C}_{free})$. There exists a unique edge or vertex Z of $\beta = cl(\mathcal{C}_{free})$ which is closer from $\mathbf{c}$ than any other edge or vertex in β. Let $p \in \beta$ be the point such that $clearance(\mathbf{c}) = distance(\mathbf{c}, p)$. If Z is an edge then $p \in Z$; if it is a vertex then $p = Z$. Consider the line L passing through $\mathbf{c}$ and p. If we move along L away from p and beyond $\mathbf{c}$, then $clearance(\mathbf{c})$ increases at maximal rates[11] until it reaches a point in $Vor(\mathcal{C}_{free})$. This point, denoted $Im(\mathbf{c})$, is the **image** of $\mathbf{c}$ on the Voronoi diagram. Passing this point, either $clearance(\mathbf{c})$ drecreases, or the direction of maximum positive increase changes. The function Im can be extended to all $\mathcal{C}_{free}$ by putting $Im(\mathbf{c}) = \mathbf{c}$ for all $\mathbf{c} \in Vor(\mathcal{C}_{free})$. It is easy to verify that the function Im is continuous. In topology a continuous function $f : E \rightarrow F \subset E$, whose restriction to F is the identity is called a *retraction* [21].

Let $\tau : [0,1] \rightarrow \mathcal{C}_{free}$ be a free path between two free configurations $\mathbf{c}_1$ and $\mathbf{c}_2$. Thanks to the continuity of Im, $Im \circ \tau : [0,1] \rightarrow Vor(\mathcal{C}_{free})$ is a path between $Im(\mathbf{c}_1)$ and $Im(\mathbf{c}_2)$. This gives the following theorem, which expresses that the connectivity of $\mathcal{C}_{free}$ and the connectivity of $Vor(\mathcal{C}_{free})$ are 'equivalent':

THEOREM 5: *Let $\mathcal{C}_{free}$ be the interior of a polygonal region with polygonal holes. Let $\mathbf{c}_1$ and $\mathbf{c}_2$ be two configurations in $\mathcal{C}_{free}$. There exists a free path from $\mathbf{c}_1$ to $\mathbf{c}_2$ if and*

[11]It is easy to verify that $\nabla clearance(\mathbf{c})$ is a vector supported by L and pointing away from p.

only if there exists a path from $Im(\mathbf{c}_1)$ to $Im(\mathbf{c}_2)$ with $Vor(C_{free})$.

The path finding algorithm known as the ***retraction method*** derives directly from the above theorem. It consists of the following steps:

1. Compute the Voronoi diagram $Vor(C_{free})$.
2. Compute the points $Im(\mathbf{c}_1)$ and $Im(\mathbf{c}_2)$ and identify the edges of the Voronoi diagram containing these two points.
3. Search $Vor(C_{free})$ for a sequence of arcs $e_1, ..., e_q$, such that $Im(\mathbf{c}_1) \in e_1$, $Im(\mathbf{c}_2) \in e_2$, and $\forall i \in [1, q-1]$ e_i and e_{i+1} have a common end-point.
4. If the search terminates successfully return $Im(\mathbf{c}_1)$, $Im(\mathbf{c}_2)$, and the sequence of arcs along the path in $Vor(C_{free})$; otherwise, return failure.

There exist algorithms for computing the Voronoi diagram in $O(n \log^2 n)$ time (see [26,28]) and in $O(n \log n)$ time (see [23]), where n is the number of vertices in $\beta = \partial(C_{free})$. The number of arcs in the diagram is $O(n)$. At Step 2, for each configuration $\mathbf{c}_i$ $(i = 1, 2)$, one can:

- First, compute both the distance of $\mathbf{c}_i$ to every edge and every vertex of β, and the point p_i of β to which it is closest,

- Second, determine the intersections of the ray drawn from p_i through $\mathbf{c}_i$ with the $Vor(C_{free})$, and select the intersection which is closest from p_i.

These computations require $O(n)$ time. Finally, the search at Step 3 can be done using the A^* algorithm and takes $O(n)$ time. Thus the overall time complexity of this path planning method is $O(n \log^2 n)$ or $O(n \log n)$. The most expensive step is Step 1, but it depends on C_{free} only. Subsequent path finding in the same free space between other pairs of configurations can be done omitting Step 1 and takes $O(n)$ time. In addition, there exists linear techniques to update the Voronoi diagram when a few C-obstacles are changed.

9 Freeway Method

The path finding method presented in this section is an ad hoc retraction-like method applicable to a robot moving both in translation and rotation in a two-dimensional workspace among fixed polygonal obstacles. The method is due to Brooks [9], whose original goal was to capture the effects of both translating and rotating an object into a more intuitive representation of free space than configuration space. In fact, since configuration space is just a general concept, the 'new' representation can easily be related to configuration space. Instead, it turns out that the actual contribution of the freeway method is that it makes it possible to build a description of the connectivity of C_{free}, known as the *freeway net*, without having to completely compute C-obstacles. The intuition behind the freeway method is similar to the intuition behind applying retraction to path finding, i.e. always keep the moving object as far away as possible

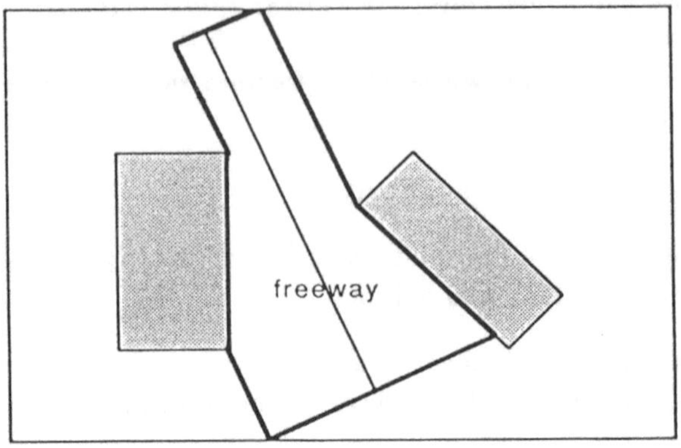

Figure 13: **Example of a Freeway**

from a pair of obstacles. Consequently, the freeway net resembles a Voronoi diagram, but its construction is based on some ad hoc assumptions and it incompletely describes the connectivity of $\mathcal{C}_{free}$, so that the freeway method itself is incomplete. Nevertheless, experiments have shown that it can solve many problems efficiently, when the workspace is not densily occupied by obstacles.

The freeway method consists of extracting geometric figures called ***freeways*** from the workspace, to connect them into a graph called the freeway net, and to search this graph. A freeway is a straight linear generalized cylinder [5] whose axis, the ***spine***, is annotated with a description of the free orientations along the freeway. Figure 13 illustrates the geometry of a freeway between two obstacles in a two-dimensional workspace bounded by a rectangle. Figure 14 shows additional freeways in the same workspace. The moving object $\mathcal{A}$ can be displaced along a freeway (or part of it) if there is a connected set of free orientations of $\mathcal{A}$ along the spine. In addition, whenever two spines intersect, $\mathcal{A}$ can transfer from one freeway to the other if the ranges of free orientations of $\mathcal{A}$ along both spines have non-empty intersection. The freeway net is a representation of the possible motions of $\mathcal{A}$ along spines and between spines. This graph describes (incompletely) the connectivity of $\mathcal{C}_{free}$. A path in this graph determines a class of homotopic free paths of $\mathcal{A}$.

The freeway method successively consists of extracting freeways from the description the obstacles $\mathcal{B}_i$ in the workspace, computing the range of free orientations at some points along the spines of the freeways, constructing the freeway net, and searching this graph for a path. We present these steps in the following subsections. We assume that $\mathcal{A}$ and the $\mathcal{B}_i$'s are modeled as polygonal regions of $\Re^2$ and that the workspace is itself a bounded polygonal region of $\Re^2$. We denote $\mathcal{E}$ the subset of the workspace not occupied by obstacles.

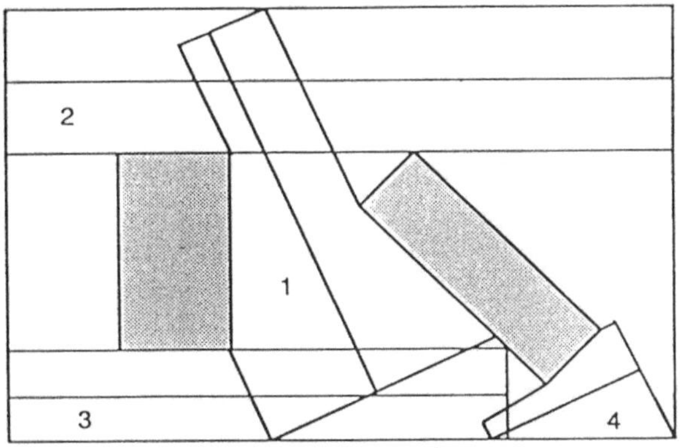

Figure 14: **Overlapping Freeways**

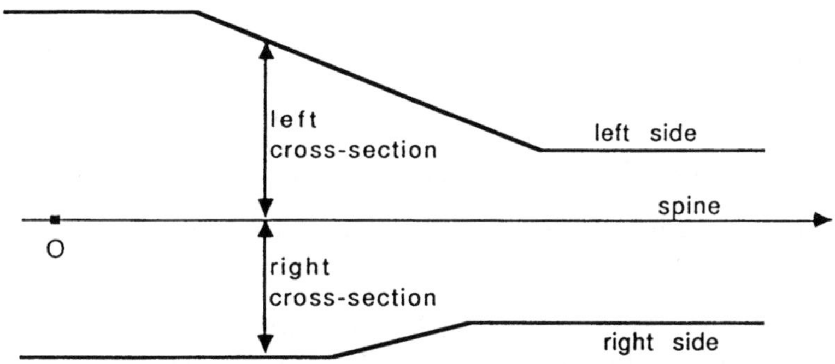

Figure 15: **Two-Dimensional Straight Linear Generalized Cylinder**

9.1 Extraction of Freeways

The geometry of a freeway in $\Re^2$ is that of a truncated two-dimensional straight linear generalized cylinder (see Figure 15).

DEFINITION 17: *A* **two-dimensional straight linear generalized cylinder** *is a region of* $\Re^2$ *obtained by sweeping a straight line segment, the* **cross-section**, *along and perpendicularly to a straight line, the* **spine**. *An origin and an orientation is defined on the spine. The cross-section is partitioned by the spine into two segments, the* **right** *and the* **left** *cross-sections. The lengths of the right and left cross-sections are continuous, piecewise linear functions of the abscissa along the spine. These functions are the right and left* **radii**. *The two lines drawn by the extremities of the cross-section are called the right and left* **sides** *of the cylinder.*

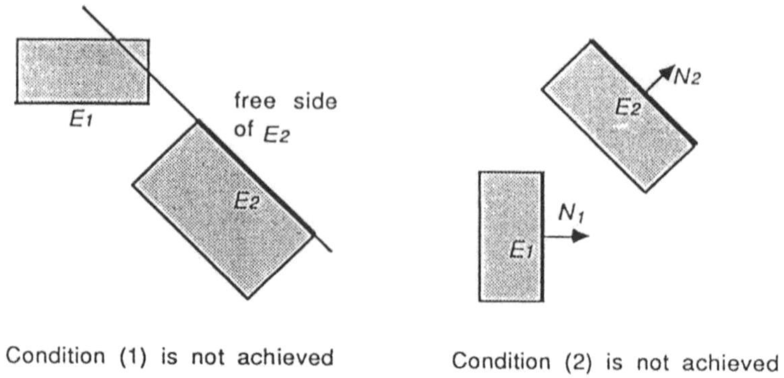

Figure 16: **Pairs of Edges Not Producing a Generalized Cylinder**

Freeways are extracted from $\mathcal{E}$ by considering all pairs of edges in $\partial(\mathcal{E})$. Any edge E is contained in an infinite line that divides $\Re^2$ into two half-planes. The outgoing normal vector of E points toward the half-plane called the *free half-plane* of E. A pair of edges (E_1, E_2) produces a generalized cylinder if and only if it satisfies the following two conditions:

(1) For $i, j \in \{1, 2\}$, $i \neq j$, one extremity of E_i is in the free half-plane of E_j.

(2) The inner product of the outgoing normal vector of E_1 and E_2 is negative.

These two conditions impose that E_1 and E_2 'face' each other. Figure 16 show examples of pairs of edges that do not satisfy these conditions.

Given a pair of edges (E_1, E_2) satisfying the above two conditions, a generalized cylinder (from now on, we omit the other qualifiers) GC is constructed as follows. GC's spine is the bisector of the angle formed by the lines containing E_1 and E_2; if these lines are parallel, the spine is parallel and equidistant to both of them. Each side of GC is made of one edge (E_1 or E_2) extended at each extremity by a half-line parallel to the spine. This construction is illustrated by Figure 17. The choice of the bisector for spine is reminiscent of the construction of a Voronoi diagram in the workspace.

GC partly lies outside $\mathcal{E}$ and we now select those pieces of GC which completely lie within $\mathcal{E}$ as illustrated by Figure 18. First, GC is intersected with the region $\Re^2 - \mathcal{E}$. The intersection is then normally projected on the spine, and the corresponding slices are removed from the cylinder. The truncated GC is thus sliced into several truncated generalized cylinders. Truncated cylinders whose sides do not include portions of both E_1 and E_2 are discarded. In the example of Figure 18, it only remains one slice denoted Φ. This way of removing slices from generalized cylinders is empirical and may seem a bit radical. However, one must keep in mind that there are usually many generalized cylinders leading to many overlapping freeways. In particular, some of the edges bounding the obstacles intersecting GC can usually be paired with either E_1 or E_2 to produce other generalized cylinders overlapping GC. The success of the freeway

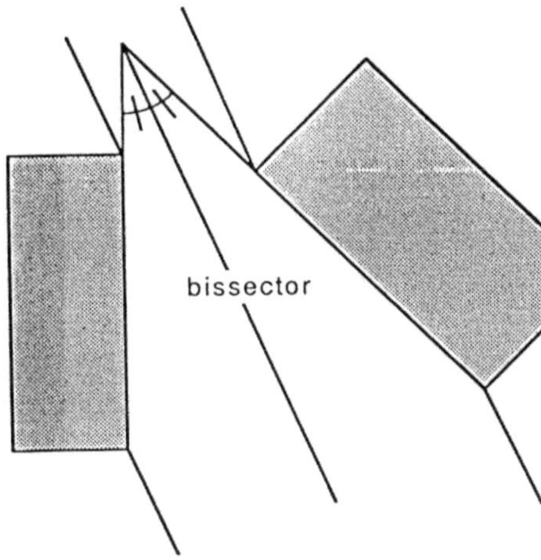

Figure 17: **Construction of a Generalized Cylinder**

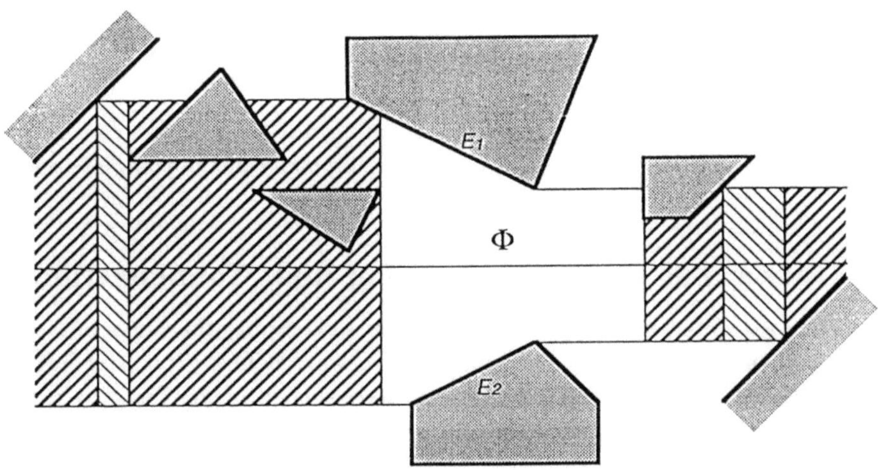

Figure 18: **Removing Slices from a Generalized Cylinder**

method derives from this multiplicity of freeways.

When all the pairs of edges of $\partial(\mathcal{E})$ have been considered, the remaining truncated generalized cylinders are the freeways to be used for path finding.

Each freeway is an instance of the general case shown in Figure 19. The spine of a freeway is oriented from the 'big' end toward the 'small' end; if the freeway has parallel sides, an arbitrary orientation is selected. The origin of spine abscissae is taken at the freeway's end (when it exists, the big end), such that any point in the freeway projects on the spine at a positive abscissa. Thus, the geometry of every freeway is completely specified by 7 parameters illustrated in Figure 19: L, B_l, B_r, S_l, S_r, ϕ and W.

Extraction of the initial non-truncated set of generalized cylinders takes $O(n^2)$, where n is the number of edges in $\partial(\mathcal{E})$. Intersecting $\Re^2 - \mathcal{E}$ with a generalized cylinder can be

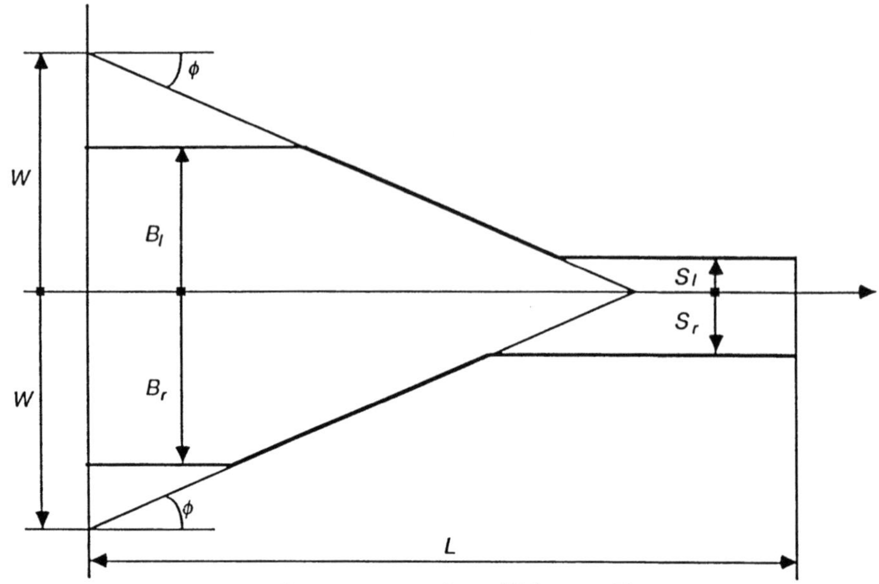

Figure 19: **Parameters Specifying a Freeway**

done in time $O(n)$. If the intersection is not empty, the intersection can be projected on the spine in time $O(n)$. Therefore, the time complexity of the complete operation is $O(n^3)$.

9.2 Freeways as Cross-Sections Through Configuration Space

After having extracted freeways, the next step is to compute the range of free orientations of $\mathcal{A}$ when it moves along the spine of a freeway. Before explaining how this can effectively be done, we must define more accurately what it means for $\mathcal{A}$ to 'move along' the spine.

We represent a configuration **c** of $\mathcal{A}$ by (x, y, θ), where x and y are the coordinates of the origin of $O_{\mathcal{A}}$ (the origin of $\mathcal{F}_{\mathcal{A}}$) in $\mathcal{F}_{\mathcal{W}}$, and $\theta \in [0, 2\pi)$ is the angle (modulo 2π) between the x-axes of $\mathcal{F}_{\mathcal{W}}$ and $\mathcal{F}_{\mathcal{A}}$.

When $\mathcal{A}$ 'moves along' the spine of a freeway, we impose that $O_{\mathcal{A}}$ stays on the spine. In other words, we constraint potential paths in $\mathcal{C}_{free}$ to be contained into planes projecting on the (x, y)-plane of $\mathcal{C}$ along the spines of the extracted freeways. This makes sense since each spine is equally distant from two obstacle edges. Determining the range of free orientations of $\mathcal{A}$ along the spine of a freeway is thus equivalent to computing the intersection of a plane parallel to the θ-axis with C-obstacles. In fact, as we will show later, the range of free orientations has to be determined only at some points on the spines.

Since $O_{\mathcal{A}}$ moves on a line which is equidistant from two obstacle edges, it is preferable to select it so that the maximum distance from $O_{\mathcal{A}}$ to the points on the boundary of $\mathcal{A}$ is the smallest possible. This is achieved by taking $O_{\mathcal{A}}$ at the center of the minimum

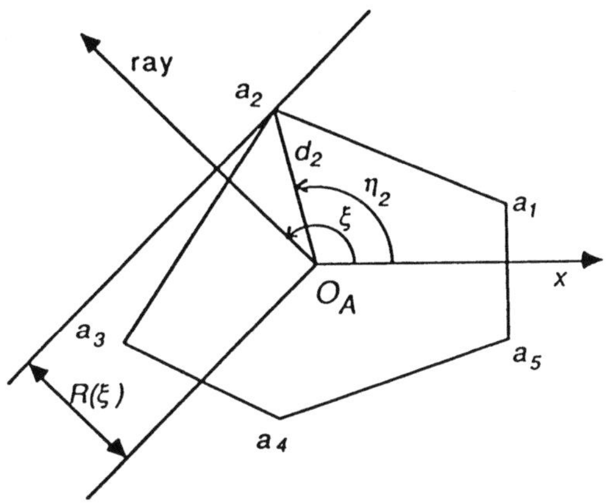

Figure 20: **Radius Function of the Moving Object**

spanning circle[12] of the set of vertices of $\mathcal{A}$. The computation of the minimum spanning circle of a set of points is a classical problem in Computational Geometry [42]. A naive algorithm does the computation in $O(n_{\mathcal{A}}^4)$ time, where $n_{\mathcal{A}}$ is the number of vertices of $\mathcal{A}$. Improved algorithms do the computation in $O(n_{\mathcal{A}}^2)$ [42], or even in $O(n_{\mathcal{A}} \log n_{\mathcal{A}})$ ($O(n_{\mathcal{A}})$ if we already know the convex hull of the points, which is the case when $\mathcal{A}$ is a convex polygon) [11].

9.3 Determination of Free Orientations

We now describe the technique proposed by Brooks for approximating the range of free orientations of $\mathcal{A}$ along a spine of a freeway. This technique makes use of a function, called the radius function of $\mathcal{A}$, and of its inverse. We first define these functions (see Figure 20).

DEFINITION 18: *A half-line issued from $O_{\mathcal{A}}$ is called a* **ray**. *The angle ξ (modulo 2π) between the x-axis of $\mathcal{F}_{\mathcal{A}}$ and the ray is called the* **angle** *of the ray with respect to $\mathcal{A}$. The* **radius function** *$R(\xi)$ of $\mathcal{A}$ is defined as the infimum of the distances from $O_{\mathcal{A}}$ to the lines which both are normal to a ray of angle ξ and do not intersect $\mathcal{A}$.*

We have (see Figure 20):

$$R(\xi) = \max_{1 \leq i \leq n_{\mathcal{A}}} \{d_i \cos(\xi - \eta_i)\}$$

where:
- d_i is the distance from $O_{\mathcal{A}}$ to the ith vertex of $\mathcal{A}$,
- η_i is the angle of the ray passing through the ith vertex of $\mathcal{A}$.

$R(\xi)$ specifies how close a line perpendicular to the ray of angle ξ can be from $O_{\mathcal{A}}$ without intersecting the interior of $\mathcal{A}$. The inverse image of an interval $[0, r]$ by R is

[12] The minimum spanning circle of a set of points is the smallest circle that encloses all of them.

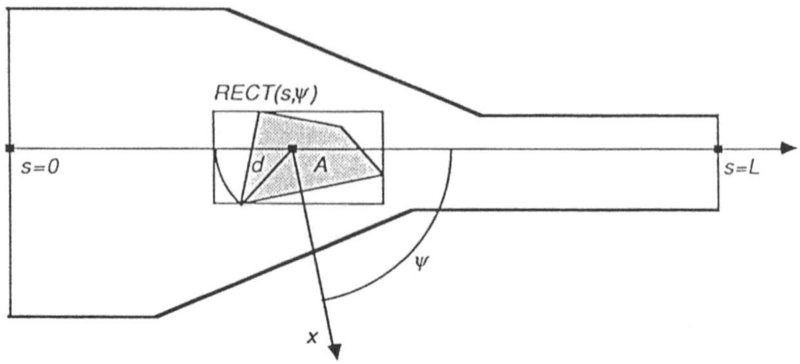

Figure 21: **Bounding $\mathcal{A}$ by a rectangle**

the function $R^{-1}(r)$ defined by:

$$R^{-1}(r) = \{\xi / R(\xi) < r\}.$$

$\forall \xi \in R^{-1}(r)$, a line perpendicular to the ray of angle ξ and distant from $O_{\mathcal{A}}$ by more than r is guaranteed to have no intersection with $\mathcal{A}$.

Let us assume that $\mathcal{A}$ is a convex polygon (otherwise, we approximate it by its convex hull). Then, $R^{-1}(r)$ can easily be computed as follows:

$$R^{-1}(r) = [0, 2\pi) - \bigcup_{1 \leq i \leq n_{\mathcal{A}}} \mathcal{I}_i$$

where:
- $\mathcal{I}_i = \emptyset$ if $r > d_i$,
- $\mathcal{I}_i = [\eta_i - |\arccos\frac{r}{d_i}|, \eta_i + |\arccos\frac{r}{d_i}|]$, modulo 2π, if $r \leq d_i$.

Let us now use the inverse radius function for computing the free orientations of $\mathcal{A}$ along the spine of a freeway Φ. In order to simplify computation, we approximate $\mathcal{A}$ by a bounding rectangle. By doing so, we obtain a subrange of the free orientations of $\mathcal{A}$.

Let us denote s the spine abscissa of $O_{\mathcal{A}}$ and ψ the angle between the x-axis of $\mathcal{F}_{\mathcal{A}}$ and Φ's spine. We enclose $\mathcal{A}$ in a rectangle $RECT(s, \psi)$ defined as follows (see Figure 21). Two sides of $RECT(s, \psi)$ are parallel to the spine, the *right* and *left* ones. The distance from $O_{\mathcal{A}}$ to these sides are $R(\psi - \pi/2)$ and $R(\psi + \pi/2)$, respectively. The other two sides are the *front* and the *rear* sides. The distance from $O_{\mathcal{A}}$ to the front side is $R(\psi)$. The distance from $O_{\mathcal{A}}$ to the rear side is conservatively $d = \max_{1 \leq i \leq n_{\mathcal{A}}}\{d_i\}$. The later choice is conservative, but the independence of d from ψ presents practical advantages as it will be appearent below. Notice that the bounding rectangle $RECT(s, \psi)$ depends on the orientation of $\mathcal{A}$ relative to the spine of Φ.

Let $\mathcal{V}(s)$, with $0 \leq s \leq L$, denote the set of orientations ψ, when $O_{\mathcal{A}}$'s spine abscissa is s, such that $RECT(s, \psi)$ is completely contained within Φ. Since $RECT(s, \psi)$ is a bounding approximation of $\mathcal{A}$, $\mathcal{V}(s)$ is a subset of the range of free orientations of $\mathcal{A}$. $\mathcal{V}(s)$ can be computed using the formulas given in the following lemma.

LEMMA 6: *For a freeway Φ constructed from two non-parallel edges, we have:*

- if $0 \leq s \leq d$, then:

$$\mathcal{V}(s) = \emptyset; \tag{1}$$

- if $d < s \leq L$, then:

$$\begin{aligned}\mathcal{V}(s) = R^{-1}(L-s)\bigcap([(R^{-1}(S_r)+\frac{\pi}{2})\cap(R^{-1}(S_l)-\frac{\pi}{2})]\\ \bigcup[(R^{-1}(B_r)+\tfrac{\pi}{2})\cap(R^{-1}(B_l)-\tfrac{\pi}{2})\cap R^{-1}_{-\frac{\pi}{2},\alpha}(\alpha s+W)\cap R^{-1}_{\frac{\pi}{2},\alpha}(\alpha s+W)])\end{aligned} \tag{2}$$

where $\alpha = \tan\phi$, $R_{\zeta,\alpha} = R(\psi+\zeta) - \alpha R(\psi)$, and $R^{-1}_{\zeta,\alpha}(u) = \{\psi \,/\, R_{\zeta,\alpha}(\psi) < u\}$.

In the particular case where Φ is a traditional 1-cylinder of radius W, relation 1 is unchanged and relation 2 simplifies to:

$$\mathcal{V}(s) = R^{-1}(L-s)\cap(R^{-1}(W)+\frac{\pi}{2})\cap(R^{-1}(W)-\frac{\pi}{2}).$$

Relation 1 is obvious, since for $0 \leq s \leq d$, the rear side of $RECT(s,\psi)$ does not lie within Φ, whatever is ψ's value. (Notice here the pratical advantage of the conservative definition of d.) The best way to understand relation 2 is to parse it:

- The first part of the main conjunct, $R^{-1}(L-s)$, expresses the constraint that the abscissa of the front side of the rectangle must be less than L.

- The first sub-expression in the second part of the main conjunct, $(R^{-1}(S_r)+\frac{\pi}{2})\cap(R^{-1}(S_l)-\frac{\pi}{2})$, expresses that the distance from the spine to the light (resp. left) side of Φ is nowhere smaller that S_r (resp. S_l). The conjunction of $R^{-1}(L-s)$ and this sub-expression is a sufficient condition insuring that $RECT(s,\psi)$ lies within Φ (for $s > d$). However, it is a too strong condition.

- The next two elements in relation 2, $(R^{-1}(B_r)+\frac{\pi}{2})\cap(R^{-1}(B_l)-\frac{\pi}{2})$, express the fact that the distance from the spine to the right and left sides of Φ cannot be greater than B_r and B_l, respectively. These two conditions are too weak and remained to be composed with conditions considering the non-parallel sides of Φ.

- The last two elements in relation 2, $R^{-1}_{-\frac{\pi}{2},\alpha}(\alpha s+W)\cap R^{-1}_{\frac{\pi}{2},\alpha}(\alpha s+W)$, express a necessary and sufficient condition that the bounding rectangle be between the two lines containing the non-parallel sides of Φ. Indeed, for each line, we can first write:

$$R(\psi-\frac{\pi}{2}) \leq \alpha(s+R(\psi))+W$$

and

$$R(\psi+\frac{\pi}{2}) \leq \alpha(s+R(\psi))+W$$

where $\alpha = \tan\phi$. The above two elements in 2 directly derive from these relations.

Using formula 2 to compute $\mathcal{V}(s)$ requires being able to compute $R^{-1}_{\zeta,\alpha}(u)$, where $R_{\zeta,\alpha}(\psi) = R(\psi+\zeta) - \alpha R(\psi)$. This can be done easily by noticing that the sum of two radius functions has the same form as a radius function, whose inverse can be computed as shown above. Indeed, if:

$$R(\xi) = \max_{1 \le i \le n} \{d_i \cos(\xi - \eta_i)\}$$

and

$$S(\xi) = \max_{1 \le j \le n'} \{d'_j \cos(\xi - \eta'_j)\}$$

then:

$$R(\xi) + S(\xi) = \max_{1 \le i \le n,\, 1 \le j \le n'} (d_i \cos(\xi - \eta_i) + d'_j \cos(\xi - \eta'_j))$$

where each term inside the 'max' can be written in the form $e_{ij} \cos(\xi - \nu_{ij})$.

Since both the right and left radii of Φ are non increasing functions of the spine abscissa, one can easily verify the important following lemma:

LEMMA 7: $\forall s_1, s_2 \in [0, L] : s_1 \le s_2 \Rightarrow \mathcal{V}(s_2) \subseteq \mathcal{V}(s_1)$.

This property is the key to the construction of the freeway net, because it will permit us to determine free orientations of $\mathcal{A}$ only at some points along freeways' spines.

Notice that $\mathcal{V}(s)$ may consist of several disconnected intervals.

9.4 Construction and Search of the Freeway Net

At this point, we have built a collection of freeways. For each freeway Φ, the expressions giving $\mathcal{V}(s)$ describe the complement of a bounding approximation of the cross-section of C-obstacles in a plane containing the spine of Φ and perpendicular to the (x, y)-plane. We thus have a partial representation of free space. We now use this representation to construct the freeway net FN, which determines all the possible ways for $\mathcal{A}$ to move along spines.

Let $\mathcal{X}$ be the set of every point P such that the spines of two freeways intersect at P inside both freeways. Since there are $O(n^2)$ freeways (n is the number of edges in $\partial(\mathcal{E})$), a straightforward computation of $\mathcal{X}$ requires $O(n^4)$ time.

The nodes of FN are created as follows:

for every freeway Φ do:
  for every point P of Φ's spine that is in $\mathcal{X}$ do:
    - let s be the spine abscissa of P;
    - create one node of FN for every maximal connected interval in $\mathcal{V}(s)$.

The corresponding interval of $\mathcal{V}(s)$ is associated to each node of FN. Since each interval is originally defined with respect to a spine's orientation, an appropriate constant angle has to be added to both extremities of the interval, so that all the intervals are defined with respect to the same reference, for instance the x-axis.

In addition, the initial and goal positions of $O_{\mathcal{A}}$ are used to create the *initial* and *goal* nodes of FN. We assume here, for simplification, that each of these positions is on a

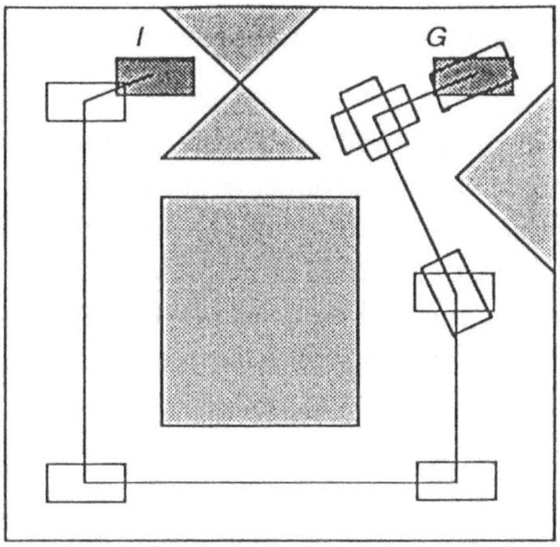

Figure 22: **A Path Produced Using the Freeway Method**

spine. The interval of $\mathcal{V}(s)$ containing the initial (resp. goal) orientation of $\mathcal{A}$ at the initial (resp. goal) position is associated with the corresponding node of FN. If the initial and goal positions of $\mathcal{A}$ are not located on freeway spines, then a technique has to be devised for moving $\mathcal{A}$ from the initial position to a spine, and from a spine to the goal position.

The arcs connecting the nodes of FN are created as follows:

(1) Let N_1 and N_2 be two nodes corresponding to two different points on the same freeway spine at abscissae s_1 and s_2. Let $\mathcal{I}_1$ and $\mathcal{I}_2$ be the two intervals of free orientations of $\mathcal{A}$ associated with these nodes. If $\mathcal{I}_1 \cap \mathcal{I}_2 \neq \emptyset$, then N_1 and N_2 are connected by an arc in FN. Indeed, thanks to Lemma 7, at any orientation in $\mathcal{I}_1 \cap \mathcal{I}_2$, $\mathcal{A}$ can move in free space from abscissa s_1 to abscissa s_2, and vice-versa.

(2) Let N_1 and N_2 be two nodes corresponding to the same point in space on two different spines. Let $\mathcal{I}_1$ and $\mathcal{I}_2$ be the intervals of free orientations associated with these nodes. If $\mathcal{I}_1 \cap \mathcal{I}_2 \neq \emptyset$, then N_1 and N_2 are connected by an arc in FN. Indeed, at the corresponding point, $\mathcal{A}$ can transfer in free space from one of the freeway to the other.

The constructed freeway net can then be searched for a path between the initial node to the goal node. A A^* algorithm can be used [36]; various heuristic functions are possible. A path in the net determines a class of homotopic free paths in C_{free}. A particular path can be selected by rotating $\mathcal{A}$, if necessary, only at the intersections between spines. Figure 22 shows a path produced using this technique (this example is drawn from Brooks's paper [9]).

Experiments with the freeway method has shown that it works fast in 'relatively uncluttered' workspace. However, the method is not complete, and may not always find a free path when one exists. Indeed, it embodies several empirical choices, whose justification is merely intuitive, for instance: the spine of each freeway is the bisector of the two lines containing a pair of edges, each freeway is extended at each vertex by parallels to the spine, $O_{\mathcal{A}}$ is constrained to remain on spines, etc. One of the major drawbacks when the workspace is densely occupied by obstacles, is that freeways are often too short to contain the rectangle enclosing $\mathcal{A}$. Even in rather uncluttered workspaces the freeway method may fail.

Some of the limitations of the method could be alleviated. For example, it is possible to construct additional freeways by introducing fictitious edges in the workspace and by considering spines which are not bisectors of the lines containing edges. But, the size of the freeway net and the cost of searching it would increase significantly. Probably, it is better to take the freeway method as it is, i.e. fast and incomplete, and not try to make it too complicate, i.e. slow and still incomplete.

10 Visibility Graph

Besides decomposition and retraction, the visibility graph method is another approach to path finding. However, unlike the other two, it is not universal, i.e. its principle is only applicable to the specific case where configuration space is made of (generalized) polygonal obstacles. Nevertheless, historically, the visibility graph method has been one of the earliest path finding methods used to control a robot (SHAKEY project [35]). Due to its conceptual simplicity, it remains quite popular for implementing mobile robot systems. Furthermore, it can be combined with other types of methods to handle more difficult cases, e.g. to deal with rotation.

We have $\mathcal{C} = \Re^2$ and the C-obstacles are possibly overlapping convex polygons. The principle of the visibility graph method is to construct a path as the concatenation of line segments connecting the initial to the final configurations through C-obstacle vertices. Therefore, it produces paths which lie in valid space[13].

Let us consider a polygonal object $\mathcal{A}$ translating among convex polygonal obstacles $\mathcal{B}_i$. We want to plan a path between any given initial and final valid configurations $\mathbf{c}_1$ and $\mathbf{c}_2$.

DEFINITION 19: *Let VG be the graph constructed as follows:*
- VG's nodes are the initial and final configurations $\mathbf{c}_1$ *and* $\mathbf{c}_2$, *and the vertices of the C-obstacles.*
- Two nodes are connected by an undirected arc if and only if they are 'mutually visible', i.e. the line segment joining the corresponding two points lies completely in valid space.
VG is called the **visibility graph** *in* $\mathcal{C}$.

Figure 23 shows the visibility graph for a simple configuration space with three C-obstacles (the graph arcs include the obstacle edges).

The visibility graph contains a path from $\mathbf{c}_1$ to $\mathbf{c}_2$ if and only if there is a valid path between the two configurations. In addition, if a valid path exists, the shortest valid path (according to the Euclidean metric in $\Re^2$) is made of line segments passing through vertices of the C-obstacles. The graph VG contains such a path, which is shown in bold line in Figure 23.

[13]One can grow C-obstacles slightly, so that the visibility graph method produces free paths.

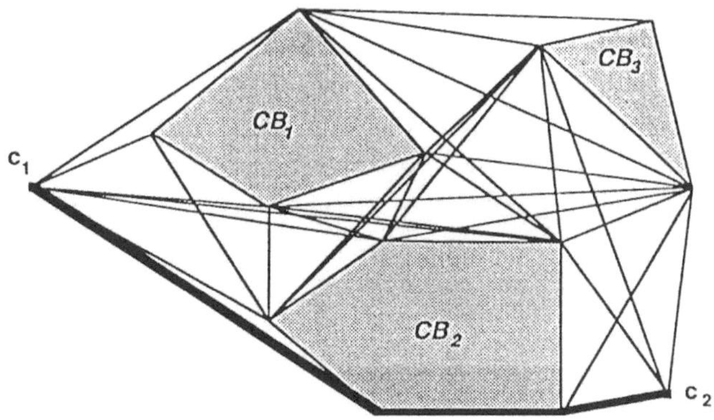

Figure 23: **Example of Visibility Graph**

The overall algorithm of the visibility graph method is the following:

1. Compute the vertices of the C-obstacles and construct the visibility graph VG.
2. Starting at $\mathbf{c}_1$, search VG for a path from $\mathbf{c}_1$ to $\mathbf{c}_2$.
3. If $\mathbf{c}_2$ is attained, then return the list of vertices along the path; otherwise indicate failure.

In order to construct VG, one may consider each node X in VG. By rotating a half-line emanating from X and using a line-sweep algorithm, one can construct the arcs connecting X to other nodes in VG.

Searching VG can be done using the A^* algorithm [36]. The heuristic function f to guide the search is a function which maps each node X into an estimate of the length of the shortest path from $\mathbf{c}_1$ to $\mathbf{c}_2$ passing through X. We may take:

$$f(\mathbf{c}_1, \mathbf{c}_2, X) = g(\mathbf{c}_1, X) + h(X, \mathbf{c}_2)$$

where $g(\mathbf{c}_1, X)$ is the length of the shortest path discovered so far between $\mathbf{c}_1$ and X, and $h(X, \mathbf{c}_2)$ is the Euclidean distance in $\Re^2$ between X and $\mathbf{c}_2$. Since we have $0 \leq h(X, \mathbf{c}_2) \leq d_{valid}(X, \mathbf{c}_2)$, where $d_{valid}(X, \mathbf{c}_2)$ denotes the (unknown) length of the shortest path between X and $\mathbf{c}_2$ in C_{valid}, the heuristic function is admissible. Then, the visibility graph method is guaranteed to return the shortest path from $\mathbf{c}_1$ to $\mathbf{c}_2$, if a valid path exists, and to indicate failure, otherwise.

LEMMA 8: ***With a*** A^* ***search algorithm using an admissible heuristic function, the visibility graph method is complete and generates the shortest valid path whenever a valid path exists.***

Let n be the number of vertices of the C-obstacles. The construction of the arcs outgoing from a node in VG using the rotating half line-sweep takes $O(n \log n)$ time, so that the total time for constructing VG is $O(n^2 \log n)$. The resulting graph contains $O(n)$ node and $O(n^2)$ arcs, so that the combinatorial search of VG using the A^* algorithm with

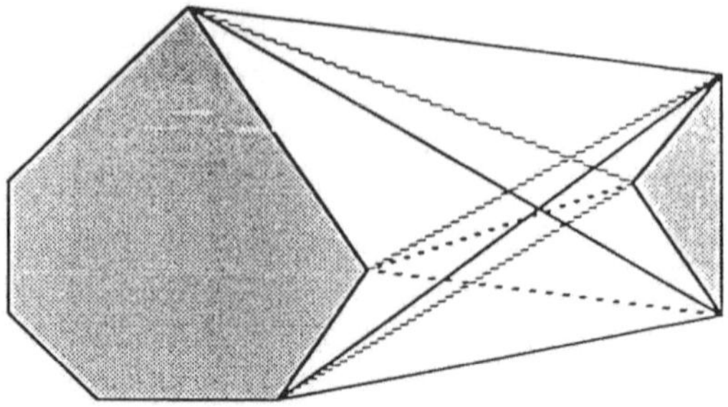

Figure 24: **Simplification of the Visibility Graph**

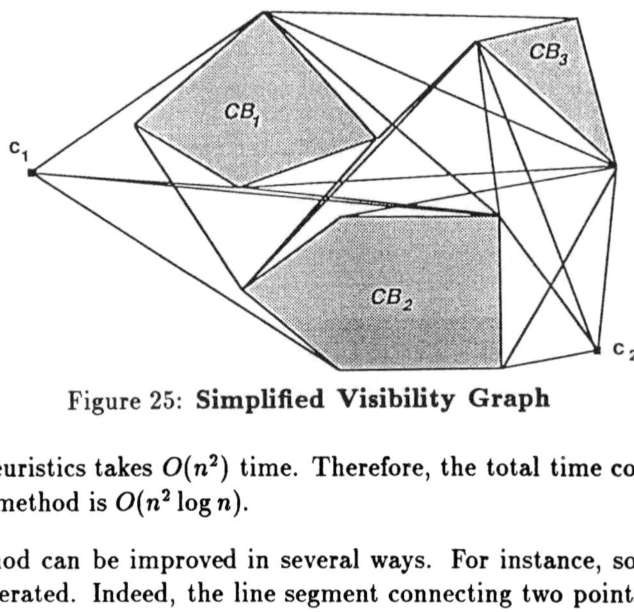

Figure 25: **Simplified Visibility Graph**

an admissible heuristics takes $O(n^2)$ time. Therefore, the total time complexity of the visibility graph method is $O(n^2 \log n)$.

The above method can be improved in several ways. For instance, some arcs in VG need not be generated. Indeed, the line segment connecting two points X_1 and X_2 in free space may be part of a shortest path, only if the infinite line containing the two points does not intersect the interiors of the C-obstacles to which these points belong in an arbitrarily small neighborhood of these points. The remaining segments are called ***supporting segments*** and there are exactly four of them among the vertices of any two disjoint convex polygons [45]. This is illustrated by Figure 24. The supporting segments are shown as plain lines, the non-supporting segment in free space are shown as dashed lines. Thus, considering only supporting segments reduces the size of the search graph (see Figure 25). In addition, more efficient graph construction techniques are applicable if the C-obstacles are known to be both convex and disjoint [53,44,3,45].

The visibility graph technique can be extended to the case where both $\mathcal{A}$ and the $\mathcal{B}_i$s are **generalized polygons** bounded by line segments and circular arcs. In this case, C-obstacles are also generalized polygons (see [25]). The 'generalized' visibility graph is built as explained above using actual vertices of the generalized polygonal C-obstacles, and is completed by fictitious vertices as follows [25]:

- Let X be a vertex (or the initial or final configuration) and E a circular arc. If there

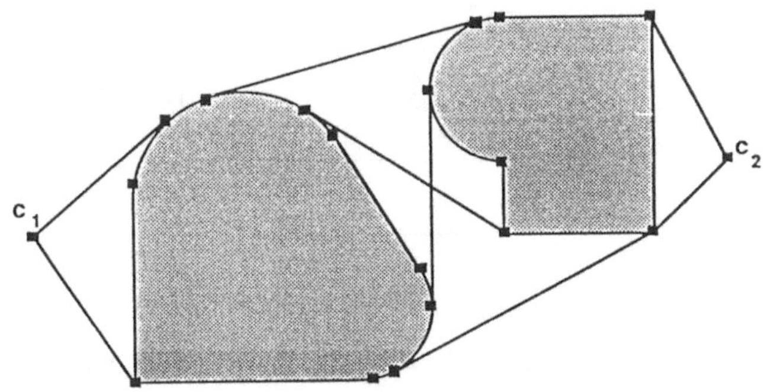

Figure 26: **Generalized Visibility Graph**

exists a point X' on E such that the open line segment joining X and X' lies in free space and the infinite line supporting the segment is tangent to E, then X' is included as a fictitious vertex in the generalized visibility graph, and X and X' are connected by an arc.

- Let E and E' be two circular edges. If there exists a point X on E and a point X' on E' such that the open line segment between these two points lies in free space and the infinite line supporting the segment is tangent to both E and E', then X and X' are included in the graph as fictitious vertices, and they are connected by an arc.

- Any two vertices X and X' located on the same circular edge of a C-obstacle are connected by an arc.

Figure 26 shows the generalized visibility graph in a configuration space with two generalized polygonal C-obstacles. Only supporting segments are shown in the figure.

The same rotating line-sweep technique as above is applicable and allows to compute the generalized visibility graph in $O(n^2 \log n)$ time.

Unfortunately, the visibility graph method is not applicable to the case of a three-dimensional polyhedral configuration space. Indeed, in such a space, there may exist no valid path consisting of a sequence of straight line segments joining the initial and final configurations through C-obstacle vertices. For instance, the polyhedral C-obstacles may form kinds of 'tunnels'. However, the problem of finding shortest valid paths among polyhedral obstacles has recently attracted interest in Computational Geometry. It can be shown that, as illustrated by Figure 27, shortest valid paths are sequences of line segments adjacent at points on polyhedron edges. Canny [10] shows that the problem of generating such paths is NP-hard in n and the best bound so far is due to Reif and Storer [44] who give a $2^{n^{O(1)}}$-time algorithm. Papadimitriou [41] gives an algorithm that finds a path which is at most $(1+\varepsilon)$ times the length of the shortest path, in time that is polynomial in n and $1/\varepsilon$. The basic idea of this algorithm is to break edges into short segments and to search a graph with these segments as nodes.

Algorithms for finding shortest paths among polyhedra are quite technical and of limited interest in robot path planning. In fact, the visibility graph method is attractive for robot motion planning in two-dimensional workspace, more because it is conceptually

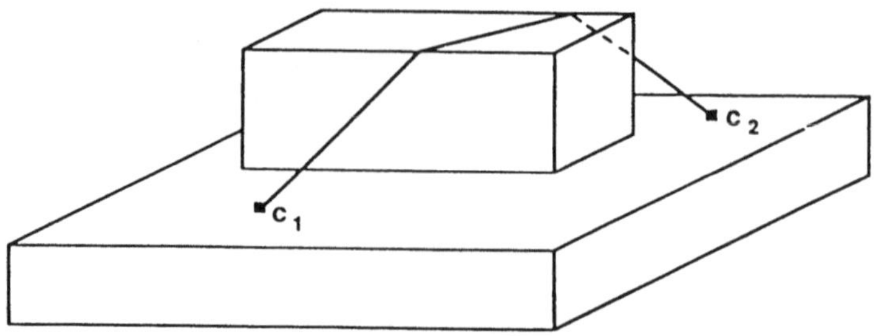

Figure 27: **Shortest Path Among Polyhedral C-Obstacles**

simple, easy to implement, and relatively time efficient, rather than because it generates shortest paths. It is more appropriate to apply other methods such as those described in the previous sections to the three-dimensional path planning problem.

The visibility graph method can neither be extended to the case of a robot translating and rotating in a two-dimensional workspace. However, it can be combined with other techniques in order to handle rotations of the robots. For instance, it can be combined with the 'orientation slicing' technique described in Subsection 7.2. A visibility graph VG_k can be built for every orientation interval $[\theta_k, \theta'_k]$, by first computing a bounding polygonal approximation of the area swept out by the moving object when it rotates from θ_k to θ'_k about the reference point, and next, considering this approximated area as a translating object, computing the polygonal C-obstacles to this object. The graphs VG_k can be combined into a larger graph VG by linking any two nodes $X_1 \in VG_{k_1}$ and $X_2 \in VG_{k_2}$ $(k_1 \neq k_2)$, whenever the two points X_1 and X_2 can be connected by a straight segment intersecting no C-obstacle in both $\mathcal{C}^{k_1}$ and $\mathcal{C}^{k_2}$, and $\theta'_{k_1} = \theta_{k_2}$ (modulo 2π). A path consisting of interweaved pure translations and pure rotations can be generated by searching VG.

11 Conclusion

Until rather recently path planning has not been considered a central and difficult problem in the development of advanced autonomous robots. It was often thought that appropriate heuristics would be sufficient to solve it in almost every case. Research on motion planning became quite active after the mid-70's, when the importance of the problem started being recognized. Theoretical studies and practical implementations showed it was also a difficult problem.

Work on path planning has attracted the interest of a many researchers in the 80's. Much progress has been done and the problem is now fairly well understood. Several approaches have been proposed and, within each of them, different techniques have been developed. Some of these methods may have practical applications, provided they are carefully engineered if order to fit the characteristics of the application tasks. Indeed, the computational complexity of both the path planning problem (lower bound) and the proposed techniques (higher bounds) has been analyzed in depth. Since this complexity increases exponentially with the number of degrees of freedom of the robotic systems, it is clear that there is no universal efficient solution. But experiments with various implementations have shown that some of the existing techniques can be very efficient in restricted domains.

It is clear that path planning for one robot among fixed obstacles is only one facet of the larger motion planning problem. Research on the other facets (dealing with multiple robots, mobile obstacles, dynamic constraints, incomplete knowledge, inexact knowledge, ...) currently attracts a lot of interest, but is still at an early stage and the application of existing approaches is more remote.

Existing path planning techniques can be applied to material movements in several ways. For example, they can be implemented as utilities in a CAD system, in order to allow the designer to anticipate the effects of his decisions on the execution process. Several software packages in the construction industry include tools for simulating the movements of material (such as pipes and beams) and machines, some have collision-checking capabilities, but none can perform automatic path planning automatically. Automatic path planning could also be useful within process planners. But it is clear that a major goal should be to make them available on-line on automatic material transportation robotic systems. In addition to making these systems more autonomous, this would greatly facilitate the integration of material physical movements with the updatings of information (e.g., where objects are at every instant, what processing steps have been performed so far), which is necessary to control material flow systems.

Acknowledgements

The author thanks Dr. Jocelyne Pertin-Troccaz who presented this paper at the NATO-Advanced Research Workshop on *Advanced Information Technologies for Industrial Material Flow Systems*, in Grenoble, France June 13-17. Support for the work was provided in part by the Stanford Institute for Manufacturing and Automation (SIMA) and in part by the Center of Integrated Systems (CIS). The author also thanks Wonyun Choi, Jérôme Barraquand, Krasimir Kolarov, Jean-François Rit, Shashank Shekhar and David Zhu for useful discussions on motion planning and/or constructive comments on earlier drafts of this paper.

References

[1] V.I.Arnold (1978). *Mathematical Methods of Classical Mechanics.* Springer-Verlag, New York.

[2] D.S.Arnon (1979). *A cellular decomposition Algorithm for Semi-Algebraic Sets.* Technical Report No. 353, Computer Science, University of Wisconsin.

[3] T.Asano, T.Asano, L.Guibas, J.Hershberger, H.Imai (1986). *Visibility of Disjoint Polygons.*, Algorithmitica 1 (1), 49-63.

[4] D.Ayala, P.Brunet, R.Juan, I.Navazo (1985). *Object Representation by Means of Nonminimal Division Quadtrees and Octrees.* ACM Transactions on Graphics, 4, 1 (January).

[5] T.O.Binford (1971). *Visual Perception by Computers.* Proc. of the IEEE Conference on Systems Science and Cybernetics, Miami, FL (December).

[6] O.Bottema, B.Roth (1979). *Theoretical Kinematics.* North-Holland Publishing Co., New York.

[7] R.A.Brooks, T.Lozano-Pérez (1982). *A Subdivision Algorithm in Configuration Space for Findpath with Rotation.* AI Memo 684, AI Laboratory, MIT (December).

[8] R.A.Brooks, T.Lozano-Pérez (1983). *A Subdivision Algorithm in Configuration Space for Findpath with Rotation.* Proc. of the Eighth International Joint Conference on Artificial Intelligence (IJCAI), Karlsruhe, FRG, August 1983.

[9] R.A.Brooks(1983). *Solving the Find-Path Problem by Good Representation of Free Space.* IEEE Transactions on Systems, Man and Cybernetics, SMC-13, No. 3.

[10] J.F.Canny (1987). *The Complexity of Robot Motion Planning.* Ph.D. Dissertation, Department of Electrical Engineering and Computer Science, MIT, Cambridge, MA (May).

[11] C.Castells, R.Melville (1985). *An Unusual Algorithm for the Minimum Spanning Circle Problem.* Technical Report, Electrical Engineering and Computer Science, John Hopkins University, Baltimore, MD.

[12] R.Chatila (1982). *Path Planning and Environmental Learning in a Mobile Robot System.* Proceedings of the European Conference on Artificial Intelligence (ECAI), Orsay, France.

[13] B.Chazelle, D.P.Dobkin (1979). *Decomposing a Polygon into Convex Parts.* Proceedings of the 11th Annual ACM Symposium on Theory of Computing, 38-48.

[14] B.Chazelle (1985). *Approximation and Decomposition of Shapes.* In [50].

[15] G.E.Collins (1975). *Quantifier Elimination for Real Closed Fields by Cylindrical Algebraic Decomposition.* In Lecture Notes in Computer Science, No. 33, Springer-Verlag, New York, pp. 135-183.

[16] B.R.Donald (1984). *Motion Planning with Six Degrees of Freedom.* Technical Report 791, Artificial Intelligence Laboratory, MIT, Cambridge, MA (May).

[17] R.L.Drysdale III (1979). *Generalized Voronoi Diagrams and Geometric Searching.* Ph.D. Dissertation, Report STAN-CS-79-705, Department of Computer Science, Stanford University.

[18] M.Erdmann, T.Lozano-Pérez (1986). *On Multiple Moving Objects.* A.I.Memo No. 883, Artificial Intelligence Laboratory, MIT, Cambridge, MA (May).

[19] F.Germain (1984). *Planification de Trajectoires Sans Collision.* DEA Report, LIFIA Laboratory, National Polytechnic Institute of Grenoble.

[20] C.L.Jackins, S.L.Tanimoto (1980). *Octrees and their Use in Representing Three-Dimensional Objects.* Computer Graphics and Information Processing, 14, 3 (November).

[21] K.Jänich (1984). *Topology.* Springer-Verlag, New-York.

[22] O.Khatib (1986). *Real-Time Obstacle Avoidance for Manipulators and Mobile Robots.* International Journal of Robotics Research 5, 1.

[23] D.G.Kirkpatrick (1979). *Efficient Computation of Continuous Skeletons.* Proceedings of the 20th Symposium on Foundations of Computer Science. 18-27.

[24] J.C.Latombe (1988). *Motion Planning With Uncertainty: The Preimage Backchaining Approach.* Technical Report No. STAN-CS-88-1196, Computer Science, Stanford University (March).

[25] J.P.Laumond (1987). *Obstacle Growing in a Non Polygonal World.* Information Processing Letters (April).

[26] D.T.Lee, R.L.Drysdale (1981). *Generalization of Voronoi Diagrams in the Plane.* SIAM Journal of Computing, 10, 73-87.

[27] D.Leven, M.Sharir (1985). *An Efficient and Simple Motion Planning Algorithm for a Ladder Moving in Two-Dimensional Space Amidst Polygonal Barriers.* Proc. of the First ACM Symposium on Computational Geometry, 211-227.

[28] D.Leven, M.Sharir (1987). *Intersection and Proximity Problems and Voronoi Diagrams.* In [50], 187-228.

[29] A.Lingas (1982). *The Power of Non-Rectilinear Holes.* Proceedings of the 9th Colloquium on Automata, Languages and Programming, Aarhus, LNCS Springer-Verlag, 369-383.

[30] T.Lozano-Pérez, M.A.Wesley (1979). *An Algorithm for Planning Collision-Free Paths Among Polyhedral Obstacles.* Communications of the ACM, Vol. 22, 10, 560-570 (October).

[31] T.Lozano-Pérez (1981). *Automatic Planning of Manipulator Transfer Movements.* IEEE Transactions on Systems, Man and Cybernetics, SMC-11.

[32] T.Lozano-Pérez (1983). *Spatial Planning: A Configuration Space Approach.* IEEE Transactions on Computers, C-32, No. 2.

[33] T.Lozano-Pérez (1987). *A Simple Motion-Planning Algorithm for General Manipulators.* IEEE Journal of Robotics and Automation, RA-3, 3 (June).

[34] D.Meagher (1982). *Geometric Modeling Using Octree Encoding.* Computer Graphics and Image Processing, 19.

[35] N.J.Nilsson (1969). *A Mobile Automaton: An Application of Artificial Intelligence Techniques.* Proceedings of the 1st International Joint Conference on Artificial Intelligence (IJCAI), Washington D.C., 509-520.

[36] N.J.Nilsson (1980). *Principles of Artificial Intelligence.* Morgan Kaufmann, Los Altos, CA.

[37] C.Ó'Dúnlaing, C.K.Yap (1982). *A Retraction Method for Planning the Motion of a Disc.* Journal of Algorithms, 6, 104-111.

[38] C.Ó'Dúnlaing, M.Sharir, C.K.Yap (1983). *Retraction: A New Approach to Motion Planning.* In Proc. of the 15th ACM Symposium on the Theory of Computing, Boston, 207-220.

[39] C.Ó'Dúnlaing, M.Sharir, C.K.Yap (1984). *Generalized Voronoi Diagrams for Moving a Ladder: I. Topological Analysis.* Technical Report No. 32, Robotics Laboratory, Courant Institute, New-York University.

[40] C.Ó'Dúnlaing, M.Sharir, C.K.Yap (1984). *Generalized Voronoi Diagrams for Moving a Ladder: II. Efficient Construction of the Diagram.* Technical Report No. 32, Robotics Laboratory, Courant Institute, New-York University.

[41] C.H.Papadimitriou (1985). *An Algorithm for Shortest-Path Motion in Three Dimensions.* Information Processing Letters 20, 259-263 (June).

[42] F,P.Preparata, M.I.Shamos (1985). *Computational Geometry: An Introduction.* Springer-Verlag, New York.

[43] J.Reif (1979). *Complexity of the Mover's Problem and Generalizations.* Proceedings of the 20th Symposium on the Foundations of Computer Science, 421-427.

[44] J.Reif, J.Storer (1985). *Shortest Paths in Euclidean Space with Polyhedral Obstacles.* Technical Report CS-85-121, Computer Science Department, Brandeis University (April).

[45] H.Rohnert (1986). *Shortest Paths in the Plane with Convex Polygonal Obstacles.* Information Processing Letters 23, 71-76 (August).

[46] H.Samet (1980). *Region Representation: Quadtrees from Boundary Codes.* Communications of the ACM, 23, 3 (March).

[47] J.T.Schwartz, M.Sharir (1983). *On the 'Piano Movers' Problem. I. The Case of a Two-Dimensional Rigid Polygonal Body Moving Amidst Polygonal Barriers.* Communications on Pure and Applied Mathematics, 36, 345-398.

[48] J.T.Schwartz, M.Sharir (1983). *On the 'Piano Movers' Problem. II. General Techniques for Computing Topological Properties of Real Algebraic Manifolds.* Advances in Applied Mathematics, 4, 298-351.

[49] J.T.Schwartz, M.Sharir, J.Hopcroft (1987). *Planning, Geometry, and Complexity of Robot Motion.* Ablex Publising Corp., Norwood, New Jersey.

[50] J.T.Schwartz, C.K.Yap (1987). *Algorithmic and Geometric Aspects of Robotics.* LEA, Lawrence Erlbaum Associates, Publishers, Hillsdale, New Jersey.

[51] M.Spivak (1979). *A Comprehensive Introduction to Differential Geometry.* Publish or Perish, Inc., Wilmington, Delaware.

[52] A.Tarski (1951). *A Decision Method for Elementary Algebra and Geometry.* University of California Press, Berkeley, CA.

[53] E.Welzl (1985). *Constructing the Visibility Graph for n line segments in* $O(n^2)$. Information Processing Letters 20 (4), 167-171 (May).

[54] C.K.Yap (1987). *Algorithmic Motion Planning.* In [50], 95-143.

THE NEED FOR A GENERIC FRAMEWORK FOR SYSTEMS INTEGRATION

R H Weston, J D Gascoigne, C M Sumpter & A Hodgson
Department of Manufacturing Engineering
Loughborough University of Technology
LOUGHBOROUGH, Leicester, LE11 3TU
ENGLAND

ABSTRACT

This paper outlines the need for a widely accepted generic framework for systems integration within manufacturing enterprises. The provisions of this framework would form a basis for the specification of integration projects and sub-components such as software packages in order to permit the interchange, reconfiguration, expansion and transferability of the whole or part solutions so generated. The overall aim is to allow a logical mapping of integration tools, implementations and solutions to physical resources at all phases from manufacturing system inception to operation so enabling rapid and cost-effective response to changes in products, markets, enabling technologies and manufacturing methodologies.

The discussion is based on the findings of systems integration studies at Loughborough University of Technology aimed at a flexible and adaptable approach to product realisation in electronic industries. More specifically, research findings are reported with respect to:

(i) the need to proceed from a position of existing installed bases of manufacturing machines and operations, both automated and manual,

(ii) the integration of the activities of contemporary proprietary software packages with specific reference to production planning and control, and

(iii) the provision of an environment for the definition and realisation of interaction, synchronisation and organisation between manufacturing entities.

The approaches studied incorporate OSI computer network methodologies and products in the role of an underlying communication "utility".

NATO ASI Series, Vol. F53
Advanced Information Technologies for Industrial Material Flow Systems
Edited by Shimon Y. Nof, Colin L. Moodie

1 RATIONALE

The concept of integrating computer based machines to improve the efficiency of manufacturing systems is widely accepted. What is less generally agreed upon is what this integration means in practice and how it can be achieved. A basic problem is that the people potentially involved have varying perceptions depending on their previous experiences and environment. Examples include bespoke system providers, the products of which are often really only 'larger machines' which would require re-integration to form part of a wider integrated system. Conversely, others see the need for system wide architectures to support the perceived end system features such as cell control, program downloads, status monitoring, workpiece tracking etc. Further, while many people intuitively foresee many benefits from integration they have great difficulty producing convincing (ie cost based) justifications for it.

We suggest that an underlying problem is that while it is inherent in the term 'factory integration' that a wide scale of implementation is involved this is not practicable as yet. The reasons for this include both technological and cost problems associated with multi-vendor information (in addition to and distinct from data) transfer, the requirement for factory systems to change reasonably slowly area by area, uncertainty as to what will work and what will be an expensive failure and disagreement as to what basis to start from.

One current approach is to start with networking and add functionality on top while another is to use well understood technology to build internally flexible but outwardly inflexible integrated sub-systems before then trying to standardise and generalise the technologies and techniques involved.

We propose the definite requirement for a 'master plan' or framework, the provisions of which would be incorporated into the basic specification of individual projects in this sphere. It is essential that such provisions would not absolutely pre-determine every component mechanism of an attempted application but would provide a model of functionality which proposed solutions should not contradict. In this way the necessity of allowing for the constraints of pre-existing system elements, gaps in technology and the inevitable learning process could be accommodated, ie solutions might well not be perfect in terms of eventual integration requirements but will meanwhile not be incompatible with the overall scheme.

Something analogous to the ISO/OSI (International Standards Organisation/ Open Systems Interconnection) reference model for communication[1] is required for open factory integration. It would probably be necessary for individual companies to derive their own 'flavours' of this model to suit their particular operations. An example of the overall concept in practice would be the incorporation of messaging functionality above the communications service without stipulating the exact messaging format to be used.

The framework proposed should exploit the results of worldwide standardisation activities and studies, aimed at evolving distributed environments and provide mechanisms for the mapping of reconfigurable logical structures onto multi-vendor hardware architectures.

A secondary objective of the framework would be to decompose the whole integration problem into regions in order that existing 'know-how' could be most productively applied to its relevant sphere of activity. The approach suggested in this paper is illustrated by Figure 1. In this way the wealth of Information Technology expertise could be directed at information architectures, manufacturing skills brought to bear upon distributed control and so on. This would help avoid the duplicate efforts currently being exerted in certain areas eg by the information technology (IT) and manufacturing communities.

The concepts presented here have been derived as a result of UK Government funded research at Loughborough University which has been conducted in collaboration with UK manufacturing companies (the primary collaborator being International Computers Limited [ICL], UK). Thus, features of the work at Loughborough will be described to facilitate an appreciation of the background thinking involved.

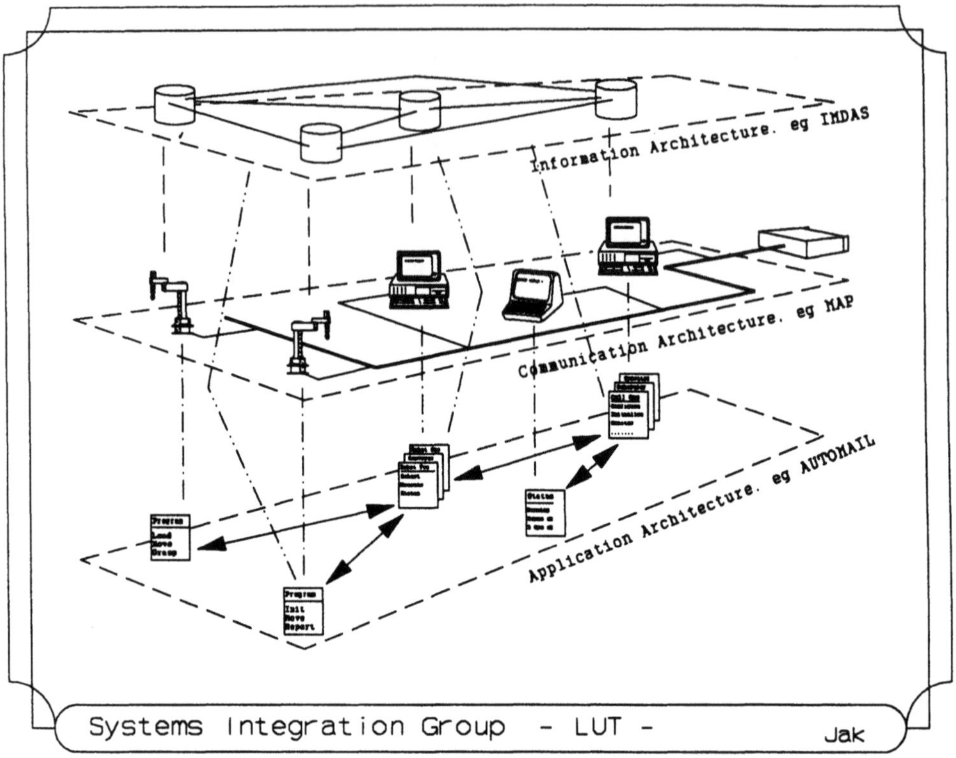

Figure 1 Architecture for Flexible Integration

2 STUDIES EVOLVING A DISTRIBUTED ENVIRONMENT FOR INTEGRATION

Many of the problems of total systems integration, which a 'generic framework for integration' can address, are evident at individual sub-system levels of integration. For example, the diversity of manufacturing activity and proprietary bespoke integration solutions found at the manufacturing cell level epitomizes the integration problem within the whole manufacturing enterprise.

Consider the manufacturing cell across a wide range of industry. Shop floor activity (eg component manufacture, assembly and test functions) is structured on a cell basis where a number of machines and personnel work closely to achieve a manufacturing function. Typically, their individual tasks will be quite different but will be required at times to interact with each other in a synchronised and ordered manner. The difficulties of managing and controlling such a cell are particularly apparent if one has to consider requirements such as dynamic work re-scheduling, exception handling or flexible cell re-configuration at this management level.

Cell supervision and control has been implemented as a computer based function by numerous systems integrators and vendors, at varying levels of complexity. However, all too often solutions are engineered to suit a specific requirement and configuration, requiring major surgery if a different configuration or functionality is later required. Open machine interconnectability based on OSI implementations such as MAP/TOP (Manufacturing Automation Protocol/Technical Office Protocol[2]), whilst providing an essential 'utility' for the integration infrastructure, is only the first step towards systems integration.

Once this 'utility' is available, potentially allowing computer based systems and machines to interchange information, questions relating to organisation, interaction, messaging, etc begin to arise.

AUTOMAIL (AUTOMAtion Integration Language) was evolved by the Systems Integration Group initially to provide a generic computer based solution for the cell control/supervisory problem allowing the systems integrator to use a standard set of tools when developing supervisory software for specific applications. Above all, applications thus implemented would be intrinsically 'flexible' in nature facilitating reconfigurability in response to changing requirements. The real benefits of utilising such tools is the promotion of a consistent and familiar approach to integration across the whole enterprise, which will remain stable over an acceptable lifetime.

This work has resulted in an implementation of AUTOMAIL suitable for cell control application programming which has supported the programming, debugging and running of supervisory applications based around the Loughborough University Flexible Assembly System[3] (see Figure 2).

2.1 The AUTOMAIL Methodology

The manufacturing enterprise structure can be broken down, in the limit, to a number of discrete manufacturing entities (MEs), ie individual personnel and machines that are organised collectively to achieve a manufacturing goal such as design, test, quality appraisal, assembly, planning etc. Although an exhaustive model for a ME is not being proposed here, AUTOMAIL evolve assuming certain ME attributes. For instance, each ME would be

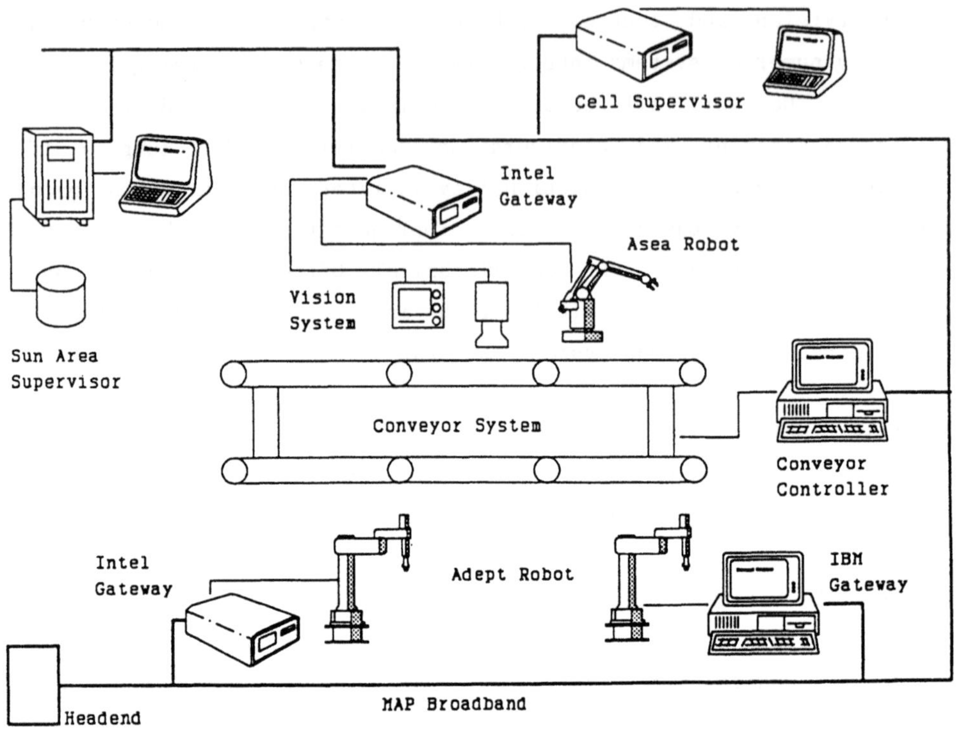

Figure 2 Loughborough Integration Proving Cell

assumed to have a communication capability pertinent to the context in which it is working, ie the ability to 'understand' and 'deliver' relevant information. In practice, this could for example require a manual machine tool operator to receive and log job information via a terminal, or an assembly robot to exchange control and data information in a specific format over a network link. The physical realisation of this will require some operator retraining and a support environment for any given machine.

For the example of the cell supervisor application, AUTOMAIL supports the programming of supervisor/ME interaction on an individual basis, ie separate program tasks defining the sequence of activities for each ME in the cell, allowing the essentially asynchronous and individual ME role to be retained in the supervisor application. This approach facilitates reconfigurability as ME control tasks can readily be replaced, inserted or modified in response to changing requirements, either dynamically or as determined by longer term strategic planning goals. As mentioned above, dialogue is built on an assumed level of communication, pertinent dialogue primitives being available for programming the ME/supervisor interactions. The importance is placed on dialogue content rather than specific syntax ie as manufacturing message communication services evolve (eg Manufacturing Message Services (MMS) plus companions[4]), inclusion within AUTOMAIL should be natural.

The AUTOMAIL environment for program task development/debugging, task management and exception handling is explained in some detail in reference 3, but for the purpose of this document, AUTOMAIL is best illustrated by way of an example.

2.2 An AUTOMAIL Cell Supervisor

Figure 3 shows a manufacturing cell that could typically be a target for integration as defined by a company's current integration strategy. In this example, human intensive functions as well as stand-alone machine operations are to be co-ordinated and monitored by a computer-based supervisory entity.

AUTOMAIL supports the development of the individual ME control tasks that at run-time execute in a multi-tasking asynchronous manner. The control tasks themselves have a sub-structure based on the concept of a 'function module' building block which typically contains a small number of program statements, sensibly grouped together to initiate an action or to attain a particular program state. Figure 4 provides a control task snapshot which shows how dialogue primitives within a function module can be used to establish communication and obtain information relevant to machine type and functionality etc. These dialogue primitives, which essentially are program function calls, initiate communication based on suitable standard network services. In this example, a one-to-one mapping onto MAP 2.1

Manufacturing Message Format Standard (MMFS) is used. It is important to stress, at this point, that the exact nature of the dialogue is application determined, and in fact industry specific application dialogue should not be precluded. However, it is essential that any such dialogue is delivered in a standard manner. This would be achieved within AUTOMAIL by providing custom dialogue functions constructed upon suitable standard network service primitives which themselves are derived, for example from MMS,

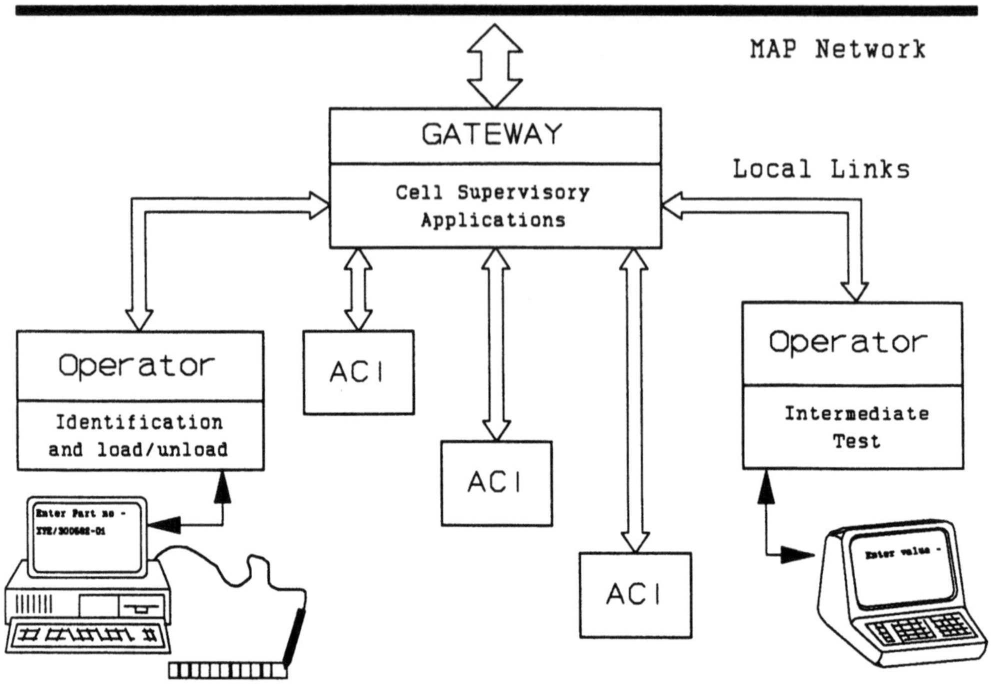

Figure 3 Example Cell Controller/Gateway Function

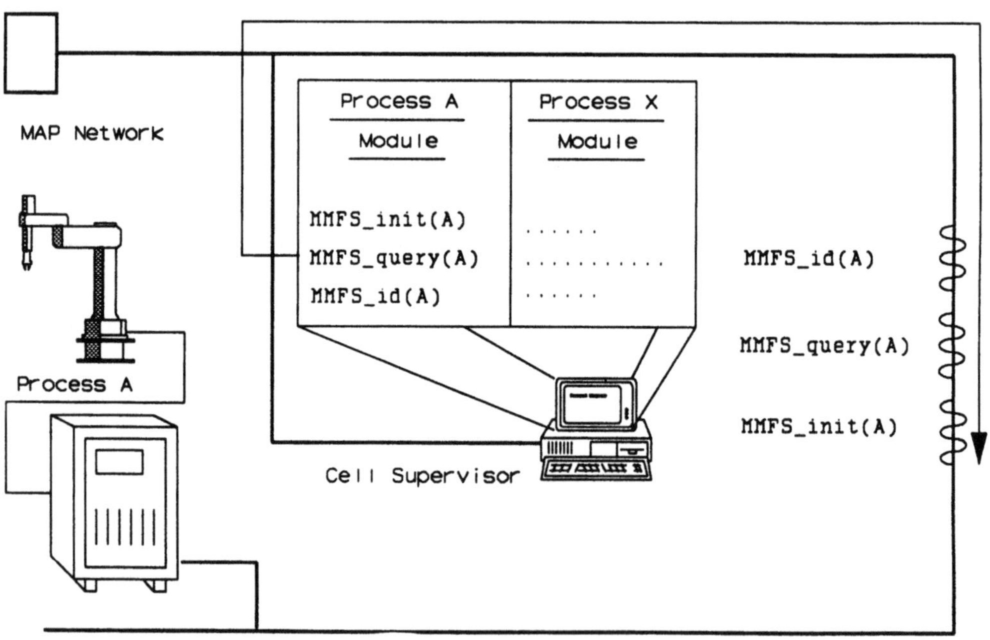

Figure 4 Establishing Dialogue Via AUTOMAIL

MHS, FTAM, etc. Figure 5 shows a snapshot of an alternative task which facilitates menu-driven input from an operator terminal. A single dialogue function call could utilise, for example, block data transfer for screen updating via MMS file transfer, followed by a MMS read variable request to ascertain operator selection.

Exception handling in AUTOMAIL is programmed away from the main control task flow. If an exception occurs during the execution of a function module, subsequent program flow is directed through user-written handlers: here control task sequencing can be modified eg individual function modules can be replaced and reorganised, or complete control tasks replaced. The approach facilitates, for example, orderly shutdown or cell re-scheduling.

The requirement of task synchronisation (eg the halting of one manufacturing process until another completes or reaches a desired state, such as work transfer) is supported by the standard mechanism of program semaphores (Figure 6).

The resulting AUTOMAIL application can be readily modified or extended as appropriate, and can exhibit features of pre-programmed decision making and scenario changing to accommodate short-term manufacturing tactical changes.

The cell supervisor application thus developed can be considered an ME in its own right, interacting in the context of the shop floor on the one hand, and with suitable dialogue, responding to higher level management directives on the other. Taken to its ultimate conclusion, AUTOMAIL should facilitate the development of specific MEs throughout the manufacturing enterprise hierarchy.

Similar attributes of sub-ordinate organisation and management, information gathering, report generation, decision taking, exception handling etc are obvious requirements, to a greater/lesser extent, of MEs throughout the hierarchy. This observation has directed current work on AUTOMAIL towards the concept of a 'configurable' AUTOMAIL that can provide the necessary support for developing control and management applications vertically within a CIM hierarchy. To crystallise thinking in this area it is anticipated that control and data models of the ME will evolve

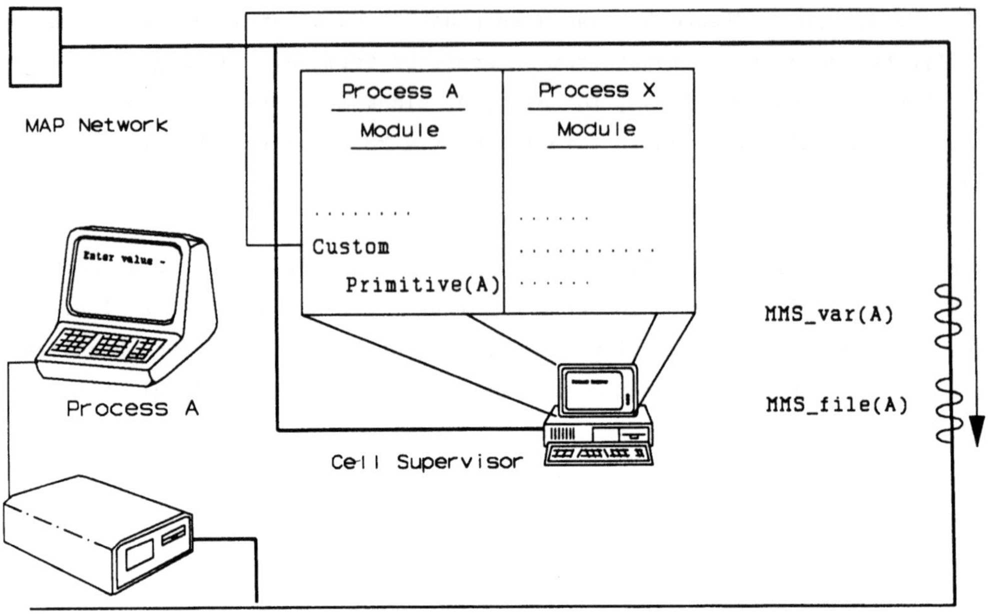

Figure 5 Example Operator Interaction Using AUTOMAIL Primitives

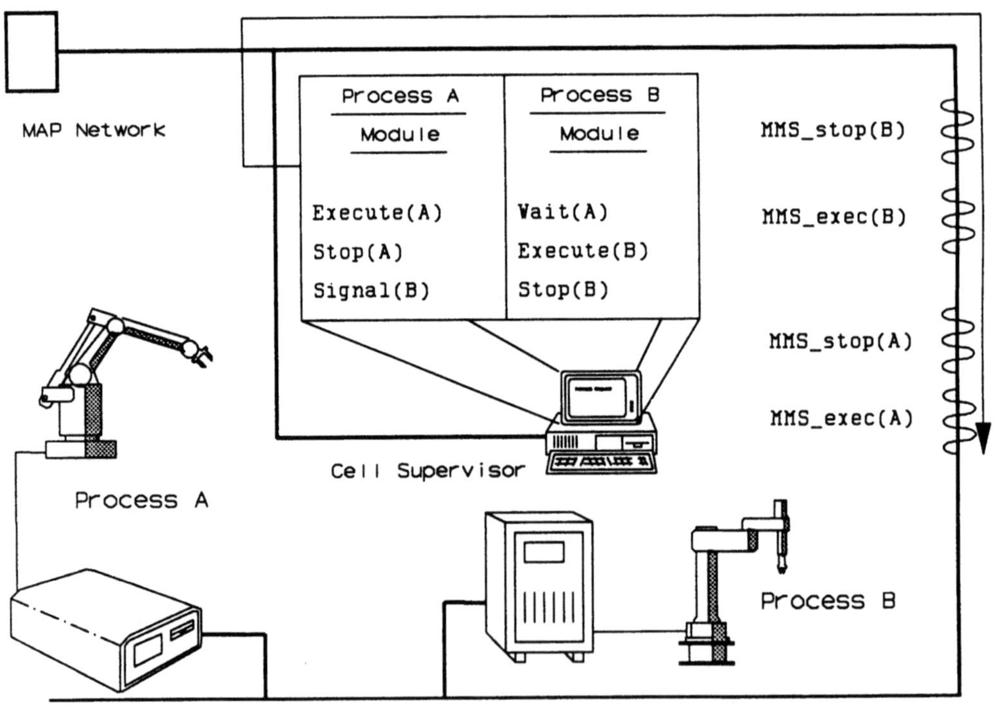

Figure 6 Task Synchronisation Features within AUTOMAIL

alongside the generic AUTOMAIL. We shall return to this issue.

3 CONNECTING THE ELEMENTS OF AN INSTALLED MANUFACTURING BASE

3.1 Shop Floor Manufacturing Entities

Any plan for the integration of manufacturing machines must assume the existence of an inter-connection facility. This is the "telephone" of the machine communication domain and as such must feature an extremely high degree of standardisation and acceptance amongst a wide range of equipment vendors. A solution, whilst technically almost trivial, has proved difficult and expensive to produce for essentially commercial reasons. However, not only is data interchange required but also meaningful information exchange, whether it be for 'action causing' or 'storage and processing' purposes. It is crucial to realise that a command and information message structure (messaging format) is required before interconnected machines can become interacting machines. However, it appears somewhat obscure as to why the microprocessor based controller of a machine tool should need a different set of messaging formats to the micro-processor based controller of a robot or programmable logic controller in order for it to be instructed to receive a program, started, stopped, interrogated etc.

The MAP/TOP specification must now be regarded as the basis for general factory data transfer of any significant scale. However, as yet it is not possible to select from a multi-vendor range of factory floor, commercial and business machines a collection which can 'plug' together and be made to interchange data, let alone purposeful information, at once or indeed cost effectively. This problem should lessen in the near future.

The two main choices for the mechanism of machine interconnection are:

(a) A 'backplane solution' which requires incorporation of a network interface card into the machine controller hardware and extensive modifications to the controller software structure (as illustrated by Figure 7).

(b) A 'gateway' solution whereby existing communications facilities of the machine controller are used to establish a local link to another

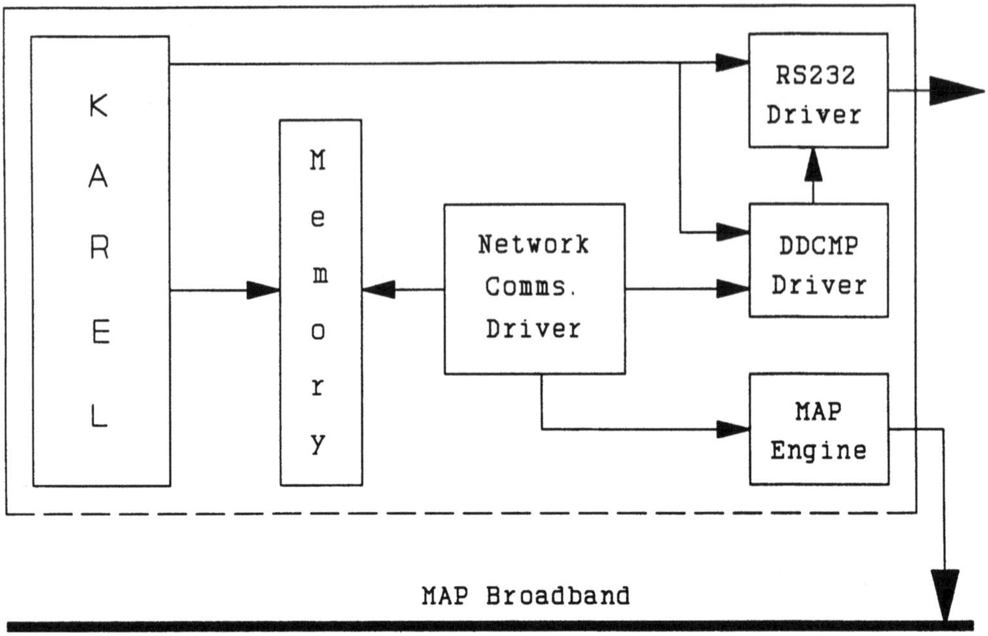

Figure 7 GMF Backplane Solution for Machine Connection

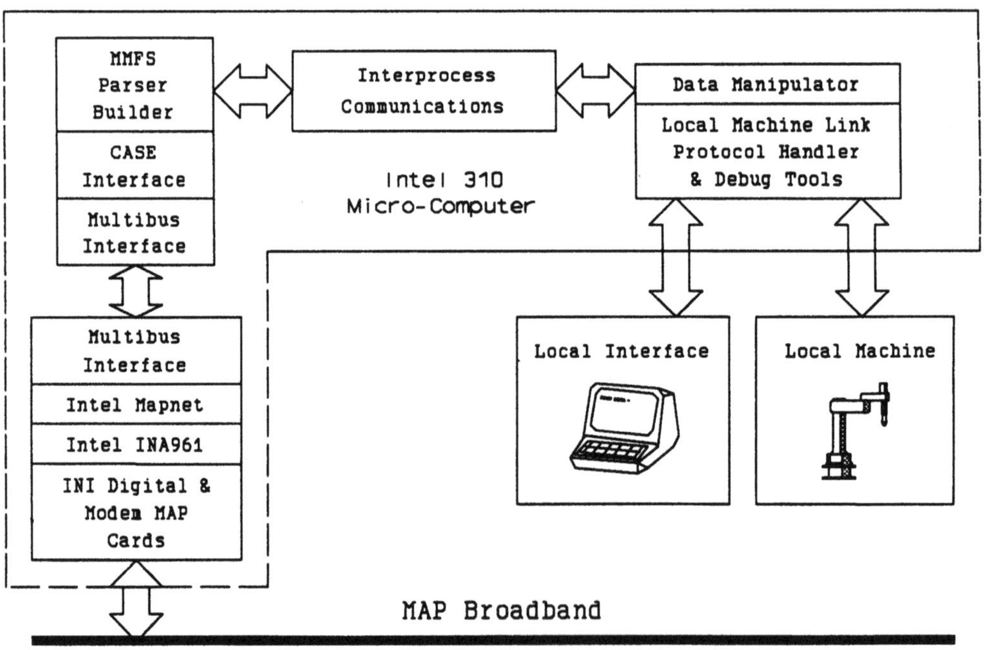

Figure 8 Gateway Solution for Machine Connection

computer which then performs the task of data manipulations and actual network connection (Figure 8).

The former method can only be effectively accomplished by the machine manufacturers. Given estimates of up to 100 man years for a new controller[5], it is clear that it will be some time before large numbers of directly MAP connectable machines appear after the demand is sufficient to prompt manufacturers to make the decision to provide such a feature.

Meanwhile, it has been and continues to be necessary to connect machines together for experimental, pilot and early generation integrated systems. At Loughborough University, gateways have been used to connect together the elements of a manufacturing cell using a MAP broadband network. This communications base was chosen as being the only conveniently available OSI implementation for shop floor use at its time of purchase. Availability led to current devices being based around Intel 310 micro-computers fitted with Intel supplied INI based MAP 2.1 interfaces and using proprietary interfaces to communicate with the local devices. In this way, Adept-One and ASEA robots, a conveyor controller and cell supervisory computers have been linked to facilitate the investigation of integration tools. Further work has led to the inclusion of a network of Sun mini-computers and IBM PCs onto the broadband network and studies continue into the incorporation of complex sensory (vision) and test equipment.

Software in the gateways provides 'intelligent' interfacing between the MAP network using MMFS (Manufacturing Message Format Standard) based messaging formats[2] and the local machine controller using custom or proprietary supervisory links. An example of this approach was demonstrated at the CIMAP exhibition in 1987 by the Loughborough Systems Integration Group (refer again to Figure 8). This software has been written in essentially two sections in order to generalise the interface as much as possible. One section codes and decodes the network messages and handles all network protocol using CASE[2] (Common Application Service Element) provided by the actual MAP interfaces. The other provides for local link protocol handling, local control and debug facilities as well as the required data manipulation to service the messaging requirements.

Discussions with collaborating partners and other industrial organisations has indicated a desire for the merging of cell control and network connection functions. This arises from the following factors:

(a) the existing installed base of machines,

(b) the lack of available machines with MAP connectability,

(c) the current prohibitive cost of network interfaces,

(d) The use of proprietary links inside existing cells,

In this manner local processing functions such as cell scheduling can be performed by the gateway computer which interacts with other machines in the manufacturing hierarchy over the network. As an example, the foregoing discussion with regard to AUTOMAIL, (refer to Figure 3) illustrates the local control of a cell of component insertion machines and manual stations. The supervisory computer is capable of breaking down manufacturing requirements delegated to it and operating the cell accordingly. Data collection and analysis into summary form can also be performed as well as the control of local operator/machine interaction.

While this system may not be totally agreeable to MAP purists it does provide further potential benefits. The cell can operate in the event of network or area manager computer failure. This allows the cell to continue performing its tasks in a reasonably flexible manner and can prevent the loss of collected data by means of local storage. In addition, maintenance, system development and debugging are simplified by the ability to localise machine interaction and control to the immediate area on the shop floor.

In addition, and possibly of more fundamental importance is the practicality of introducing this type of system into an existing factory environment area by area but without limiting the potential for future wider scale integration, ie provided the sub-system is implemented in conformance with an appropriate integration framework it will be upwardly integratable by reconfigurability rather than by major modification.

One aim of the work at Loughborough is to enable the use of the same integration tool (AUTOMAIL) to configure, control and monitor the interaction between manufacturing entities and hence provide a configurable gateway function. This will provide consistency between the different levels of system control and help present a uniform operator/development interface at each level.

3.2 Manufacturing Organisation, Administration and Design Entities

It is clearly important to place research within the context of a structured set of long-term goals, though these may change in the light of experience. The current approach of the Loughborough Systems Integration Group places work on the development of higher level information and control systems in the context of three major stages, each increasingly long-term.

(i) A standard command (action causing) and information (storage and processing) message structure, utilising an OSI interconnect mechanism, between manufacturing cells and conventional proprietary packages. A limited range of transactions and system flexibility will be available from this.

(ii) Three-schema systems, providing independency between a range of user processes and proprietary data repositories. These will use (semi- or defacto) standard query languages and have local command and data translators to interface to each data repository.

(iii) Three-schema systems providing direct independence between data repositories, functional packages and user processes. Such systems will consist of:

(a) Database repositories based on appropriate data models, utilising standard relational database information storage/retrieval mechanisms (eg based on ANSI/SQL[6]). This represents the central core of the 'information architecture'.

(b) Separate proprietary parameter-driven packages which act on these relational database systems to provide a range of typical manufacturing information system functions eg MRP, scheduling, capacity planning. The packages or functions implemented will reflect the manufacturing methodology and organisational structure.

(c) Enabling tools for decision support. These will include rule-based systems with learning ability, capable of using the facilities described in (a) and (b) above (which represent the functions of traditional information systems). Additional 'soft' data will be available, held as rules.

Currently, the group is developing a detailed understanding of the background and software tools required for stage 3, and has started some initial work on a demonstration system for stage 2. However, significantly more work has been carried out towards defining and implementing a stage 1 system. This can be regarded as a relatively short-term 'achievable', which could produce significant benefits for small-to-medium batch manufacturing companies. The above 'stages' are described in more detail in the following sub-sections.

3.2.1 Standard Messaging between Cells and Conventional Packages

This approach recognises that, whether interfaced to each other or not, there currently exist a number of clearly identifiable information (sub-) systems, ie

- Design for manufacture systems, eg CAD, CAM, CAPP systems.
- Factory organisation and control systems, eg production planning and control, quality control systems.
- Enterprise administration and control systems, eg payroll, accounts, financial planning.

A number of functions and transaction types can be identified in each which are particularly relevant to cell and work-centre planning and control, for example:

- Work-to-list detail requests
- Part list/works order allocation requests] and any
- Component insertion data requests] resulting
- Works order progress updates] transactions
- NC program downloads

A set of 'standard' messages can be defined in order to implement transactions such as those above. Ultimately, by implementing an agreed message format, a proprietary package manufacturer could gain a low-cost generic entry to CIM systems. This would in turn enable small to medium sized companies to implement simple CIM systems without much of the current major software costs and delays.

MMS currently has a number of developing companion standards. An additional companion standard could provide an effective vehicle for a set of production planning and control messages. Standards such as IGES and EDIF could be utilised as the basis for appropriate areas of CAD information.

A set of production planning and control transaction types has been implemented by the Loughborough Systems Integration Group (see Figure 9). These transactions currently provide interconnection between a Sun workstation configured either as a cell or area controller and a microcomputer-based proprietary production planning and control package. This work has already demonstrated the simplicity of the concept, and also some of the limitations and difficulties of interfacing to low-cost packages without in-built transaction processing or comms facilities. Figure 10 illustrates the requirements placed on a package designed by stage 1.

3.2.2 Currently Developing Hierarchical Three-Schema Systems

A considerable amount of IT research activity is currently taking place in the design of distributed database management systems[7] and in data modelling[8]. The methodologies developed here can be utilised in devising flexible integration systems as witnessed by research initiatives, including the NBS-AMRF[9] and the USAAF-IISS[10] programmes.

This approach recognises the continuing presence into the future of fragmented proprietary data repositories and their associated update, retrieval and processing systems. To overcome the potential problems of multiple interfaces between users (workstations, cell supervisors, etc.), and data repositories, an intermediate buffer or conceptual schema is introduced. Only one interface is now needed, to this conceptual schema, for each user entity and for each data repository. In addition, a

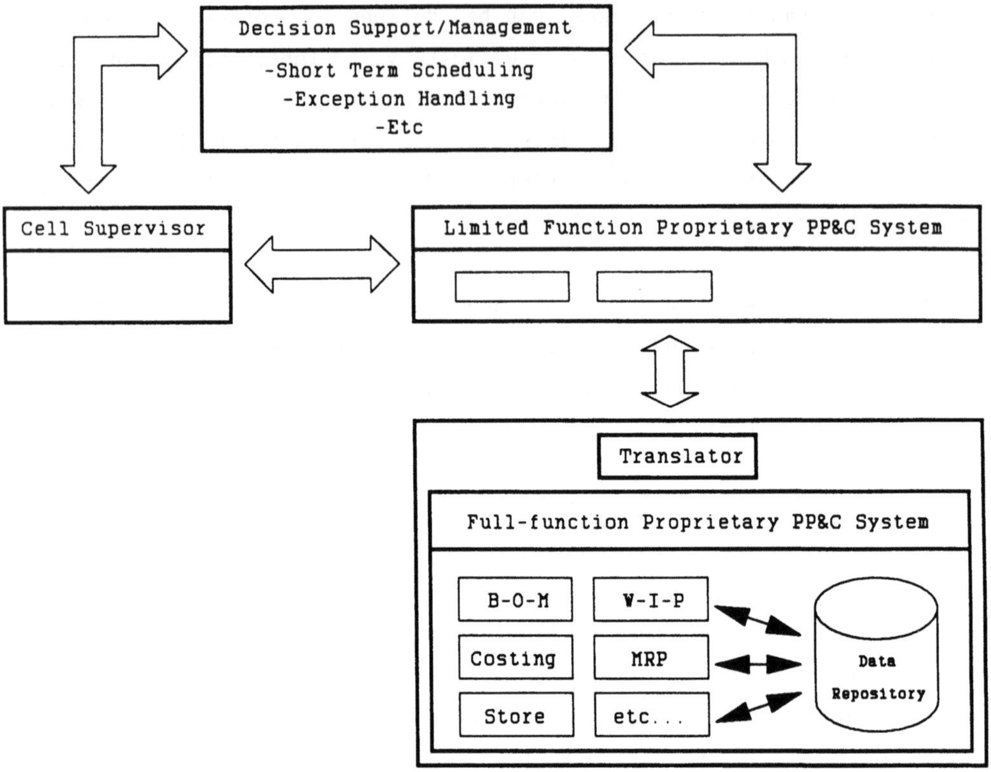

Figure 9 Conceptual View of Stage 1 Implementation at Loughborough

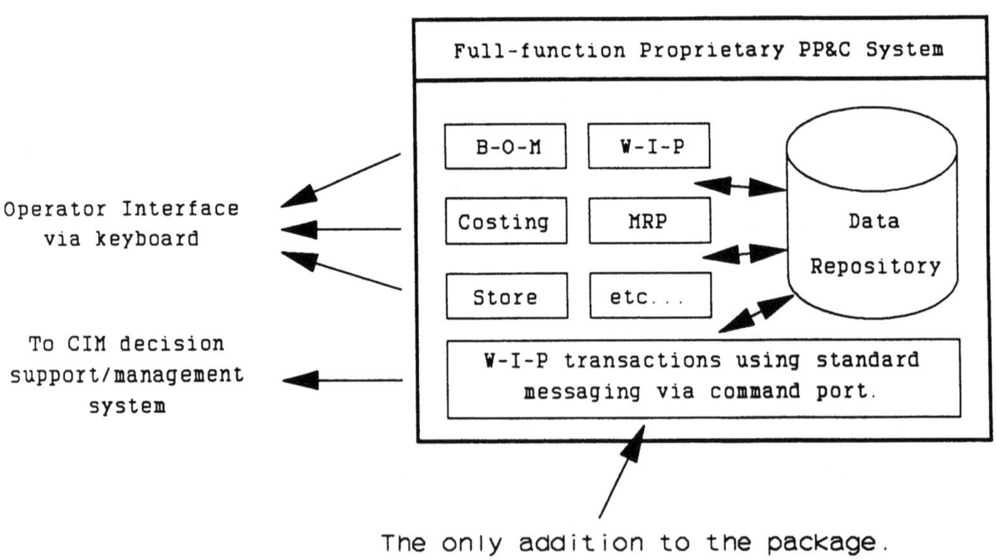

Figure 10 Requirements Placed on Package Designer to Achieve Stage 1

hierarchical control structure is typically imposed, each level (e.g. factory, shop, cell, ...) in this structure dealing with different levels of aggregation and time horizons. Additional problems relating to data distribution and co-ordination of activities impose further complexity. Figure 11 illustrates this approach.

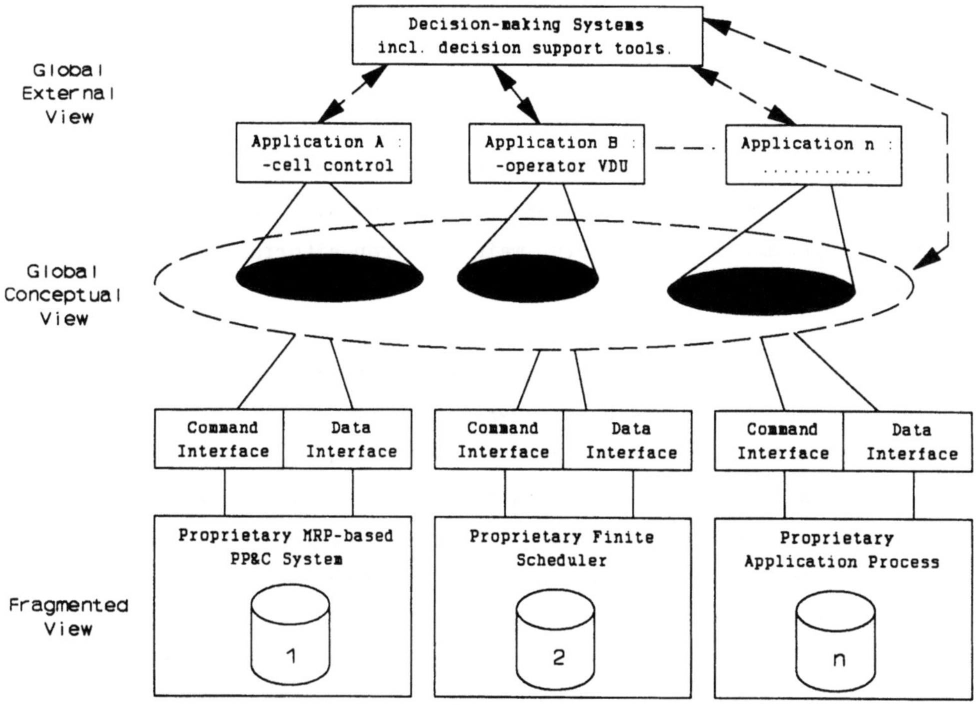

Figure 11 Typical Stage 2 Three Schema Approach

The work done on this approach so far by the Loughborough Systems Integration Group centres mainly on some exploratory programming of part of the three-schema architecture in Prolog.

3.2.3 Three-schema Systems with Independent Data Storage Systems

This approach represents an 'ideal world' solution, but, with the currently emerging standards (eg ANSI/SQL, MAP/MMS, IGES[11], EDIF[12], PDES/STEP[11], etc), it is becoming an increasingly realistic long-term goal.

The first and most important requirement is the separation of the major data items, as a resource accessible via standard retrieval and update mechanisms, from the manipulative processes provided by proprietary packages. The emerging SQL standards are a significant step in this direction. In addition to standard storage and retrieval mechanisms, standard data types and must be defined. These data types may include additional complex types, eg for co-ordinates in space, etc.

The second requirement is for packages to provide highly modular functions, with standard 'hooks' to allow a common format for initiating processes, passing values to, and subsequently receiving values from the functions

As in stage 2, the decision-making systems (both human and computer-based) will utilise a range of decisions-support tools. These will include rule-based systems with access to main data repositories and application processes and simulation systems.

Results from work on stage 2 should largely be transferrable to stage 3. One of the major differences between stages three and two is the reduction of translation/interface requirements in the information architecture. Little, if any, change is required in the network architecture. The easier direct access to, and manipulation of, information resources may encourage a less rigid hierarchy of command and feedback in the application architecture. The separation of data repositories from proprietary application processes will enable the location and organisation of data to be based much more closely on a realistic model of the actual physical systems and their functional requirements. Figure 12 illustrates the approach of stage 3.

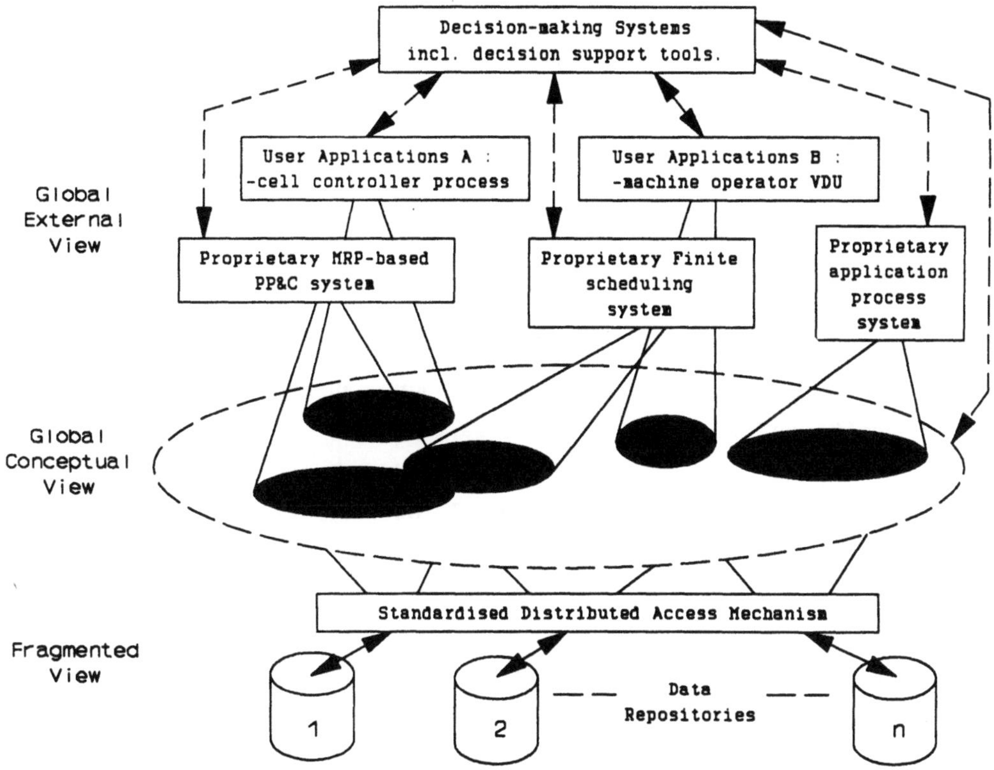

Figure 12 A Conceptual View of Stage 3

Migration from stage 2 to stage 3 will be an incremental process as more proprietary packages utilise standard database systems.

4 THE FEATURES OF A GENERIC FRAMEWORK FOR INTEGRATION

We have stressed the need to avoid bespoke customised solutions so that the benefits of integration can be realised without attendant disadvantages of inflexibility. Almost invariably, however, todays factory integration schemes incorporate 'hard' applications software, this software being written in accordance with some previously determined specification which describes a specific relationship between the manufacturing entities involved. Using such an approach, integration benefits can accrue but will generally result in the creation of a more complex standalone manufacturing entity, the characteristics of which cannot easily be modified. Such solutions may be regarded as fixed or hard integration as opposed to flexible integration where individualentities remain distinct but interact in a reconfigurable manner as desired by the user. Even so engineering a hard integration solution will involve very high costs.

In summary hard integration solutions will demonstrate major limitations resulting from:

(i) the need for high cost, 'major surgery' if a different configuration or functionality is required,

(ii) the need to 're-invent the wheel' at very high cost for the wide variety of manufacturing systems and industry requirements that can be encountered, and

(iii) the need to restrict the range and source of entities involved to those with which the system builder has expertise. Hence the facilities and functionality included in a system can be severely restricted by the experience of the system builder.

Thus a reliance on hard integration solutions cannot represent a long term prospect so that the need for flexible integration methods and tools will be of increasingly vital importance to all manufacturing organisations.

Despite worldwide standardisation activities facilitating communication between heterogeneous computer systems, truly flexible integration is not yet with us. However, those activities have and will continue to provide a major 'stepping stone' in the right direction. Furthermore, the results of major manufacturing systems integration and information technology research programmes (particularly in North America and Europe) are evolving integration tools and methods which can provide the next step forward.

The authors believe that it is now time to build on these initiatives to specify a framework for flexible integration. The necessary provisions laid down by this framework (or 'master plan') could then be incorporated within any specific integration scheme, thereby allowing it to be expandable over an acceptable time frame.

Clearly, the evolution of a generic framework is by no means a trivial task. However, it is possible to suggest certain features of that framework.

As previously discussed, the collaborative research studies at Loughborough University have served to highlight the definitive need to separate the integration problem into at least three architectures as illustrated by Figure 1. Through separation maximum flexibility can be achieved and the results of worldwide initiatives can be most effectively harnessed. These facts can be best illustrated by considering existing and developing integration tools and methods relating to those architectures. To aid the reader's understanding, simply stated the application and information architectures address organisation, interaction and messaging issues relating to 'action causing' and 'storage' functions respectively.

4.2 The Network Architecture

The anticipated availability over the next decade of various implementations of the ISO/OSI reference model such as broadband TOP, MAP, MAP/EPA and Fieldbus, will provide a limited range of widely accepted network architectures. These standard interconnect mechanisms will allow multi-vendor distributed manufacturing entities to be connected together with Figure 13 depicting their possible use. Thus shop floor, engineering office and business computer systems can be interconnected with a particular OSI option being chosen based on criteria such as cost, data rates involved, quantities of data transferred and the reliability with which the timely arrival of data can be ensured.

The ISO/OSI reference model implementations can provide seven layers of communication protocols which at the topmost layer (ie the application layer or layer 7) provides support functions for the users application software so that virtual connection between distributed computers can be achieved.

Currently much activity within the MAP/TOP initiative is concentrated on evolving application layer support functions including network management,

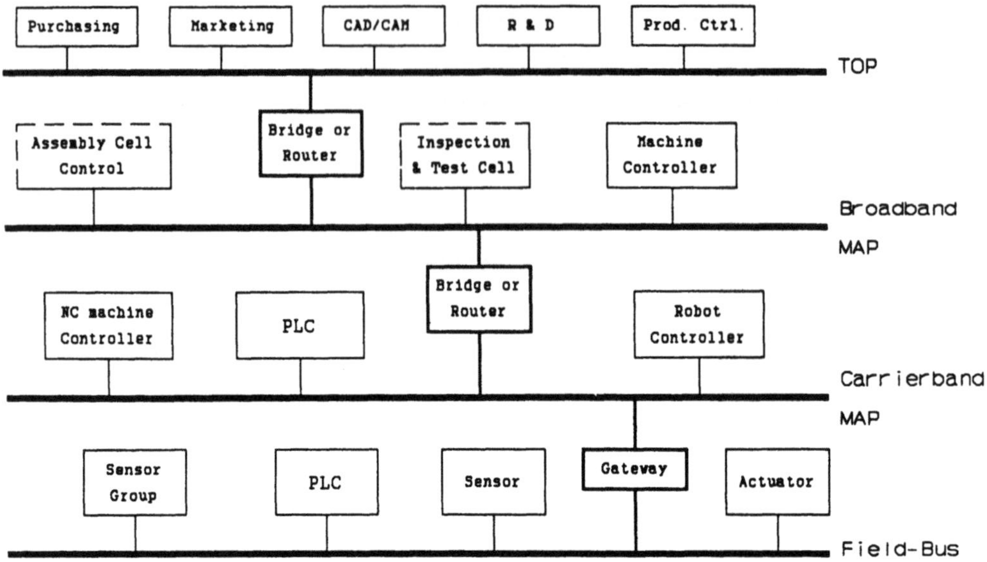

Figure 13 Example Enterprise Communications Structure

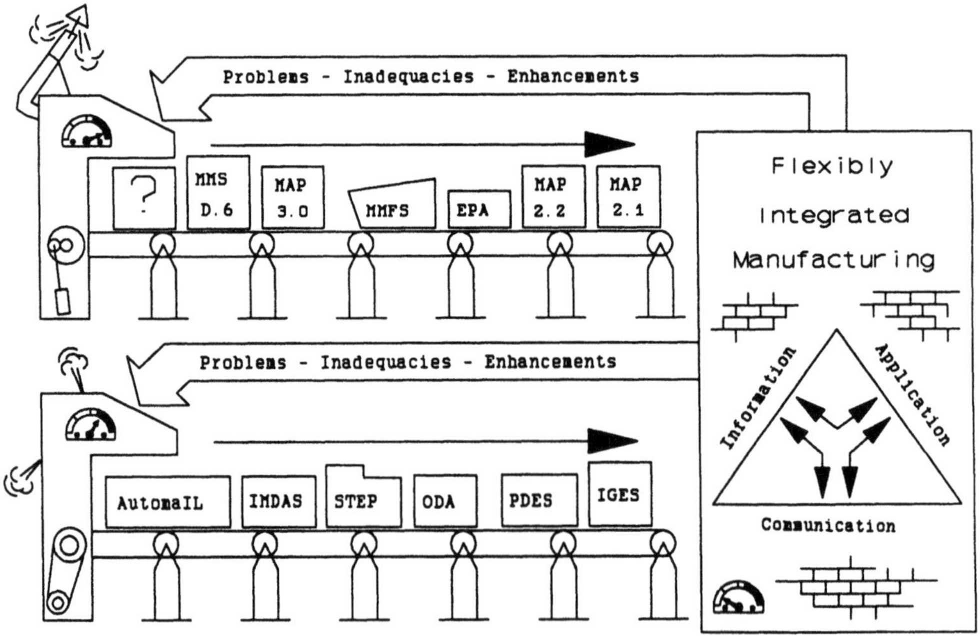

Figure 14 Proliferation of Standards and Concepts

file transfer and management (FTAM), manufacturing message services (MMS) and message handling services (MHS).

The evolution of these layer 7 application software support functions was initiated following the realisation within the MAP/TOP community that interconnection is not integration and that solutions to the organisation, interaction and messaging problems should be facilitated and where possible regularised. In general, however, MAP/TOP layer 7 support functions are growing a little like 'topsy' (as illustrated in Figure 14).

The authors believe that conceptually at least certain of the application support functions (eg MMS, MHS) should be considered to reside outside the scope of the network architecture. By separating these functions greater flexibility is facilitated eg running MMS over TOP or MHS over MAP would be considered to be natural and would have advantages in certain application scenarios.

4.3 The Information Architecture

As previously stated much progress has been made by the IT community in the areas of distributed database technology and data modelling. Furthermore, the USAAF and NBS programmes have illustrated that those concepts can be translated into specific areas of manufacturing organisations.

Also the MAP/TOP community have approached this problem from the 'bottom up', aiming to include data modelling standardisation activities such as PDES, IGES, GKS, Electronic Mail, Virtual Terminal and possibly EDIF within the TOP specification.

Those different initiatives serve to illustrate the need to separate organisation, interaction and data modelling issues relating to manufacturing information (ie data storage and processing functions). Such issues require subtly different solutions at different hierarchical levels and in different types of manufacturing environment. Through separationthe problems of providing high performance, reconfigurable information architectures can be properly assigned to the IT community. However, the functionality required from and the interfaces to the information architecture must be clearly defined so that the needs of manufacturing organisations can be served over an acceptable time frame.

The existence of appropriate information architectures will grow in importance as closer customer/supplier relationships need to be formed when seeking higher efficiency levels and opportunities for company growth.

4.4 The Application Architecture

In many ways, the functionality of the application architecture can be considered to mirror that for the information architecture. However, here the purpose is to provide support functions for action causing events. Many of the issues herein have been discussed with respect to the evolution of the generic AUTOMAIL.

The MAP/TOP initiative has also addressed certain problems in this area through activities in regard to MMS. However, currently MMS and it's companions are somewhat specific in nature and directed mainly at the shop floor environment.

Like the information architecture, the authors believe that the complexities involved in providing an application architecture, suitable for the various application scenarios found in manufacturing organisations, requires a separate top-down approach which treats emerging OSI network architectures as a utility. Again issues such as determining the interfaces to the network and information architectures need to be addressed.

4.5 Conclusion

As yet, truly flexible integration schemes do not exist which can have a sufficiently long time frame to accommodate unforseen changes in:

(i) products and processes,
(ii) manufacturing practices, and
(iii) enabling technologies.

However, the results of worldwide initiatives can be utilised in formulating a master plan or framework for flexible integration. The specification of this framework would need to be an interactive process and can only be evolved by decomposing the problem so that appropriate expertise and emerging integration tools can be effectively harnessed.

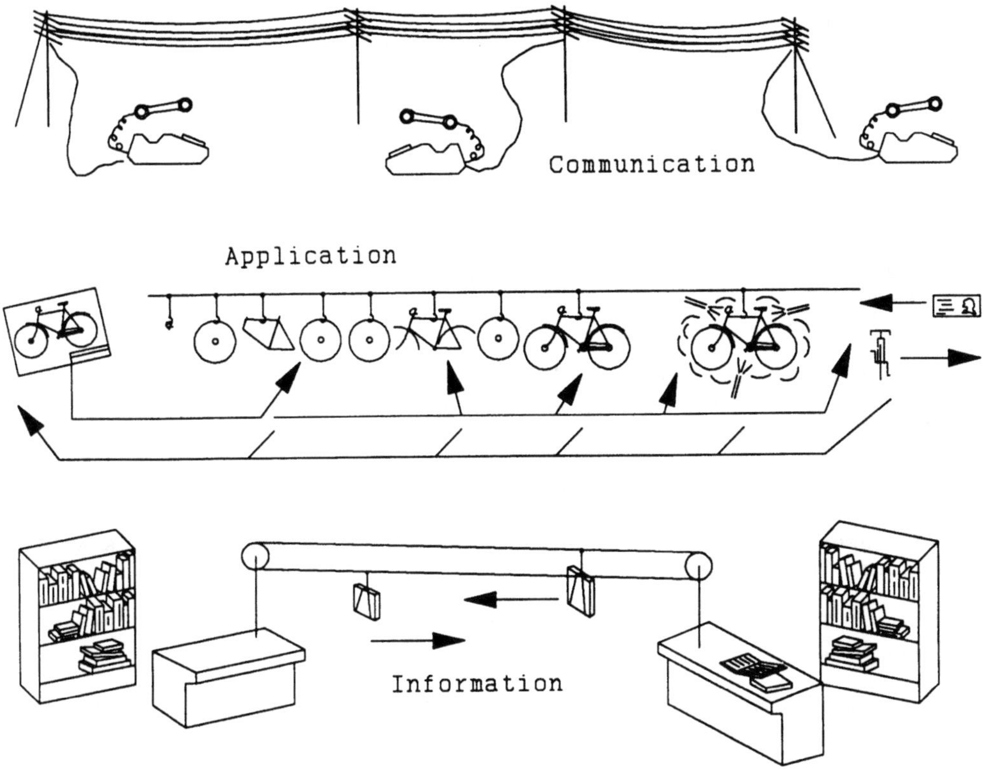

Figure 15 Interconnect, Control and Information Support

Initially decomposition should result in three separate architectures which subsequently would themselves be decomposed into more containable sub-problems. As illustrated by Figure 15, the three configurable architectures so formed can be loosely categorised as follows:

Network Architecture	-	The Interconnect Mechanisms
Application Architecture	-	The Manufacturing "Actions" and their Support Environments
Information Architecture	-	A Library of Manufacturing Information and Support Services

5 REFERENCES

1 Day, J D, and Zimmerman, H,: The OSI Reference Model: Proceedings of IEEE, 1983, 71, (12)

2 MAP and TOP Version 3.0 Specifications: Obtainable from the Society of Manufacturing Engineers, Detroit, Michigan, USA

3 Weston, R H, Sumpter, C M, Gascoigne, J D, and Hodgson, A,: MAP and Integration Architectures in the Flexible Manufacture of PCB's: Proceedings of CIRP Int Conf on Production Research, Cincinnati, Ohio, USA, June 1987

4 Manufacturing Message Specification, Part 1: Service Specification, Appendix B: Guidelines for Writing Companion Standards: EIA Project 1393A Draft 6, May 1987

5 Personal Communication: Cincinnatti Milacron, Ohio, USA, August 1987

6 ANSI American National Standard Database Language SQL: American National Standards Institute Inc, New York, December 1986

7 Teorey, T J, and Fry, J P,: Design of Database Structures: Prentice Hall, 1982

8 Vernadat, F,: Artificial Intelligence in CIM Databases, IFS Artificial Intelligence: Impliciations for CIM, 1988

9 Su, S Y W, Lam, H, Khatib, M, Krishnamurthy, V, Kilmar, A, Malik, S, Mitchell, M, and Barkmayer, E,: The Architecture and Prototype Implementation of an Integrated Manufacturing Database Administration System: Proceedings of Spring COMPCON Conference, San Francisco, CA, USA, Dec 1986

10 ICAM Project: Computer Program Development Specification for ICAM Integrated Support Systems (IISS) Configuration Item: Precomputer prepared by Control Data Corporation and D Appleton Company, USA, Dec 1983

11 Rui, A, Weston, R H, Gascoigne, J D, Hodgson, A, and Sumpter, C M,: Automating Information Transfer in Manufacturing Systems: Computer Aided Engineering Journal, June 1988

12 Hillawi, J I, and Bennett, K R,: EDIF - An Overview: Computer Aided Engineering Journal, March 1987

Modeling Issues of Specific Material Flow Systems Functions

SOME ISSUES IN CONTROLLING INTEGRATED WIP MATERIAL HANDLING SYSTEMS

Leon F. McGinnis

School of Industrial and Systems Engineering
Georgia Institute of Technology
Atlanta, GA 30332-0205

ABSTRACT

In an automated factory, material handling will be but one element of a hierarchical control system. Experience shows, however, that the computer controls for some highly automated work-in-process (WIP) handling systems are not consistent with strictly hierarchical architectures. In this context, we propose an approach to integrating the control of automated WIP handling within hierarchical process control and explore some of implications for the design of interfaces and coordinating mechanisms.

1.0 INTRODUCTION

As computer integrated manufacturing draws closer to a commonplace reality, a great deal of attention is being focused on manufacturing control systems, their logical architectures, and the underlying communication networks and database systems. While there certainly are no standards in this arena, there are some reasonably well-defined themes.

In terms of logical architecture of the control system, clearly the dominant theme is hierarchical control. As promulgated by the National Bureau of Standards [1, 6, 8], the hierarchical architecture embodies the "black box" principle of structured analysis/structured design. Conceptually, the discipline of the hierarchical organization should simplify the designing and coding of the control software.

NATO ASI Series, Vol. F53
Advanced Information Technologies for Industrial Material Flow Systems
Edited by Shimon Y. Nof, Colin L. Moodie

Hierarchical architectures appear to work quite well for controlling manufacturing processes, at least discrete processes. However, we have observed that highly automated material handling systems for work-in-process material often do not seem to fit well within the strictly hierarchical paradigm. Our objective in this paper is to explore some of the reasons for this apparent lack of fit, and to suggest an approach to resolving the problem.

The paper is organized as follows. In the remainder of this section, we shall summarize the key relevant aspects of the hierarchical paradigm, describe two examples of automated handling systems for WIP, and discuss the issues arising from the two examples. In the following section, we propose a slightly modified paradigm that seems to overcome many of the problems, and discuss some implementation issues. In the final section, we attempt to place this issue in the context of current research activities.

1.1 Hierarchical Control in Manufacturing

The National Bureau of Standards (NBS) has been quite active in promoting hierarchical control, both through publications and in developing a hierarchical control system in their demonstration facility, the Automated Manufacturing Research Facility, or AMRF [6]. The NBS model of a hierarchical control system is shown in Figure 1 (see, e.g., [1], or [8]).

According to the NBS model, each level of the hierarchy has a different type of task to manage. The facility level is concerned with very aggregate and long term issues, such as information management, manufacturing engineering, and production management. The shop level allocates resources to tasks, e.g., assigning jobs to machine tools, or allocating storage space, and schedules tasks, e.g., job orders or maintenance, over a much shorter horizon, e.g., a day.

In the NBS model, the cell level manages the details of monitoring job progress, dispatching resources, and scheduling operation start times. The work station level directs and

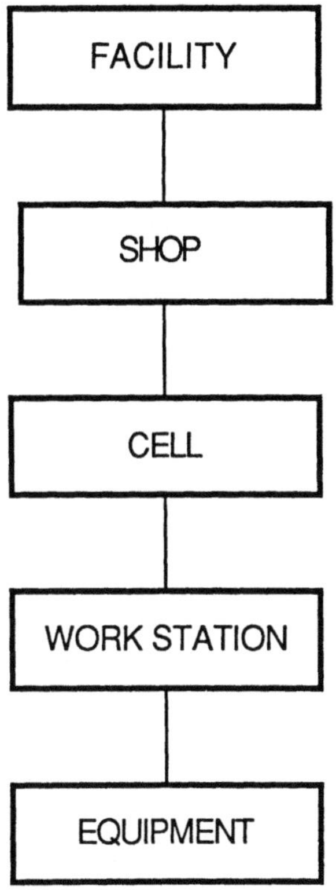

Figure 1. NBS Hierarchy

coordinates specific production resources, such as machine tools, robots, storage systems, or guided vehicles. Finally, the equipment level represents proprietary controllers for specific equipment, e.g., the end-of-aisle controller for an automated storage system.

A slightly different hierarchy has been proposed by Digital Equipment Corporation and Philips [9], and is shown in Figure 2. Essentially, the hierarchy in Figure 2 allows more levels of control, for example, having several "factories" under one roof. Implicit in both the NBS and DEC/Philips models is the notion of a "command unit," i.e., any one of the boxes in Figure 2 represents a self-contained control system having a single interface to the "outside world." Thus, for example, a particular cell controller represents the only avenue for commands that will eventually result in the activation of devices within its command unit.

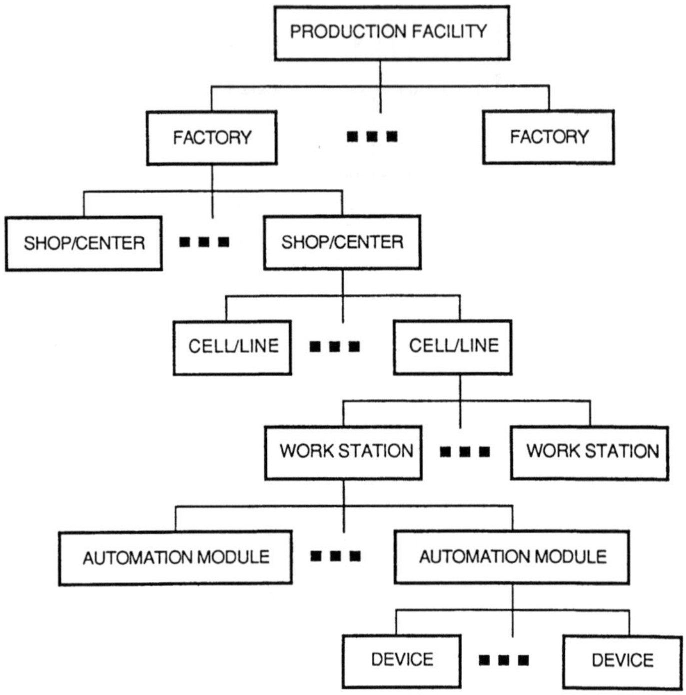

Figure 2. DEC/Philips Control Hierarchy

A reference model for the structure of a command unit controller also has been proposed in [9], and is shown in Figure 3. The reference model distinguishes between commands and their status, environmental data about the domain of the command unit, and global data about the system outside the command unit. It incorporates three functions: the H function decomposes a given command or task into subcommands which are transmitted to lower level controllers. THe G function condenses and processes environmental data, e.g., from limit switches, barcode readers, etc.. The M function is a "world model" which may be used by the H function in making the task decomposition.

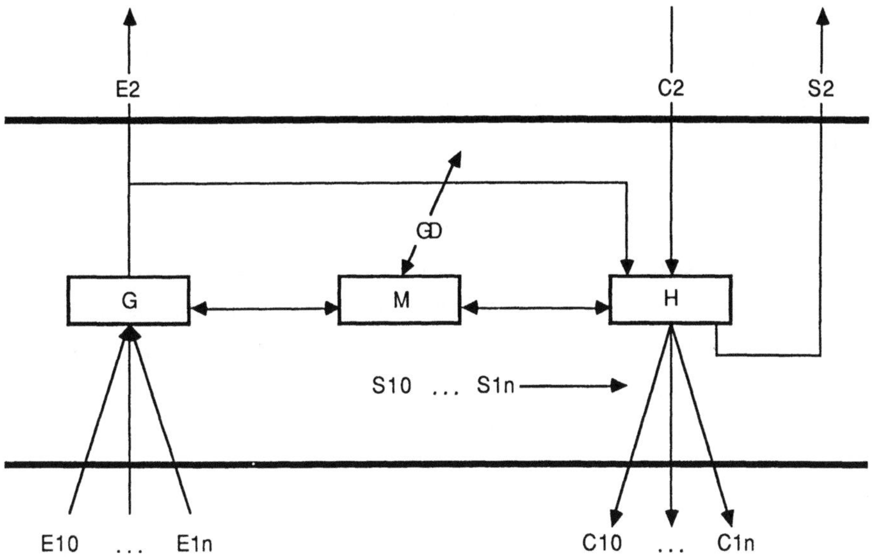

Figure 3. General Model of a Controller

As an example, consider a robotic assembly station consisting of a pallet positioner, parts feeders, part registration stations, grippers, robot arm, force and torque sensors, and vision inspection. A task for the assembly station controller might be to add three specific components to an assembly. The H function might decompose this into a series of specific operations, e.g., register assembly pallet, visually verify, move gripper to first component feeder, acquire component, move to registration station, register component, move to component assembly location, assemble component, etc. As each of these operations is completed successfully, the H function initiates the subsequent operation.

A failure of any operation requires a contingency action to be determined by the H function, perhaps in consultation with the M function. For example, the M function might estimate the probability of successfully completing a retry of a failed operation.

A bar code reader might read a bar code on the assembly pallet, and relay this information through the G function.

Similarly, sensors in the part feeders might be used through the G function to updarte the global database, so that a higher level controller might issue a replenishment command when a parts feeder nears an empty condition.

In the command unit controller reference model, a controller communicates upward only with its immediate superior controller, and downward only with its immediate subordinate controllers. Recognizing that some of the functions shown in Figure 3 may not be present in a particular realization of the reference model, it is a very general way of viewing the structure of a control unit in the manufacturing control hierarchy.

1.2 Automated Storage/Retrieval System for WIP Handling

Microload automated storage/retrieval systems, often referred to as "totestackers," have been used widely for movement and storage of work-in-process (see [7] for an extended discussion and references). The totestacker is simply a small unit load automated storage/retrieval system which permits loads to be removed from the system at certain storage locations, processed, and returned to the system at other storage locations. Figure 4 illustrates a plan view for a WIP application.

Consider the following typical application as an example [7]. A totestacker is used to provide WIP storage and transport between 20 different workstations in an extrusion finishing cell. Extrusions enter the totestacker in totes which have unique identifying bar codes. Each workstation has one or more input ports and one or more output ports. Each port consists of a rack location and a segment of unpowered roller conveyor, so each port has its own capacity for buffering totes.

Jobs entering the totestacker have both the parts to be processed, and any tooling or instructions required. Whenever the storage/retrieval (s/r) machine picks up a tote, its bar code is scanned and transmitted to the s/r controller, which then communicates with a shop host computer to determine the

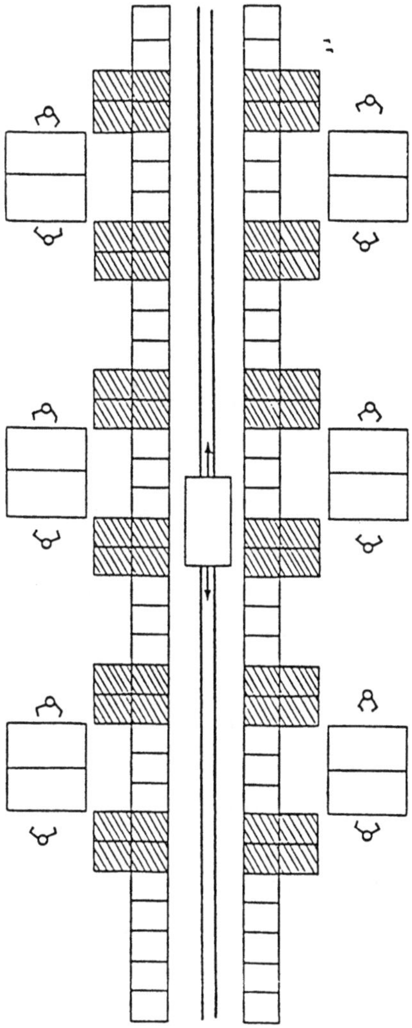

Figure 4. WIP AS/RS Application

next operation/workstation. The s/r machine then delivers the tote to the storage zone for the destination workstation. If there is space available in that workstation's input port, the s/r machine takes the oldest job from the storage zone and delivers it to the input port. If there is a job in the output port, the s/r machine picks it up, and delivers it to its destination. Each port has sensors which signal the s/r controller whenever a tote is placed in the or removed from the port.

The cycle continues until there is no space in an input port, or no job in an output port for some workstation. At that point, the s/r controller looks for the "oldest" request for service (either a delivery to an input port or a pickup from an output port) and begins the cycle again at the corresponding workstation. Obviously, this is not the only way to operate such a system. Nevertheless, it represents a fairly typical type of control strategy.

1.3 Automated Guided Vehicle System for WIP Handling

Automated guided vehicle systems (AGVSs) are another very popular technology for moving material on the factory floor. AGVs find application for moving assemblies, for moving parts in fabrication, and for supplying parts to the shop floor and removing dunnage. The current technology in AGVS employs self-powered vehicles which follow either an active (radio-frequency) or passive (retroreflective strip) path which is layed out on the factory floor.

Figure 5 illustrates a layout for a large distribution warehouse. The facility contains a number of activities, including receiving, shipping, bulk storage, rack storage, some material processing, bagging, overpacking (placing several smaller boxes in a larger box) and shrink wrapping. The guidepath is uni-directional, which is typical for this type of application. In this particular application, the guidepath is a wire in the floor, emitting a radio frequency signal. To avoid collisions, the guidepath is divided into control segments, and only one vehicle is allowed on a control segment at one time.

The AGVS has a dedicated controller, which assigns load movements to vehicles. A number of algorithms have been proposed for this type of controller, including such rules as: oldest load to first available vehicle; closest load to an available vehicle; and when a vehicle becomes available, search a specified area for a load, if none is found assign the oldest load in the system. For more details on vehicle dispatching and references, see [10].

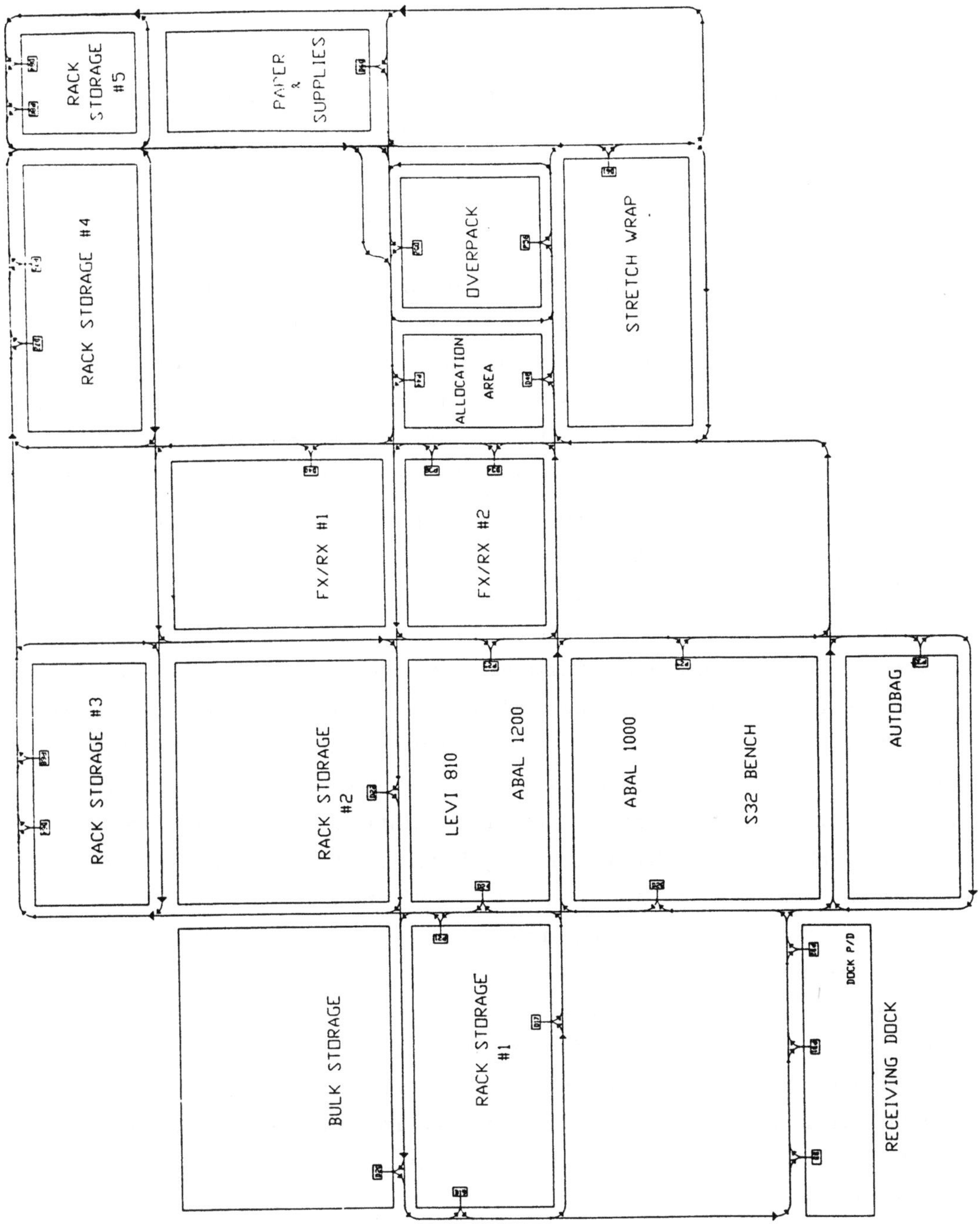

Figure 5. AGVS Application Example

1.4 Problems

These two examples of automated material handling for work-in-process present a number of problems for purely hierarchical control system design. In some cases, the problems arise from ambiguity in specifying the control hierarchy.

In the totestacker system, for example, how does one represent the hierarchy? It seems reasonable that there should be a controller for the workstations, and a controller for the AS/RS. In fact, there usually is a proprietary controller for the s/r machine. However, the workstations effectively "command" the s/r machine to deliver or pick up totes, based on the state of their input and output ports. As illustrated by Figure 6, no matter how we try to represent this relationship, we do not have a strict hierarchy.

By their very nature, AGVS control systems cannot be represented by a pure hierarchy. The central AGV system controller assigns tasks to individual vehicles. However, it is not at all clear how we should represent the relationship between the vehicles and the path segments. If, for example, we want to make the segments subordinate to the vehicles (arguing that the vehicle "assign" themselves to the segments), then the fact that the segments are shared by all vehicles leads to the diagram in Figure 7. Again, we do not have a strict hierarchy.

Another difficulty with AGVS is that the vehicles may interface with manufacturing at different levels of the hierarchy. A vehicle may interface with a particular workstation within a cell, or it may interface with a cell which has its own internal material handling system, or it may interface with a shop which has its own internal material handling system, all in the same factory.

For example, in an electronics assembly plant, a vehicle may deliver WIP to an automated board testing system which is logically a cell in the hierarchy. It also may deliver material directly to workstations in a through-hole insertion cell which moves circuit cards on a pallet conveyor within the

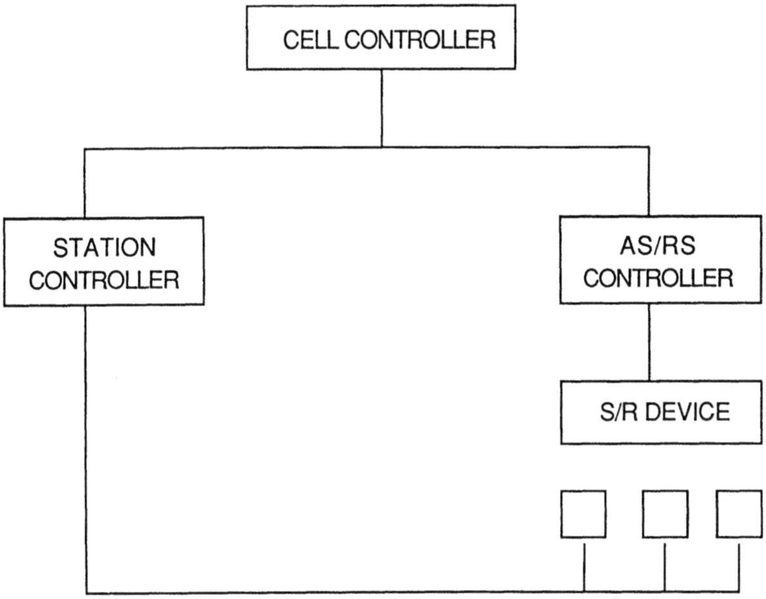

(a) S/R Device Controls Workstations

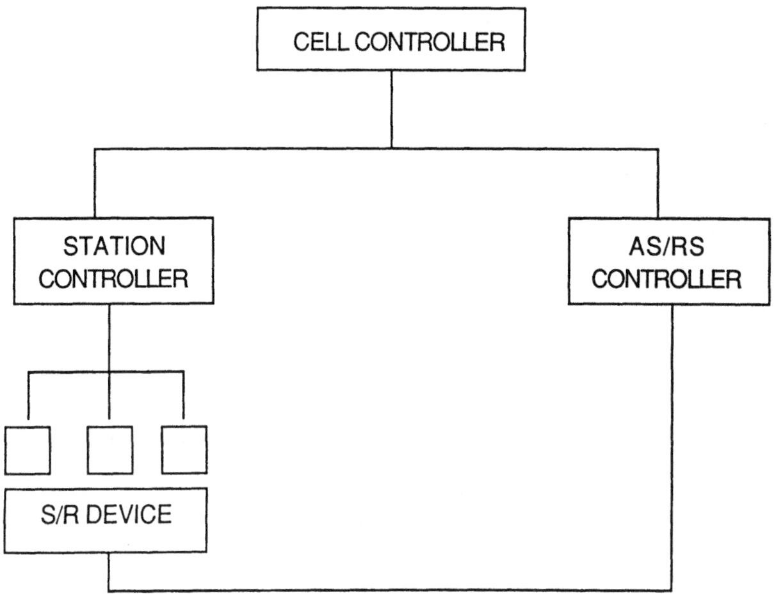

(b) Workstations Control S/R Device

Figure 6. First Example of Ambiguous Control.

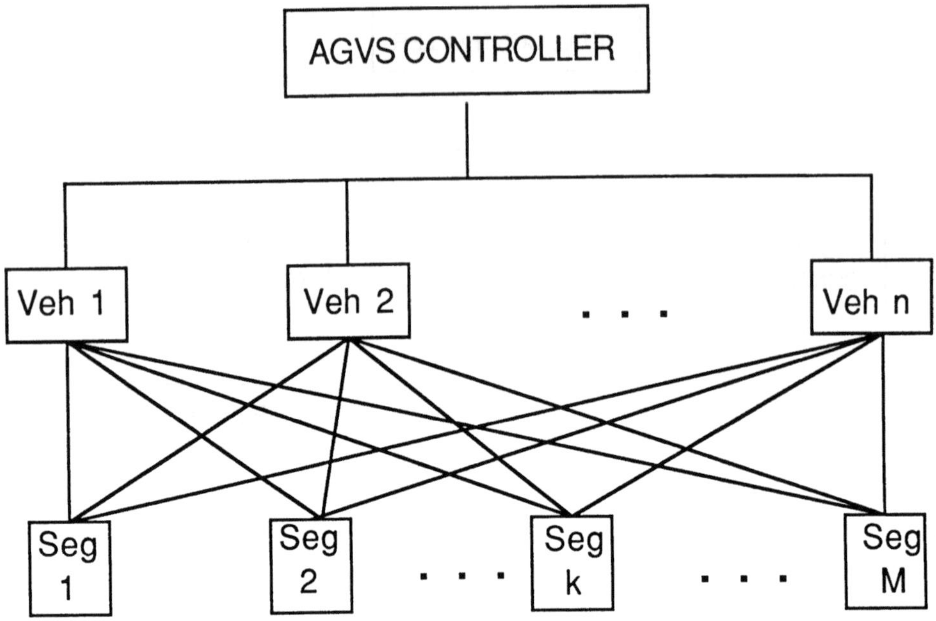

Figure 7. Second Example of Ambiguous Control Hierarchy

cell. The parts kitting area might be treated as a separate shop, where the vehicle picks up kits and delivers empty containers.

In this case there are two problems. First is the ambiguity; where should the AGVS controller be placed in the hierarchy, i.e., who are its "peers"? The second problem is really a practical problem. In order for a load to come to the attention of the AGVS controller, messages may have to be transmitted through several levels of the hierarchy. In practice, this communication across levels can introduce some delay into the system. For example, in the totestacker system described earlier, this delay turned out to be quite significant for overall system performance, even though it was only a few seconds for any one load movement.

Thus, strict hierarchies present problems for controlling WIP material handling systems. In the first place, by their very nature, such systems are "shared servers", i.e., they respond to requests for service from many different locations. In many cases, it is difficult to find a way to cast this relationship into a strict hierarchy. Even when it is possible, the result often is a requirement for commands and responses to traverse several levels of the hierarchy, potentially leading to significant delays.

2.0 PROPOSED PARADIGM

Integrating automated WIP handling systems into the currently accepted hierarchical paradigm presents two types of problems. One is the conceptual problem of specifying the structure of the hierarchy, and affects primarily the system designers and developers. The other is the potential problem of communication delays when several command control units are involved in what is essentially a single material handling transaction.

In attempting to resolve these problems, we should bear in mind that the hierarchical paradigm functions quite well for the non-material handling functions. Thus, we might not want to consider changes to the paradigm that jeopardize those other functions. What we propose, in fact, are some relatively modest changes to the hierarchical paradigm.

The first proposal is to modify the structure of the task decomposition function. In the past, the task decomposition function has been viewed as one of determining which subtasks must be performed by subordinates in order to perform the task assigned by the superior, scheduling the subtasks, and assigning them to subordinates.

For example, to manufacture a printed circuit card assembly, a cell controller might assign the various assembly operations to appropriate workstations and the corresponding move operations to the material handling system. However, the cell controller would issue the commands in a sequence that corresponds to their performance. The next command in sequence would be issued only upon completion of the previous command. Thus, the cell controller would issue a command to the material handling system to move the card to the workstation for the first operation. Upon completion of the move, the cell controller would issue a command to the workstation to perform the appropriate assembly operation. Upon completion of the assembly operation, the cell controller would issue another move command, and so on, until the card assembly was completed.

As pointed out earlier, the need to issue a move operation and get the corresponding operation complete message may lead to "dead time" for either, or both, the material handling device and the workstation. This dead time results from a purely serial synchronization of the move and assembly operations, i.e., the assembly workstation can't begin its opearation until the operation is assigned to it by the cell controller, but the cell controller can't assign the assembly operation until it knows that the move operation has been completed.

An alternative way to view task decomposition is to separate the task activation aspect from the task assignment, as illustrated in Figure 8. We may view a subtask as consisting of two elements: an operation assigned to a particular workstation; and an activation key. The assigned workstation must have both the operation and the key in order to execute the task.

Continuing the circuit card assembly illustration, suppose the cell controller simultaneously issues two commands, one a move command to the material handling system, and one an operation command to the receiving workstation. The key for the move command would be issued along with the command, and the key for the operation command would be the circuit card itself. Upon completion of the assembly operation, the cell controller would be notified, and it could determine the next operation and issue the appropriate move and operation commands. Note that the structure for issuing commands would continue to be hierarchical in this example.

If we separate task assignment from task activation, we can identify three modes of activation. A task is activated by the superior controller when the subordinate is given both the task and the key and has no alternative task to execute. A task may be activated by the assigned controller when that

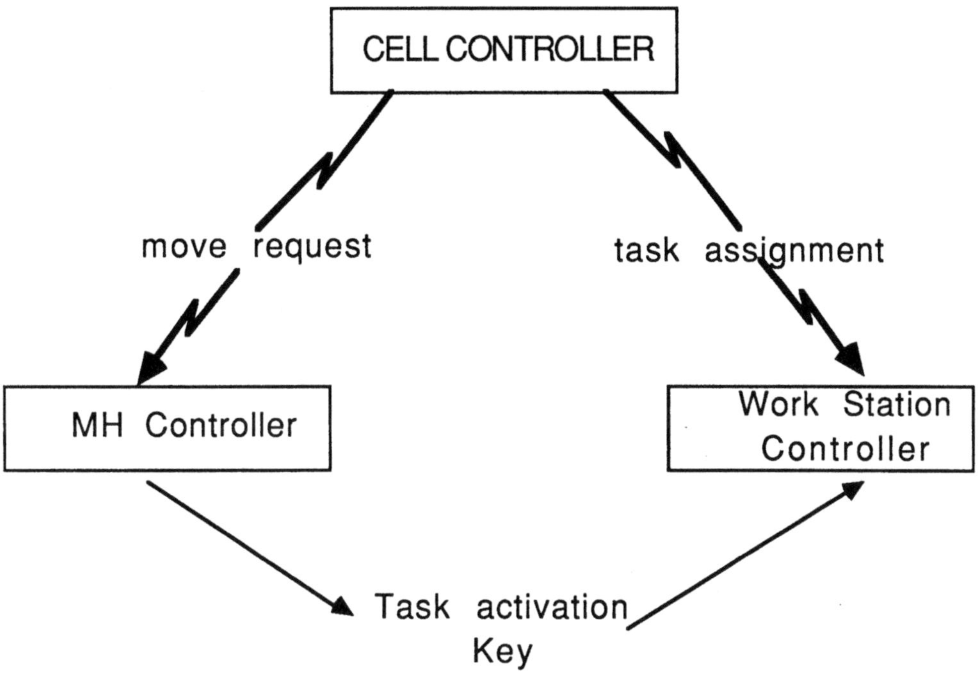

Figure 8. Alternative Task Decomposition

controller has several tasks and their keys, and must choose which one to execute. A task may be activated by a "peer" when the assigned controller has the task, and then receives the associated key from a peer.

Superior activation corresponds to the type of stepwise control typical of, say, a robotic assembly station, where a station controller is coordinating such diverse devices as parts feeders, grippers, x-y tables, and vision registration. Self activation corresponds to the situation of a WIP material handling system controller, say, a totestacker controller or an AGVS controller. Peer activation, as we have described it, currently is not consistent with the accepted hierarchical paradigm.

In order for the revised definition of task decomposition to be useful, we also must revise our view of the command hierarchy. Specifically, we must permit direct peer-to-peer communication for the purpose of transmitting activation keys, as illustrated in Figure 9. For the purposes of designing and developing a control hierarchy for a particular application, the essential requirement is to identify those command units that will be allowed (or required) to exchange activation keys. The reference model for a command unit must be revised to reflect this added interaction, which is similar, conceptually, to the interaction of the world model with global data.

As an aside, we have attempted to avoid confusing the discussion of the control hierarchy by simultaneously considering the communication network, as we consider them

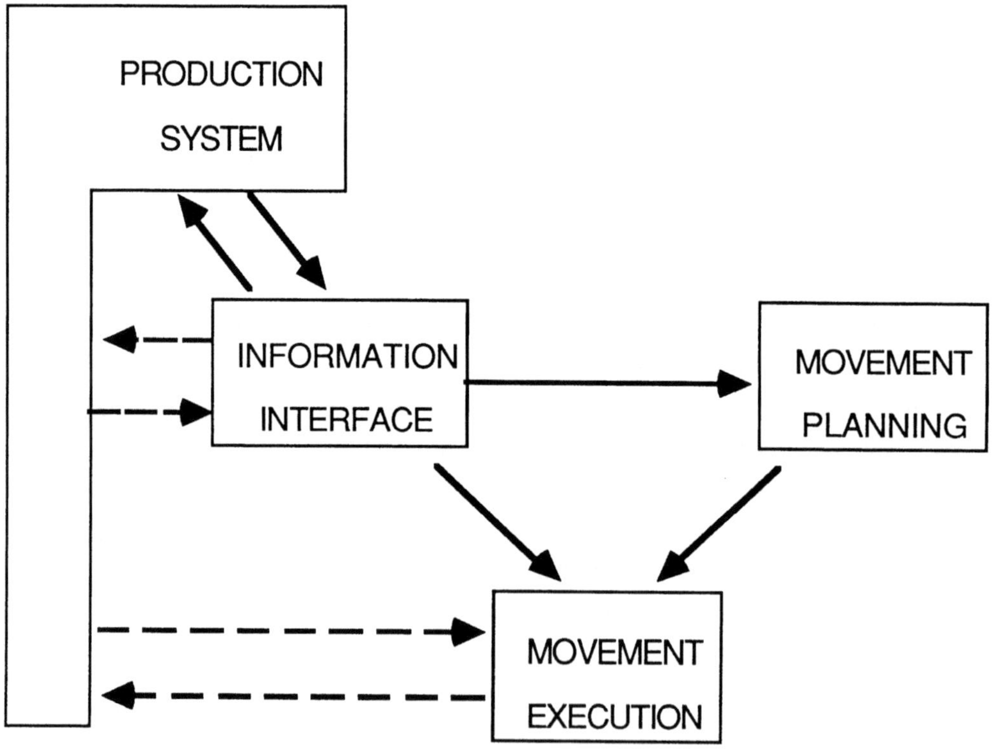

Figure 9. Peer-to-Peer Communications

distinct issues. The implementation of the control hierarchy certainly will depend upon the structure of the communication network, but the logical structure of the control hierarchy should not depend upon the physical communication network.

These two structural changes to the hierarchical paradigm should be coupled with two changes in "standard" practice. The first practice is to pair a move command with the subsequent operation command, and have the item being moved carry the activation key for the subsequent operation. This is a change in implementation, rather than a change in the hierarchy paradigm.

The second suggested change in practice is to treat the WIP material handling system as a black box entity in the hierarchy. That is, the control hierarchy should not attempt to represent the details of the internal structure of the WIP material handling command unit. There are several justifications for this. First, as we've discussed, there are some material handling systems whose controls simply don't fit a hierarchical model. Second, most automated WIP material handling systems have their own proprietary control systems, which, hierarchical or not, are not easily accessible to the systems designer. Finally, if we adopt the notion of "peer-to-peer" exchange of activation keys, there is no reason for us to look inside the WIP material handling systems's control structure.

3.0 IMPLICATIONS

There are two types of implications for adopting the proposed changes to the structure and practice of hierarchical control of integrated manufacturing systems. On the practical side, there may be an impact on the design of lower level command units. For example, the ability to store and randomly retrieve commands (or match up commands with their activation keys) may represent a higher level of functionality than many such controllers now exhibit. In a similar fashion, allowing peer-to-peer communication may involve revising communication protocols for some systems.

There also are some important implications for manufacturing systems research. Figure 9 illustrates the conceptual structure of the WIP material handling system within the proposed hierarchical paradigm. As the figure shows, if we could adopt the proposed paradigm, we could begin to focus on the three distinct elements: (1) the interface between the production system (the manufacturing operations) and the WIP material handling system; (2) the material handling system control; and (3) the physical components of the material handling system.

A critical question from the production system design perspective is, "How reliable will the material handling system be?" The distribution of the delay in executing move commands will impact the need for a WIP buffer at the workstations. The more unreliable, i.e., the higher the variance, the greater the need for WIP buffers, with the attendant disadvantages. Thus, a fundamental research problem is to develop methodology for predicting not only the average execution time, but also the distribution of execution times, in order to better design the production workstations.

From the WIP material handling system perspective, a critical issue is to maximize the throughput capacity (for a given hardware configuration and system cost). In order to most efficiently utilize the material handling equipment, the material handling system controller would prefer to have a large number of move commands to choose from. The selection method is sometimes referred to as a "command arbitration scheme," and there has been some recent research interest in command arbitration, especially for AGVS [10] and AS/RS [2, 3, 4, 5].

Obviously, the larger the number of move commands that are "pending," the larger the average execution delay. Moreover, for most command arbitration schemes, we also could expect a higher variance. Thus, another research problem, which has attracted less attention is the problem of devising command arbitration schemes which can simultaneously optimize the material handling system and consider the urgency of specific move commands, so that the variance in execution times does not lead to much higher buffer requirements.

The move command execution time clearly depends upon the capability of and configuration of the material handling hardware. While there continues to be active research in developing new technology, e.g., autonomous vehicles, better sensors, better communication, etc., there appears to be relatively less research interest in the fundamental problems of facility layout or configuring the material handling system. At least there is little specific research which focuses on the layout or configuration problems as they impact the requirement for workstation buffers.

4.0 CONCLUSION

Automated WIP material handling systems are not easily represented in the accepted hierarchical control system paradigm. We have proposed two modifications to the hierarchical paradigm, and two changes in the practice of control system implementation. We have argued that these changes will clarify the relationship between manufacturing operations and material handling operations, thus simplifying the tasks of designing and implementing integrated control systems. In addition, we have identified some implications of these changes for research on manufacturing systems.

Obviously, our proposal is not the only approach to resolving the problem. It represents only one possibility, and we hope it can be the impetus for others to be developed.

REFERENCES

1. Albus, J. S., C. R. McLean, A. J. Barbara, and M. L. Fitzgerald: An Architecture for Real-Time Sensory-Interactive Control of Robots in a Manufacturing Facility. Proceedings of the 4th IFAC Symposium on Information Control Problems in Manufacturing Technology. Gaithersburg, MD. pp. 81-90: 1982.

2. Han, Min-Hong, L. F. McGinnis, J. S. Shieh, and J. A. White: On Sequencing Retrievals in an Automated Storage/Retrieval System. IIE Transactions, 19, 1, pp. 56-66 (1987).

3. Han, Min-Hong, and L. F. McGinnis: Automated Work-in-Process Carousels: Modeling and Analysis, Georgia Institute of Technology working paper. 1987.

4. Han, Min-Hong, and L. F. McGinnis: Rotary Rack Operation. Material Flow, 4, pp. 283-293 (1987).

5. Han, Min-Hong, and L. F. McGinnis: Control of Material Handling Transporter in Automated Manufacturing. IIE Transactions, to appear.

6. Jones, A. T., and C. R. McLean: A Cell Control System for the AMRF. Computers in Engineering, 2. The American Society of Mechanical Engineers, pp. 353-359 (1984).

7. McGinnis, L. F.: Evaluating Integrated WIP Handling Systems. Proceedings of the 3rd International Conference on Automated Materials Handling (Prof. R. H. Hollier, ed.) IFS (Publications) Ltd. (1986).

8. McLean, C., M. Mitchell, and E. Barkmeyer: A Computer Architecture for Small Batch Manufacturing. IEEE Spectrum, 20, 5, pp. 59-64 (1983).

9. Reference Model of Production Systems, Version 1.0, Digital Equipment Corporation and Nederlandse Philips Bedrijven BV (1987).

10. Tanchoco, J. M. A., and C. L. Moodie, eds.: Automated Guided Vehicle Systems, (Special Issue of) Material Flow, 4 (1987).

ACKNOWLEDGEMENTS

Many of the ideas described in this paper are the result of an ongoing project to develop a manufacturing system emulation in the Material Handling Laboratory at Georgia Tech. This project has been supported by three organizations within Georgia Tech: the Computer Integrated Manufacturing Systems Program, the School of Industrial and Systems Engineering, and the Material Handling Research Center. I especially want to express my gratitude to my students, Debbie Dutton and Curt Ide, whose creativity and enthusiasm have continually pushed me to elaborate and clarify the notions of control and hierarchies in manufacturing logistics systems.

TOOL MANAGEMENT AND ITS IMPLICATIONS FOR MATERIAL FLOW

A S Carrie and U S Bititci
Manufacture and Engineering Management Division
Department of Design, Manufacture and Engineering Management
University of Strathclyde
GLASGOW G1 1XJ
Scotland

ABSTRACT

Tool Management is a topic receiving increased attention, although diverse and narrow concepts of what is involved are still current. This paper reviews the aspects of tool management and presents three brief cases. The first illustrates the complexity of the logistics problem; the second the difficulties in planning and support despite automated tool handling; the third the operational problems in achieving communication links. The paper identifies areas where further research is needed, namely, the tool management practices, communication standards and effective modelling of tool management systems.

INTRODUCTION

Tool Management is an issue now receiving much attention in industrial practice. It is recognised that the lack of proper tool management procedures may jeopardise the efficiency and effectiveness of the manufacturing system, whether conventional or computer integrated.

In the research community there has been some interest in the field, but the work is not sufficiently comprehensive and not related to technological practicalities.

It is contended that the effectiveness of tool management is restricted by several factors:

(1) inadequate shopfloor discipline
(2) inadequate conception of functions involved
(3) complexity of the logistics which may be involved and inadequacy of current models
(4) restricted tool supply technology

NATO ASI Series, Vol. F53
Advanced Information Technologies for Industrial Material Flow Systems
Edited by Shimon Y. Nof, Colin L. Moodie

(5) difficulties in the specification and implementation of the information system necessary to support the complex logistics and integrate various functions

These topics are discussed and the research needs are identified with reference to several uses.

SHOP FLOOR DISCIPLINE

Because computerisation of tool management is at an early stage, even in some automated manufacturing systems, in many cases shop floor discipline is inadequate. Procedures for return of tools for refurbishment are poorly defined and inadequately controlled. Operators may hoard tools at the work centre because they do not trust the tool room to refurbish and return them in time. Tool stores may not commence refurbishment of tools until they are already needed back at the machines. Companies which are investing millions of dollars in just-in-time manufacturing systems leave tooling to be ordered by discredited stock control systems.

In many respects the situation resembles that which existed in the early days of computerised inventory management for manufactured items and materials, prior to the concept of inventory control data accuracy being established as they are to-day.

A similar situation may exist with fixtures and other manufacturing aids. Where fixtures are shared between components, or are built up from modular components and may be stripped down between batches, the situation is additionally complex.

In flexible manufacturing cells and systems the need for clear procedures is obvious, and great progress is being made. There is a need to extend this approach throughout the company.

Further research is needed to evaluate advantages and limitations of various shop floor procedures whether manual or automated. Standard models are required which define shop floor practices under several configurations and constraints for use as building blocks to define efficient, effective shop floor procedures for tool management and control within a specific application.

FUNCTIONS OF TOOL MANAGEMENT

Tool Management involves both planning and control functions, including:

Tool Inventory Control

Inventory control of consumable items. This is not merely a matter of maintaining a stock of parts, but of tools ready for use, in use, or requiring refurbishment. Furthermore, since many tools are built up from several parts, Bill of Materials (multi-level) considerations have to be considered.

Tool Requirements Planning

This involves looking at the forward production schedule and determining what tools will be needed. It is necessary to know not only what tools are available at machines, but also their remaining tool lives. There are technical aspects to this.

Tool Preparation

This is the assembling of tools from holders, sleeves and inserts or other consumables according to the requirements list, their presetting, measurement and recording of offsets. This is at present a manual operation. However, there is much scope for research into the automation of this area.

Tool Supply

Delivery of the tools to the machine, by manual or automated methods. If automated, supply may be by equipment which also handles workpieces or by dedicated equipment. The technology of tool supply is rapidly developing, and it is a fruitful area for research.

Tool Condition Monitoring

It is necessary to monitor the residual life on the cutting tool to know whether it remains in working condition for subsequent operations. The technology is developing; a major limitation is the accuracy of tool life predictions. Normally the life is expressed as a cutting time which is decremented by the cutting duration of an operation. Alternatively, it may be decided to replace every tool after each operation.

Tool Offset Data Entry

The exact pre-set dimensions of the tool must be communicated to the machine controllers. This may be done manually, automatically or by a variety of intermediate methods such as bar codes or embedded memory chips.

Perhaps because errors in tool identification and offset definition can have serious consequences, the perception of tool management is frequently restricted to comprise the presetting and offset entry stages. Inventory control and build-up are becoming more accepted as major components of tool management systems. Fully integrated systems encompassing requirements planning are not yet common.

The major problem in achieving integrated tool management systems is that, at present, the various elements come from different types of company or enterprise. Thus machine tool suppliers may provide facilities for monitoring tool wear and remaining tool life, and may provide some means of communicating this data from the machine's CNC controller; it may not be possible to obtain this data in useful form due to communication system limitations. Software from system suppliers usually provides a means for capturing and transmitting this information as well as downloading offsets, but system suppliers seldom take interest in tool inventory control and requirements planning. Tool pre-setters often come with quite sophisticated software for their specific tasks; again, little thought has been given to standardisation and integration. Tool suppliers can offer quite sophisticated inventory control software, perhaps coupled by hotline to their sales offices. The emphasis is on getting a handle on the tool inventory and purchasing problem, rather than system integration. Tool requirements planning is often the missing link; it has to bring together factory scheduling, tool inventory control and current system status.

There is a substantial need for further work in this area. This involves software specification and design, communication techniques and system integration. When standards such as MAP and TOP become common, the problem may be simplified; at present it is a problem to most companies.

LOGISTICS OF TOOL MANAGEMENT

There are two major decision areas concerning the logistics of tool management:

(1) The logistics of tool supply. How frequently should individual tools be replenished as they wear out? Should sets of tools be replaced together at scheduled intervals?
(2) Work allocation. What operations and tools should be allocated to each machine? This question arises when there are similar or alternative machines in a system.

The first problem area is concerned with the operation of the system on a minute-by-minute or hour-by-hour basis. It depends largely on the decisions made on work allocation. Some work allocation decisions are made on a day-by-day or week-by-week basis, but others are semi-permanent decisions taken when the system is being designed.

The main parameters of the work allocation decisions are the variety of parts to be produced, the number of tools they require and the capacity of the tool magazines in the machine tools. As the variety of parts increases so will the total number of tools required to produce them. At some point it will no longer be possible to mount all the tools for all the parts permanently in magazines. At this point work scheduling decisions have to be taken jointly with tool allocation decisions. A subset of parts in current schedule will be selected, and the tools required mounted in the machines. After some time other parts will be added to the subset, new tools loaded into the magazines and any tools no longer needed for the revised subset removed. In the limit, every time a part is launched tools would be exchanged. At the other end of the spectrum it may be possible to maintain an allocation of tools to machines for a complete week's operation. In that case the interchange of tools due to tool wear may be of more significance than tool changes due to product variety.

In almost every FMS, planning decisions are taken to reduce the complexity of the problem. These limit the flexibility of the system. These may be organisational "soft" decisions or they may be technical "hard" decisions.

The "soft" decisions arise because it is necessary to simplify the planning process. For example, a system may have several similar machines. It may be decided to allocate certain operations to some machines and certain other operations to other machines. This gives the process planner a framework within which part programs may be developed. Such a framework is necessary, even although as a consequence, the load on the machines cannot be easily balanced on a day-to-day basis.

The "hard" constraints arise most frequently when a system is controlled by combining existing and new facilities. As a result, some machines may have smaller or larger tool or workpiece capacity.

Carrie and Perera (1) have given a review of the factors influencing the work allocation decision, examples of systems with simple and complex problems, and a critique of models in the literature. A substantial number of models has been published.

Unfortunately, most are too simplistic and make unrealistic assumptions, such as that all tools are of equal size. Most are static allocation models; very few recognise the trade-off between the selection of a subset of parts and the length of time for which the decision is applied.

Carrie and Perera (2) give a detailed discussion of a particular case in which work scheduling was greatly influenced by tooling constraints.

This is a field where there is a need for more research. This work should not be aimed at developing bigger or better mathematical models, but rather the development of practical heuristics. Simulation Modelling will be an important tool.

TOOL SUPPLY METHODS

Tool handling may be by manual or automatic methods. If manual methods are used, it is necessary to have some means of checking that the correct tools have been loaded into the correct tool pockets. Alternatively, there must be a means of identifying which tool is in each pocket. The technology of the machine and its controller may provide fairly rudimentary or quite sophisticated facilites. This influences the rigour which is required of human operators. Clearly if human intervention is required, then there is scope for error. Machines and controllers should provide the maximum of flexibility and facilities to monitor what the operator has done rather than rigidly prescribe what he has to do. The technology has developed significantly in recent years.

Another area in which technology is developing rapidly is that of automated tool supply methods. Two basic approaches to automated tool supply exist. Tools may be transported by either:

(1) Devices dedicated to tool supply. Two examples are AGVs for tool handling and mobile robots which exchange tools between a central store and machine magazine.
(2) Devices which are also used for workpiece transportation, such as AGVs for material and tool transportation.

If the vehicles are shared between tool and workpiece transportation, logistical planning becomes more complex. While the use of shared devices should result in fewer devices, thus saving on capital cost, there are several counter arguments. For example, the method of locating and holding tools during transport and the method for loading tools into the machine magazine from the vehicle, may introduce additional costs. Another problem is that since each operation requires both tools and workpieces, the use of shared facilities may introduce delays, waiting for tools to be supplied after the workpiece has been loaded, or vice versa. There are interesting questions to be answered on this topic, and some simulation studies have been reported, e.g. PATHSIM **(3)**.

There are opportunities for research in both the logistics of tool supply and handling, as well as in the engineering of improved designs.

INFORMATION SYSTEMS FOR TOOL MANAGEMENT

Various research papers **(4,5,6)** have been published which emphasise the use of distributed hierarchical control systems for achieving integrations, improved management information and better control of various manufacturing functions.

Various functions within a manufacturing organisation may be divided into various levels according to the nature of their operations - rather like the management structure of an organisation, where each function is autonomous, it receives instructions from a higher authority, issues instructions to a number of level lower functions and receives feedback. It performs its task based on this information and reports back to its controlling/managing function as appropriate.

Recent developments towards the standardisation of industrial communication protocols reinforce this argument of hierarchical management and control systems with the introduction of various specifications for Broadband MAP (Full MAP), Carrierband MAP and Mini MAP (EPA) which are designed to operate at various levels of an organisation, horizontally integrating the functions at each level as well as providing a means for inter-level communications through gateways and bridges.

Evans **(4)** defines various functional levels that may exist within an organisation as the Plant, Centre, Cell, Station and Sensor levels.

Plant - At this level, plant wide, financial and long-term planning decisions are taken. Market and Production plans are also generated.

Centre - Supervise production and provides management information by monitoring and supervising lower levels.

Cell - Co-ordinates production flow between various stations. Integrates individual stations into an automated system.

Station - Performs the production task by coverting information from lower lines to output commands based on directions from above. A typical CNC machine tool will be classified as a Station, therefore, the CNC controller is the station controller.

Sensor - Encompasses all sensors and activators of encoders and motors of the machine tool

None of the papers referred to earlier **(4,5,6)** attempt to define a relationship between the control hierarchy explained and tool management systems for FMS.

For this purpose we believe that the centre, cell and station levels are directly relevant. Fig. 1 shows various functions at each level and the information flow between functions.

At centre level less time-critical functions reside such as tool requirements planning and tool inventory control. These functions are comparatively less time-critical as they may be performed weekly or daily depending on the application. At the other end of the spectrum functions which reside in the station level are highly time-critical, i.e. tool force and torque need to be monitored at all times and tool life must be updated in real time in order to provide a realistic picture for those cell and centre level functions.

CASE STUDIES

In this section, we present three cases, each presenting different forms of tool management problem, different constraints and different approaches to managing the problem.

Case 1

This company has an FMS, for producing major castings, which consists of six machining centres, of which one is a special purpose "facing head" machine, the remaining machines are similar, as shown in Fig. 2. The installation has been described in several publications, e.g. Carrie et al **(7)**. The workpieces are complex requiring operations in typically three fixtures, with approximately 20 operations in all. The operation times vary between 10 minutes to 2 hours, although, on the facing head machine, allowing for manual intervention for tool changes, etc., the machine occupancy may be several hours.

This application largely employs manual tool inventory control and tool requirements planning systems. Tool preparation is carried out on a tool presetting machine where offsets are manually entered into the executive computer. A manual tool supply system is in operation. DNC links are available for communication of tool offsets.

The executive computer records the wear incurred by each tool during the execution of the part program. Estimates of tool wear are contained in the tooling required file for each operation. When the cycle start command is issued, the estimated values are added to the wear total of each tool. The machine tooling status file also contains Warning Point Limit and Maximum Wear Limit which can be used for scheduling tool changes and allocation of parts to machines.

Due to the complexity of parts a large number of cutting tools is required, varying between approximately 140 and 240 in total, averaging 18 per operation. The machines have tool magazines with pockets for 100 tools. However, since some tools are large and occupy more than one pocket, the effective capacity is less than 100.

In planning the operation of the system, two types of decisions are required:

(1) Should some strategic allocation of types of operations to machines be decided upon, which could be used when process planning and preparing NC tapes?
(2) Which tools should be allocated to which machines on a day-to-day or hour-by-hour basis?

In this case, it was decided that two of the machining centres should be used for roughing work, two should be used for semi-finishing, and one for finishing. As a result, a library of the tools which are nominally available on each machine was created, which then became available to NC planners for use in developing part programs for new workpieces. In making this decision the system designers sought to achieve:

(1) An approximately-balanced workload across the five machining centres.
(2) A roughly even allocation of tools to machines, taking tool size and sister tooling into account.

It has been hoped that the 100 tool magazines would be large enough to accommodate the tools needed for the first set of parts to be produced on the system, but it was recognised that once new parts were introduced, the capacity would soon be exceeded. In the event, the first seven parts required more than 100 tool pockets on all but one of the machines.

When the tool requirements exceed the magazine capacity then it becomes necessary to consider tool availability when scheduling work. The demand for tool pockets will vary from week to week with the production schedule. Carrie and Perera **(2)** gave an assessment of the problem in this case.

It is essential to get a feel for the magnitude of the problem. Since in this case tool supply was manual and had to be carried out while the machines were idle, it was considered important to minimise tool changes. The literature now contains many operations research models for this work allocation problem. Frequently they make unrealistic assumptions, for example, assuming that all tools are of the same size, or omit important practical details. Carrie and Perera **(2)** discovered that the number of tool changes due to tool wear would be about 10 times the number due to workpiece variety. Thus, any model which sought to minimise changes due to variety, but ignored changes due to wear, was unrealistic. Strictly, as more workpieces have been introduced to the system, the situation has changed.

O'Grady and Menon **(8)** developed a multi-objective linear programming model for allocating work and tools to the machines, which appeared to be highly appropriate for this system. Unfortunately, according to Carrie and Perera, it suffered from several defects. Firstly, it was based on the concept that some tools would be standard and kept in the magazine, others would be non-standard and changed according to workpiece requirements. In this case the decision whether a tool was standard or non-standard seemed arbitrary.

Secondly, because several criteria were incorporated in the model, it was necessary to assign weighting factors for each criteria. While, in principle, it should be possible to give a financial measure to each criteria, in practice these weights are rather subjective. Thirdly, because of the size of the model, and the volume of data required, capturing the status of the system to load into the model, and running the model, posed practical difficulties. Fourthly, this model (in common with many) made once-off decisions, but did not adequately take the dynamic nature of the schedule into account. One important decision is how far into the future the planning horizon should be set, and how large a volume of work should be considered. Fig. 3 from Perera **(9)** shows the effect of the volume of work considered on the expected number of tool changes, as well as on the start-up and tail-off effects on machine utilisation. It was found that there was a trade-off between these effects and a workload level which gave a good compromise. Perera and Carrie **(10)** found a tool flow simulator of value in analysing the situation and in developing heuristic scheduling rules.

This case testifies to the complexity of the work allocation problem, even in a situation where manual tool supply avoids any need for scheduling of vehicles or robots for tool supply. Many megabytes are required just to store the basic data.

Case 2

This system (Fig. 4) consists of two machining centres linked together by a railcar, which also services two load/unload stations and six buffer stations. The two machining centres differ in the size of workpieces which can be accommodated. The larger of the two machines accommodates an 80-tool chain, whereas the smaller machine houses two tool chains of 40 and 80 tool capacities. Large tools occupying 2 or 3 pockets are not uncommon, reducing the effective tool capacity of the system. The system requires over 300 tools in total.

A tool buffer storage area is provided at the rear of the machines which is supplied manually. A gantry robot is used to supply tools to the 80-tool chains from the buffer area. The 40-tool chain on the smaller machine must be loaded manually, as it is not accessible by the robot.

The machine tool supplier has quite sophisticated tool management software available, built up through several key projects. Fig. 5 gives an indication of the options then available for tool pre-setting, tool storage, tool transportation and tool handling into the machine magazine, with the method in use in this system shown.

The tool requirements function is carried out manually. The information available to the system manager is the "tool list per part program" and "current tool list" which gives a snapshot of the tools currently on the machines and associated data such as tool lives. This information is obtainable from the cell computer. Using this information the system manager then works out, manually, a tool requirements list and issues appropriate tool preparation instructions to the tool room.

The prepared tools are manually supplied to the buffer area, and the relevant data, i.e. the position in buffer, position on machine, offsets, etc., are manually entered to the cell computer. As tools are loaded into the machine tools, the relevant offsets and other relevant data are downloaded through DNC links. During processing, tool life is monitored in a similar fashion to the previous case. The spindle torque is also monitored to detect any sudden tool breakages.

The production schedules are received and revised weekly. The re-tooling exercise has to be carried out once a week, which leads to very low utilisation of the expensive tool handling system. Meanwhile, the supervisors and the operators painstakingly employ manual methods to carry out the tool planning procedure, which is inevitably prone to errors. The lack of computer-based support at this level is felt to a great extent, as the whole of the tool requirements planning exercise is demanding of supervisors' and operators' time. The company now recognises this shortfall and are at present specifying an appropriate tool requirements planning package.

This case serves to show that, even if an integrated tool supply system is available, the effectiveness of tool management can be limited due to the inability of the user to extract data from the machines.

Case 3

This case consists of a nine-cell FMS shown in Fig. 3. Cells 1 and 2 are flexible machining cells. The remaining are robotic assembly, inspection and finishing cells. The company has decided to implement a tool management system for cells 1, 2 and 4/5, as they are relatively tool-intensive.

The company is currently installing a computer-based tool requirements planning system which is capable of generating the following output:

Tool preparation list.
Tool pick list - tools to be removed from machines.
Tool kitting list - tools to be supplied to the machines.

Fig. 7 shows the architecture of the tool management system adopted in this application with specific reference to various levels of control.

The line controller is responsible for functions such as tool requirements planning and tool inventory control. At the cell level, the cell control computers are responsible for monitoring and control of tool supply to individual machines using tool data generated by tool preparation. Tool life, force and torque monitoring is carried out at the station level by the CNC controllers. This data is communicated to cell level through the communication links provided.

Tool requirements planning is carried out by comparing the tool requirements based on the oncoming production schedule and comparing this gross tool requirement with what is already available in the machines and at the "prepared tools" store. This effectively gives net tool requirements per machine. The tool preparation list is passed to tool pre-setting where each tool is pre-set and identified by a unique bar code. The offsets at this stage are logged on to the cell controller via the communication link provided between the pre-setter and the cell controller. Tools are supplied to the machines manually by the operators according to the tool kitting instructions supplied from the line controller. The tool kitting instruction identifies those tools which need to be removed and those which need to be supplied to each machine. The operator first identifies the pocket, then the tool, using the bar code, i.e. the pockets on each machine are also identified by a unique bar-code. This information from the bar code readers is automatically relayed to the appropriate cell controller, which then informs the CNC controller of the tool identity and its offsets, as well as other relevant data such as tool life, maximum torque and force limits.

Although a scheduled tool supply procedure is dominant, a tool condition monitoring system based on detecting a cutting code will also be used. However, supply of individual tools will be minimised due to extensive use of duplicate tooling.

Although not yet operational, this tool management system shows more promise than the previous cases. A great deal of time was spent on the specification of the system in terms of its detailed operation and hardware and software requirements, including details of communication protocols and telegram formats. A series of discussions between the user and the vendor of the tool management system as well as the machine tool vendors, has lead to development of a system which is no more complex than it need be, but is of good functional quality.

However, the specification of this system was not totally painless. Various difficulties were encountered due to the different protocols used and non-standard telgrams required. A survey of off-the-shelf tool management packages was made and found to be generally inadequate due to limited functionality. One off-the-shelf package was considered to be acceptable on its functionality, but was not capable of handling the different communication protocols used by the machines in Cell 1 and Cell 2 which employ different types of CNC controllers. It was possible to implement this package only by using different versions and integrating them at a higher level. This option was considered to be unacceptable. The company decided to implement a custom-made application where the final specification was agreed after lengthy discussions between the user, the system builder and the machine suppliers. This overcomes the problems associated with protocols. However, some difficulties were encountered with reference to the telegram formats. Although the information required, e.g. machine magazine configuration, was available locally at the machine controller, the facility to transmit that data was not available. A series of discussions with the machine tool suppliers was necessary in order to accurately specify the modifications required to the machine tool controller to enable it to communicate the desired information.

CONCLUSIONS

The major problem in achieving integrated tool management systems is that, at present, the various elements come from different sources with different objectives and information. Thus machine tool suppliers provide major facilities for monitoring tool wear and tool life, and may provide some means of communicating the data from the machine's controller, although it may not be possible to obtain the data in a useful form due to limitations of the communication system. Software from systems suppliers usually provides means for capturing and transmitting this information, as well as downloading offsets. Tool pre-setters often come with sophisticated software for their specific task, but again little thought has usually been given to standardisation and integration. Tool suppliers can offer quite sophisticated inventory control software, perhaps coupled by a hotline to their sales offices. The emphasis is on getting a handle on the tool inventory and purchasing problem, rather than system integration. Tool requirements planning is often the missing link. It has to bring together factory scheduling, tool inventory control and current system status, which are often implemented in incompatible packages.

We believe that there is a substantial need for further work on:

(1) The establishment of good planning and shop floor practices in connection with tool management.
(2) The definition and standardisation of telegram formats and communication interfaces, to achieve adequate information exchange between machine tool controllers and host computers at a reasonable cost.
(3) Realistic studies on the logistics of tool management.

REFERENCES

1. Carrie, A S and Perera, D T S: Work allocation in flexible manufacturing systems, Proc. Conference on Computer Aided Engineering, April 1986, pp. 91-95, Mechanical Engineering Publications Ltd.
2. Carrie, A S and Perera, D T S: Work scheduling in FMS under tool availability constraints, Int. J. of Production Research, 24, 6, 1299-1308, 1986..

3. Crite, G D, Mills, R I and Talavage, J J: PATHSIM: a modular simulator for automatic tool handling, J. Manufacturing Systems, Vol. 4, No. 7, 1985.
4. Evans, A: Cell controllers in action, Proc. 2nd Conference on Machine Control Systems, 55-64, IFS Conferences, Birmingham, 1987.
5. Bonner, D T: Traditional information technology: PLCs, Proc. 2nd Int. Conference on Machine Control Systems, 129-138, IFS Publications, Birmingham, 1987.
6. Kehoe, D F and Bititci, U S: Implementing manufacturing strategies through the application of programmable control, Proc. 2nd Int. Conference on Machine Control Systems, IFS Publications, Birmingham, 1987.
7. Carrie, A S, Adhami, E, Stephens, A and Murdoch, I C: Introducing a flexible manufacturing system, Int. J. of Production Research, 22, 6, 907-916, 1984.
8. O'Grady, P J and Menon, U: Loading a flexible manufacturing system, Int. J. of Production Research, 25, 7, 1053-1068, 1987.
9. Perera, D T S: Tool availability strategies for flexible manufacturing systems, International Industrial Engineering Conference, Orlando, US, 1988.
10. Perera, D T S and Carrie, A S: Simulation of tool flow within a flexible manufacturing system, Proc. 6th Int. Conference on Flexible Manufacturing Systems, IFS Publications Ltd, pp. 211-223, Turin, Italy, November 1987.

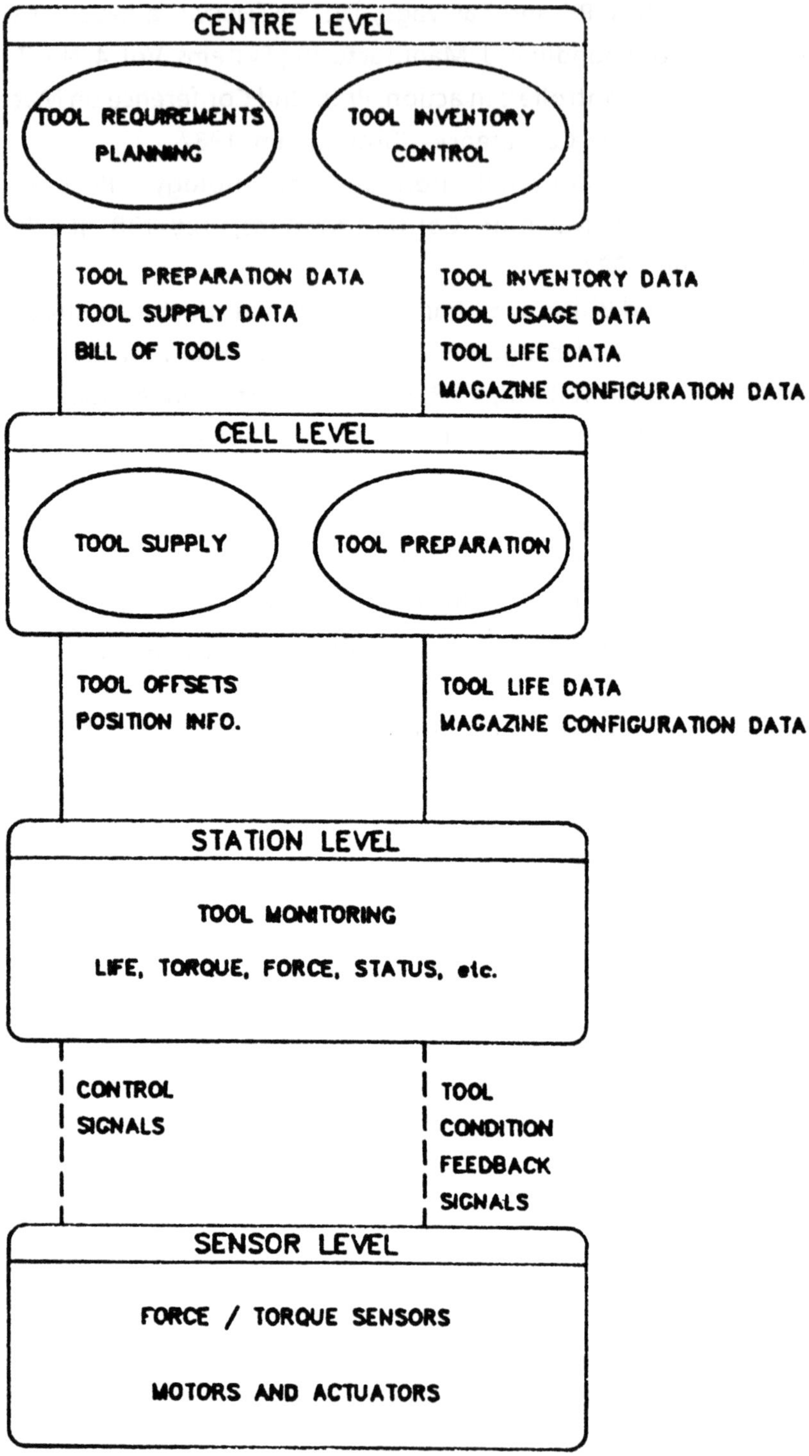

FIGURE 1. INFORMATION SYSTEM FOR TOOL MANAGEMENT

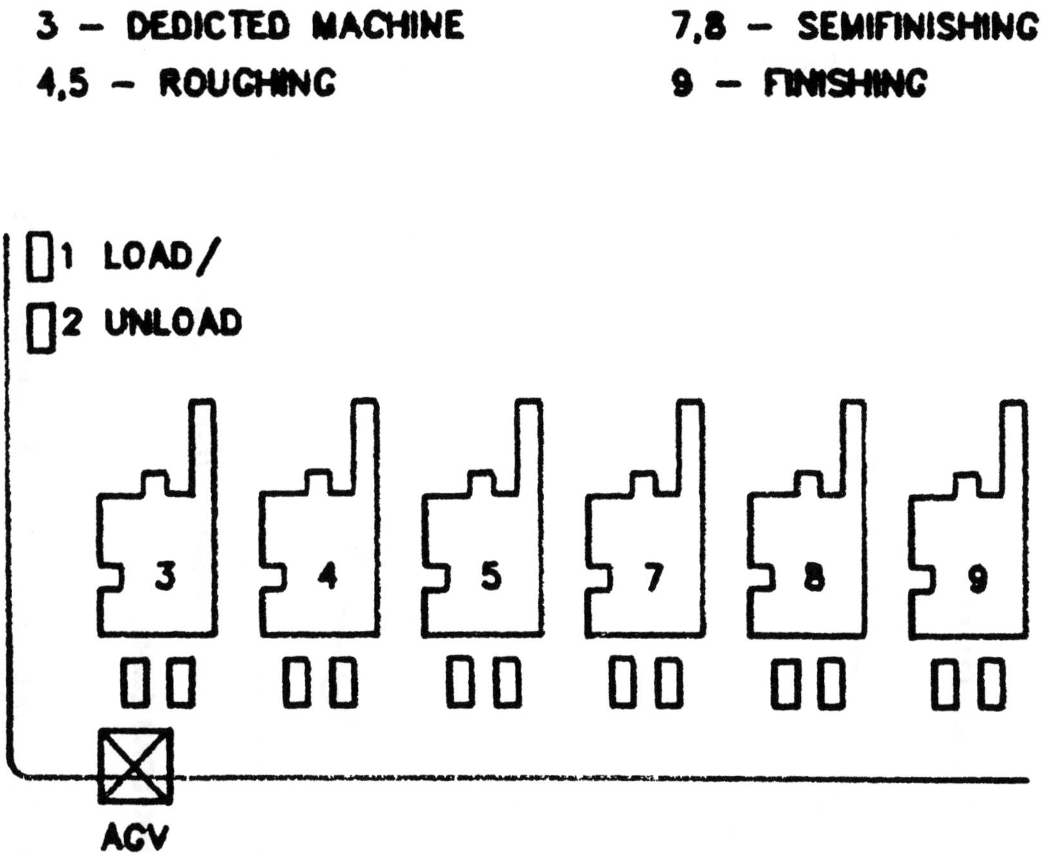

FIGURE 2. LAYOUT OF FMS PRESENTED IN CASE 1.

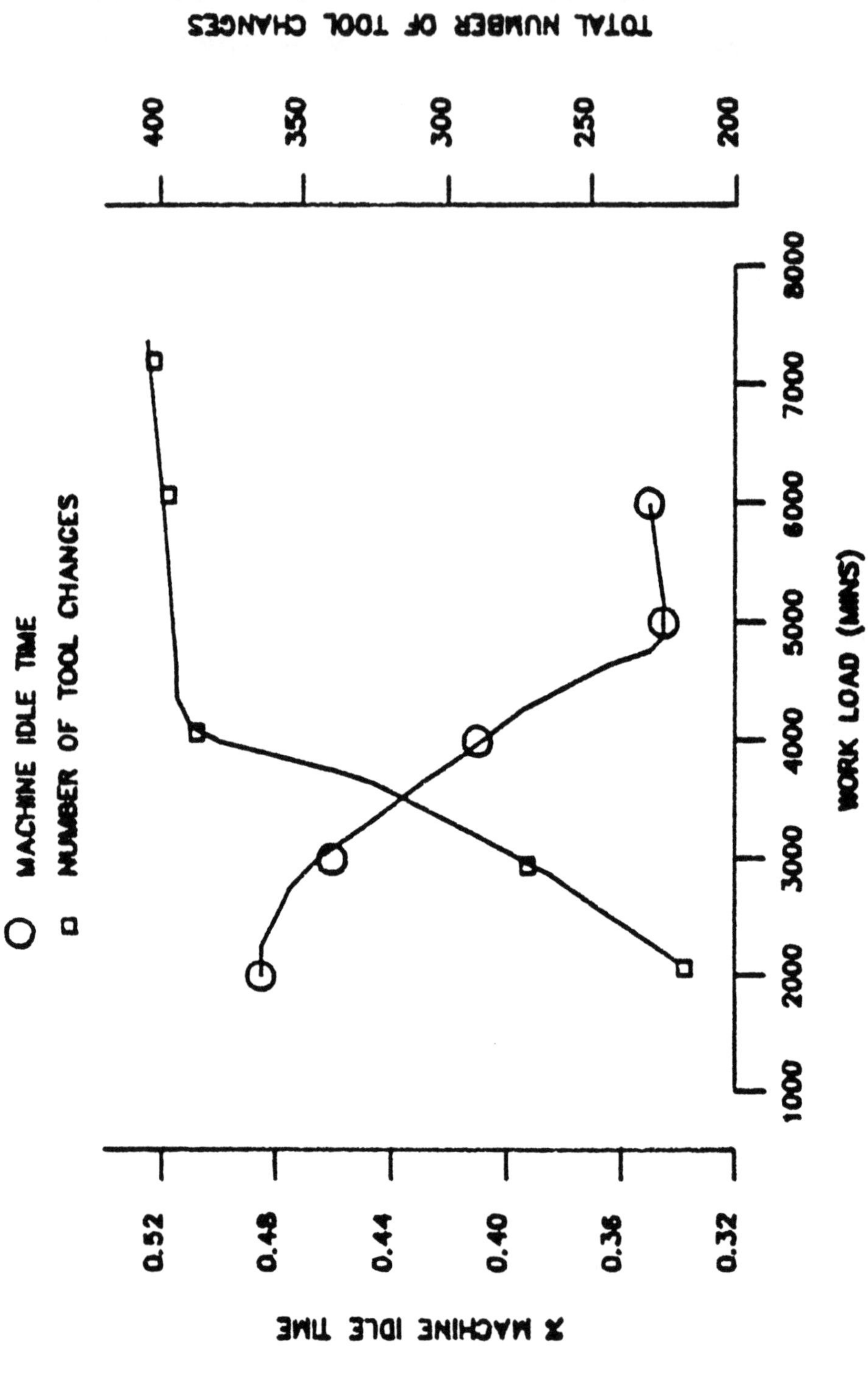

FIGURE 3. RELATIONSHIP BETWEEN MACHINE LOAD, IDLE TIME AND NUMBER OF TOOL CHANGES

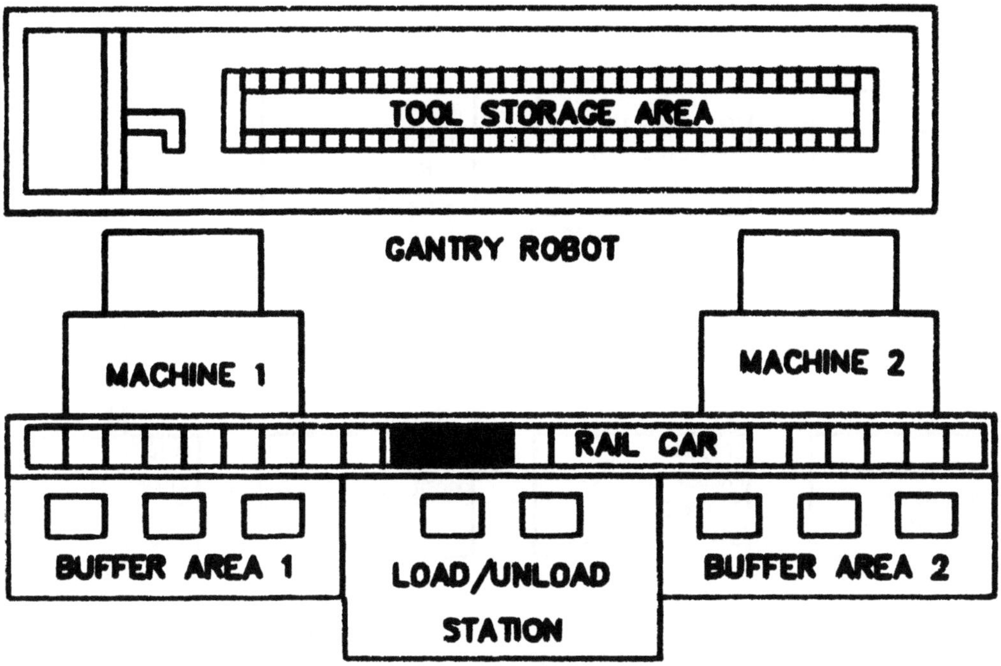

FIGURE 4. LAYOUT OF FMS IN CASE 2.

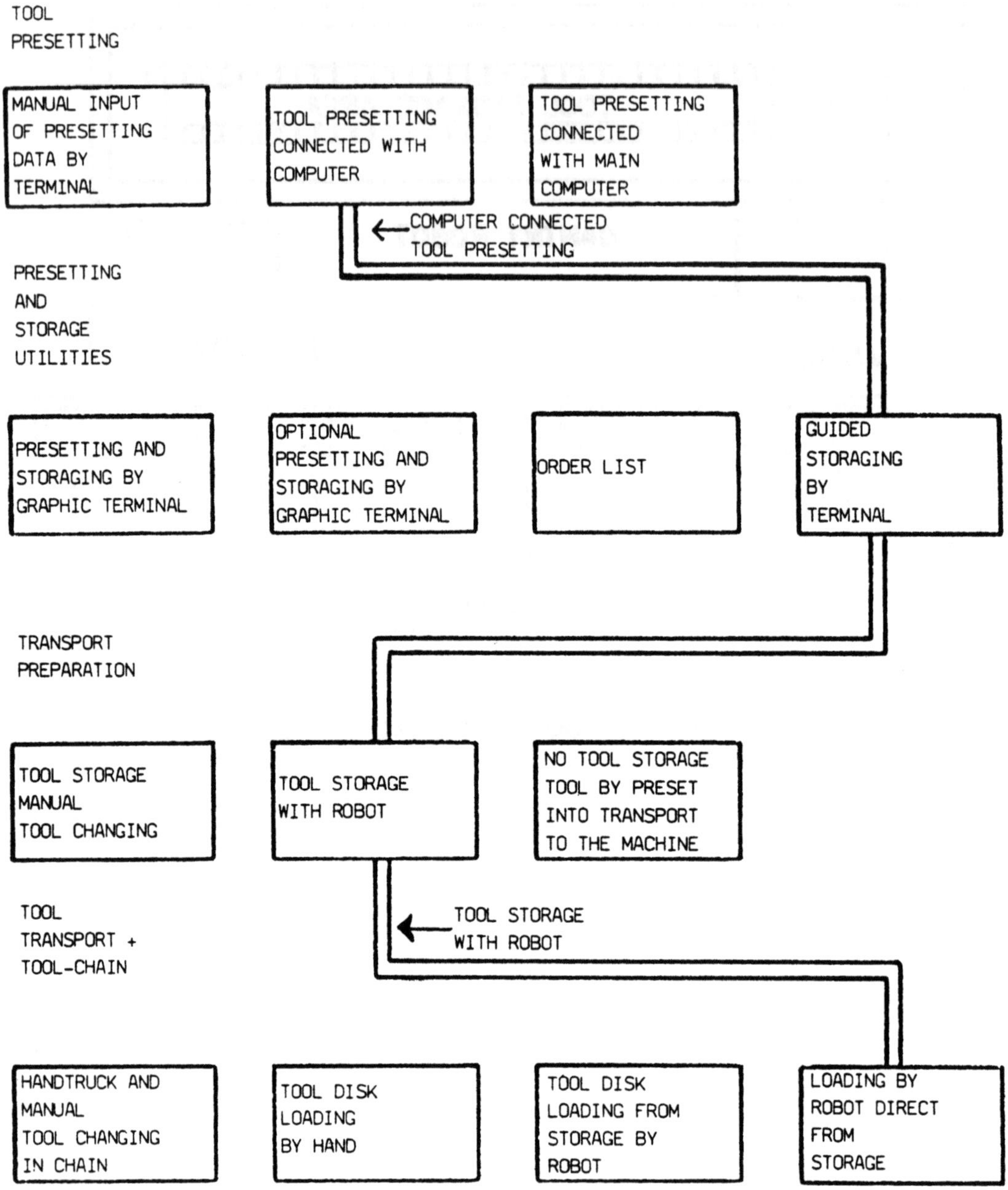

FIGURE 5. OPTIONS OFFERED BY MACHINE TOOL SUPPLIERS FOR VARIOUS TOOL MANAGEMENT FUNCTIONS

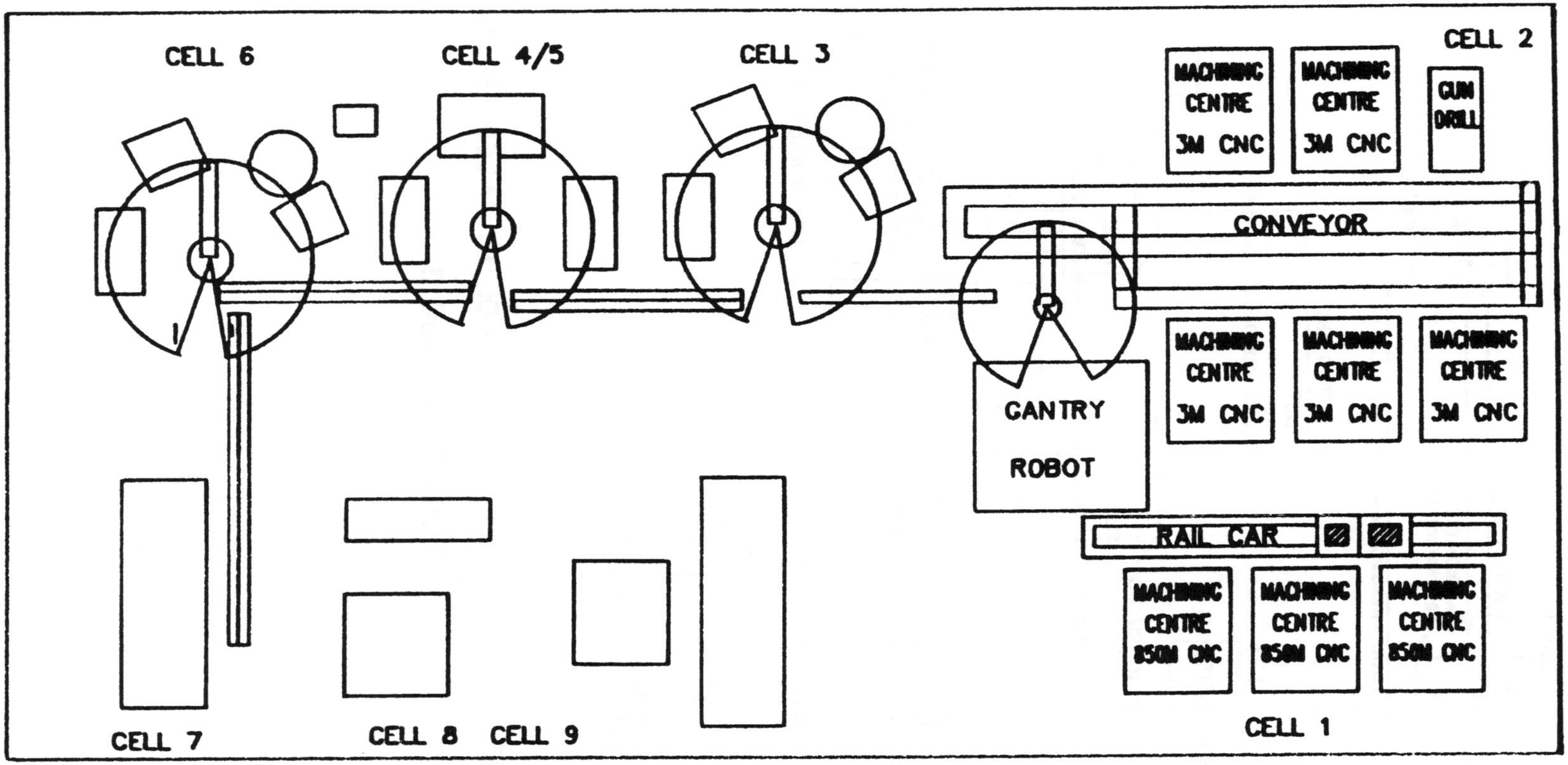

FIGURE 6. LAYOUT OF THE FMS DISCUSSED IN CASE 3.

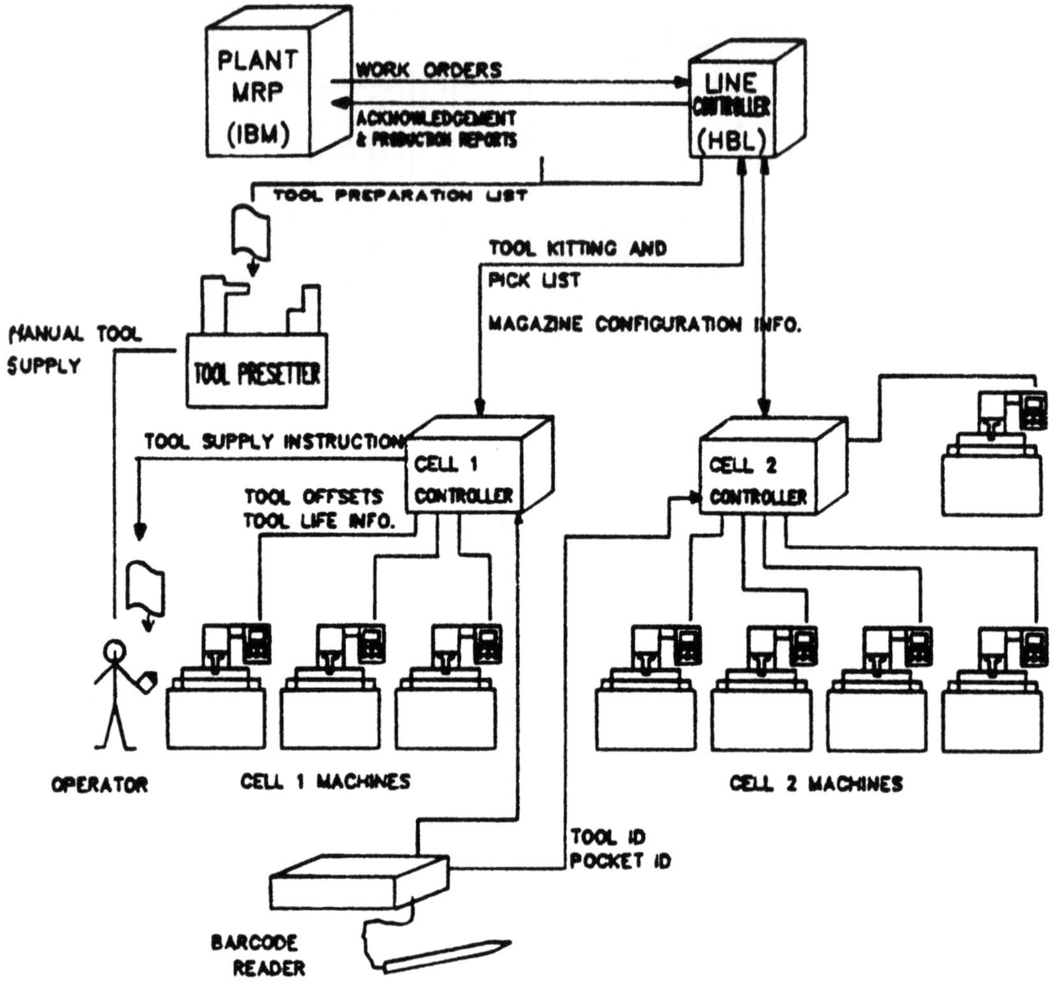

FIGURE 7. THE ARCHITECTURE OF THE TOOL MANAGEMENT SYSTEM PRESENTED IN CASE 3.

Issues in Design and Control of AS/RS

Meir J. Rosenblatt
Faculty of Industrial Engineering and Management
Technion - Israel Institute of Technology
Haifa 32000, Israel

Abstract

In this paper we deal with issues of design and control of AS/RS. Specifically, we focus on some recent developments in this area, which were part of four different studies that may affect the design and control of AS/RS. The first issue is concerned with the problem of which items (and at what levels) should be allocated to the AS/RS. The second issue deals with the problem of using class-based storage policies and dividing the items into groups so that the expected one-way travel time is minimized. The third issue deals with the actual design of the AS/RS, studying the effects of having smaller number of cranes than aisles. The fourth issue is concerned with the best container size in a warehouse. Although each one of these issues can be viewed as a separate entity (study), they are all connected with the important problem of better designing and controlling AS/RS.

NATO ASI Series, Vol. F53
Advanced Information Technologies for
Industrial Material Flow Systems
Edited by Shimon Y. Nof, Colin L. Moodie

1. Introduction

Automated storage and Retrieval Systems (AS/RS) are being introduced into industry and warehouses at an increasing rate. Forecasts indicate that this treat will continue for the forseeable future (I.E.,1986). Some of the characteristics of AS/RS are large variety of items and customers (orders), large throughput, short processing time and reliability in identifying items and managing inventory. Some of the advantages of AS/RS, compared to a conventional warehouse, are: Increased throughput and higher utilization of floor space and storage volume and workforce; Improvements in safety and working conditions; Decreases in losses, thefts and damages; Energy savings; Decrease in idenfication errors and a better control of the workforce in the warehouse. Obviously, introducing AS/RS is costly and involves quite a substantial investment. Therefore, issues of design and control in AS/RS are of considerable importance.

In this paper we briefly present some of the results of four studies that were carried out at the Technion - Israel Institute of Technology. All the topics deal, in one way or another, with issues of design and control of Automated Storage/Retrieval Systems (AS/RS).

The first study deals with the situation where a new storage system, e.g. AS/RS or carousel, is introduced. However, given budgetary and other constraints, it is impossible to accommodate all the currently stored items. Thus, the question is which items should be stored at this preferred area and at what volumes. A solution procedure is provided for the static problem and , as demonstrated, the amount of information needed is quite managable.

The second problem addressed in this paper is concerned with managing the items once they are allocated to the preferred area. It is assumed that items' demand is acording to an ABC curve. In order to minimize the expected one-way travel time a class-based approach is used. Under this policy, items are grouped into classes and the warehouse is divided into areas, and the problem is that of deciding on the boundaries of the various classes (areas) to which the warehouse should be divided. Previously, this problem was solved for two and three classes in a squared in time warehouse. Here, a solution procedure is provided for the general case of n classes. Managerial implications of dividing items into groups and the savings (in time) obtained by continuing dividing the warehouse into classes is provided.

The third topic deals with deciding the best layout design of a warehouse. In considering the various cost components of an AS/RS, it is evident that the number of cranes if of significant importance. As a result, a new approach to the problem is considered here, which affects the design, control, information and technology needed to operate such a system. This sytem allows the transfer of any crane from one aisle to another, and, obviously, requires additional equipment and controlling machanisms. An iterative solution procedure is provided. In each iteration optimization and simulation models are solved and output of one model is used as input into the other model.

Finally, the fourth subject to be discussed here is the optimal container size. The importance of the container size and its effects on the total cost is realized. A set of assumptions allows one to determine the optimal size of the container.

2. Allocating Items to AS/RS

We consider the problem of assigning items to a preferred area within a large warehouse, for example, a large distribution center in which a new AS/RS is installed. The new AS/RS reduces the costs of managing an item. By assumption the capacity of the preferred area is not large enough to accummodate all of the items in the warehouse; hence, an allocation problem exists, i.e., which items should be managed by the preferred area?

If an item is allocated to the preferred area, all customer requests for that item will be issued from the preferred area. All receipts of stock for items located within the preferred area, however, are stored in a secondary location within the main warehouse. Replenishments of stock for items within the preferred area are issued from the secondary location.

We assume that during each period items depleted due to demand are immediatley replenished. Further, the lead time is assumed to be negligible - which is a reasonable assumption if requests for issue are not to frequent. Hence, if an item i is allocated to the preferred area, it is managed by a $(0,Q_i)$ inventory policy; where Q_i - the economic order quantity (representing the number of units of item i in the preferred area) - has to be determined too. It should be noted that the allocated space to each item can be viewed as a dedicated storage system, although the actual storage location of the items within the warehouse may change during the period.

Throughout this study we assume a deterministic environment. More specifically, we assume that the periodic demand of each item is known and constant. This can be supported by the fact that in certain manufacturing organizations the actual product mix and consequently the demand for items is fixed for the planning period.

The optimal storage assignment problem in automatic warehousing systems has been treated by Hausman et al. (1976). There, the authors compare full turnover-based assignment with class-based turnover assignment, where the objective function used is to minimize the expected one-way travel time per pallet. The study by Hausman et al (1976) can be viewed as the second phase of our problem. Namely, once it is decided, based on economic grounds, which items are assigned to the AS/RS and at what levels, then one is concerned with the actual assignment (location) of each item. This is further discussed in the next section.

2.1 The Model

Each item assigned to a preferred storage area accrues a periodic (annual) profit. This profit is the difference in costs between the old and new way of material handling. Under the old system, the annual (periodic) material handling cost is given by the product of the number of requests per period, R_i, and the cost of satisfying each such request from the general warehouse, e_i^o . When the preferred storage area is introduced, the cost of satisfying customer requests is now $e_i^n R_i$, where $e_i^n \le e_i^o$. However, an additional cost exists due to the "double-handling" of products between the general warehouse and the preferred storage area. This cost is the product of the number of times required to replenish the desired level of item i (at the preferred storage area) and the cost of each repplenishment. Obviously, the number of replenishments in a given period is intimately related to the economic order quantity, Q_i. Benefits of joint replenishment are not considered. Let

e_i = "savings" per request for item i if stored in the preferred area $\left(e_i^o - e_i^n\right)$.

c_i = cost per replenishment for item i.

R_i = the periodic number of requests of item i.

D_i = the periodic demand for item i converted into units of volume.

z_i = the economic order quantity (continuous decision variable) converted into units of volume, i.e. $z_i = v_i Q_i$ where v_i is the volume of one unit of item i.

x_i = binary (0 or 1) decision variable determining if item i is assigned to the preferred area.

$f_i(z_i)$ = profit per period for item i.

N = number of items in the warehouse.

V = a predetermined volume of the preferred area.

All parameters are assumed positive. With the above notation and assumption

$$f_i(z_i) = \begin{cases} e_i R_i - \dfrac{c_i D_i}{z_i} & z_i > 0 \\ 0 & z_i = 0 \end{cases}$$

This formulation models the profit for item i as the total savings in costs of all the requests under the old and new system minus total replenishment cost (between main warehouse and preferred storage area) under the new system. To simplify notation, let $a_i = e_i R_i$ and $b_i = c_i D_i$ resulting in

$$f_i(z_i) = \begin{cases} a_i - \dfrac{b_i}{z_i} & \text{if } z_i > 0 \\ 0 & \text{if } z_i = 0. \end{cases}$$

The allocation model is formulated as follows:

$$\text{(P)} \qquad \max \sum_{i=1}^{N} f_i(z_i)$$

$$\text{s. t.} \qquad \sum_{i=1}^{N} z_i \leq V$$

$$z_i \geq 0.$$

Problem (P) arises in a variety of applied contexts (see Zipkin (1980) and Vidal (1987) and references therein). It is typically assumed that each f_i is differentiable and strictly concave over its domain, which allows a straightforward application of the well-known Kuhn-Tucker conditions. When a closed form expression for the concave conjugate of each f_i exists, Vidal (1987) provides a graphical approach to solve (P). Our problem differs from previous applications. Although each of our f_i's is strictly concave, increasing and differentiable on the open interval $(0, \infty)$, each is discontinuous at the origin, and hence not differentiable over the entire domain. Motivated by a different application, Srikantan (1963) solves a similar formulation of our problem when there exist positive lower bounds on the allocations, thereby ensuring differentiability over the entire domain. In our problem, however, each item i has a minimal threshold value (volume) of $\frac{b_i}{a_i}$ below which it does not pay to assign item i to the preferred area, and we allow the possibility of not assigning an item to the preferred area. Therefore, a different analytical approach needs to be developed.

2.2 A Heuristic for Solving Problem (P)

Let $z^* = (z_1^*, ..., z_N^*)$ denote and optimal solution to P and let $S^* = \{i: z_i^* > 0\}$ denote the corresponding optimal set of items allocated to the preferred area. Given S^*, the optimal stocking levels z_i^* for $i \in S^*$ may be determined by solving the following sub-problem:

$$P(S^*): \qquad \max \sum_{i \in S^*} f_i(z_i)$$

$$\text{s. t.} \qquad \sum_{i \in S^*} z_i \le V \tag{2.1}$$

$$z_i^* > 0, \qquad i \in S^*.$$

Problem $P(S^*)$ is a concave program which clearly possesses a unique optimal solution. By the well-known Kuhn-Tucker conditions, the optimal solution to $P(S^*)$ necessarily satisfies:

$$\lambda(S^*) = \frac{df_i(z_i^*)}{dz_i} = \frac{b_i}{(z_i^*)^2} \tag{2.2}$$

where $\lambda(S^*)$ denotes the optimal Lagrange multiplier for constraint (2.1). From (2.2) we deduce that

$$z_i^* = \sqrt{\frac{b_i}{\lambda((S^*)}}. \tag{2.3}$$

Since the f_i's are increasing it follows that the $\sum_{i \in S^*} z_i^* = V$; hence,

$$\lambda(S^*) = \left(\sum_{i \in S^*} \sqrt{\frac{b_i}{V}} \right)^2 \tag{2.4}$$

and

$$z_i^* = \left(\frac{\sqrt{b_i}}{\sum_{i \in S^*} \sqrt{b_i}} \right) V. \tag{2.5}$$

Thus, it is sufficient to find an optimal set S^* from which the determination of the corresponding allocation vector z^* is given by (2.5). Note that the value of z_i^* is independent of $a_i = e_i R_i$, the total savings of all the requests in the period. The actual values of the a_i's are essential to determine which items should be stored in the preferred area. Once this decision has been made, the actual level of each z_i^* is dependent only upon the b_i's. Furthermore, the following ratio between the allocation to the different items can be easily established:

$$\frac{z_i^*}{z_j^*} = \frac{\sqrt{b_i}}{\sqrt{b_j}} \text{ or } z_i^* = z_j^* \sqrt{\frac{b_i}{b_j}}$$

However, as yet, an S^* has not been found. Suppose, on the other hand, that the value of $\lambda(S^*)$ was given. If item i were to be allocated to the preferred area, its optimal stock level z_i^* would be given by (2.3). Let $\lambda^* = \lambda(S^*)$ and define

$$z_i^*(\lambda^*) = \sqrt{\frac{b_i}{\lambda^*}}$$

Given the knowledge of λ^* the optimal solution to problem P may be obtained by solving the following knapsack problem:

$$P_N: \qquad \max \sum_{i=1}^{N} x_i \cdot f_i\left(z_i^*(\lambda^*)\right)$$

s. t.

$$\sum_{i=1}^{N} x_i \cdot z_i^*(\lambda^*) \le V$$

$$x_i = 0 \text{ or } 1.$$

Problem P_N is NP-complete. A well-known heuristic for solving problem P_N is to rank the items from highest to lowest according to the ratio

$$\frac{f_i\left(z_i^*(\lambda^*)\right)}{z_i^*(\lambda^*)} \tag{2.6}$$

and sequentially fill the knapsack until adding the next item on the list exceeds the capacity V .While this procedure does not guarantee optimality it generally performs well in practice. The ratio (2.6) is of course a function of λ^* - which has yet to be determined. Substituting the definition $z_i^*(\lambda^*)$ from (2.3) into ratio (2.6) and rearranging terms) the ratio becomes

$$\frac{f_i\left(z_i^*(\lambda^*)\right)}{z_i^*(\lambda^*)} = \frac{a_i}{\sqrt{b_i}}\sqrt{\lambda^*} - \lambda^* \tag{2.7}$$

It is clear from (2.7) that for any value of λ^* item i out-ranks item j if and only if $\frac{a_i}{\sqrt{b_i}} \ge \frac{a_j}{\sqrt{b_j}}$.

The heuristic algorithm can now be stated as follows:

<u>The Heuristic Algorith</u>

Step 1: Rearrange the items so that $\frac{a_i}{\sqrt{b_i}} \geq \frac{a_{i+1}}{\sqrt{b_{i+1}}}$, $i = 1,2,\ldots, N-1$.

In case of a tie, put the item with the highest b_i first.

Step 2: For each ordered set of items, $S_k = \{1,2,\ldots k\}$, $1 \leq k \leq N$, use eq (2.5) to compute z_i^* for problem $P(S_k)$.

Step 3: Compute the value of $v(S_k) = \sum_{i \in S_k} f_i(z_i^*)$.

Step 4: Select the minimum cardinality set with the maximum value of $v(S_k)$ over $1 \leq k \leq N$.

Normally, one would have to check N subproblems in step 4. However, a proposition , which is proved in the Hackman and Rosenblatt (1988), shows that $h_k = v(S_k)$ is unimodal with respect to k . Hence, a bi-section search reduces the number of subproblems that need to be solved to $O(\log_2 N)$.

Finally, Hackman and Rosenblatt (1988) developed sufficient conditions for the optimality of the heuristic algorithm. The solution procedure was tested on data from a Naval Supply Center and it was shown that usually the sufficient conditions do hold which guarantee the optimality of the heuristic procedure.

3. Deriving the Optimal Boundaries for Classed-Based AS/RS.

In a study by Hausman, Schwarz and Graves (1976) storage policies based on the level of demand of each item are suggested. They considered two such policies: Full Turnover-Based assignment and Class-Based Turnover assignment. Under the full turnover-based assignment the distance (in time) of an item from the Input/Output (I/O) point in the warehouse is proportional to its demand level. Thus, an item with a higher demand rate is stored closer to the I/O point. However, given the large number of items that need to be stored, the implementation of such a policy can be an enormous task to perform. Therefore, they recommend the more practical class-based approach. Under this policy, the warehouse is divided into several (L shaped) areas and the items are divided into the same number of groups. Those groups of items are asigned to the various levels, based on their demand. This policy is easier to implement than is the full turnover policy.

In this section, a simple solution is developed for finding the optimal boundaries of each of the storage areas. This procedure enables us to extend the results of Hausman et al for two and three class-based turnover assignment policies to any number of classes. We show that using a relatively small number of classes (storage regions) can result in average travel times which approach travel times obtained for the full turnover-based assignment.

3.1 Assumptions and Definitions

The following assumptions are made:

1. All storage locations are of equal size and each location holds only one type of item.
2. Each crane is dedicated to a specified aisle and is capable of moving simultaneously both vertically and horizontally.
3. The warehouse system is "square" in time, i.e., the cranes' vertical and horizontal speeds are such that the time to reach the most distant row from the I/O point equals the time to reach the most duistant column.

4. The I/O point is in the corner point of the aisle.
5. The demand, or turnover frequency, of each item is known and constant through time.
6. Items are ordered according to the simple EOQ model.

In addition, it is assumed that items are ranked according to their contribution to total demand using an ABC curve. The ABC curve is a plot of ranked cumulative percentage demand, G(i), versus percentage of inventoried items, i. Using the notation of Hausman et al (1976), we represent the ABC curve by:

$$G(i) = i^s = \int_o^i D(j)dj \qquad \text{for } o < s \leq 1 \tag{3.1}$$

where: D(i) is the demand rate of item i.

Then, the actual demand of item i is obtained by taking the derivative of G(i), namely:

$$D(i) = si^{s-1} \qquad 0 < i \leq 1 . \tag{3.2}$$

3.3 Preliminaries - The case of two and three classes.

Consider the case of two classes first:

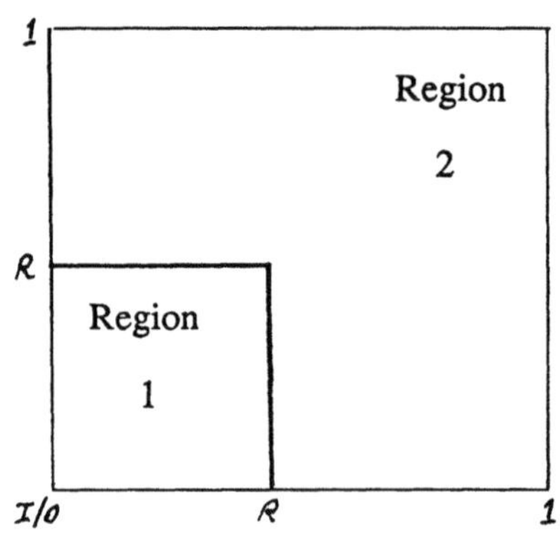

Fig. 1: Warehouse divided into two regions

The warehouse is square (normalized) in the horizontal and vertical dimensions and R_1 represents the distance from the I/O point to the boundary between regions 1 and 2. We need to determine the value of R_1 and the two groups of items belonging to each class in order to minimize the expected one-way travel time, T. Let t_i represent the average one-way travel time to region i, and G_i the cumulative percentage of items up to region i. Then, in the case of two regions:

$$T = t_1 \cdot G_1 + t_2(1 - G_1) \tag{3.3}$$

It can be easily shown that the values of t_1 and t_2 are $2R_1/3$ and $2(1-R_1^3)/3(1-R_1^2)$, respectively. As previously stated, we will assume that items are ordered according to the simple EOQ model, $Q_i^* = \sqrt{2KD(i)}$, where K is the ratio of ordering to holding cost, and for the sake of simplicity K is assumed constant for all items. Then, given a large number of items in the warehouse, on average the total number of required storage places is: $\sum_i \frac{Q_i^*}{2}$.

Let $(A_1)^2$ be the total area (region 1) required for the first group. Then, since each location has the same size, we get

$$(A_1)^2 = \int \frac{Q_i^*}{2} = \int \frac{\sqrt{2KD(i)}}{2} \tag{3.4}$$

$$i \in \text{1st group}$$

substituting D(i) from (2), we get:

$$(A_1)^2 = \frac{\sqrt{2Ks}}{s+1} \cdot (\alpha)^{\frac{s+1}{2}} \tag{3.5}$$

where: α is the last item (%) in the first group.

In a similar manner, substituting $\alpha = 1$, the total area, A, required for the whole warehouse is:

$$A^2 = \frac{\sqrt{2Ks}}{s+1} \tag{3. 6}$$

Therefore, the boundaries of A_1 are given by:

$$(R_1)^2 = \frac{(A_1)^2}{A^2} = (\alpha)^{\frac{s+1}{2}} \quad \text{and}$$

$$R_1 = (\alpha)^{\frac{s+1}{4}} \quad \text{or} \quad \alpha = (R_1)^{\frac{4}{s+1}} \tag{3. 7}$$

Now from (1) it is known that $G_1 = \alpha^s$, substituting the value of α from (3.7), we get

$$G_1 = (R_1)^{\frac{4s}{s+1}} \tag{3. 8}$$

Substituting the value off G_1 into (3.3) and using the values of the t_i's, T can be represented as a function of only one variable, R_1. Indeed, in this case, similar results to those obtained by Hausman et al (1976) are derived for the various values of s.

Let us now consider the case where the warehouse is divided into three regions. Here, T is given by:

$$T = t_1 G_1 + t_2 (G_2 - G_1) + t_3 (1 - G_2) \tag{3.9}$$

Using a similar logic to the one used for the two class system, we obtain the following values for the three class system:

$$t_1 = 2R_1/3;\ t_2 = 2\left(R_2^3\right)/3\left(R_2^2 - R_1^2\right);\ t_3 = 2\left(1 - R_2^3\right)/3\left(1 - R_2^2\right),$$

and $\quad G_1 = (R_1)^{\frac{4s}{s+1}}$ and $G_2 = (R_2)^{\frac{4s}{s+1}}$.

Substituting these values in (3.9), T is obtained as a function of R_1 and R_2 only, $T = f(R_1, R_2)$. To solve this problem, Hausman et al (1976) suggested a grid-search approach for arriving at the optimal values of R_1 and R_2.

However, as is shown here, the value of T can be represented as a function of only one independent decision variable. This simplifies the solution procedure, and enables one to derive accurate results.

3.4 Analysis

Consider again the ABC curve, where the items are divided into three groups.

The optimal values of R_1 and R_2 for the three class case can be easily derived if the values of ρ_1 and ρ_2 are known, where ρ_1 is the % of items in the first group and ρ_2 is the cumulative % of items in the first two groups. Given the values of ρ_1 and ρ_2, the values of $G_1 = G(\rho_1)$ and $G_2 = G(\rho_2)$ are obtained. Thus, similarly to (3.7), the values of R_1 and R_2 are:

$$R_1 = (\rho_1)^{\frac{s+1}{4}} \quad ; \quad R_2 = (\rho_2)^{\frac{s+1}{4}} .$$

The following simple property, for this specific demand function, should help us in developing a mechanism to simplify the solution procedure.

Property 1

For the cumulative (ABC) demand function $G(i) = i^s$, $o < s \leq 1$, and for any c, $c \neq o$, the following holds:

$$G\left(\frac{\rho_2}{\rho_1}\right) = \frac{G(\rho_2)}{G(\rho_1)} . = \frac{G(c\rho_2)}{G(c\rho_1)} . \tag{3. 10}$$

Thus, if the value of ρ_2 is known, there is only a need to solve the problem of dividing the warehouse into two groups. This problem can be easily solved using a simple one-dimensional search. In general, let:

$R_i(j)$ = the value of R_i, $i = 1,2,...,j-1$ where the warehouse is divided into j regions (groups).

Then,

$$R_1(2) = \left(\frac{\rho_1}{\rho_2}\right)^{\frac{s+1}{4}} = \frac{(\rho_1)^{\frac{s+1}{4}}}{(\rho_2)^{\frac{s+1}{4}}} . = \frac{R_1(3)}{R_2(3)}$$

Thus, $R_2(3)$ can be represented as a function of $R_1(3)$ and $R_1(2)$. As $R_1(2)$ is already known, the value of T can be represented as a function of only $R_1(3)$, which can be easily solved. Note also that the proportion of $R_1(3)$ to $R_2(3)$ is equal to $R_1(2)$ or the proportion of $R_1(2)$ to $R_2(2)$, where $R_2(2) = 1$.

In extending this approach to the general case of n class-based system, we use the following theorem.

<u>Theorem 1</u>.

For any i, j and n, where $i < j \leq n$, the following general relationship holds:

$$R_i(j) = \frac{R_i(n)}{R_j(n)} \tag{3. 11}$$

The proof is given in Rosenblatt and Eynan (1987).

Now, for the special case of $i = 1$, the following corollary is obtained:

<u>Corollary 1:</u> $$R_j(n) = \frac{R_1(n)}{R_1(j)} \qquad \forall\, j < n. \tag{3.12}$$

Whereas the theorem gives us the general relationship between the various parameters i, j and n, the corollary shows that for any value of n, the different R_j's can be expressed as a function of $R_1(n)$ and the previously computed R's. Thus, there is no need to perform an n-1 dimensional search for the n-class based system, but rather a simple one dimensional search.

The expected one-way travel time for the n class based system, T_n, is represented by:

$$T_n = \sum_{i=1}^{n} t_i(G_i - G_{i-1}) \tag{3.13}$$

where $G_0 = 0$ and $G_n = 1$.

Substituting the general values of the t_i's and the G_i's, the general expected one-way travel time can be represented by the following:

$$T_n = \frac{2}{3} \sum_{i=1}^{n} \left[\frac{R_i^3(n) - R_{i-1}^3(n)}{R_i^2(n) - R_{i-1}^2(n)} \right] \left((R_i(n))^{\frac{4s}{s+1}} - (R_{i-1}(n))^{\frac{4s}{s+1}} \right) \tag{3.14}$$

Finally, the expected one-way travel time for the full turnover case (and as previously obtained by Hausman et al (1976)), T_∞, is given by:

$$T_\infty = \frac{4s}{5s+1} \tag{3.15}$$

Even though practically it may be infeasible to implement a full turnover system, this value can serve as a lower bound on the expected one-way travel time. Furthermore, as is illustrated, even a modest number of regions can achieve most of the savings (in time) obtained by the full turnover system.

The solution procedure for a given ABC curve and any desired number of regions or groups, n, is divided into n stages. At the k-th stage, k = 2...n, the various values of $R_j(k)$ are represented as a function of $R_1(k)$ using equation (3.12). Using the value of T_k from equation (3.14), we derive the optimal value of $R_1(k)$ by a one-dimensional numerical search procedure. Note that $R_1(i)$ is smaller than $R_1(i-1)$, which has already been found. Thus, the search procedure for $R_1(i)$ is bounded by the value of $R_1(i-1)$, which also saves some computational effort. Given the value of $R_1(k)$, the solution proceeds to the k+1 stage until it derives the different $R_j(n)$ values for stage n.

3.5 Computational Results

Consider the following results for the 20/80 demand curve (s=.139) case.

n	T_n	$R_1(n)$	% of Absolute Improvement	% of Relative Improvement
1	0.6667	1.000	0	0
2	.4277	.2838	35.85	70.63
3	.3754	.1331	43.68	86.08
4	.3557	.0773	46.64	91.91
5	.3462	.0505	48.07	94.73
6	.3408	.0359	48.87	96.31
7	.3375	.0266	49.37	97.28
8	.3354	.0205	49.69	97.92
9	.3339	.0163	49.92	98.36
10	.3328	.0134	50.08	98.68
11	.3320	.0110	50.20	98.92
12	.3314	.0093	50.29	99.10
13	.3309	.0078	50.37	99.25
14	.3305	.0068	50.42	99.36
15	.3302	.0058	50.47	99.45
16	.3299	.0052	50.51	99.53
17	.3297	.0046	50.54	99.59
18	.3295	.0040	50.57	99.65
19	.3293	.0036	50.59	99.69
20	.3292	.0033	50.61	99.73

Table 1: Improvements in expected one-way travel time for various number of classes for the 20/80 demand curve.

The value of the percentage of the absolute improvement is derived for each T_n relative to the value of T_1. Thus, Percentage of Absolute Improvement = $\frac{T_1 - T_n}{T_1} \times 100\%$

The percentage of relative improvement is derived by the improvement in travel time due to dividing the warehouse into n regions relative to the maximum improvement obtained by using a full turnover approach. Thus,

$$\text{Percentage of Relative Improvement} = \frac{\frac{T_1 - T_n}{T_1}}{\frac{T_1 - T_\infty}{T_1}} \times 100\% = \frac{T_1 - T_n}{T_1 - T_\infty} \times 100\%$$

where T for the 20/80 demand curve is 0.3280.

The percentage of relative improvement can be used to determine the desired number of regions in the warehouse such that most of the benefits of the full turnover case are gained. In this case, for example, 99% of the potential gain of full turnover is obtained by a 12-region (class-based) warehouse. It is further seen that even for four classes, over 90% of the potential gain is obtained. This result was found to be consistent for various patterns of demand.

Finally, this study is now extended to the case of class-based policies in rectangular (time-wise) AS/RS.

4. Warehouse Design - A Combined Approach of Optimization and Simulation.

This section presents a heuristic procedure for determining the physical layout of an automated warehouse. The procedure is an iterative one and is based on solving optimization and simulation models. In recent years there have been several attempts to solve the warehouse desigh problem using optimization techniques, see Ashayeri et al (1985), Bassan et al (1980), Karasawa et al (1980), and Ashayeri et al (1983), Chisman (1983), Gelders et al (1983), Perry et al (1983), respectively. Also, Zollinger (1975) has offered a method for arriving at a realistic budget price for proper evaluation of storage methods based on data assembled from existing warehouses. However, the main contribution of this section is in incorporating simulation results into our optimization model from the technical point of view, and by considering the case where the number of cranes may be smaller than the number of aisles from the design point of view.

In the optimization model the objective function is to minimize total cost which is composed of the various physical components of the AS/RS. The basic idea is to find the best layout design given a number of constraints on the capacities of warehouse and cranes. The major advantage of the optimization model is that it provides an optimal solution for the various physical parameters while minimizing total cost. However, the major disadvantage is that it ignores the stochastic nature of the system and does not guarantee the satisfaction of certain service levels.

On the other hand, the simulation model deals with the stochastic nature of the problem. For a specific layout and design characteristics, it can answer questions regarding service levels of the system. However, the simulation model must be run each time for a different layout. Thus, theoretically, in order to determine the optimal layout with the minimum total cost that satisfies service level requiirements, a large number (possibly infinite) of experimentations needs to be made.

In this section, a combined (optimization and simulation) procedure is developed in order to determine the design of an AS/RS. The optimization model minimizes the total cost of the system subject to constraints on service times and service levels. The main purpose of the simulation model is to provide information on the daily behavior of the system and to relate the physical layout with its measures of efficiencies The iterative procedure which is presented here uses the output of one model (simulation or optimization) as input into the other model until it converges to the desired solution.

4.1 The Optimization Model

The objective function of the optimization model is composed of initial investment costs and annual operating costs which are discounted throughout the life of the project. The decision variables include: The number of levels (rows) in the warehouse, M, the number of columns along each rack, P, the number of aisles in the warehouse, R, the number of cranes, S, the number of workers, W, the length of a conveyor, Q, and number of storage spaces, B, both in pallet units. The constraints relate to the maximum dimensions of the warehouse and the minimum number of workers and cranes. The objective function is composed of the following costs:

Costs of Walls and Ceiling:

$$C_c[2(R(2L+a) + Pw + K_2)(Mh + K_1) + R(2L + a)(Pw + K_2)]$$

where L , w and h are the length, width and height of a standardized pallet, respectively; K_1 and K_2 are the required space between the top pallet and the ceiling, and between the end of a rack and the end of the warehouse, respectively; a is the width of an aisle and C_c is the average cost per squared meter of walls and ceiling construction.

Cost of Land and Floor:

$$(C_f + C_L)R(2L + a)(Pw + K_2)$$

where C_f and C_L are the cost of one squared meter of floor and land, respectively.

Cost of pallets:

$$C_p(2RPM)$$

where C_p is the cost per pallet.

The cost for cranes, storage workers, conveyor and storage space is:

$$C_s \cdot S + C_w \,.\, W + C_q \cdot LQ + C_b \cdot B$$

where:

C_s and C_w are the annual costs per one crane and one worker (work station), respectively; C_q is the cost per one meter of conveyor and C_b the cost per one unit of storage (pallet).

Cost per aisle in AS/RS can be divided into two, one which is fixed per aisle (C_a') and one which is dependent on its length. Thus, total cost per aisles is:

$$R(C_a' + C_aP)$$

In warehouses where the number oc cranes is smaller than the number of aisles ($S < R$), additional equipment is needed to transfer and control the shift of cranes from one aisle to another. Then, the total cost of this equipment is given by:

$$\delta(K_3 \cdot R \cdot C_a')$$

where: K_3 is the additional cost (in terms of percentages) on top of the original aisle cost, with δ being a binary decision variable with a value of 1 if $S < R$ and 0 if $S = R$ The operating cost in the warehouse is mainly composed of salaries for the operators. Thus, if D is the annual salary of a worker and Z is the present worth factor for the entire life of the project, then, the total operating cost (wages) is:

$$WDZ$$

Finally, it is assumed that annual maintenance costs is represented by a prespecified percentage (I) of the initial investment in cranes, conveyors and storage spaces. Thus, the present worth of maintenance costs is:

$$I \cdot Z \cdot (C_s \cdot S + C_q \cdot Q \cdot L + C_b \cdot B)$$

<u>Constraints</u>:

In addition to the physical constraints, the model includes the following constraints:

$$(\delta-1)(R\text{-}S) \geq 0 \quad , \ \delta = 0,1$$

This constraint guarantees that $\delta = 1$ if $R < S$, namely, that the number of cranes is less than the number of aisles and therefore additional cost is needed. Also, throughout the iterative process additional constraints relating the physical characteristics of the warehouse with performance measures will be added.

4.2 The Simulation Model

The model was written using SLAM II. The input for the system included the inter-arrival time for orders of pallets (products) in the warehouse; the processing times of the orders; operating decision rules, and other technical parameters of the system (e.g. crane's speed (vertical/horizontal), conveyor's speed, etc.).

The physical layout of the warehouse include data regarding the dimensions (length/width/height) of each pallet; number of pallets/rack, number of cranes and aisles, storage spaces (in pallets) near the aisles and work stations, length of conveyor (in pallets) and number of work stations, where it is assumed that each work station is manned by one worker.

The output of the simulation model was measured in two ways:

1. Average service time per order. This was measured as the elapsed time from receiving an order for a certain pallet till it was completely processed in the work station
2. Service level in the warehouse (%). This was measured as the ratio between the actual number of orders processed throughout the first shift in the warehouse and the total number of orders received.

4.3 Solution Procedure

An iterative solution procedure is proposed. First the optimization model is solved. Its output is used as input for the simulation model. Given the output of the simulation model, the two measures of effectiveness are compared with the predetermined desired levels of service (e.g., an average service time of one hour). If the result of the simulation model satisfy these desired levels, the procedure is terminated. Otherwise, the following experimental design is carried out. For each physical input of the warehouse (obtained from the optimization model), an upper and lower bound values is derived. Then, the theoretical number of combinations is 2^7, since the number of variables in the optimization model is 7 . Then, out of the theoretical number of combinations, choose the relevant (feasible) combinations and run the simulation model for them.

Based on the simulation output and using multiple linear regression techniques, we determine the relationship between each measure of performance and the various variables (representing the actual characteristics of the warehouse). The next step is to incorporate the linear equations into the optimization model as inequalities, with the right hand side (R.H.S.) of these inequalities being the prespecified levels. The iterative procedure continues until the results of the optimization model are the same as those obtained in the previous stage, and are satisfying the required measures of performance. If, however, in one of the iterations, the simulation results do not comply with the prespecified service levels, then, again, we need to find the various combinations of variables for which to run the simulation models and obtain the new linear inequalities.

Several examples with different warehouse characteristics were studied. In all the examples the procedure converges very rapidly (few iterations) to its final solution.

5. Determining the Size of a Warehouse Container

Several researchers addressed the problem of the optimal handling unit (pallet or container) size, to be used in material handling and warehousing systems. Steudel (1983), Tanchoco and Agee (1981), Tanchoco et al (1983) and Grasso and Tanchoco (1983) studied various aspects of this subject. The last two references incorporate the size of the pallet, or unit load, in evaluating the optimal lot sizes for multi-item inventory systems with limited storage space. In a report on a specific case, Normandin (1967) has demonstrated that using the "best-size" container can result in considerable savings. A simulation model combining container size and warehouse capacity considerations, in an AS/RS environment, was developed by Kadosh (1987). The general results, reflecting the stochastic nature of the flow of goods, are similar to those reported by Rosenblatt and Roll (1988). Nevertheless, container size was found to affect strongly overall warehousing costs.

In this section, we suggest a procedure for determining the optimal size of a warehouse container. First, the type of data necessary for analyzing this problem is considered. Then, an approximate solution, based on a series of underlying assumptions, is proposed. For cases where the approximate approach does not seem to be valid, a simulation model should be used in which all pertinent variables are accounted for, at the desired level of detail

5.1 Physical Characteristics

Determining the optimal container size is strongly related to the overall physical solution proposed for a given situation. The important factors under this heading are the material handling equipment (with the resulting requirements on the type of container to be used), the degree of mechanization, and the overall size of the planned warehouse. Other physical aspects pertain to the general style of storage facilities to be installed. This includes the type of warehouse furniture (racks, etc.) and the material and structure of the container to be used.

Another set of relevant data comprises the proportional dimensions of the container. These proportions are affected by the proposed work organizations and the ergonomic requirements stemming from it, the general storage style and the nature, size and quantities of items to be stored. This set of data is required for obtaining an initial rough estimate of the range within which the container size will finally be. Such an initial estimate is necessary for determining the general characteristics of the solution. This may also affect our storage decisions, e.g. whether each item should be stored in a separate container or if the containers should be partitioned into compartments each holding (possibly) different items. In the former case, the proposed analysis will yield the best container size, while in the latter, the outcome may be the appropriate compartment size.

5.2 Assumptions and Implications

The following general assumptions will be used throughout this section,

(1) There is a relatively large number of items to be stored.

(2) One container size is used.

(3) Container size (volume) is such that:

(a) The volume of a single unit of any item is considerably smaller than the container volume.

(b) The average total volume held of each item is larger than the container volume.

(4) Each container holds units of one type of item only,

(5) A random storage policy is used.

(6) A strict FIFO policy is not imposed.

(7) The average storage equipment cost can be expressed as a function of container volume.

The implication of the first assumption is that we can make use of the "law of averages" in our approximating model. The assumption of a single container size is applicable to situations where the items are similar in their physical characteristics (e.g. weight, colume, shape, etc.). However, if this is not the case, a procedure for dividing the item population into groups possessing similar characteristics is suggested in Roll et al (1988). Each such group is, then, treated separately. The implication of (3a) is that items can be arranged within a container in a manner leading to a satisfactory volume utilization. On the other hand, (3b) is necessary in order to ensure that a high proportion of containers are utilized to their full capacity. This is related to assumption 6 regarding a non-strict FIFO policy.

Assumption 4 does not preclude the possibility that storage units (containers) are partitioned into comparrtments, each holding units of different items. In such a case, the model can be used for finding the best compartment size. The random storage policy is assumed in 5 in order to allow any empty container (compartment) to be used for the storage of any incoming shipment of items. Assumption 6 deals with the practiced storage/retrieval policy, and is related to the number of non-full containers at any point in time. Under strict FIFO the number of non-full containers will generally approach the number of different lots (of items) held in storage. On the other hand, when a strict FIFO policy is not imposed, then at any point in time only one container (at most) for each item type will not be full.

The last assumption deals with the annual cost of storage space, furniture, equipment, containers, etc. These costs may vary widely for different storage styles. Nevertheless, in a series of real-life cases, see Kadosh (1987), it was found that these costs can be closely approximated by a 3-arameter expression of the kind

$$T_v = \alpha + \beta v^{\gamma} \tag{5. 1}$$

where T_V denotes the average cost per container of volume v and α, β and γ are parameters specific to the particular style of storage.

In addition, a set of container-dependent operating costs should be considered. Typical among these is the cost associated with moving to a single location in the warehouse. The smaller the container, the larger the number of locations at which a batch of incoming items would be stored, and the higher the probability that a withdrawal of a required quantity will have to be done from more than one location. Thus, there is a relationship between container size and the number of movements. Other cost parameters within this group are those associated with external storage, see Rosenblatt and Roll (1984). The larger the containers, the larger would be the space required on average for a warehouse and the lower the probability of having to recourse to costly external storage of incoming shipments.

5.3 Modeling and an Approximate Solution

The total storage capacity, K_v, needed on average, for storing N different items in containers of size v is:

$$K_v = v \sum_{i=1}^{N} \left[s_i / [v/m_i]^- \right]^+ \qquad (5.2)$$

where: m_i is the volume of item i and s_i average quantity held of item i, $[v/m_i]^-$ is the maximum number of units of item i that can be stored in one container; $[s_i/[v/m_i]^-]^+$ represents the minimum number of containers needed to store, on average, the units of item i.

In order to get the total cost for a given container size, equation (5.2) should be multiplied by (5.1). However, to get the optimal v, given the formulation of (5.2) may be very complicated, and therefore some simplifications and approximating are introduced.

Let u_{iv} denote the container's utilization factor, when item i of volume m_i is stored in a container of capacity v. This factor accounts for the fact that, due to the different geometric properties of the stored items, the shape of a particular container v is seldom perfectly suitable for storing each of the items. Hence, the number of units of item i that can be stored in one container is given by:

$$\left[\frac{v}{m_i} \right]^- = \frac{v u_{iv}}{m_i} \qquad (5.3)$$

Defining e_i as the unutilized proportion of the "end container" in the group of containers holding item i, and using (5.3), the number of containers necessary for storing a quantity s_i, can be written as:

$$\left[s_i / \frac{v u_{iv}}{m_i} \right]^+ = \frac{m_i s_i}{v u_{iv}} + e_i \qquad (5.4)$$

Substituting (5.3) and (5.4) into (5.2), the total storage capacity needed is:

$$K_v = v \sum_{i=1}^{N} \left(\frac{m_i s_i}{v u_{iv}} + e_i \right) \tag{5.5}$$

To simplify the solution procedure, the following assumption will be made: Let the various items be sufficiently similar in character so that m_i, s_i and u_{iv} can be replaced by m, s and u_v, respectively. This is a rather strong assumption and should be used only in situations where it is justified. Here it is presented mainly for illustrative purposes. Also, given the assumption of large N, $\sum_{i=1}^{N} e_i$ is approximated by $N/2$. Thus, the required capacity of a warehouse can be approximated by

$$K_v = Nv \left(\frac{ms}{u_v v} + 1/2 \right) \tag{5.6}$$

Combining expressions (2), (5) and (7), the average container dependent cost, C_v per item stored is given by:

$$C_v = \left(\frac{ms}{u_v v} + 1/2 \right) \left(\alpha + \beta v^{\gamma} \right) \tag{5.7}$$

Finally, if we assume that, for any specific application, u_v exhibits little variation so that it can be replaced by u, then we can proceed and try to obtain the optimal container size.

Taking the derivative of C_v with respect to v, the optimal value of v must satisfy the following equation:

$$\frac{u\beta\gamma}{2sm} v^{\gamma+1} + \beta(\gamma - 1) v^{\gamma} - \alpha = 0 \tag{5.8}$$

As in general, $0 < \gamma < 1$, the value of v is obtained numerically. For the trivial case of $\gamma = 1$, a closed form solution for v is given by:

$$v^* = \left(\frac{2\alpha sm}{u\beta}\right)^{1/2} \tag{5.9}$$

Finally, the following property provides the worst case (lower bound) utilization factor values dependent only on the relationship between v and m_i.

Property 1: The utilization factor is bounded below (worst value) by $k/(k+1)$ where $[v/m_i]^- = k$. (Proof is provided in Roll et al (1988)).

The following table presents the lower bound values of the utilization factor as a function of k.

k	1	2	3	4	5	10	20	30
u_{iv}	.5	.667	.75	.8	.833	.909	.952	.967

Note, that even for the small ratio of $k = 10$, a utilization factor of over 90% is guaranteed. Although this utilization factor ignores issues like the shape of the various units, for a large ratio of v to m_i, as assumed in 3(a), this property still holds.

An extensive simulation study is reported in Roll et al (1988) taking into consideration the stochastic nature of the problem. The input to the model include the characteristics of the items to be stored, the inventory policy, retrieval characteristics, external storage rules, container size, overall warehouse capacity and relevant cost data. A series of runs was made to cover the entire range of possibilities. The outcome of these runs were the overall costs for each capacity-container combination, as well as the corresponding service levels. When no limit is imposed on warehouse capacity, the relationships between container size and required capacity for a specific service level, is established. It is found, among other results, that the larger the container size is, the more capacity is needed for maintaining a required service level. Finally, a detailed solution procedure is provided there for determining the optimal container size, or whether multiple container sizes are needed.

REFERENCES

"AS/RS are expected to rise 500% by 1994", **Journal of Industrial Engineering,** 18, No. 7, p. 6, 1986.

Ashayeri, J., Gelders, L.F. and Van Looy, P.M., "A simulation package for automated warehouses", **Material Flow**, 1, pp. 189-198, 1983.

Ashayeri, J., Gelders, L.F. and Van Wassenhove, L., "A microcomputer based optimization model for the design of automated warehouses", **Int. J. Prod. Res.**, 23, 4, pp. 825-839, 1985.

Bassan, Y., Roll, Y. and Rosenblatt, M.J., "Internal layout design of a warehouse", **AIIE Transactions**, 12, 4, pp. 317-322, 1980.

Chisman, J.A., "Anatomy of a simulation model", Institute of Industrial Engineers, 1983, **ICAW Proceedings**, pp. 51-53, Atlanta, Georgia.

Gelders, L.F., Ashayeri, J. and Van Looy, P.M., "Microcomputer assisted design of an automated warehouse", Institute of Industrial Engineers, 1983, **ICAW Proceedings**, pp. 43-46, Atlanta, Georgia.

Grasso, E.T. and Tanchoco, J.M.A., "Unit load and material handling considerations in material requirement planning systems", **Material Flow**, 1, p. 79, 1983.

Hackman, S.T. and Rosenblatt, M.J., "Allocating items to a preferred storage area", Working Paper, Faculty of Industrial Engineering and Management, Technion, Haifa, Israel, March 1988.

Hausman, W.H., Schwarz, L.B. and Graves, S.C., "Optimal storage assignment in automatic warehousing systems", **Management Science**, 22, No. 6, pp. 629-638, 1976.

Kadosh, D., "Determining the optimal size of a warehouse container", M.Sc. thesis, Faculty of Industrial Engineering and Management, Technion, Haifa, Israel, 1987.

Karasawa, Y., Nakayama, H. and Dohi, S., "Trade-off analysis for optimal design of automated warehouses", **Int. J. Systems Science**, 11, 5, pp. 567-676, 1980.

Normandin, G., "Computer picks the one "Best" container size", **Modern Materials Handling**, p. 60, May 1967.

Perry, R.F., Hoover, S.V. and Freeman, D.R., "Design of an automated storage/retrieval system using simulation modeling", Institute of Industrial Engineers, 1983, **ICAW Proceedings**, pp. 57-63, Georgia.

Roll, Y., Rosenblatt, M.J. and Kadosh, D., "Determining the size of a warehouse container", Working Paper, Faculty of Industrial Engineering and Management, Technion, Israel, May 1988.

Rosenblatt, M.J. and Eynan, A., "Deriving the optimal boundaries for class-based automatic storage/retieval systems", Working Paper, Faculty of Industrial Engineering and Managment, Technion, Haifa, Israel, June 1987.

Rosenblatt, M.J. and Roll, Y., "Warehouse capacity in a stochastic environment", **International Journal of Production Research**, forthcoming, 1988.

Rosenblatt, M.J. and Roll, Y., "Wartehouse design with storage policies considerations", **International Journal of Production Research**, 22, p. 809, 1984.

Srikantan, K.S., "A problem in optimal allocation", **Operations Research**, 11, pp. 265-273, 1963.

Steudel, H.J., "Determination of economic pallet size for in-plant handling", **Material Flow**, 1, p. 71, 1983.

Tanchoco, J.M.A. and Agee, M.H., "Plan unit loads to interact with all components of warehouse system", **Journal of Industrial Engineering**, 13, No. 6, p. 36, 1981.

Tanchoco, J.M.A., Davis, R.P., Egbolu, P.J. and Wysk, R.A., "Economic unit loads (EUIL) for multi-product inventory systems with limited storage space", **Material Flow**, 1, p. 141, 1983.

Vidal, R.V.V., "A simple method to solve some simple allocation problems", **IIE Transactions**, 19, No. 2, pp. 234-237, 1987.

Zipkin, P., "Simple ranking methods for allocation of one resource", **Management Science**, 26, No. 1, pp. 34-43, 1980.

Zollinger, H.A., "Planning, evaluating and estimating storage systems", **Advanced Material Handling Technology Seminar**, Purdue University, Indiana, 35, pp. 1-35, 1975.

ACKNOWLEDGEMENT

The studies reported in this paper are a summary of work done at the Technion jointly with the following: Prof. Steven T. Hackman, Mr. Amit Eynan, Mrs. Vered Ziser, Prof. Y. Roll and Mr. David Kadosh.

KNOWLEDGE MODELS FOR THE DESIGN AND MANAGEMENT OF AUTOMATED STORAGE AND RETRIEVAL SYSTEMS

Abraham Seidmann*

Department of Industrial Engineering

Faculty of Engineering

Tel-Aviv University, Tel-Aviv, 69978, Israel

Abstract

The successful operation of every manufacturing facility or distribution center depends to a large extent on the efficient hadling, storage and flow controls of materials. A key component in many material flow activities is the Automated Storage and Retrieval System (AS/RS) which serves as an unmanned warehouse for raw material, work in progress inventories and finished goods. In this paper a new design and management methodology for AS/RS's is developed and examined. It is based on knowledge structures, mathematical models and on the state-operator framework for problem solving. An industrial case study is used to illustrate the applicability of these concepts to the operational controls of a unit-load AS/RS. The impacts of recent developments in related information technologies, data representation schemes and industrial communications are also examined.

---

* Currently an Associate Professor of Computers and Information Systems and Operations Management with William E. Simon Graduate School of Business Administration at the University of Rochester, Rochester, NY 14627, USA.

NATO ASI Series, Vol. F53
Advanced Information Technologies for Industrial Material Flow Systems
Edited by Shimon Y. Nof, Colin L. Moodie

1. Introduction

The introduction of innovative information technologies to the material flow application area has had a dramatic impact on storage and warehousing. A case in point is the numerous installations of automated Storage and Retrieval System (AS/RS). These systems combine equipment and controls to automatically handle, store and retrieve materials with great speed and accuracy. (Tompkins and White (1984)). Such complex systems may incorporate laser beam scanners, automated rollers, chain or overhead conveyors, computer controlled palletizers, weight and dimension checking stations, in-floor towlines, driverless tractors, or other automated links to the manufacturing or distribution facilities. (White(1987)). The entire operation is overseen by a supervisory computer which monitors the storage location of each item and its movements from one location to another (Hill (1980)). These systems are used for storing raw parts, tools, in-process inventories and finished goods in conventional and Flexible Manufacturing Systems (Sellers and Nof (1986), Arbel and Seidmann (1984)).

In recent years, such systems have reached a point of maturity and economy (with rack supported structures depreciable as equipment). There are additional benefits to such systems: High floor and cube space utilization, improved material flow and inventory control and substantial saving in labor cost wherein AS/RS are justifiable as an alternative to conventional pallet rack warehousing (Rygh (1981)). The AS/RS consists of storage racks, storage/retrieval crane, and input/output (I/O) pick-up and deposit station. Each crane operates in a single aisle with storage racks on either side. The crane has three mechanical drives: The vertical drive which raises and lowers the cargo, a horizontal drive, which moves the frame back and forth along the aisle, and a shuttle drive, which transfers the cargo load between the crane's carriage and either side of the aisle. Typically, the horizontal and the vertical drives operate simultaneously in order to reduce the travel time.

A crane cycle generally begins with the crane at the I/O point; it picks up a load, travels to the storage location, deposits the load, travels empty (interleaves) to the retrieval location, retrieves the load, travels to the I/O point and deposits the load. The efficiently with which an AS/RS operates is influenced by the storage assignment, and by the picking policies. Storage policies refer to the assignment of arriving loads to an empty rack location. Picking rules consist of the determination of the load to be picked in a given cycle.

Previous research work in automated warehousing systems includes computer simulations, exact and approximate mathematical models. Bozer et al (1982) proposed a six-phase design process for such systems:

1. Problem definition,
2. data base development,
3. generation of design alternatives,
4. evaluation of alternatives,
5. selection of the preferred system and,
6. system implementation.

It is well known that **dedicated** storage, based on activity rates, may maximize the crane output, but it also maximizes the storage space requirements (White (1980)). On the other hand, **randomized** storage minimizes storage space, but reduces the crane throughput (Francis and White (1974)). A common compromise between the random assignment and dedicated turnover-based assignment is the **class-based** assignment. In the class-based assignment rule, rather than allocating each equidistance group of racks locations to a narrow range of turnover times, the rack locations are divided into several classes with respect to the time from the I/O point. For example, in a four-class storage assignment system the locations are divided into four classes such that and location in class II is closer to the I/O point than class III. The stored items are then divided into the same number of classes based on their turnover time-volume rates. Using this approach the total crane time is reduced compared to that of the random assignment while sufficient storage flexibility is still provided for new arrivals. The selection of the appropriate storage method depends upon the weight given to storage space economy versus that given to throughput. The seminal paper of Hausman et al (1976) formulates the merits of class-based storage assignment rules for single-aisle, no interleaving AS/RS. Later studies by these authors (Graves et al (1977) and Schwarz et al (1978)) examine several heuristic storage assignment interleaving policies assuming knowledge of the turnover time for the various items. The major performance measure investigated was the crane cycle time with mandatory interleaving.

The cube per order index rule (COI) for storage assignment was presented by Kallina (1977) and later studied by Wilson (1977). In two subsequent studies Bozer et al (1982) and White and Bozer (1984) extend the results presented by Graves et al (1977) for a rack that is not necessarily square-in-time. They computed the crane throughput for two alternative operating policies and for three alternative I/O configurations. These studies were later used by Han et al (1987) to develop lower bounds on the dual command cycle times and an effective Nearest-Neighbor heuristic. Other analytical studies are presented by Karasawa et al (1980),

Matson and White (1982), Ratliff and Rosenthal (1983), and by Kusiak et al (1985) . The paper by Kusiak (1985) used dynamic programming formulations and a heuristic model to study the single-command order picking policy with due-dates.

The complexity of the AS/RS is such that exisiting analytical procedures do not have the capabilities to adequately model all their operational features, (Maxwell (1981)). As a result, simulation models are used to verify system specifications and to study system behavior at stochastic dynamic environments. Bafna (1981) used simulation search technique to determine a minimum cost system design. A Generalized Manufacturing Simulator (GEMs) for design verification is presented by Sathayi and Phillips (1979). Another simulator for automatic warehousing systems was developed by Ulgen and Elayet (1981). The optimal determination of the items to be picked during a single crane trip is presented by Elsayed (1981). Azadivar (1984) developed a simulation optimization approach aimed at finding the best way for allocating storage racks among items with various turnover times. During the search process the simulation model is used as a mean for evaluating the value of the objective function for the level of the decision variables determined by the search procedure. The objective function used was the mean time for a crane cycle. The actual decision variables used are the number of storage classes and the boundaries of each class.

The simulated impact of seasonal trends on the crane activities was recently studied by Linn and Wysk (1987). They stress the importance of using a Pareto analysis for storage allocation at heavily utilized AS/RS. Moreover, they introduce a "pursuit" mode for the crane by which it remains idle at the termination position of the last task waiting for a new command. Improved crane performance following this mode is shown. A combination of simulation and optimization techniques was also used by Roll and Rosenblatt (1981) to determine the desired storage capacity, configuration and storage policies in a conventional warehouse.

Most of these studies were based on historical averages rather than on instantaneous system feedback. For example, they assume static demand and arrival rates for all product types and that the stored pallets are retrieved based on a strict First In First Out (FIFO) basis to ensure proper age control. In reality, the instantaneous and the long-run arrival or the demand rates fluctuate due to various short term random effects, seasonality effects, and changes in the corporate marketing strategy (Seidmann et al (1985)). Relaxing the stringent FIFO policy constraint, when operating a real AS/RS, might also prove useful. For instance, one may search for retrieval candidates among all products having the same age or expiry date. Moreover, with very few exceptions, these technical papers don't describe actual real-life applications comparing their data, assumptions and results with the current industrial practice and objectives.

The principal concern of this paper is to outline and examine the development of several knowledge models used for the design and management of AS/RS's. A general design procedure for AS/RS is discussed and then followed by a brief presentation of a control scheme for an AS/RS as developed for a national distribution center of a certain producer. This control scheme is based on the state-operator framework for controlling intelligent systems (Nilson (1980), Barr and Feigenbaum (1982)). Using this approach one evaluates states of data and use operators which convert one state to another in order to reach a goal state. A state, in this case, is the data defining the state of the AS/RS at any general instant. The structure of the paper is as follows: Section 2 presents the AS/RS design process. The development of the AS/RS operating rule is given by section 3 and section 4 outlines the knowledge structure for the system controls. Section 5 presents a brief summary of some experimental evaluations. Key information technology issues related to AS/RS's are discussed in section 6 and section 7 concludes the paper.

2. The AS/RS Design Process

2.1. Design Objectives

The major objective of an AS/RS design is to maximize the effective use of capital resources while satisfying the operational requirements. Based on this major objective Tompkins (1982) defined the following design objectives:

1. maximize the effective use of storage volume,
2. maximize the effective use of transport hardware,
3. maximize the effective use of labor,
4. maximize the accessibility of all goods stored, and
5. maximize the protection of all items in storage.

The AS/RS planning methodology starts by analyzing the space requirements and by characterizing the appropriate storage technology for all items. For instance, some items require unit-load transfer while others are handled on mini-load totes.

2.2. The Part Selection Process

The design process starts with identifying the parts to be stored in the proposed AS/RS. Several data bases are designed to help in characterizing the parts and the parts flow process. Table (1) illustrates the corporate material flow database to be established. In addition a product database is established to handle the various product descriptions to be used later as well. This database is illustrated by Table (2). Various "expert" models can then be implemented in order to help in analyzing the above data. For example, the SRAC(Shipping and Receiving Analysis Chart) is used to define the type of specifications of interest by classifying the materials into generic categories and then completing the SRAC analysis for each category (Tompkins (1982)). These "expert" models are then used in allocating the various items among several potential storage options as pallets, totes, cartons or cases.

2.3. The Volume and Space Determination

The storage volume requirement of an AS/RS depends on four major factors:

1. the attributes of the products to be stored,
2. the storage method,
3. the expected inflow and outflow rates, and
4. the desired "service-level" of the AS/RS.

Determining the storage capacity it is common to assume Poisson Batch arrival process. It has been shown that an AS/RS can be modeled as an $M/G/\infty$ queue (Schwarz et al (1978)). In a recent study Keilson and Seidmann (1987) have shown that for random batch arrival, the number of items in storage is compound poisson with a mean of:

$$\bar{x} = \lambda \alpha \mu \qquad (1)$$

and a standard deviation given by

$$\sigma(\bar{x}) = \lambda \alpha \int_0^\infty [1- F(s)]\; F(s)\, ds \qquad (2)$$

$$+ \lambda (\alpha^2 + \beta^2) \int_0^\infty [1-F(s)]^2\, ds,$$

where

λ = mean batch arrival rate,

α,β = mean and standard deviation of the batch size, and

$F(s)$= the batch size pdf.

These results can be used to estimate directly the required storage volume under various safety factor considerations.

Item

- Raw Materials
- Inprocess Item
- Finished Goods
- Delivered Item
    - Number of Items Volume
    - Seasonalities
    - Special Sales Drives

Customers/Vendors

- Domestic
- International
- Geographic Distributor
- Shipping Methods
- Containers Usage / Dispatch Policies
- Pallets Used
- Delivery Requirements
- Special Priorities
- Return Policies

Facilities

- Parts flow data
- Time Losses
- Rework and Returns
- Batching and Scheduling
- Layouts, Obstructions, Limitations
- Existing Hardware
    - Within Aisels
    - Between Aisels
- Current Information Infrastructure
- Data Collection Points

Table 1. The Material Flow Data Base

Physical Properties

- Storage State
- Shelf Life
- Dimensions
- Weight
- Storage Requirements
    - Temperature
    - Light
    - Separation
    - Static Electricity
    - EMI

SKU

- Quantity of Items
- Dimensions
- Volume
- Conveyability
- Weight
- Number of Items per Pallet/Tote
- Storage Date
- Supply Sources
- Lead Times
- Input Units
- Return Units
- Common Outlets
- Shipment/Picking Units
- Special Handling Requirements

Table 2. The Product Descriptors Data Base

2.4. The AS/RS Layout Parameters

The efficient operation of an AS/RS depends on the generation of an effective layout that can handle both current needs as well as an acceptable migration path plan.

Major layout decisions include:

1. the number of storage aisles,
2. the relative dimensions (height & length) of each aisle,
3. Storage allocation policies
   - items to classes
   - classes to aisles
   - duplication (backup) allocations,
4. receiving and shipping area
5. crane speed and hauling capacity, and
6. service layout for the (Conveyer) I/O system.

Various models were developed in order to address these issues. For example Bozer, Shiek and White (1982 a) studied the impact of alternative I/O points from a throughout perspective. Using cost estimates, developed by Zollinger (1975), While and Bozer (1980) developed an optimization model to determine the number, length, and height of aisles to minimize an after-tax economic cost measure. Tompkins and Reed (1976) developed the COFAD (COmputerized FAcilities Design) model which jointly determines the facility layout and transportation hardware combination in order to minimize the aggregate material handling costs.

2.5. The Design Process

The design process integrates the product and material flow databases with a facilities design knowledge base and contemporary business plan information (Figure 1). This integration process results in a tentative layout structure with same "basic" or primitive operating rules. Rapid performance analysis techniques are applied to the proposed layout in order to derive a set of lexicographic performance measures. The rapid analysis is done by rough-cut approximate mathematical models combined with simulation and perturbation analysis. The resulting performance measures are evaluated by an expert assessment mode (human or programmed) in order to select/modify the approved design. The acceptable design is then evaluated and the details of a more sophisticated operational procedure are then determined.

The major advantage of the proposed methodological process is in its ability to focus on the key issue in an orderly manner and to iterate among design and management/operations decisions until an acceptable performance level is derived. The rest of the paper will focus on the development and assessment of such an operational system which was actually developed for a national distribution center of a certain food producer.

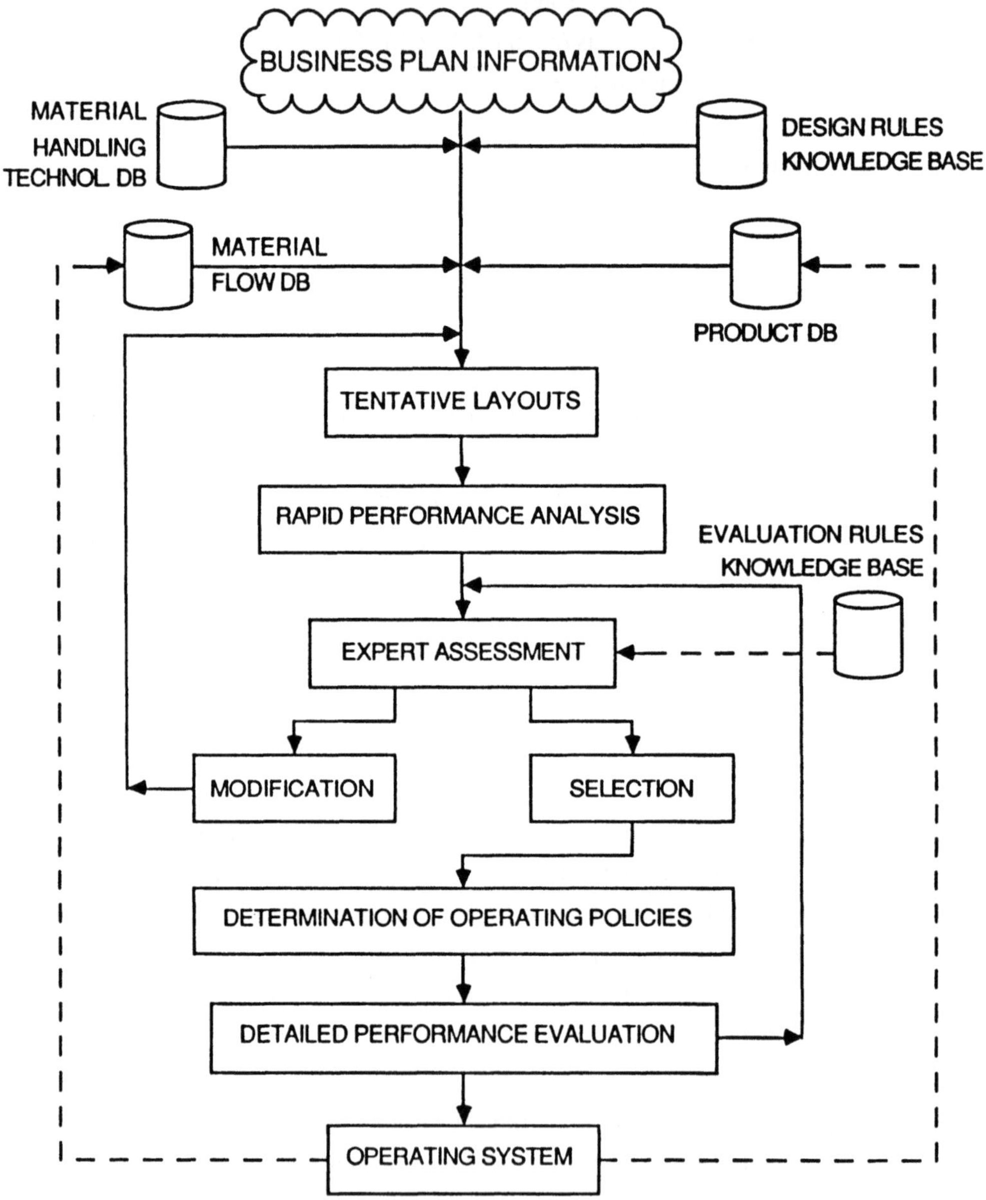

Figure 1. The Knowledge-Driven AS/RS Design Process

3. Developing the Operating Rules: A Case Study

The AS/RS discussed here operates in a national distribution center of a major dried food and pasta products manufacturer*. Finished products are containerized into special unit loads (UL) which are then hauled from the various production lines to this central distribution center. Each UL consists of single product type. Following strict shelf-life controls the current managment policy is to store these UL's for about 10 working days, and then to retrieve them back according to the standing orders of the various wholesalers. The AS/RS facility itself was designed and constructed by Demag AG (W. Germany) . It consists of a single aisle stacker with storage racks located at both sides. Figure 2 depicts a schematic structure of this system. For further details the reader is referred to Seidmann (1988). All storage and retrieval requests are initiated with the cranes at the I/O point. Each crane cycle consists of a storage task (ST), idle travel to the retrieval point (IT) and then a retrieval tast (RT) terminating back at the I/O point. This operational mode is called "dual-command" or "interleaved" cycle. It is used during peak turn-over periods, which are at the focus of our study.

The size of the AS/RS and the crane speed are given by.

Vh = horizontal speed= 180 (fpm)

Vv = vertical speed=90(fpm)

H = rack height = 63 (ft)

L = rack length =285(ft)

$$th = \frac{L}{Vh} = \text{time to reach the end of the rack} = 1.58 \text{ (min)}$$

$$tv = \frac{H}{Vv} = \text{time to reach the top of the rack} = 0.7 \text{ (min)}$$

Each crane cycle starts and terminates at the I/O point which has coordinates (0,0). The three serial legs composing the cycle are :

1* Some minor technical details of this system were modified at the request of the above manufacturer which prefers to remain unidentified.

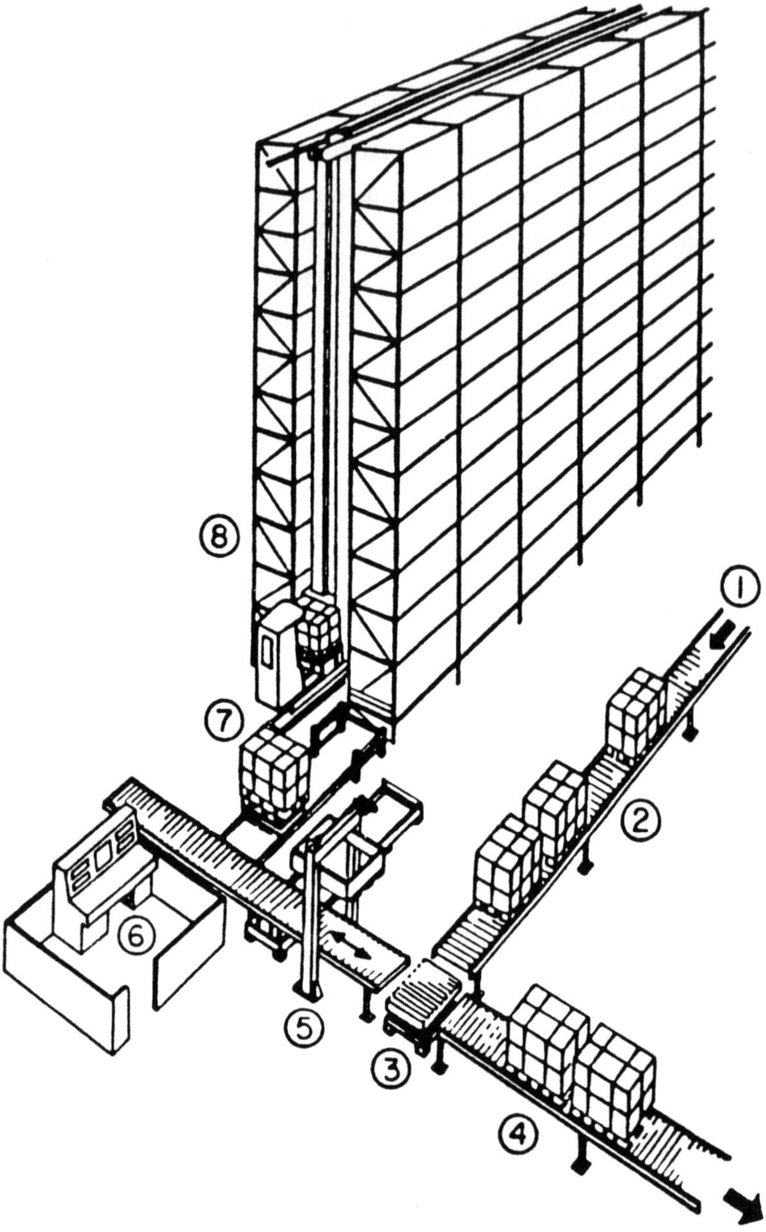

Figure 2. The distribution center AS/RS studied in this paper: 1-input conveyor from plant, 2-accumulation conveyor, 3-turntable, 4-output congeyor from storage, 5-UL checking device, 6-contral desk, 7- crane, 8-racks.

I) delivering an arriving UL from the I/O point to an empty bin (with coordinates [a,b])

II) Travelling from the storage bin to the required retreival bin (with coordinate [c,d]), and

III) Delivery of the retrieved UL back to the I/O point.

The cycle travel time CT is computed as (Bozer and White (1984)):

$$CT = t_{ST} + t_{IT} + t_{RT} \tag{3}$$

with each leg given by:

$$t_{ST} = \text{Max}\left\{\frac{a}{Vh}, \frac{b}{Vv}\right\} \tag{4}$$

$$t_{IT} = \text{Max}\left\{\frac{|c-a|}{Vh}, \frac{|d-b|}{Vv}\right\} \tag{5}$$

$$t_{RT} = \text{Max}\left\{\frac{c}{Vh}, \frac{d}{Vv}\right\}. \tag{6}$$

Using FIFO the expected value of CT can be computed from:,

$$E(CT) = \left[\frac{5}{3} + \frac{4}{6}B^2 - \frac{1}{15}B^3\right] \quad \text{Max (th, tv)} \tag{7}$$

where,

$$B = \min\left(\frac{th}{\max(th,tv)}, \frac{tv}{\max(th,tv)}\right). \tag{8}$$

In our case we get that:

B=0.44, and E(CT) = 2.84 (min).

This FCFS cycle time might be reduced through the use of a more intelligent control scheme which considers the relative locations of the empty bins and the various candidate UL's to be retrieved. The actual time required by the crane to seize or to deposit a UL is in the order of ten seconds per cycle, and slightly reduces the crane capacity. Being a constant it is not affected by the AS/RS control scheme.

The current practice at the AS/RS (denoted as DC-Default Control) was to store arriving UL's at the **closest** available (empty) bin and then to **retrieve the first** UL type from the computerized retrieval list. UL's were retrieved according to a FIFO scheme in order to minimize the variance of the storage times. Since several UL's of type k typically arrive at the same day one can permute the retrieval sequence within the same age group without violating the age control policy. During our study, management policy was even modified allowing for the retrieval logic to scan the retrieval candidates among the oldest UL's as well as those arriving one or two days later than the oldest item in storage.

The relative reduction in IT using the shortest distance among ri candidates retrieval locations for UL type i and y empty bins-rather than FIFO is explored next.

In an earlier study Bozer and White (1984) show that the distance between two randomly selected locations is a random variable, Z, with pdf f(z) and cdf F(z):

$$f(z)= \begin{cases} (2-2z)\left[2(\frac{z}{B}) - (\frac{z}{B})^2\right] & 0 \leq z < B \\ + (2z - z^2)\left[\frac{2}{B} - \frac{2z}{B^2}\right] & \\ (2(1-z) & B < z \leq 1) \end{cases} \tag{9}$$

$$F(z)= \begin{cases} (2z-z^2)\left[2(\frac{z}{B}) - (\frac{z}{B})^2\right] & 0 < z < B \\ 2z-z^2 & B < z \leq 1. \end{cases} \tag{10}$$

At each cycle there are r_i, type i UL's for retrieval. Searching among all UL's type i with the same arrival date means that r_i has a Poisson distribution with mean $p_i\lambda$. The expected value of $Z_1^{r_i}$ - the smallest order statistic (reflecting the distance to the nearest-neighbor bin) of a sample of size r_i is (Hogg and Tanis (1977):

$$E(Z_1^{r_i}) = \int_0^1 z(r_i)[1-F(z)]^{r_i-1} f(z)dz. \tag{11}$$

Unconditioning on r_i leads to

$$E(Z_1^i) = \int_0^1 \sum_{r_i=0}^{\infty} z\ (r_i)[1-F(z)]^{r_i-1} \frac{(\lambda p_i)^{r_i}}{r_i!} e^{-p_i\lambda} \quad f(z)dz, \tag{12}$$

and clearly,

$$E(t_{IT}) = \sum_{i\epsilon I} p_i\ E(Z_1^i). \tag{13}$$

Note that our (12 - 13) extends the expression given by Han et al (1987) to handle the case of a **randomly distributed** sample size r_i. This equation was numerically evaluated by us (for finite r_i values) using IMSL/DMLIN routine for gaussian tensor product integrals.

Comparing the $E(t_{IT})$ results against the FCFS values of $E(t_{IT})$ indicated an overall potential for crane cycle time reduction of about 18% to 25%. These estimates were then validated against detailed simulations (Zilbiger and Seidmann (1984)). Realizing these savings means that for each of type i UL to be retrieved one must look for all the UL type i in storage having the proper age for retrieval and then find the closest empty bin for the storage leg of the crane. Furthermore, the system dynamic is such that the sample size (being the number of empty bin plus the ri value)is typically 50 or above. When operating under heavy storage load there are many type i UL's in storage (with only few empty bins); on the other hand, when the storage load reduces then there are many more empty bins (with fewer type i candidates). In those cases one may search through the retrieval file looking for other retrieval requests that may result in improved cycle design. Those observations formed the basis of the retrieval rules described in the next section.

4. The System Controls

Given the enormous computational requirements posed by standard techniques as applied to AS/RS an effective (but not guaranteed to be optimal) AI approach has been selected. The crane cycles are planned one-at-a-time in a real-time basis. The goal state as well as the initial state are unkown before every crane cycle and a forward search determines the decision variables. In this case the decision variables are the storage bin (i), the UL type to retrieved (k), and the actual retrieval bin (j)- when several type k UL's are available in storage.

This problem is a discrete multi-stage decision problem. The values of the decision variables define the state of the system. Thus system state is constantly being monitored during the solution process (and the AS/RS operations) by changing some of these variables. This characteristic of changing the system state through the changes of variables is basic to the state-operator scheme. In classical AI approaches, the state description is given formally in first order logic. The problem is then solved in state-operator framework using PROLOG to represent the elements, the attributes and their relationships in the AS/RS.

The stage search is directed by three control functions which are discussed next. Going from the first function to the second and then to the third one increases the knowledge level of the controller (and the computational effort per state) as well as the resulting AS/RS effectiveness.

Feasibility is ensured by a Rule-Based system which enforces certain management constraints (Kusiak (1987)),

For example:

Rule R_1 : If: item k has to be retrieved
Then the candidate UL's are those stored more than SL^k days (SL^k is an age-control management parameter).
If: there are no candidate UL's stored more than SL^k days
Then: search among all other type k UL's.

Defining the storage data structures let:

T_j = the UL type in bin j

B_k= the set of all bin locations of UL's type k.

R = the ordered retrieval set

Y = the set of empty bins.

4.1 The Basic Control function: BC

The basic control function has 3 components. Each one of those components affects the system performance in a distinct manner. The linear combination of these compact performance measures attempts at producing the required long-term response pattern. If the crane stores a UL at i and retrieves one from j then the basic control function (BC) is:

$$\min_{T_j \varepsilon R,\ i \varepsilon Y} \{F_I (D_j, S_j, E_{ij})\}$$

where

$$F_I (D_j, S_j, E_{ij}) = \alpha_i D_j + \alpha_2 S_j + \alpha_3 E_{ij} \qquad (14)$$

S.T:

$$S_j \geq \max_{k=T_j} \left\{ S_{B_k} \right\} + 2 \qquad (15)$$

$O \leq \alpha_z$, z=1,2,3 are the weighting coefficients used in order to track the relative significance of the control variables.

D_j = the distance of (candidate) retrieval bin (j) from the I/O point,

S_j = the arrival time to storage of the UL in bin j, and

E_{ij} = the cycle "effectiveness" measure if the crane stores a UL at i and retrieves one from j.

The procedure used for computing D_j, S_j and E_{ij} are described next.

The retrieval distance (D_j) has an immediate effect on the cycle length attempting at minimal values to reduce the crane travel time. Computing D_j recall that the travel times at the two axes are independent of each other since each axis has an independent motor. The arrival time to storage (S_j) measures the shelf life of the produce in storage. An important consideration in our system is the elimination of those instances where UL assigned to remote storage bins are left over there for extended time periods. It means, therefore that certain retrieval priority is assigned according to the UL's current storage time. Hence higher priority is given to UL's with earlier arrivals to the storage bins. The lower bound (15) on S_j is enforcing the management age control policy as described earlier. Strictly following the minimal S_j rule for selecting the retrieval bin means using a FIFO logic. The measurement unit used for S_j is storage time in days.

The long term effects of the crane tours are measured by the cycle effectiveness measure (E_{ij}). Its objective is to drive the crane into making the most out of each cycle travelled. Doing so it strives at storing and retrieving a UL from closeby bins. Whenever the crane travels to store a UL at a remote storage bin it attempts at trying to combine that long travel leg with a UL retrieval from that remote neighborhood.

The controller computes the values of $F_I(D_j, S_j, E_{ij})$ for all the feasible cycles for each one of the items in the retrieval file R at time t. These computations are carried out as a function of the UL type to be stored. Specifically,

D_j - is given by the bin being evaluated,

S_j - is given by the arrival time of the UL to bin j, and

E_{ij} - is computed by ray searching among all the closest empty bins around the candidate bin (j) for retrieval.

4.2 The Adaptive Control Function: AC

Increasing the information level of the system controller on Adaptive Control Function (AC) was explored. It is designed to better operate the AS/RS in a time-varying environment. If the internal parameters of the system are fixed, as in the case of the BC function, the system might operate quite differently in one environment than it should in another. The AC function is designed to compensate for some dimensions of the changing environment by monitoring the AS/RS performance and altering, accordingly, some parameters of its control functions to gain improved performance. This control scheme changes the policy decision on a real time basis as a function of the number. of orders on the retrieval queue at the time t, N(t), N(t) =0,1,2,... . Adding this knowledge component means that the relative importance of the retrieval distance component increases with the increase in the retrieval queue length; this results in the AS/RS attempting at reducing N(t) by retrieving UL stored closer to the I/O point. Consequently, the crane cycle times are shortened leading to the desired transient decrease in N(t).

On the other hand, at times when N(t) is relatively small the immediate cycle performance is given inferior priority - as compared with the other operational parameters (i.e., D_j and S_j) which pander to the long-term objectives of system effectiveness.

The AC function is

$$\underset{T_j \varepsilon R,\ i \varepsilon Y}{\text{Min}} \{F_{II} (D_j, S_j, E_{ij}, N(t)\} \tag{16}$$

Where:

$$F_{II} (D_j, S_j, E_{ij}, N(t)) = \alpha_1 N(t) \cdot D_j + \alpha_2 \ S_j + \alpha_3 E_{ij} \tag{17}$$

S.T.:

$$S_j \geq \underset{k=T_j}{\text{Max}} \left\{S_{B_k}\right\} +2 \tag{18}$$

Experimental evaluations of this function are provided in the next section.

4.3 The Dynamic Adaptive Control Function: DAC

The dynamic adaptive control structure (DAC) is designed in order to handle another fundamental AS/RS management issue. This is the storage assignment problem where different product groups have different arrival rates and, moreover, these rates tend to vary dynamically as a result of several reasons. These reasons include seasonal changes in the customer's taste, introduction of new products varieties and special sales campaign typically aimed at increasing the sales level of slow moving items. The proposed Dynamic Adaptive Control (DAC) scheme delineated below is designed to capture these structural changes in the probability density function of the arrival process and to reflect them in the operational node of the AS/RS. The basic idea is to modify the control structure is such a way that UL's of products having higher traffic intensity will be stored closer to the I/O point (Hausman et al (1976)).

The average storage time for UL type k is computed dynamically whenever such a type k UL leaves the AS/RS. The time average for storage of type k UL's at time t is denoted as $A_k(t)$. Fast moving UL's will have smaller values of $A_k(t)$.

It has been shown by Hardy, Littlewood and Polya (1949) that the cross product of two series (in our case the travel time to bin i, ($i \varepsilon Y$) which is the storage location for an arriving type k-D_i^k and $1/A_k(t)$) is minimized when these series are montonic in antithetical senses. (i.e., one series is non increasing and the other is non decreasing). Hence, the time average ratio of $[D_i^k/A_k(t)]$ is minimized when slow moving items (smaller $1/A_k(t)$) are stored further away from the I/O point (larger D_i^k) and the fast moving items (smaller $A_k(t)$) - closer to the I/O point.

These foregoing arguments lead to the following dynamic adaptive control (DAC) function:

$$\text{Min} \{F_{III}(D_i^k, D_j, S_j, E_{ij}, N(t), A_k(t))\} \qquad (19)$$
$$T_j \varepsilon R, i \varepsilon Y$$

Where:

$$F_{III} = \alpha_1 N(t) D_j + \alpha_2 S_j + \alpha_3 E_{ij} + \alpha_4 \left[\frac{D_i^k}{A_k(t)} \right] \tag{20}$$

$$\text{S.T. } S_j \geq \text{Max} \left\{ S_{B_k} \right\} +2 \tag{21}$$

$k = T_j$

Note that the forth component of F_{III} facilitates the dynamics of the storage allocation decisions.

5. Experimental Evaluations

A simulation model (ASIM), that allows for detailed examination of those parameters was developed. The results of the simulation runs were analyzed using multivariate analysis of variance program (MANOVA) and an ordinal ranking procedure (RANK). This procedure was applied first to evaluate the parameters of the heuristic solution algorithm and then to compare the various control schemes. Only significant differences ($\alpha < 0.05$) are presented here.

Next few definitions of the major performance measures are introduced:

1. The maximal number of items on the retrieval queue, MQOUT.
2. The average waiting time per item on the retrieval queue, AQOUT.
3. The average travel time for the crane from the I/0 point to the storage bin, TRAVC.
4. The aymptotic percent idle-time of the crane, PIDLE.

Table 3. Comparing BC and AC vs DC performance.

Performance Measure	Improvement (%)	
	BC/DC	AC/BC
1. MQOUT - Maximal retrieval queue	29.2	32.9
2. AQOUT - Average waiting time for retrieval	51.4	41.6
3. TRAVC - Crane travel time from I/0 to storage	-3.6	8.2
4. PIDLE - Crane idle time	273.2	75.0

Table 3 presents the improvement gained when using BC over DC and then further increasing the knowledge level using AC instead of BC. The increase in crane idle time, as given in this table is a result of improved system schedule using the BC. This also leads to reductions in the maximal retrieval queues and the average waiting times or retrieval. The slight increase (3.6%) in the crane travel time from I/0 to storage is expected since the DC directs the crane at storing UL's in the nearest empty bin.

Clearly, using AC further improves the system performance through reductions in the retrieval queues. This means improved logistical service. The increased crane idle times mean that the same crane can handle augmented traffic volume.

Finally, the performance of the Dynamic Adaptive Control (DAC) function was investigated for a case where the arrival rate of the UL's was dynamically modified. Our objective was to detect the sensitivity of the DAC scheme to changes in the steady-state arrival rates of the various product types. Properly operating, DAC should respond by changing the storage allocating scheme such that the faster moving items are stored closer to the I/O. Doing so the system was first simulated using the original arrival pattern. Following a simulation of 400 operating hours the arrival pattern was reversed; the p_i values and the storage times of class A ("fastest moving items") were assigned to D ("slowest moving items") those of B to C and so on. This new arrival rate structure was used for a similar time period. Implementing the DAC scheme leads to a new storage allocation mode of the UL's as they get into the system.

The impact on the system performance is illustrated by Table 4. The results underscore the superiority of the "knowledge based" DAC over the default control (DC) control or over the adaptive control (AC) schemes which ignore the dynamics of the underlying changes in the relative arrival rates of the various UL types. In fact, using the DC the crane was utilized at about 97% of the time. Such a high utilization rate indicates that there is no spare capacity for future increase in the UL traffic intensity.

Table 4. Comparing DAC vs DC and AC Performance

Performance Measure	Improvement (%)	
	DAC/DC	DAC/AC
1. MQOUT	47.1	22.8
2. AQOUT	70.7	50.5
3. TRAVC	11.7	6.9
4. PIDLE	670	437

6. Research and Development Issues

Some of the more difficult issues in modern material handling systems are associated with the introduction of new control schemes that are information intensive. As such, these systems require advanced modelling tools for diserete parts flow optimization in realtime as well as the support of reliable data collection and communication technology. Major progress has been made in bringing about an integrated communication system in the material handling environment over the last few years. In particular, the MAP and TOP efforts have resulted in protocol stacks which allow diverse equipments to intercommunicate, thus allowing for the interchange of information amongst the various functional units of a manufacturing company. Over the same time period, proprietary systems have continued to evolve (Relton (1986)). The potential for effective communications has been demonstrated, but much remains to be done (Schaffer (1987)). In the highly competitive manufacturing environment, efficiencies must continue to be improved and costs reduced. Advances in communications can help to bring about such continued improvement. Both development issues, which relate to the application of currently available technology, and research issues, which relate to the solution of fundamental problems in designing communaction networks, bear heavily on the potential for continued improvement.

One can begin to appreciate the complexity of the material handling systems in modern manufacturing plants by realizing that such a system is essentially a complicated distributed processing system. Since even the most simple distributed processing systems are very difficult to understand, it is easy to see that understanding a complex distributed processing system in an (Hammond (1986)). A possible key to advancing the state of the art might be the formulation of a language having sufficient descriptive power to describe the manufacturing environment. In essence, what this means is that the tools should be able to describe parallelism, correlation, and coordination in both real-time control and non time-critical environments, and in addition, should allow for multiple levels of abstraction. Further, the language should be able to describe the impact of exceptions on operation of the system. Once the environment can be described symbolically in a language commonly understood by researchers and developers, there would be an increased likelihood that key certainly lead to progress. The key point here is that the fragmented research which is going on in the factory communications and logistics area will likely lead to a thorough understanding of system components, but it is unlikely that such efforts would lead to an understanding of the relevance of those results to manufacturing enterprises.

Experimental work in realistic factory environments is likely to be a strong contributer to advances in these major issues (Daigle et al 1988). The results of such experimental efforts can be of great value to advancing the state of the art. In particular, researchers might work with developers in defining the types of measurements which are appropriate, and may help to design statistical test and measurement programs which might yield useful data for analysis purposes. It seems unlikely that equally useful results could be obtained in a pure laboratory environment without direct input from some experimental system. What is needed is a cooperative effort between the experimenters and the analysts, both within corporations and throughout the research communities. More frequent use of nondisclosure agreements, which protect the property rights of corporations while providing valuable data to researchers, could go a long way towards improving the quality of both research and development in this area. Such models of cooperative work are not new, but there do not seem to be many such cooperative efforts in the factory communications area.

A serious impediment to the implementation of advanced information technologies for CIM support is the lack of adequate tools to support cost justification. The lack of tools results primarily from the difficulty of assessing the strategic value of moderization expenditure and the integration they bring about. Management expectation for short term retrun on investment impose economic barriers to improvement of productivity which will be obsolete in a period of less than five years, having essentially no value by the end of that time (Kaplan (1986)). Such tactical underpinnings of accounting and the resulting investment practices do not appear to be the appropriate framework for evaluating investments relative to CIM.

Industrial systems integration includes everything from the interfaces to the individual devices at the lowest level through the highest level financial systems of the corporation. Such an integration aims at providing the end user with application solutions, while hiding the means through which that integration is achieved. Binding of major technological advances in three areas: communication technology, knowledge engineering and data management, is required to achieve these objectives. While the OSI MAP/TOP and data management, is required to achieve these objectives. While the OSI MAP/TOP reference models have extensively addressed the communication technology issues, minimal progress has been made in the development of standards and specifications for the application to application communication logic or to standardise knowledge representation schemes.

Another important missing element beyond current standardization effort is a global database scheme for creating data communication applications and data management models. What is needed is a conceptual solution to the problem of effective storage, retrieval, manipulation, and management of large qunantities of text, graphical, audio, and video data in a dynamic, distributed, multi-application and multi-vendor environment. Such a system will result in uniform data dictionaries, data structures, storage modes and display procedures.

7. Conclusions

The research results presented in this paper demonstrate that proper extensions of constructive information models can be implemented in designing and later controlling of automated material handling system in a dynamic operational environments. The system design process presented here is based on integrating several operational databases and two design knowledge bases through the use of some rapid (mathematical) evaluation rules. An industrial case study is used to show that the proposed approach - which uses generalized goal functions for state transition evaluations is superior to the one currently proposed by various system vendors.

Attempting at improved AS/RS operation three goal functions were developed step by step, while increasing the information level: The first function is a weighted combination of variables that indicates the correspondence of a specific cycle to the desired policy. The second one is a generalization of the first one: It dynamically updates coefficient weights in the goal function, utilizing more information about the system status to actually adapt the policy in real time. The third goal function further generalizes the second model: It includes a variable argument that deals with dynamic location assignment of items in the warehouse. Location assignment means remote storage of items that have a slow turnover rate, and storing fast moving items in the vicinity of the I/O port. These goal functions use continuous statistical interpolation (dynamically produced), to make the proper rack assignments and retrievals. Each increase in the information level lead to an additional improvement in the AS/RS performance as measured for instance, by the crane throughput and the retrieval gueue length. With the advent of Computer-Integrated-Manufacturing, and its accompanying technological advantages one can use this scheme in a hierarchical control context for related operations and controls of other material handling systems such as carousels and automated guided vehicle systems.

Acknowledgements

This study has benefited from the useful comments of Professors Y.A. Bozer, A. Kusiak and S.Y. Nof.

References

1. Arbel, A., Seidmann, A., (1984), "Performance Evaluation of Flexible Manufacturing Systems," IEEE Transactions: System, Man and Cybernetics, Vol. 14, No. 4, pp. 606-617.

2. Azadivar, F., (1984), "A simulation Optimization Approach to Optimum Storage and Retrieval Policies in an Automated Warehousing System," Proceedings of the 1984 Winter Simulation Conference, S. Sheppard, U. Pooch, D. Pegden (Editors), pp. 207-214.

3. Bafna, K.M., (1981), "Use of Computer Simulation in Designing Complex Material Handling Systems, " Winter Simulation Conference, Oren, T.I, Delfusse, C.M., and Shub, C.M. (Editors), pp. 181-185.

4. Barr, A., and Feigenbaum, E., The Handbook of Artificial Intelligence, Vol 2, Kaufman Publishers, Los Altos, CA.

5. Bozer, Y.A., Shiek, J.S., and White, J.A., (1982 (a)), "Modelling Automated Storage/Retrieval Systems," TIMS/ORSA Joint National Meeting, Detroit, MI.

6. Bozer, Y.A., Branigan, M.J., and Mullens, M.A., (1982) "Designing Warehousing Systems," Presented at the TIMS/ORSA Conference, Detroit, MI.

7. Bozer, Y.A. and White, J.A. (1984), "Travel Time Models for Automated Storage and Retrieval Systems, " IIE Transactions, Vol. 16, No. 4, pp. 329-338.

8. Bullers, W.I., Nof S.Y., and Whinston, A.H. (1980), "Artificial Intelligence in Manufacturing Planning and Control," AIIE Transactions, Vol. 12, No. 4, pp. 351-363.

9. Daigle, J.N., Seidmann, A., and Pimentel, J. R., (1988), "Communication for Manufacturing: An Overview," to appear in IEEE Networks.

10. Elsayed, E.A., (1981), "Algorithms for Optimal Material Handling in Automatic Warehouse System, "International Journal of Production Research, Vol, 19, No. 5, pp. 99-109.

11. Francis, R.L., and White, J.A., 1974, Facility Layout and Location: An Analytical Approach, Prentice Hall, Englewood Cliffs, NJ.

12. Graves, S.C., Hausman, W.H. and Schwarz, L.B., (1977) "Storage-Retrieval Interleaving in Automatic Warehousing Systems," Management Science, Vol. 23, No.9, pp. 935-945.

13. Hammond, J. L., (1986), "Performance Analysis of Local Computer Networks," Addison-Wesley, Reading, NJ

14. Han, M.H., McGinnis, L.F., Shieh, J.S., and White, J.A., (1987), "On Sequencing Retrievals in Automated Storage/Retrieval Systems, "IIE Transactions, Vol, 19, No. 1, pp. 56-66.

15. Hardy, J.A., Littlewood, A.B., and Polya, A. (1949), Inequalities, Prentice-Hall, Englewood Cliffs, NJ.

16. Harris, R.J., (1975), A Primer of Multivariate Statistics, Academic Press, NY.

17. Hausman, W.H., Schwarz, L.B., and Graves, S.C. (1976), "Optimal Storage Assignment in Automatic Warehousing System, " Management Science, Vol. 22, No: 6, pp. 629-638.

18. Hill, J.M. (1980), "Computers and AS/RS Revolutionize Warehousing," Industrial Engineering, Vol. 12, No. 6., pp. 34-45.

19. Hogg, R.V., and Tanis, E.A., (1977), Probability and Statistical Inference, MacMillan, NY.

20. Kallina, K., (1976), "Application of the Cube Per Order Index, " Interfaces, Vol. 7, pp. 19-25.

21. Kaplan, R.S., (1986), "Must CIM be Justified by Faith Alone," Harvard Business Review, March - April.

22. Karasawa, Y, Nakayama, H., and Dohi, S., (1980), "Trade-Offs Analysis for Optimum Design of Automated Warehouses," International Journal of Systems Science, Vol. 11, No. 6, pp. 567-576.

23. Keilson, J. and Seidmann, A., (1987), "M/G/∞ with Batch Arrivals," working paper WP1932-87, A.P. Sloan School of Management, Massachusetts Institute of Technology, Cambridge, Mass., to appear in Operations Research Letters.

24. Kleinrock, L., (1976) Queueing Systems: Vol 2, Computer Applications, Wiley, NY.

25. Kumara, S., Soyster, A.L., and Kashyap, R.L., (1986), "An Introduction to Artificial Intelligence," Industrial Engineering, Vol 18, No. 12, pp. 9-20.

26. Kusiak, A., "(1987) "Artifical Intelligence and Operations Research in Flexible Manufacturing Systems, " INFOR, Vol. 25, No. 1, pp. 2-12.

27. Kusiak, A., (1987 a) , "Designing Expert Systems for Scheduling Automated Manufacturing," Industrial Engineering, vol. 19, No. 7, pp.42-46.

28. Kusiak, A., Hawaleshka, O., and Cormier, G. (1985), "Order Picking Policies in Automated Storage System," Proceedings of the 6th International Conference on Automation in Warehousing, IFS Publications, London, UK.

29. Linn, R.J., and Wysk, R.A., (1987), " An Analysis of Control Strategies for An Automated Storage/Retrieval System," INFOR, Vol. 25, No. 1, pp. 66-83.

30. Matson, J.O. and White, J.A., (1982), "Operational Research and Material Handling," Journal of Operational Research, Vol.11, pp. 309-318.

31. Maxwell, W.L., (1981), "Solving Material Handling Design Problems with OR," Industrial Engineering, Vol. 13, No. 4, pp. 58-69.

32. Morrison, D.F., (1967), Multivariate Statistical Methods, McGraw-Hill, San-Francisco, CA.

33. Nilsson, N.J., (1980), Principles of Artificial Intelligence, Toga Publishing Co., Palo Alto, CA.

34. Pelton, G.A., "An Application Architecture and Application Interface for Manufacturing Floor Environments," MAP Application Review, SME Publication, Dearborn, MI, pp. 119-128.

35. Ratliff, H.D., and Rosenthal, A.S., (1983), "Order Picking in a Rectangular Warehouse: A solvable case of the traveling salesman problem," Operations Research, Vol. 31, pp. 507-521.

36. Roll, Y., and Rosenblatt, M.J., (1981), "Optimal Warehouse Size, Configuration and Storage Policies," Proceedings of the VI International Conference on Production Research, Novi Sad, Yugoslavia.

37. Rygh, O.B. (1981), "Justify An Automated Storage and Retrieval System," Industrial Engineering, Vol. 13, No. 7, pp. 20-24.

38. Sathaye, S., and Phillips, D.T. (1979), "Generalized Manufacturing Simulator: GEMS, "Texas A&M University, Research Report RF-3539, College Station, TX.

39. Schaffer, G., (1987), "MAP Climate: Cloudy but Clearing," American Machinist, pp. 75-78.

40. Schwarz, L.B., Graves, S.C., and Hausman, W.H. (1978), "Scheduling Policies for Automatic Warehousing Systems: Simulation Results, " AIIE Transactions, Vol. 10, No. 3, p. 260-270.

41. Seidmann, A., Schweitzer, P.J., and Nof, S.Y., (1985), "Performance Evaluation of a Flexible Manufacturing Cell with Random Multiproduct Feedback Flow, "International Journal of Production Research, Vol. 23, No. 6, pp. 1171-1184.

42. Seidmann, A., (1988), "Intelligent Control Schemes for an Automated Storage and Retrieval System," to appear in the International Journal of Production Research.

43. Sellers, C.J., and Nof, S.Y., (1986), "Part Kitting in Robotic Facilities," in Robotics and Material Flow, S.Y. Nof (Editor), Elsevier, NY.

44. Tompkins, J.A., and Reed, R., (1976), "An Applied Model for the Facilities Design Problem," International Journal of Production Research, Vol. 14, pp. 583-595.

45. Tompkins, J.A., (1982), "Facility Design," in Material Handling Handbook, R.A. Kulwiec Ed., Wiley & Sons, NY, pp. 654-707.

46. Tompkins, J.A., and White, J.A., (1984), Facilities Planning, Wiley, NY.

47. Ulgen, O.M. and Elayat, H. (1981), "GENSAWS: A General Simulator for Automatic Warehousing Systems, "Proceedings of the VI International Conference on Production Research, Novi Sad, Yugoslavia.

48. White, J.A. (1987), "Automation: A New Perspec'ive," Modern Material Handling, Vol. 42, No. 1, pp. 27-29.

49. White, J.A., and Bozer, Y.A., (1980 (a)), "Optimum Design of Automated Storage/Retrieval Systems," TIMS/ORSA Joint National Meeting, Washington, D.C.

50. White, J.A., (1980), "Randomized Storage or Dedicated Storage ?,"Modern Material Handling, January, pp. 19-25.

51. Wilson, H.G., (1977), "Order Quantity Product Popularity and the Location of Stock in a Warehouse," AIIE, Vol 19, No. 3, pp. 25-33.

52. Wysk, R.A., and Linn, R.J., (1987), "A Knowledge Based Intelligent Control System for Automated Storage/Retrieval System Operation," Working Paper Series, Dept. of Industrial and Management Systems Engineering, Pennsylvania State University.

53. Zollinger, H.A., (1975), "Planning, Evaluating and Estimating Storage Systems,"Advanced Material Handling Technology, Seminar, Purdue University, West Lafayette, IN.

54. Zillbiger, Y., and Seidmann, A., (1984) "Operational Controls of AS/RS Systems, "Proceedings of the International Conference on the Factory of the Future." CASA/SME and ITIM, Tel-Aviv, November 4-7.

GUIDED VEHICLE SYSTEMS: INFORMATION/CONTROL SYSTEM

IMPLICATIONS OF ALTERNATIVE DESIGN AND OPERATION STRATEGIES

Yavuz A. Bozer
Industrial and Operations Engineering
The University of Michigan
Ann Arbor, Michigan 48109-2117
USA

ABSTRACT

Automated Guided Vehicle (AGV) systems continue to be a viable and flexible handling alternative for many low to medium flow, discrete parts handling requirements. Today, in addition to the towing vehicle, AGVs can be configured as unit load carriers, automatic guided pallet trucks, automatic guided fork trucks, or light load carriers. The automatically positioned stock selector (used for in-the-aisle picking) is also generally accepted as an AGV.

In this paper, assuming a pick & drop system based on unit load carriers, we describe current control methods for vehicle dispatching, and we examine their relationship to various empty vehicle dispatching policies developed in the AGV literature. Subsequently, we discuss further work needed in the area along the following directions: developing standard problems for evaluation purposes; analyzing the interaction between the system layout and the dispatching rules; analyzing the impact of machine processing times and the routing matrix; integrating or interfacing the AGV control system with higher level information systems; and evaluating the impact of self-guided vehicles.

NATO ASI Series, Vol. F53
Advanced Information Technologies for Industrial Material Flow Systems
Edited by Shimon Y. Nof, Colin L. Moodie

I. INTRODUCTION

According to the Material Handling Institute (MHI), "automated guided vehicle (AGV) systems feature battery powered driverless vehicles with programming capabilities for path selection and positioning" [8]. Although the above definition does not fully account for rail guided AGVs (which have relatively limited applications), it does emphasize two distinct characteristics of an AGV; namely, the fact that it is driverless, and that it can be programmed (or directed) to follow a specific (guide) path in moving material between any two points.

Furthermore, the guide path itself - which typically consists of energized wire buried in the floor or an optical/chemical path applied to the floor - does not obstruct personnel and other material flow in the facility. It is also relatively easy to install or to rearrange (compared to, say, conveyors) and it is relatively inexpensive. Hence, coupled with the aforementioned characteristics, AGV systems continue to be a viable and flexible handling alternative for many low to medium flow, discrete parts handling requirements. Naturally, conveyor systems (including smart monorails which have characteristics somewhat similar to those of AGVs) remain to be appropriate handling alternatives for larger flow volumes.

The earliest applications of AGV systems were based on the towing vehicle which is an AGV capable of pulling several trailers. Today, in addition to the towing vehicle, AGVs can be configured as unit load carriers (or transporters), automatic guided pallet trucks, automatic guided fork trucks, or light load carriers (or transporters). The automatically positioned stock selector (used for in-the-aisle picking) is also generally accepted as an AGV.

Note that, except for automatically positioned stock selectors, the vehicle types mentioned above are designed and used primarily to pick-up a unit load and deliver it to its destination. Such systems are also known as "pick & drop" systems. More recently, however, the unit load carrier has been used as an assembly process vehicle where the main body of the product to be assembled (along with some of the other components arranged in kits) remains on the AGV while it moves through all the workstations. Once the assembly operations are completed, the finished product is removed from the AGV and a new component is installed and the process is repeated. With increasing accuracy in vehicle stopping tolerances and the use of on-board robots (for pick-and-place purposes), the above application has also been extended to fabrication operations. Since the vehicle stays with the product throughout the production process, such AGV applications are generally known as "stop & go" systems.

To facilitate subsequent discussions, it is assumed that the system in question is a pick & drop system with unit load carriers. Such a system has proven to be effective in many manufacturing applications, including Flexible Manufacturing Systems. However, most of the issues addressed are equally applicable to automated pallet trucks and fork trucks. For a more detailed description of various AGV types and their application areas, the reader may refer to Koff [5] or MHI [8], among others.

II. THE CONTROL SYSTEM

From a conceptual standpoint, two types of control functions are essential for the successful operation of an AGV system. The first control function is primarily concerned with satisfying the move requests. That is, it is concerned with the following questions: how does a vehicle know when and where to pick up a load? How does a vehicle know where to deliver a load? How does a vehicle know which route to follow in reaching its destination?

The second control function is concerned primarily with the uninterrupted and unobstructed flow of vehicles through the system. Similar to managing overall traffic flow in a town, this control function is responsible for avoiding collisions, including those that would occur at intersection points and merge/diverge points as well as pick-up/deposit (P/D) points (where the vehicles stop for loading and unloading).

This paper is primarily concerned with the first control function. More specifically, we are interested in the first two questions which are concerned with recognizing and serving the move requests, or "vehicle dispatching" in short. The last question (concerned with routing a vehicle with a known destination) is beyond the scope of our interest. For further reading on this question, as well as the second control function responsible for traffic management, the reader may refer to [5], among others.

The remainder of this section is devoted to briefly describing current control methods used for vehicle dispatching. Generally speaking, vehicles may be dispatched through:

- a control panel onboard the vehicle
- a fixed station programming panel
- a remote dispatch terminal
- central computer control

The above methods are ordered from the simple to the sophisticated; that is, the list reflects an increasing level of automation. Also, some systems may use a combination of these methods for various reasons including those cases where a second method is made available for backup purposes only.

Under the first method, the vehicle is dispatched to one or more P/D stations by the operator who programs the AGV through the control panel mounted on the vehicle. Note that the operator must have direct access to the vehicle and he/she must be physically present when the vehicle arrives in order to program its next instruction(s). If manual load transfer is used and if the vehicle serves a few P/D points located far apart, this control method can be used successfully.

However, if load transfer is automated (which implies that no operator presence is required during P/D operations), or if the vehicle serves many P/D points and makes frequent stops over shorter distances, then programming the vehicle through the onboard panel will not be an efficient method in most cases. Also note that, after executing the last programmed instruction, the vehicle will remain idle at its present location until a new set of instructions are manually entered through the control panel.

The fixed station programming panel eliminates some of the shortcomings of the first method. Under this method, the operator places the loads in a waiting line (typically a conveyor) and enters the destination of each load through a programming panel mounted at the pick-up station. Subsequently, the control system (driven by the local programming panel) does not only "flag down" a vehicle but it also automatically transfers the destination of each load to the vehicle. Hence, no operator presence is required for vehicle programming at P/D stations.

Obviously, there is not much incentive to use this method if load transfer is not automated. Furthermore, note that the vehicle obtains the destination of a load only after it stops at the pick-up station. In fact, in some systems using the fixed station programming panel, the vehicle does not even know if there is a load waiting to be picked up until it approaches the station or actually stops in front of it.

Under the third control method, using the remote dispatch terminal, an operator may control all the vehicles through a central location. In one sense, the third method represents a centralized version of the previous one since all vehicle programming is done by a single operator at a central location.

Obviously, in dispatching the vehicles, the operator must know the status of each pick-up station as well as the location and status of each vehicle. This may be accomplished through a real-time graphic (or tabular) display on a CRT which shows the location and status of all the vehicles. A less sophisticated approach is to use a "locator panel" [5] which essentially shows the presence (or absence) of a vehicle in pre-defined segments of the path.

The fourth method is a computerized version of the third one. The operator is replaced by a computer and the dispatching decisions are automatically made by the program. Naturally, these systems allow the operator to interfere and manually dispatch the vehicles through the central terminal if necessary. Under this approach, system monitoring is greatly facilitated since the computer may now keep track of each vehicle individually (including their performance statistics) and alert the operator only when error or abnormal conditions are detected. With central computer control, it is customary to display the system status on a CRT as described earlier.

Note that, under the last two methods, since the vehicles are dispatched from a central location, some form of communication must be established between the vehicles and the dispatching center. For wire guided systems, this communication can be established through the guide wire. However, in those cases where other high-energy wires are present in the floor, some interference problems may be encountered.

An alternative approach is to use radio frequency (RF) signals. Under this approach, each vehicle and the dispatching center are equipped with a transmitter and receiver. Although magnetic fields generated inside the facility and/or complex building shapes (with many interior walls) may cause some difficulties, like the previous approach, using RF signals is a proven technology. In fact, it has been successfully used in dispatching operator driven lift-trucks in warehousing applications as well.

The above control methods are intended only as an outline of basic approaches presently available for vehicle dispatching. Naturally, AGV vendors show certain variations in how they approach the control issue. Consequently, a variety of techniques and capabilities exist in the market place.

For example, some control packages allow the vehicles to communicate not only with a dispatching center but with eachother as well. Therefore, some deviations are to be expected between the above descriptions and specific control packages offered by the vendors. In the following section, we will discuss various vehicle dispatching rules within the framework of the control methods described above.

III. VEHICLE DISPATCHING RULES

Vehicle dispatching rules are concerned with deciding which load (or move request) to serve next. That is, when a vehicle delivers its current load and becomes available, one must select the next task, if any, to be assigned to that vehicle. (One may also wish to consider the opposite question: when a load is ready to be moved, among all the vehicles that are available at the time, which vehicle should be assigned to that load? As shown empirically by Egbelu and Tanchoco [2], with reasonably high AGV utilization levels, the above question seldom arises during system operation since the number of loads waiting to be moved generally outnumbers the number of available vehicles).

The following list is not intended to be an exhaustive one; however, it shows a wide range of possible dispatching rules:

- Random-workstation-selection (RAND)
- First-come-first-served (FCFS)
- Modified-first-come-first-served (MODF)
- First-encountered-first-served (FEFS)
- Shortest-travel-time-first (STTF)
- Longest-travel-time-first (LTTF)
- Maximum-outgoing-queue-size (MAXQ)
- Minimum-remaining-queue-space (MINQ)
- Load-shop-arrival-time (LSAT)
- Most-desirable-station-first (MOST)

With the exception of FEFS, the above rules apply only if one or more move requests are present when a vehicle becomes available. Otherwise, one must also specify what to do with an idle vehicle; that is, a vehicle that becomes available when no move requests are present.

Until the next move request is received, an idle vehicle may either remain at its last delivery point, or it may be instructed to travel along a pre-specified closed loop (defined over the existing network). Under the former approach, say, the "stay put" strategy, the vehicle may block other (loaded or empty) vehicles unless side spurs are provided at each P/D point. With the latter approach, say, the "travel idle" strategy, the vehicle may needlessly interfere with other (loaded or empty) vehicles at intersection and merge/diverge points (unless the traffic management system has the capability to recognize idle vehicles and assign them a low priority).

Some of the above dispatching rules are self-descriptive. However, for clarification purposes, each one is briefly described below:

RAND - The next load to be moved is selected randomly.

FCFS - All the move requests are placed on a conceptual first-in-first-out (FIFO) queue and the next load to be moved is the "oldest" request in the system; that is, the one that has been waiting for an AGV for the longest duration.

MODF - Similar to FCFS except that each pick-up station may have at most one request represented in the conceptual FIFO queue at any point in time. If a second load is waiting at a particular pick-up station, the request associated with it is placed at the end of the conceptual queue when the load in front of it is picked up by an AGV.

FEFS - Regardless of the number and location of the outstanding move requests, an available vehicle is instructed to follow a pre-specified closed loop (defined over the existing network) until it encounters the first move request. Although this rule may seem identical to the travel idle strategy described above, there is indeed a subtle difference. With FEFS, the first move request to be served is the first one encountered by the vehicle, regardless of the "age" of the request. With travel idle, however, no other move requests are available and the vehicle will be assigned to the first one to arrive. Hence, even if the same loop is specified for both FEFS and the travel idle strategy, the load that is eventually picked up is not necessarily the same load.

STTF - The next load to be moved is the one closest in distance (or travel time) to the current location of the available vehicle. This rule is also known as vehicle-looks-for-work (VLFW).

LTTF - The next load to be moved is the one farthest in distance (or travel time) to the current location of the available vehicle.

MAXQ - The next load to be moved belongs to the pick-up station which has the maximum number of move requests. Note that the total number of move requests (or loads) that can be accumulated at the pick-up station are not taken into account.

MINQ - The next load to be moved belongs to the pick-up station which currently has the minimum remaining space for accumulating additional move requests. Note that a processor may get blocked if the corresponding move request queue is full.

LSAT - The next load to be moved is selected according to the time each load originally entered the shop. A load that has been in the shop for the longest time (that is, the load that has the earliest shop arrival time among all those waiting to be moved) is the one that will be moved next.

MOST - The AGV is assigned to the pick-up station with the highest score which is computed as a weighted combination of the following factors: mean arrival rate of move requests, proximity to current AGV location, and whether another request is awaiting at the destination point of the move request located at the station for which the score is being computed. Hence, in general, if a station accounts for many move requests, if it is close to the current location of the AGV, and if the destination of its move request takes the AGV to a station where another move request is present, then such a station will receive a high score. The AGV is assigned to the station with the maximum score. The score is recomputed at each P/D point even if the AGV has been assigned to a particular request at a previous stop. Consequently, while proceeding towards its assigned move request, an empty vehicle may be reassigned to another move request.

The following rules were tested by Egbelu and Tanchoco [2] via a simulation model applied to a particular layout: MODF, FEFS, STTF, LTTF, MAXQ, and MINQ. Performance bounds for the FEFS rule were derived by Bartholdi and Platzman [1] for a simple, closed-loop configuration. The STTF rule (or the VLFW rule) was compared with the FCFS rule by Newton [7] using an approximate simulation model. The MOST rule was developed and tested (via simulation against STTF) by Hodgson et al. [3] and King et al. [4]. Also, the role of vehicle dispatching in system design was discussed and demonstrated (to a limited extent) by Maxwell and Muckstadt [6].

IV. DISCUSSIONS

In the above studies, the impact of each dispatching rule of interest is evaluated by certain performance measure(s) selected by the respective authors. For example, Egbelu and Tanchoco [2] use shop throughput measured in unit loads moved (per unit time or per simulation run) as their performance measure.

The bounds derived by Bartholdi and Platzman [1], on the other hand, are based on the number of vehicle revolutions required (around a closed loop) to complete a set of move requests. Newton [7] offers certain (qualitative) arguments on the trade-offs between VLFW and FCFS. Although he provides a listing of his simulation code, no simulation results (quantitative or otherwise) are presented.

In their simulation model, Hodgson et al. [3] use throughput as the primary measure. For a particular layout with seven P/D stations, they report the simulation results by tabulating the mean and variance of the move request queue for the system. In a subsequent study, King et al. [4]

further evaluate the MOST rule via simulation and compare it with STTF for particular layouts with seven, 10, 14, and 17 P/D stations. As before, the authors tabulate the mean and variance of the move request queue for the system.

Given the above results, one is still unable to identify the "best" dispatching rule(s). (The reader may refer to each article in question for further details). Even if no single policy may be consistently the "best", it is still not possible to determine which policy would be the preferred one under which conditions because:

- A significant portion of the results are empirical and they are derived for specific layouts with a fixed number of vehicles:

Egbelu and Tanchoco keep the number of vehicles constant at six vehicles while Hodgson et al. and King et al. consider systems only with one or two vehicles. Naturally, one is inclined to speculate that the performance of these rules depend on the number of vehicles. For Bartholdi and Platzman the number of vehicles is irrelevant as long as the vehicles are not allowed to pass one another.

- Different experimental conditions apply to each study:

The layout studied by Egbelu and Tanchoco consists of 13 departments (with multiple processors in each department) where the deposit point of a department does not necessarily precede the pick-up point which may be located on another segment of the path. Hence, a vehicle may deposit a load and NOT remove one from that department even if a load is present at its pick-up point. This situation, although it may be unavoidable due to flow patterns that exist within a department, often leads to "locking" or a "grid lock" with certain dispatching rules.

Hodgson et al. and King et al. use layouts where each P/D station essentially represents a single processor with its deposit point immediately preceding its pick-up point. (In fact, in their studies, after making a delivery a vehicle always first checks the corresponding pick-up station of that P/D point). Consequently, although [2] and [3]/[4] have both considered the STTF (or the VLFW) rule, due to the above differences in layout, the actual move requests served under identical conditions in the system are not necessarily the same.

The layout considered by Bartholdi and Platzman [1] consists of a single closed-loop with no shortcuts. Similar to [2], the deposit point of a station is not necessarily adjacent to its pick-up point.

- None of the studies explicitly model the processors themselves:

In a manufacturing environment, when a load (ie. job) is delivered by an AGV, it goes through one or several processes and eventually it returns to the pick-up point of the same station where it waits to be moved to the next station. Clearly, the mean and variance of the above processing time may have a profound impact on the queue statistics associated with the move request queue of that station. (The mean and variance of the processing time at each station or department is not mentioned in [2], [3], or [4]; hence, one is forced to conclude that they were not explicitly represented in the simulation model).

- Performance measures are not consistent or standardized:

Although measures such as throughput or mean/variance of the move request queue for the system may at first seem attractive, one must ultimately face the following question: is the rule that maximizes throughput (or minimizes the mean/variance of the request queue) is the most preferred rule? The answer is a resounding no because one must also consider: the relative amount of improvement obtained (compared to other rules), the control system requirements associated with that rule, and whether or not the additional throughput capacity obtained through that rule is indeed required.

Consider, for example, the results reported by Egbelu and Tanchoco. With finite queue capacities, the only two rules for which locking is not encountered are MODF and MINQ. The maximum percent difference observed in throughput between these two rules is 1% (in favor of MODF). With infinite queue capacities, the authors report relatively larger percentage differences between the "best" rule (MODF) and the others (MAXQ, STTF, LTTF and MINQ). However, no explanation is offered as to why MODF outperforms MAXQ and MINQ. Furthermore, STTF still leads to an infeasible situation where one of the deposit queues continues to grow indefinitely (due to no pick-ups from that department).

According to King et al., for a single AGV system, the maximum percent difference observed between the mean queue lengths of STTF and MOST is slightly over 10%. Likewise, the percent difference in queue length variance is at most 16%. (For the two vehicle case, the maximum improvement obtained in queue length mean and variance is approximately 14% and 21%, respectively). Given the amount of improvement obtained, and given that MOST is a considerably more complicated rule, it is not clear that MOST is indeed the preferred rule (as long as both rules meet the required throughput).

In addition to the above four factors, differences in control system requirements has largely been ignored by the studies performed to date. More specifically, the relationship between the above dispatching rules and the four control methods described in the previous section has not been explicitly taken into account in identifying the most preferred rule.

Note that, with the exception of FEFS, all the dispatching rules listed earlier require at least the third control method; namely, a remote (centralized) dispatching terminal. Recall that, under this control method, the operator must have full knowledge of pick-up station status and vehicle location and status. Although it is theoretically possible to implement any of the rules under the third control method, STTF, LTTF, MAXQ, and MINQ can be effectively managed by an operator only if all the relevant information is displayed in a very convenient and clear fashion. Due to the amount of record keeping involved, the remaining rules (FCFS, MODF, LSAT, and MOST) actually require the fourth control method; that is, central computer control.

Although FEFS represents conceptually the simplest rule and it can be implemented with the first two control methods, determining the best path to be followed by the available vehicle is far from straightforward for reasonably complicated networks. Solving a Traveling Salesman Problem to minimize total distance traveled (see Egbelu and Tanchoco [2]) will not necessarily minimize expected empty vehicle travel; it also does not assure that the empty vehicles will visit each P/D station with proportionally equal frequency. That is, the layout and the above path may be such that some P/D stations may require considerably longer queue capacities.

Since considerable cost and flexibility differences exist between the four control methods, the evaluation of any dispatching rule must take into consideration the minimum acceptable control method required by that rule. As remarked earlier, with the exception of FEFS, all the dispatching rules studied to date require centralized control. On the other hand, according to Koff [5], the first control method, that is, programming the AGV via an onboard control panel, is "the most common form" and "usually the most flexible type."

Obviously, this does not necessarily imply that future installations would continue to use the first control method. However, there is incentive to keep the control system as "simple" as possible, and use sophisticated control methods only if the overall benefits outweigh the high cost and significant installation and maintenance efforts associated with such methods.

Lastly, one must also address the overall impact of vehicle dispatching in system cost and performance. Although the above cited studies provided much needed insight, there remains considerably more to be learned to design efficient AGV systems from a control and information requirements standpoint. Listed in increasing order of significance and/or complexity, some issues to consider are:

- Standardization or consistency of the layouts that are being selected to evaluate the dispatching rules.

Some consensus must be reached in selecting the layouts to be simulated. Perhaps a "standard" set of layouts used by all the researchers in the field would serve as an excellent test bed for future efforts. This will also enable one to compare the conclusions offered by different studies - something which is not possible today.

- Analyzing the interplay between the system configuration (guide path layout) and the dispatching rules, as well as the impact of alternative workstation arrangements for a given guide path layout.

Some of the studies have addressed this issue (see, for example, Maxwell and Muckstadt [6]); however, more research is needed to examine potential interactions. That is, certain rules may perform well only with certain types of layouts. If so, could one characterize such layouts? Are there certain "robust" layouts that perform almost equally well with all (reasonable) dispatching rules? If so, can one characterize their properties?

- Analyzing the impact of processing time distribution and the routing matrix for move requests.

As remarked earlier, in simulating a manufacturing system, one must include the processors (and the corresponding processing times, break-down frequencies, etc.) in the model. The variance of the processing time is likely to have a significant impact on queue capacity requirements; which in turn will affect the AGV system performance. The routing matrix, on the other hand, interacts with the layout to generate certain amounts of loaded vehicle travel in certain segments of the path. Future studies may consider exploiting the structure of the routing matrix in system layout and vehicle dispatching.

- Integrating or interfacing the AGV control system with higher level information systems in the facility.

In studies performed up to date, as well as in current installations, the AGV system operates essentially on a "reactive" basis as opposed to a "proactive" basis. That is, at present it is assumed that no information - other than the move request queue - is available to the AGV system. Given that manufacturing operations run according to a pre-specified schedule (that may change over time), one might be able to develop "smart" control systems for AGVs that actually "anticipate" some of the flow requirements in the shop and manage the vehicles accordingly.

One example is "dynamic fleet size" management where certain vehicles are "parked" to reduce congestion and wear and tear if a period of low demand is anticipated by the system. Another example would be the "pre-positioning" of certain vehicles to avoid congestion and delays later. More importantly, the AGV control system may feed vital information back to the shop floor control system and increase the visibility of the jobs. Hence, developing "smart" and "adaptive" AGV control systems that operate on limited information about the future may have a dramatic impact on the design and performance of AGV systems.

- Evaluating the impact of self-guided vehicles (SGVs).

SGVs do not require a guide path. Instead, they rely on external devices placed at critical locations in the facility such as walls and ceilings and/or they are equipped with additional navigational devices or sensors used for determining position, location, and distance traveled. Such systems offer considerable flexibility in those cases where P/D station locations or material flow patterns are subject to change over short periods of time. Since no guide path relocation is required, such changes can be conveniently implemented by reprogramming the "road map" stored in vehicle memory.

Most of the results derived for AGVs would also be applicable to SGVs. However, control algorithms that are developed for AGVs may not fully utilize the additional flexibility (including routing) offered by SGVs. Hence, further work is required to define and analyze system control issues in SGV systems. (Currently, only two vendors offer SGV systems).

V. CONCLUSIONS

Although the studies performed to date on AGV control issues provided some insight into the empty vehicle dispatching problem, more research is needed to generalize the existing results and to standardize the problem environments to be used in deriving future results. More importantly, future research in AGV controls must go beyond empty vehicle dispatch and consider the overall benefits to be derived from making the AGV control system an integral part of higher level information systems (primarily the scheduling system and the inventory control system). Such an approach does not only take full advantage of the flexibility offered by AGV systems, but it allows one to develop a radically different material handling system; namely, a "proactive" handling system rather than the traditional "reactive" one.

BIBLIOGRAPHY

1. Bartholdi, J. J., III and L. K. Platzman, "Decentralized Control of a Fixed Route Automatic Guided Vehicle System," TR-85-05, Material Handling Research Center, Georgia Institute of Technology, Atlanta, GA, February 6, 1985.

2. Egbelu, P. and J. M. A. Tanchoco, "Characterization of Automatic Guided Vehicle Dispatching Rules," The International Journal of Production Research, Vol. 22, No. 3, pp. 359-374.

3. Hodgson, T. J., R. E. King, S. K. Monteith, and S. R. Schultz, "Developing Control Rules for an AGV using Markov Decision Processes," Material Flow, Vol. 4, 1987, pp. 85-96.

4. King, R. E., T. J. Hodgson, and S. K. Monteith, "Extracting Heuristic Control Rules for AGV's Using Markov Decision Processes, Unpublished Working Paper, Department of Industrial Engineering, North Carolina State University, Raleigh, NC.

5. Koff, G. A., "Automatic Guided Vehicle Systems: Applications, Controls and Planning," Material Flow, Vol. 4, 1987, pp. 3-16.

6. Maxwell, W. L. and J. A. Muckstadt, "Design of Automatic Guided Vehicle Systems," IIE Transactions, Vol. 14, No. 2, 1982, pp. 114-124.

7. Newton, D., "Simulation Model Helps Determine How Many Automated Guided Vehicles Are Needed," Industrial Engineering, Vol. 17, No. 2, 1985, pp. 68-78.

8. Considerations for Planning and Installing Automatic Guided Vehicle Systems, Prepared by the AGVS section of the Material Handling Institute, Document #101, 1980.

Integration Models and Techniques

MATERIALS HANDLING AND STORAGE IN LOGISTICS SYSTEMS

R H Hollier
Manchester School of Management
UMIST, Manchester, UK

The aim of this paper is firstly (in Part A) to set AHS in the context of the overall logistic chain and to highlight some aspects which have not received much attention from researchers. It will then (in Part B) explore one particular aspect in more depth, the development of guidance systems for automated guided vehicles (AGVs) with special reference to free-ranging capability. The paper deals only with unit load systems and will stress the linkages necessary between the physical material and information flows.

INTRODUCTION

Over the past decade or so there has developed a much greater recognition that handling and storage are important activities in material flow systems, perhaps equally important as the material transformation processes and transportation activities in a total logistic chain. This has come about during a period in which enormous changes have taken place in the management of supply, manufacture and distribution and in the technology involved. On the management side there has been the revolution in thinking brought about largely by the Japanese approach and on the technological front, the rapid development of computer-aided systems. Such concepts as the 'systems approach', emphasis on flexibility and more recently the theme of integration, leading to the computer-integrated business, have come to the fore. In the realm of handling and storage, we now consider much more complex systems comprising mechanical equipment together with associated computer hardware and software for control and management information purposes - the advanced handling and storage system (or AHS).

NATO ASI Series, Vol. F53
Advanced Information Technologies for Industrial Material Flow Systems
Edited by Shimon Y. Nof, Colin L. Moodie

PART A

LOGISTICS AND AHS

Before considering the role of the AHS, attention needs to be given to some of the trends that are apparent in logistic systems as a whole (see Figure 1). Foremost amongst these trends is the continuing development of information and communications technology. This, in turn, is changing the nature of the processing stages of supply, manufacture and distribution, as well as their integration, to provide better customer service, if possible at lower cost. For example, many manufacturing companies are now re-appraising their policies concerning make-or-buy, while in distribution there has been a rapid growth in third party warehousing and transport. The wholesaler may re-emerge as a broker, holding information on stock location and availability, not stocks themselves.

Within this context, the need for an AHS is driven by market forces - to be more competitive, to give better service to the customer, to reduce costs and increase productivity, to improve quality, to give enhanced management control, etc (see Figure 2). To maximise these gains demands a thorough knowledge of the business, its strategic objectives, its products, the market place, the needs of customers and the role of technology. For example, products today are often more complex, with shorter life cycles, made in smaller batches, in reduced lead times, to give faster response to market demands (see Figure 3). An AHS must be used only to assist in achieving the strategic objectives, not as an end in itself.

In spite of general improvements in efficiency, it is still true that the material or product often spends only 5-10% of the total time being processed or transported, the remainder being occupied in handling or storage between stages of the flow chain. So the less the material or product is handled, the more efficient the handling methods and the less time is spent in storage (implying lower stocks and increased turnover), the more cost effective the overall process is likely to be. A faster flow rate brings with it the need to track the material or product right through the flow chain, since searching and identification activities waste significant time and give rise to error. Thus handling and storage have to be considered as an integrated whole, interfacing to the machine and information systems through a hierarchy of control involving connections to a computer network.

However, the concept does not necessarily imply full automation; it is sensible to automate only to the degree which is both practical and economic. Even where conventional methods are retained, the use of up-to-date control systems can provide the means to gain full utilisation of equipment, a point which is of considerable importance to the smaller company who may not be able to afford major capital cost of a complete AHS. Hence equipment suppliers are developing 'bolt on' systems which can be retrofitted to conventional equipment to achieve a low cost automation facility.

SYSTEM DESIGN ISSUES

One aspect affecting all handling and storage designs is the question of layout and location. One view of the relationship between the two is shown in Figure 4 (see Muther and Webster, 1985). Apart from systematic methods of layout planning as propounded by Muther, much research has been carried out on more analytical techniques drawn from Operational Research. Predominant amongst these has been the development of various algorithms for computer-aided layout planning, based either on construction of an initial layout or improvement of an existing one (see Driscoll and Sangi, 1986). Virtually all of these algorithms make assumptions about the nature of the handling costs involved in the movement of materials between locations which do not take into account the variety of handling methods which might be employed. Usually handling costs are assumed to depend purely on intensity of flow and distance travelled, perhaps also considering frequency of movement. Thus the interaction between layout and handling is an area which does warrant further research.

Design of handling and storage systems still relies heavily on the experience of the analyst and system design tools are often fairly rudimentary with little formal modelling. Mention was made above of simulation as a frequently used technique, particularly now that powerful graphic displays and interactive computing facilities are readily available. Other Operational Research techniques, such as queueing, inventory and scheduling theory, mathematical programming, etc, have been used, but with very limited success. Perhaps the skill of the system designer will be captured in a series of expert or decision-support systems in the future, although little has yet been done apart from some computer-aided design packages used by a number of suppliers mainly as marketing tools.

It was stated at the beginning of this paper that an AHS comprised mechanical equipment together with associated computer hardware and software for control and management information purposes. The major classes of unit load handling and storage equipment in which significant changes have taken place in the last twenty-five years are listed in Figure 5. In some cases (eg AGVS) development has been very rapid, in some gradual refinement has taken place (eg conveyors), while in others very slow progress has been made (eg vehicle loading/unloading). During this period a number of new equipment concepts have emerged which have either gone no further than the drawing board, or have not gone beyond an initial trial installation.

As far as computer hardware and software are concerned, the changes that have taken place have been dramatic, both with regard to the scope of the system and to its reliability. It is now common for there to be a control hierarchy, starting with PLCs for control of individual pieces of equipment, through minicomputers for area control, to mainframe computer for overall control of the entire system. This gives rise to the need for communication networks between the various computers and the use of communication protocols. Increasingly, such systems for AHS will have to interface to other machine and management systems, as progress is made towards the computer-integrated business.

PROJECT MANAGEMENT

AHS projects range from relatively simple problem solving or improvement studies to full scale design of a completely new facility. The methodology involved across the range is not fundamentally different (see Figure 6), but the emphasis on the various stages of the project will vary. In all cases there needs to be a clear statement of objectives, time for an adequate feasibility study and thorough system design, a full specification of the design and a well-managed project team to carry through the implementation (see Figure 7). Much has been learned about these aspects of AHS project management over the past few years from bitter experience, but the mistakes go on happening (see Hollier and Shimmings, 1988). Figure 8 summarises some of the lessons, based on White (1986).

During the project various design options will be considered, some of which can be eliminated as infeasible, leaving a few which are worthy of detailed analysis. Comparison may first be made on the basis of capital and operating costs and any obvious difference in the more intangible benefits, such as flexibility, expansion capability, ease of control, etc. More sophisticated modelling may well be called for, especially using computer simulation, but the technique needs to be used carefully to focus on those aspects of the total system where performance is unpredictable, rather than attempting comprehensive model development.

From a research viewpoint, there are many facets of the overall project management process that require further study. Such research is of interest to equipment suppliers who claim that many companies do not understand AHS and thus find themselves involved in much unpaid initial design work and to company users who do not have the internal expertise and must rely too heavily on consultants or suppliers. For example, the whole question of economic justification for AHS is a matter of concern, since there is often difficulty in measuring the intangible benefits as input to traditional methods for evaluating capital expenditure and assessing sensitivity. Frequently the major intangible benefits only appear when a degree of system integration has been achieved and after a series of investments have been made, making management cautious in their willingness to accept that such benefits will be forthcoming.

AUTOMATIC IDENTIFICATION

Automatic identification is an important, closely related technology, linking the physical identity of goods (or people) with the computer processing of information for planning and control, using various types of equipment for reading/writing. Public awareness of the technology has been raised due to the near-universal adoption of bar coding by the major supermarket groups, but other forms of technology are now becoming available which further extend the application of the basic principles (see Figure 9).

The market for automatic identification systems has grown at a very rapid pace over the last few years. It has been estimated that the European market alone is already worth over £1000 million per year and is forecast to triple by the mid-1990's. Similar growth is evident in the United States and Japan. American companies are the dominant suppliers in the Western World at present, but there are signs that European and Japanese companies are beginning to compete through product innovation.

In manufacturing and distribution, the principal benefit of automatic identification is the ability to track accurately every transaction from receipt to despatch. Indeed, one of the major advantages is the dramatic decrease in the occurrence of errors. This should provide reduced stocks and lead times, better control of quality and rework, less plant downtime waiting for materials, improved information for monitoring operations, etc, thereby giving enhanced service to the customer with greater profitability. Time and attendance recording, control of access to secure areas or to information sources are other industrial applications.

In the service sector, automatic identification has been used for some years in the financial world, in postal services, in cataloguing work and in transport for labelling, vehicle or container identification.

One of the earliest technologies used in automatic identification was optical (or magnetic ink) character recognition in which stylised alpha-numeric characters are read by machine. Sometimes information is encoded before reading is carried out, as is the case with postcodes in the UK.

Magnetic stripe systems were also early on the scene and are still widely used. However, it was the development of equipment to read bar codes remotely and accurately, without undue need for precise location of the code on the item being identified, which lead to the rapid growth mentioned earlier. The other important factor was the need for agreement on standard symbologies, sometimes achieved by international bodies, but often either by an industry group or by a major customer, such as a military procurement agency. There now exist numerous ways in which bar codes can be generated and subsequently read by many different types of scanner with its associated decoder.

Radio frequency tags (or transponders) are a rapidly growing sector of the automatic identification industry. They show great promise for the future as their memory capacity increases and their physical size and unit cost decrease. Tags need to have data entered into them which can then be read by sensors. They can be either passive (read only) or active (read/write). Another new development is the so-called 'smart card', in which a microprocessor is imbedded in a plastic card. This again has considerable memory capacity and can be interrogated in read/write mode.

To complete the picture, mention should be made of machine vision and voice recognition/response. Machine vision requires image processing which is much more complex than bar code scanning, but as microprocessors of greater power become available, applications will certainly increase. Voice recognition/response is already a reality and is growing in acceptance as equipment becomes easier to use. It is particularly applicable where an operator's hands and eyes are fully occupied with other tasks, as in sorting operations.

Automatic identification is a fast-moving technology which promises major benefits to the user. It has a vast field of application and will surely develop strongly in the years ahead.

PART B

AGVS GUIDANCE

The AGV is an example of a relatively new piece of materials handling equipment whose application has grown rapidly in the last ten years. One of the most important aspects of AGV design is the question of the guidance system to be used. This can vary from rail guidance which is completely constrained, to a totally free-ranging capability which may be achieved in a variety of ways and is still subject to much research. For the purpose of this discussion, it will be assumed that guidance determines the path of travel of the AGV (and its speed), but not other forms of communication.

Boegli (1985) suggests that a guidance system should meet the following objectives:

1) Guide the vehicle reliably from A to B within the working surroundings with a precision sufficient to avoid the known obstacles.

2) Provide for precise manoeuvring of the AGV in the vicinity of load pick-up and deposit stations.

3) Provide a base for efficient traffic control, possibly involving a large number of vehicles.

4) Fulfil safety requirements for protection of people, facilities and other material.

Point 2) may not depend purely on the guidance system, but may be achieved by finally locating on floor-mounted cones or through some form of 'hand-shaking'. Point 4) is usually partly fulfilled by ultrasonic or infrared sensors and impact bumpers. Boegli then goes on to compare the following seven guidance techniques in terms of a number of evaluation criteria:

* Mechanical or rail
* Optical, chemical or magnetic
* Inductive wire
* Dead reckoning or odometry
* Beacon system, with stationary rotating beacons scanning the area to be controlled (Note that AGV-mounted beacons are also used)
* Inertial navigation
* Direct imaging systems, with optical or ultrasonic sensors.

The last four guidance techniques are suitable for free-ranging AGVs. The evaluation criteria used were:

* Range
* Accuracy
* Flexibility
* Reliability
* Controllability
* Vehicle cost
* Fixed equipment cost

The main conclusion of the paper is that no single guidance system is dominant, but that an assessment of performance versus cost would rule out beacon systems, inertial navigation and direct imaging on cost grounds. A rail guidepath is not easily combined with other guidance systems, and three systems are left which might be used in various combinations.

Off-guidepath movements of AGVs have been available for some time, using dead reckoning or odometry. Some AGV manufacturers are developing programming aids to allow users to generate these off-guidepath movements easily without extensive software changes. However, interest in complete free-ranging capability is still strong.

Premi and Besant (1983) describe different forms of guidance for free-ranging AGVs. In addition to the last four methods mentioned above, they consider stereoscopic vision. They conclude that dead reckoning, combined with position referencing beacons, provides the best approach to free-ranging guidance.

Other research on the use of machine vision has been reported by Inigo et al (1985) using linear image arrays. Similar work is currently taking place in France also using linear arrays to analyse a painted path one metre ahead of the AGV. The binary video signal is processed to give information on detection quality and offset between camera axis and the painted path for steering control. If a detection failure occurs for a short distance, then an odometry method takes over to estimate the required path.

Miura et al (1985) and Tsukagoshi et al (1987), both from NEC, describe a guidance system using magnetic floor tiles. The earlier paper discusses the sensors used and vehicle control, while the later work covers a novel lattice lane guide consisting of magnetic-rimmed floor tiles in which the path is determined by counting the number of intersections traversed. An experimental vehicle has been built for use in office environments.

Kamimura et al (1986) have tested fixed overhead optical transceivers arranged to view the whole floor area within which movement can occur. The National Engineering Laboratory in the UK and a team at Warwick University have experimented with fixed overhead video and infra-red sensitive cameras respectively, the AGVs being fitted with emitters. In all cases positional information is transmitted to the vehicle via a central computer, but available resolution limits the positional accuracy achievable.

Although inertial guidance has been used in aviation and marine applications for many years, its use in AGVs has been limited by cost. Turpin (1986) reviews the general issues and mentions one vendor in the US who has supplied a system for a clean room application in the semi-conductor industry.

Dead reckoning or odometry combined with external inputs for positional correction, mainly due to drive wheel slip, is being actively studied. Examples are Komatsu et al (1987), Kopp (1987) and work at Purdue University and ISW Stuttgart. Whereas Komatsu et al propose a physical guidepath to correct for odometry errors, Kopp uses image processing and the work at Purdue University and ISW Stuttgart mentions global or absolute coordinate referencing. Yet another possibility is the use of radio-frequency coded transponders embedded in the floor in a grid pattern; such a system called 'FROG' (free-ranging on grid) is under development at Louvain University and by a group in The Netherlands. This latter group are also trying steel-rimmed concrete slabs (inductive) and two-coloured floor tiles (optical).

Ultrasonic guidance from solid objects, such as walls, combined with on-board storage of a route map up-dated from a central computer is described by Murata et al (1987). Similar concepts are being used by another research team at Louvain University to develop a guidance system called 'MORON' and by another Belgian group at CRIF/WTCM working on local guidance for obstacle avoidance.

Laser triangulation, with the laser mounted on the vehicle scanning bar-coded markers on the walls, combined with a route map held on a central computer, is the guidance method used on the GEC-Caterpillar AGV. Research on the combination of laser triangulation with odometry to avoid the necessity for a route map is currently being undertaken at Lulea University. It is also conceivable that a laser could be used for direct distance measurement by interference, but this would be expensive.

To date the vast majority of AGV applications have been in factories or warehouses with flat and relatively clean floors. Travel outside between buildings over rougher surfaces is known, but is still largely experimental. It raises different guidance problems, since many of the techniques discussed above are not applicable. Optical guidance and odometry over reasonably short distances have been used, and the latter could be extended if radio-frequency transponders were to be employed.

CONCLUSION

Much research and development is currently in progress on AGV guidance techniques, much of it directed towards achieving free-ranging capability. For this requirement, most of the techniques being studied are used in combination, usually as primary and secondary guidance. The alternative systems currently being investigated are summarised in Figure 10. It is difficult to judge which of these various combinations show most promise, but the 'FROG' system seems to be an interesting development. So far as commercial application is concerned, there is only one installation of the GEC-Caterpillar AGV in the UK so far, besides the use in Caterpillar's own plants in the USA.

The industrial need for free-ranging capability is primarily to achieve flexibility by being able to make rapid path changes. The lack of any physical guidepath means cheaper and quicker installation, while the possibility of distributed control and autonomous operation may be attractive. Certain applications, such as in clean rooms or travel outside, may also require free-ranging capability. However, it is clear that inductive wire (and to some extent optical) guidance is well accepted and there will have to be a strong market demand if free-ranging AGVs are to become widely available.

REFERENCES (PART A)

Driscoll, J and Sangi, N A (1986), The Development of Computer-Aided Facilities Layout Systems - International Survey. University of Liverpool.

Hollier, R H and Shimmings G F (1988), The Specification of Advanced Handling Systems. Proceedings of Advanced Handling Systems Conference, Institution of Mechanical Engineers, London.

Muther, R and Webster, D B (1985), Relationship between Materials Handling and Plant Layout. Chapter 2 in Materials Handling Handbook, Ed by Kulwiec, R A, Wiley.

White, J A (1986), Warehousing in the New Age. Proceedings of Seventh International Conference on Automation in Warehousing, San Francisco, IFS.

REFERENCES (PART B)

All the papers references appear in the following four Proceedings volumes, all published by IFS (Publications) Ltd:

Warnecke, J H (Ed), AGVS-2, Stuttgart, 1983.

Andersson, S E (Ed), AGVS-3, Stockholm, 1985.

Hammond, G C (Ed), AGVS-4, Chicago, 1986.

Takahashi, T (Ed), AGVS-5, Tokyo, 1987.

BIBLIOGRAPHY

Hammond, G C (1986), AGVS at Work. IFS (Publications) Ltd.

Hollier, R H (Ed), (1987), Automated Guided Vehicle Systems. IFS (Publications) Ltd.

Muller, T (1983), Automated Guided Vehicles. IFS (Publications) Ltd.

INFORMATION & COMMUNICATIONS TECHNOLOGY

CHANGING PROCESS STAGES & THEIR INTEGRATION

- Manufacturing
- Storage
- Transport
- Transactions

CIM	-	smaller batches, lower stocks
	-	sub-contracting - more or less
STORAGE	-	centralised, lower stocks
	-	wholesaler becomes broker
	-	automated pick, sort, load
TRANSPORT	-	along major routes, modes
	-	direct deliveries
	-	more frequent deliveries, smaller quantities
	-	third party distribution
	-	traffic planning & control
TRANSACTIONS	-	communications & data banks
	-	paperless systems
	-	automatic identification
SERVICE	-	shorter lead times
	-	deliveries to customer order
	-	produce & transport to need

Figure 1 LOGISTIC TRENDS

WHAT ARE YOUR BUSINESS OBJECTIVES?

WHAT ARE YOUR COMPETITORS DOING?

- national and international
- what trends in cost structure are apparent?
- labour, materials, energy, equipment, computing

PRESSURES FOR AUTOMATION

- cost reduction
- improved productivity
- improved service
- faster response
- better control
- increased output capacity
- lower stocks
- higher quality
- better working environment
- improved safety
- increased security
- reduced space

Figure 2 WHY AUTOMATE?

- competitive international markets
- greater emphasis on service, quality and product design
- shorter product life cycles and greater product variety
- shorter lead times in design, manufacture and distribution
- changing supplier/customer relationships
- changing process/equipment technology
- changing information technology
- need to reduce unit costs (eg stock reduction)
- greater complexity of investment decisions
- changing expectations of workforce
- changing regulations

Figure 3 PRESSURES ON MATERIALS MANAGEMENT

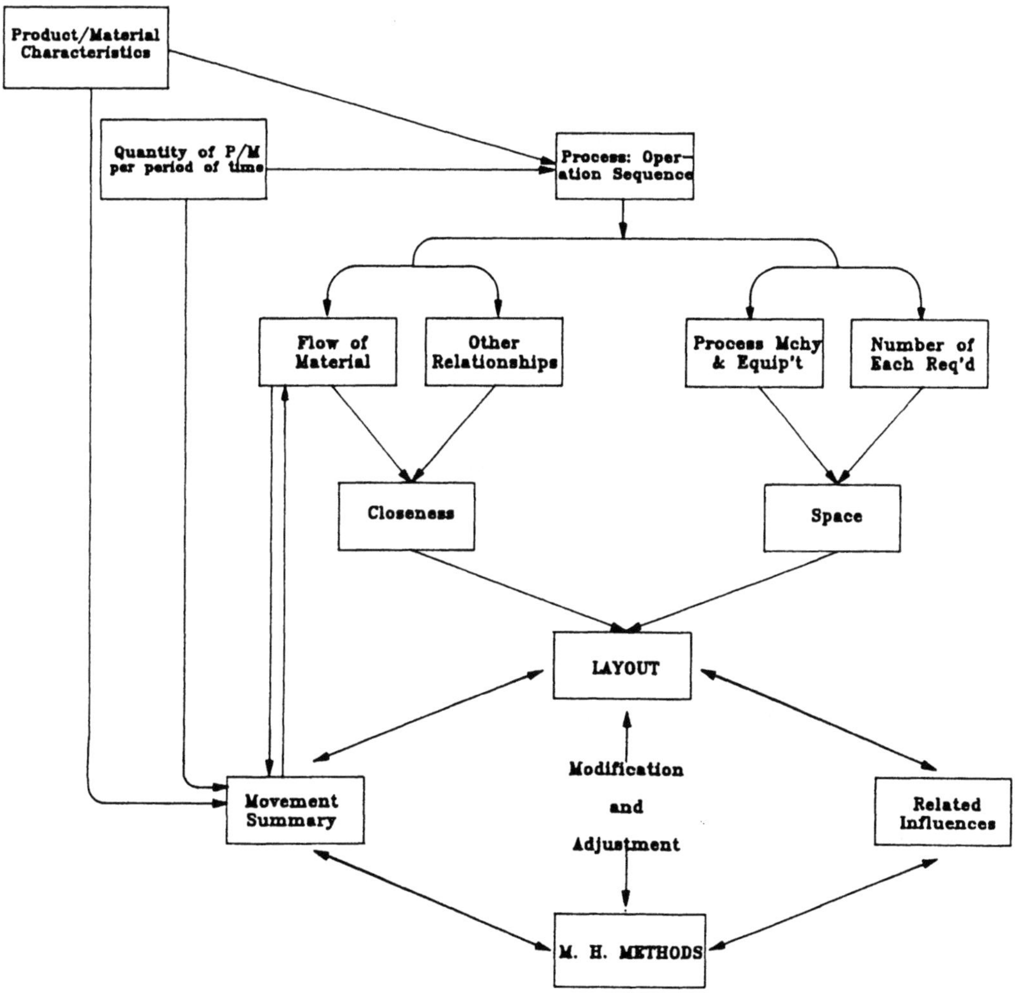

FIGURE 4 RELATION OF LAYOUT PLANNING AND M. H. ANALYSIS

VNA trucks

Automated guided vehicles

Conveyors - power and free

Monorails

Stacker cranes

Automated storage and retrieval

Miniloads-totestackers - small parts

Carousels - vertical and horizontal

High density storage

Air film

Robots

Picking

Sortation

Vehicle loading/unloading - dock operations

Ancilliary equipment - positioning devices

Packaging

Automatic identification

Figure 5 HANDLING AND STORAGE EQUIPMENT TRENDS

DEFINITION

Identification of problems
Objectives and constraints
Feasibility study - cost and time

INVESTIGATION

Work plan and schedule
Collect and analyse data
Redefine problems

SYNTHESIS

Alternative systems - degree of mechanisation
Supplier selection
Performance prediction - modelling, simulation
Evaluate and select system

DESIGN

Develop and design system components
Specification of proposed system
Supplier bid evaluation
Justification report - cost/benefit

IMPLEMENTATION

Contracts
Detailed design
Installation schedule
Training
Test, start-up and acceptance
Maintenance
Audit performance

Figure 6 PROJECT MANAGEMENT FOR SYSTEMS

FIGURE 7 **Influence of key activities on the cost of a project**

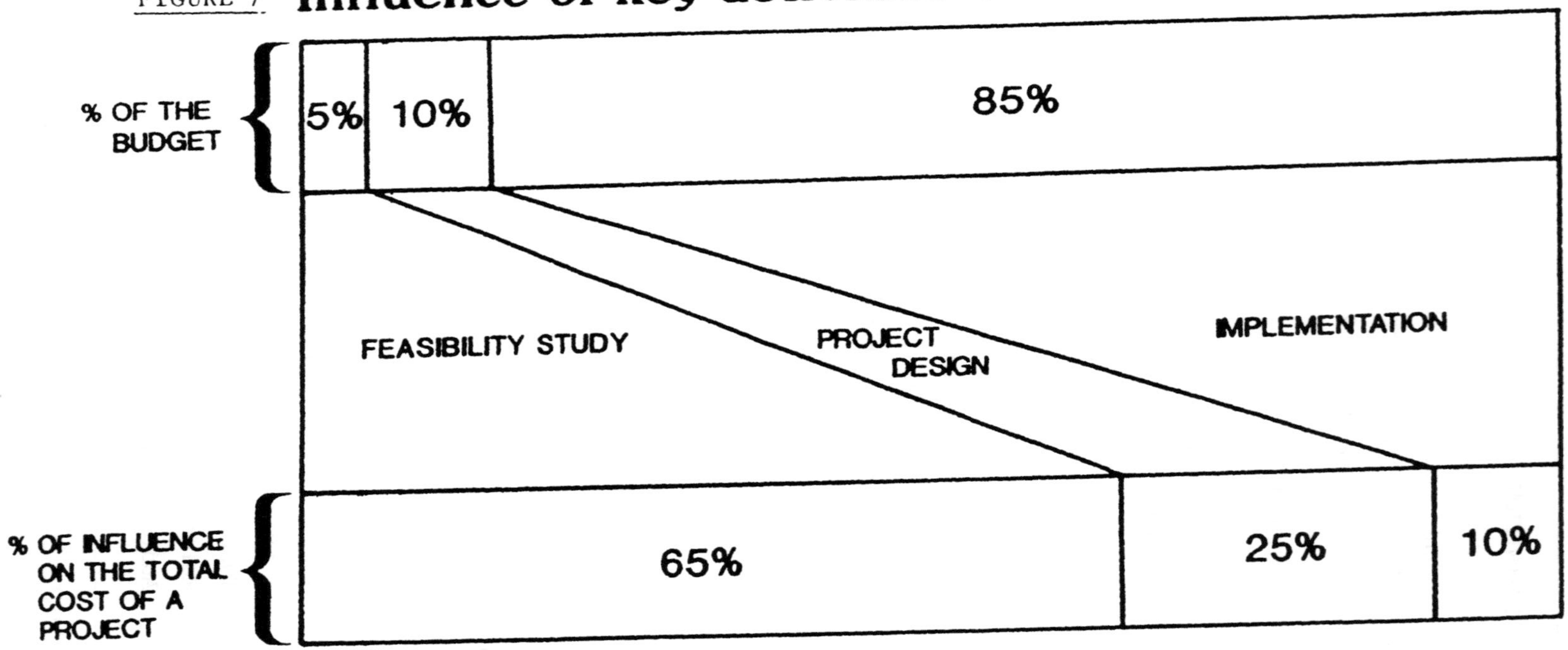

Technology - usually works, must be appropriate

Requirements driven, not solution driven

Requirements not defined

Inadequate time for design

Design the whole, implement the pieces

Failure to freeze design, changes cause chaos

Systems too extensive and complex

Software the major bottleneck, not hardware

Premature hardware specification

Unnecessary and expensive failure recovery

Primary benefit is system discipline

Cost benefit - for show not dough

- over promised, under delivered

Figure 8 OBSERVATIONS ON AUTOMATION

Punched cards/tags

Optical (or magnetic ink) character recognition

Special encoding (eg UK postcodes)

Magnetic stripes

Bar codes - symbologies

RF transponders - passive & active

Smart cards

Machine vision

Voice recognition/response

Figure 9 AUTOMATIC IDENTIFICATION

Odometry/physical guidepath (partial)

Odometry/beacons

Odometry/machine vision

Odometry/global coordinate referencing

Odometry/transponder Grid

Machine vision/odometry

Magnetic, inductive or optical grid

Overhead optical transceivers

Overhead video or IR cameras

Inertial

Ultrasonic (or laser)/route map

Laser beacons/route map

Laser beacons/odometry

Figure 10 FREE-RANGING METHODS

GRAI Approach to Designing and Controlling Advanced Manufacturing System in CIM Environment

Guy DOUMEINGTS
Deputy Director
GRAI Laboratory
University of Bordeaux I

The design of Advanced Manufacturing System (AMS) is today one of the most complex task in the field of Computer Integrated Manufacturing (CIM). This design must fit with several constraints (costs, quality, flexibility, time) and must allow the best efficiency in the control of Manufacturing System (MS). The choice of equipment and software depends strongly from the result of this design activity and the consequences of a wrong choice could have disastrous consequences if we consider the amount of investment. For fifteen years, GRAI Laboratory has been developing methods to design and to control AMS. The objectives of this paper is to give the progress of this research and to show the recent development. We will organize our presentation in four parts. The first one presents the problematic of the research work, the second details the GRAI Method, the third shows how we can use the approach on an experimental way and the last describes the new researches.

1 - PROBLEMATIC

The design of an AMS is the phase which allows to get the detailed specifications of the whole system. Traditionally in the life cycle of an AMS, we find the Analysis phase in which we analyse the problem, try to point out the main features, constraints and objectives attributed to the potential solution.

The design phase allows to determine the functional specifications of a solution, the basic architecture and the main features of the components. The result of the phase is a specification book which is given to the potential suppliers of the solution. This specification book is used as a reference in order to make a choice among the various proposals.

The next phase is the development phase : based on the specifications included in the specification book, the supplier (often an engineering company) chooses the various components (machines, transport systems, local area networks, quality control equipment, software packages...), develops the specific software. Usually this phase needs a complementary analysis in order to obtain more detailed specifications and the use of software engineering tools for the development of the software.

NATO ASI Series, Vol. F53
Advanced Information Technologies for Industrial Material Flow Systems
Edited by Shimon Y. Nof, Colin L. Moodie

The last two phases are the implementation phase : integration of various elements, tests, transitional period and the operating phase : to keep the objectives, to control the system and feedback in case of disruptions.

It is well known that a flaw in the initial phases of the life cycle may have serious consequences on the other phases, in particular, it may increase the development cost by ten, a hundred or over.

But the design phase is a very complex process.

First, design is a typical activity performed by men. We know how difficult this task is [SIMON 69]. First of all, the design of an artificial object (artifact) must take into account the complexity of the environment. Second, there is no appropriate educational system because of a lack of scientific research on the topic. This last point is related to the difficulties in designing the internal system in agreement with the environment : the usual theories such as control theory, linear and dynamic programming ... are not sufficient. SIMON suggests to analyse by dividing into "actions and goals". We will show later how to use this rule in our approach.

In our field, the difficulties increase with the specific conditions :

- knowledge about manufacturing system,
- capability of designing methods,
- process of designing a concrete system initially based on concepts.

As far as manufacturing is concerned, the knowledge does not exist in one mind, as the above-mentioned manufacturing systems have numerous functions which require some skills from various sectors. Usually an AMS has a high degree of automation. The design becomes more and more difficult as we have to design new systems. We cannot always automatize the human gesture as it is very complex ; we must therefore modify the product. We have then the new situation of manufacturing system design : the close relationship between the products to be manufactured, the facilities and means necessary to manufacture them and the management of the whole system. We can sum up this first point and say that the knowledge is not in one mind, but in several : we must collect and organize it.

The second point is related to the design method. Unfortunately the designing process is learnt from the experience [YOSHIKAWA 81.09]. It is difficult to abstract the process itself from the practical design activity even for skilled designers. There are some design theories for the design of parts but not for manufacturing system. The use of computers in the design will no doubt improve the process by making it possible to memorize the various steps.

The last point deals with the process of designing a concrete system starting from concepts. We have seen that we start from the knowledge on manufacturing systems. This knowledge is related to the type of industry. The process often exists but has to be changed when automation is introduced. We think that it is essential to analyze this current process in terms of concepts. This analysis can be made from an existing workshop. When buiding a new factory, we advise to use first the knowledge of potential users and to derive a "pseudo model" which is represented by a workshop able to manufacture products with traditional facilities.

In both cases, this first step, called the analysis phase, enables us to determine the main characteristics of the problem and to obtain an initial model which will help start the design phase.

From our point of view, the design phase deals with :

- the determination of the functional specifications of the system to be built,
- the global architecture,
- the detailed specifications of the components.

The three steps of the design phase may be linked with three abstraction levels of the design process which enables us to work from the conceptual level towards the concrete level :

- the conceptual level which enables us to represent the system mainly with concepts and functions ;

- the structural level which enables us to group the functional specifications and define an architecture : for example at this level, we have to determine the manual and automatic procedures ;

- the operating level which, from the detailed specifications of the hard components, enables us to determine the hard components, the software packages and the specific software. This step must take into account the realizational constraints.

We truly think that the design must follow the three previous steps if we want to be able to start from a conceptual point of view and reach a concrete realization.

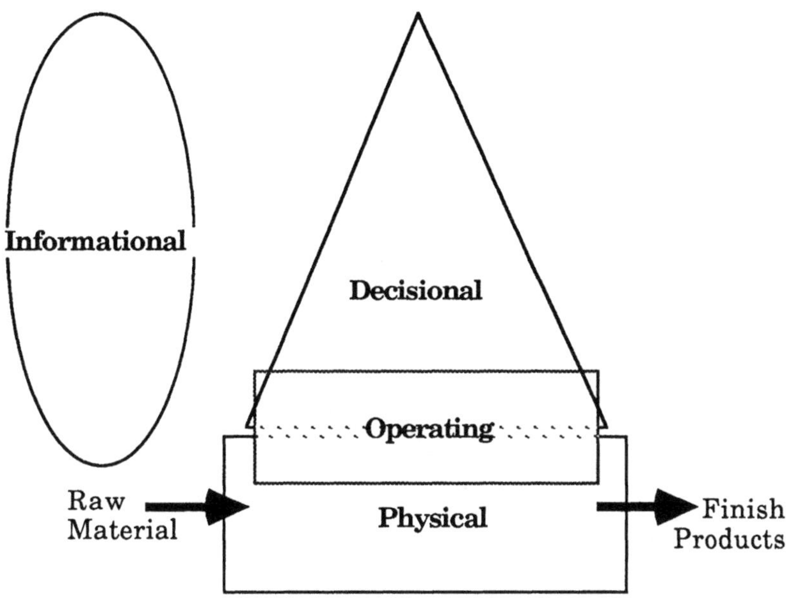

Figure 1: GRAI Conceptual Model

Finally, further to our research in this field (GRAI conceptual model : Figure 1) a manufacturing system can be split up into three subsystems [DOUMEINGTS 84.11] :

- Physical resource sub-system : this subsystem includes men, machines, material flow, techniques ; its role is to transform raw materials into end-products ;

- Decisional sub-system : it aims to control the physical sub-system in order to reach the economic and social targets while taking the constraints into account. We have improved our model lately by introducing an operating level which links the decisional and physical parts (this level includes the control of machines and all procedures to be controlled: security, quality, maintenance) ;

- Informational sub-system which links the two previous sub-systems.

To sum up, we can determine three axes of subdivision :

- one axis for the life cycle,
- one axis for the level of abstraction,
- one axis for the various sub-systems.

If we combine these three subdivisions (Figure 2), we can determine a model for each case (for instance the decisional model at the conceptual level in the analysis phase). The various boxes are linked and we must determine the transformations (T) to go from one model of a sub-system to another (decisional model at conceptual level towards decisional model at structural level, or decisional model in the analysis phase towards decisional model in the design phase) and interfaces (I) between the sub-systems.

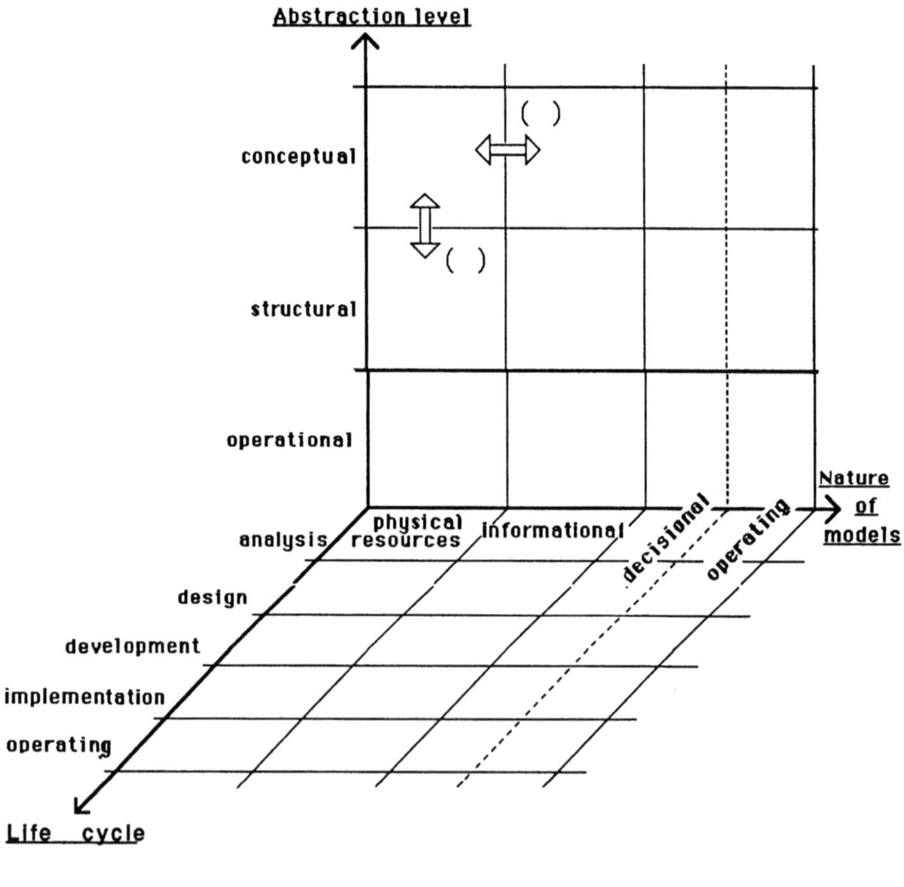

Figure 2 : Various kind of models for Manufacturing Systems

In order to build CAD systems for advanced manufacturing systems, such models, tools and methods are necessary.

Our experience in the field of methods for designing manufacturing systems leads us to search for methods, first in order to describe manufacturing systems and also to formalize. The difficulties arise from the various features of the manufacturing systems :

- the need to modelize : from a global viewpoint through a top down approach (because it is an organizational system with several aspects) and in details through a bottom up approach (in order to obtain accurate running) ;

- the need to study, not only the static but also the dynamic behavior, particularly when a disruption occurs. This comment emphasizes the role of simulation in the design process. We think it is necessary first to design the model from a static point of view (STRUCTURING) then to study the behavior over time from a dynamic point of view (SIMULATION) ;

- the last point deals with the kind of model : a continuous model is not sufficient, a discrete model lacks sometimes theory. It appears that the basic concept of activity is adopted by some designers. Some ambiguities remain especially as far as simulation is concerned : We can describe the model with activities, events or processes, but in order to improve the design process it is necessary to have a compatible model between STRUCTURING and SIMULATION.

Our conclusions are that :

- we need models to describe and formalize manufacturing systems in the design process. These models can be used as a reference and a guide all along the design process. Such models are called CONCEPTUAL MODELS because they describe the manufacturing systems in terms of concepts and relationship ;

- these models can no doubt generate tools helpful during the design process. Such tools must be supported by computers, that is why a formalization is necessary ;

-finally some methods are necessary in order to organize in a structured way the data acquisition during the design process.

We have given in the previous lines some conclusions about the design needs. We will show now the GRAI Method which is well adapted to the decisional system in a Production Management environment.

2 - THE GRAI METHOD

GRAI Laboratory developed the GRAI Method ten years ago. The first application was performed in 1980 in workshop of Telemecanique Electrique located in South East of France. We will describe successively the Conceptual Model, the tools and the structured approach. But before we want to explain the means of the four letters GRAI. At the origine GRAI was the name of our Research group: Groupe de Recherche en Automatisation Intégrée.

Later on, we were known by the first research works and we kept the letters : Graphs with Results and Activities Interconnected.

2.1 - Conceptual Model

GRAI Method use a Conceptual Model which could be considered as a general reference model. This model is built on three theoretical bases :

- basic concepts
- controlling rules for Manufacturing systems
- hierarchical theory

• **Basic concepts**

The basic concepts are supported by H. SIMON's works and J.L. LEMOIGNE's in France "Théorie du système Général".

The conceptual model uses a global approach. It defines the structure of the object, its operating, and its evolution with time. Conceptual model is open on the environment (Figure 3).

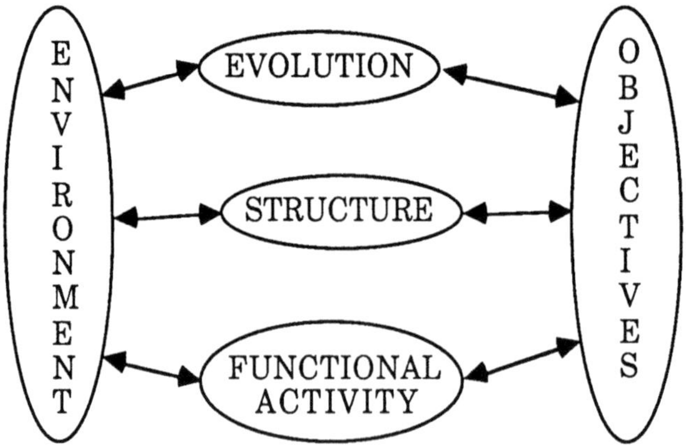

Figure 3 : Conceptual model open on the environment

The basic element of the conceptual model is the process: any change with time, of material, energy and information. This concept of process allows to identify the objects. We distinguish the events representing the change in the state of the object and the activity representing this object between two states. We have used the concept of activity to make this representation easier.

We have looked for a precise representation of the objects and we have chosen K. BOULDING's classification. Within this classification we have chosen an active object which can be co-ordinated and presents the concepts of information, decision, and physical system. We have also selected the grouping of various activities (Figure 4).

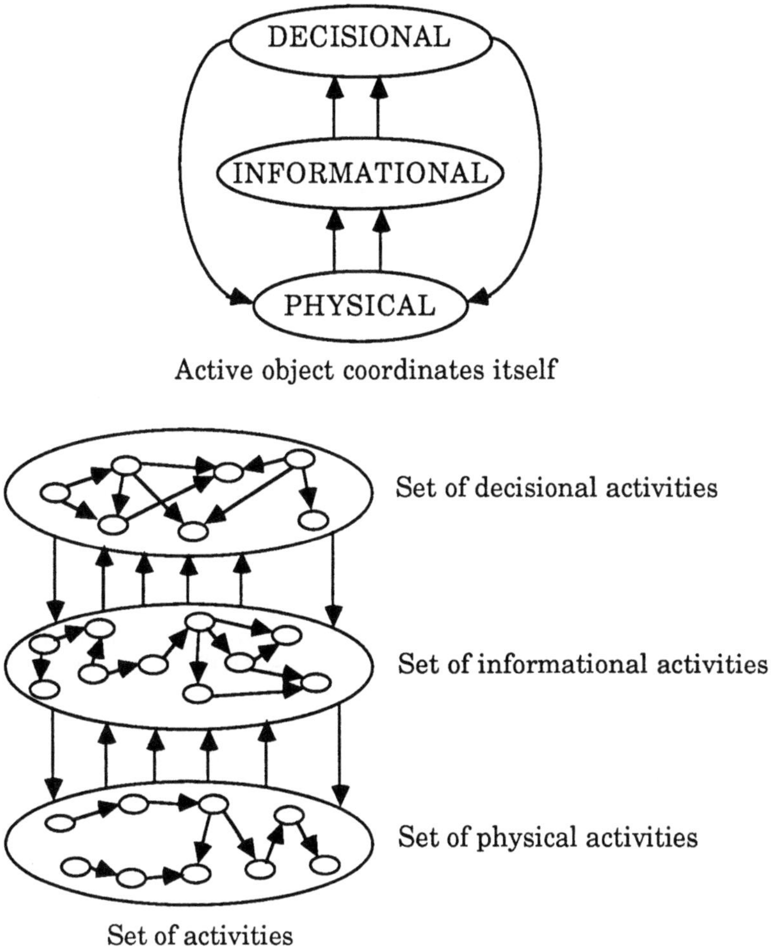

Figure 4 : Boulding classification

- **Rules to control Manufacturing System**

Controlling a system is setting one or several goals to each part of the system. By this way the system must have a finality. We define controlling activities as activities resulting in guiding each part of the system towards its goals. The system has monitoring, feedback and fitting devices which allow it to compare the results and the goals initially set.

The extend of the control of a MS Controlling depends on the ability for guiding a PMS.

This notion is similar to the one set out by ASHBY through the rule of the required variety [ASHBY 52]. It is shown that a system with a variety V cannot be completely controlled by another system unless the variety of the latter be at least equal to V.

Melese has shown that the rule of the required variety could lead to a "controlling barrier" [MELESE 72] which reverses the control. The selected example deals with the relation between factory management, and deadlines, costs, inventory. Manufacturing may set some unsatisfying deadlines whereas these deadlines are usually required by the commercial department.

This situation is due to the great variety of systems to be driven. We all know that the uncertainties in the knowledge of this system increases as we pass from the operating of a machine to the operating of a set of machines and to the operating of one or several workshops.

In order to solve this situation a relevant structure has to be drawn up, particularly breakdown into levels of decision making, so that when going from one level to a lower level some new sources of variety may come up.

This is only possible if the decision making is decentralized: each level will have some freedom of decision making : a condition is to provide this level with a decision frame and necessary information.

In order to solve this situation we must add the concepts of adaptation and learning.

Structures fitted to driving must be some hierarchical and decentralized structures. We have drawn these techniques from MESAROVIC's hierarchical decomposition techniques [MESAROVIC 70].

- **Hierarchical decomposition of PMS**

The lack in a coherent structuring of PMS doesn't allow to get an efficient driving of MS. We have derived our theory from the one developed by MESAROVIC who has considered three ways of splitting : into strata, layers and echelons.

When splitting up into strata, the aim is to describes a model with various levels of abstraction according to the considered point of view.

In our study, we can find 2 examples of splitting into strata :

- distinction between PMS and Physical System (MS) (Figure 5),
- breakdown according to abstraction levels of PPS : factory, workshop, workload, manufacturing cell. This splitting has an inclusive feature (Figure 6).

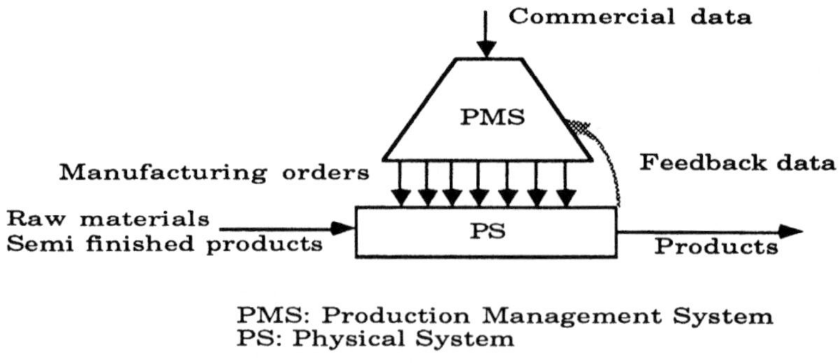

Figure 5 : Distinction between Physical System and Production Management System

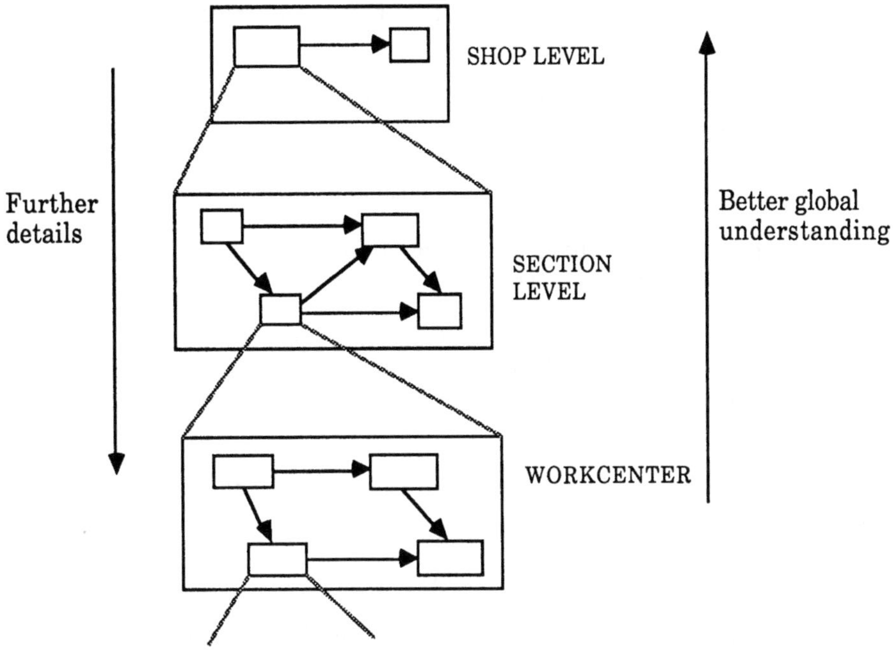

Figure 6 : Splitting up into abstraction levels of the physical system

As regard breakdown into layers, we focus on decision making. The splitting levels are (Figure 7) :

- level of decision choice,
- level of learning,
- level of self-organizing.

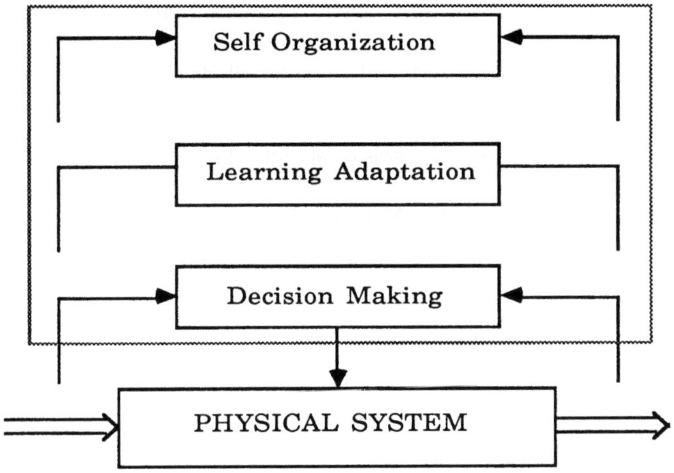

Figure 7 : Layers of decomposition

In the decomposition into levels we combine both previous decomposition. The rules are as follows :

- the system to be split up includes a family of sub-systems reacting upon one another and explicity idendifiable ;

- some of these sub-systems are defined as decision making unit ;

- decision making units are organized into a hierarchy so that some of these units act upon the management of other units.

From basic concepts, controlling rules and decomposition methods, we have built our conceptual model GRAI which has the distinctive feature of allowing decision making. We will describe this model in the next section : we will show the decomposition into MESAROVIC's echelons.

It should be noted that MESAROVIC has listed along with decomposition methods 10 rules detailing splitting conditions.

So the GRAI conceptual model consists in 2 parts :

- a macrostructure showing the global structure of the system and describing the links between its components. It is mainly oriented towards the decisional structure ;

- a microstructure which allows to identify each component of the macrocomponents determined in the macrostructure.

• *The macrostructure*

The macrostructure is decomposed according to 3 criteria :

- nature of the activities : informational, decisional, and physical ;

- time aspects of decision making :
  . notion of horizon which is a time interval through which decisions are valid ;
  . notion of period, a time interval at the end of which decisiosns are revised,

- functional nature of decisional activities.

We get thus the representation shown on Figure 8.

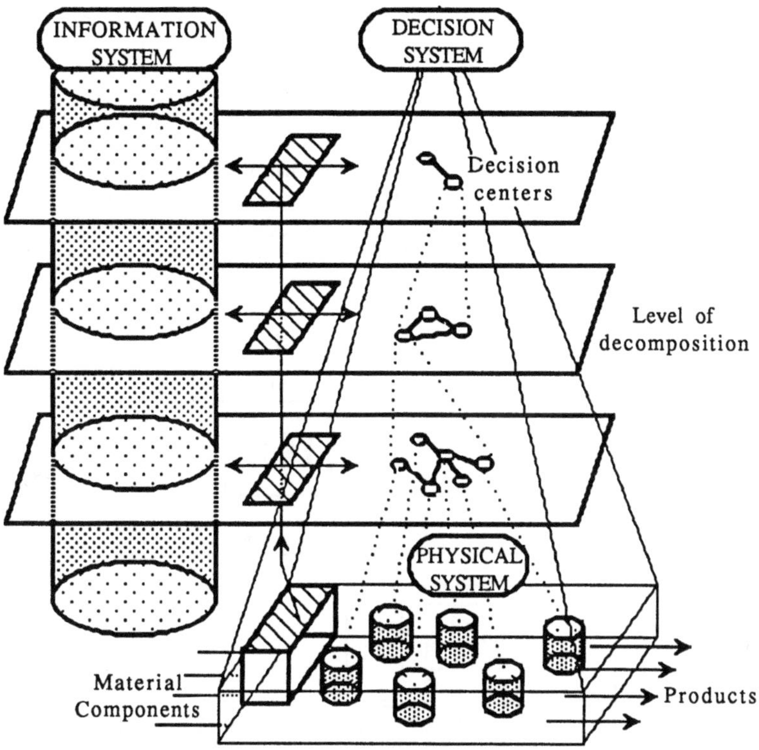

Figure 8 : Macro structure GRAI

We have already described the decomposition into decisional, informational and physical activities in the basic concepts.

The second criterium deals with decisional activities. Manufacturing system is designed in a moving external world for which the states are only known on a limited span of time, therefore decision can be made only on this horizon which caracterizes its lifetime. The notion of period is linked with the notion of adaptation. A regular revision of decisions made is necessary.

The third decomposition criterium regards also decisional activities: they are brought together on an horizon/period according to their functional nature: planning, acquisition, management of resources, design, delivery, commercial or services (tooling equipment, maintenance, handling, storage ...).

The operating of the macrostructure is as follows :

- to each decision center of the structure corresponds a representation of the PS which we call an image and which is a more or less aggregated representation of some activities of the PS. The controlling system comprises the model of the PS and decision systems which are also controlled by the considered decision center ;

- there is a link between the decomposition into Decision Centers (DC) of the decision system and the decomposition into workcenters of the PS. Both decomposition have an inclusive feature (Figure 9) ;

- the links between DC are made of information of various kind :
  . decision frame : information providing the coordination : goals linked with the notion of required performances, the orders which allow the DC toreach the goal, decision variables
  . service information : information coming from or arriving to the DC.

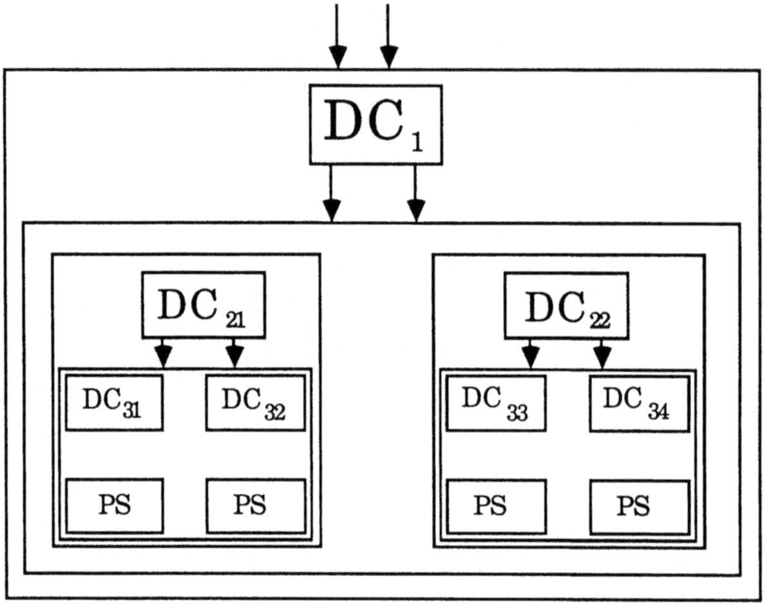

Figure 9 : Inclusive decomposition of decision centers

The structuring of decisional system allows to decentralize decision making and to provide at the same time a coherence towards the goals previously set. Actually we cannot decentralize decision making without making sure that all DC lead the PS towards the selected goal.

• *The microstructure*

For the DC we have chosen SIMON and J.L. LEMOIGNE's model: "Intelligence, modelling, choice". The decision process is developed iteratively according to three separate stages:

- the stage "Identification or Intelligence": to investigate the environment in order to identify the situations calling for a decision ;

- the stage "Modelling or Designing": the decision maker gathers some information, he puts them together on a model in order to find out several potential solutions;

- the stage "Choice": the decision maker picks out one of the solution according to the elements he has been given (criteria, constraints, decision frame).

This procedure is iterative when we don't get any satisfactory solutions.

The structure of the center is shown on Figure 10.

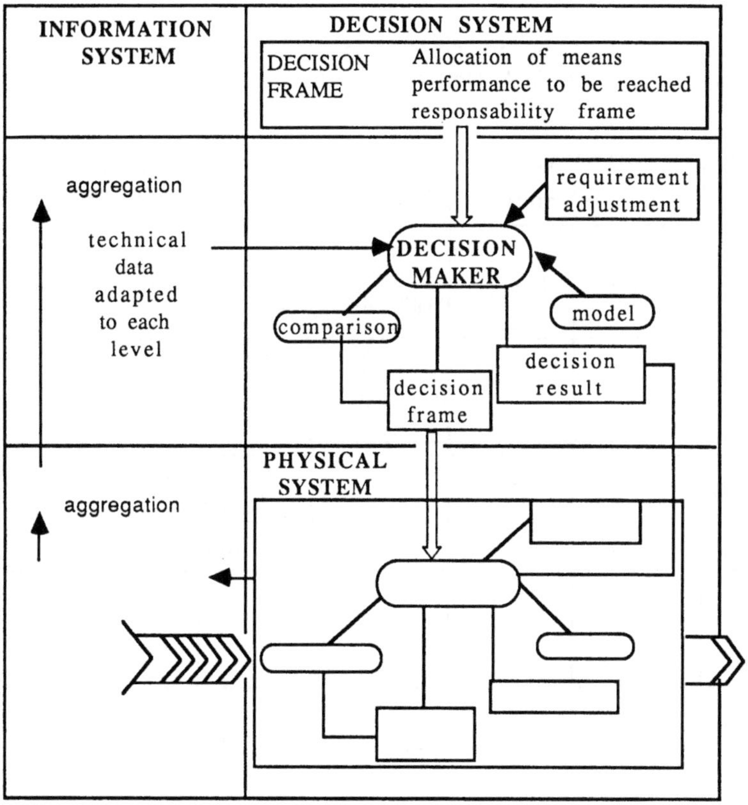

Figure 10 : Conceptual Model of decision center

In order to analyze in details the structure of a DC, we have looked for a tool able to represent the activity concepts. We have thus developed the theory on discrete activities.

An activity "a" is an operation δ (arrow) which changes an initial state q_o (circle) into another state called final q_1 (circle), this transformation is performed through a support x (square).

The mathematical definition is : an activity is an automate : $a = (q_o , \delta , x , q_1)$

in which: $q_o, q_1 \in Q$: given set of states,

$x \in X$: given set of supports,

$\delta \in \Delta$: given set of operations.

δ : $q_o \wedge x \dashrightarrow q_1$

Δ : $Q * X \dashrightarrow Q$

q_o and q_1 are vector variables, x is a vector parameter.
We have also define some basic structure.

2.2 - GRAI Tools

When applying GRAI Method, graphic tools derived from conceptual models are used : the GRAI Grid (Figure 11) and the GRAI Nets (Figure 12).

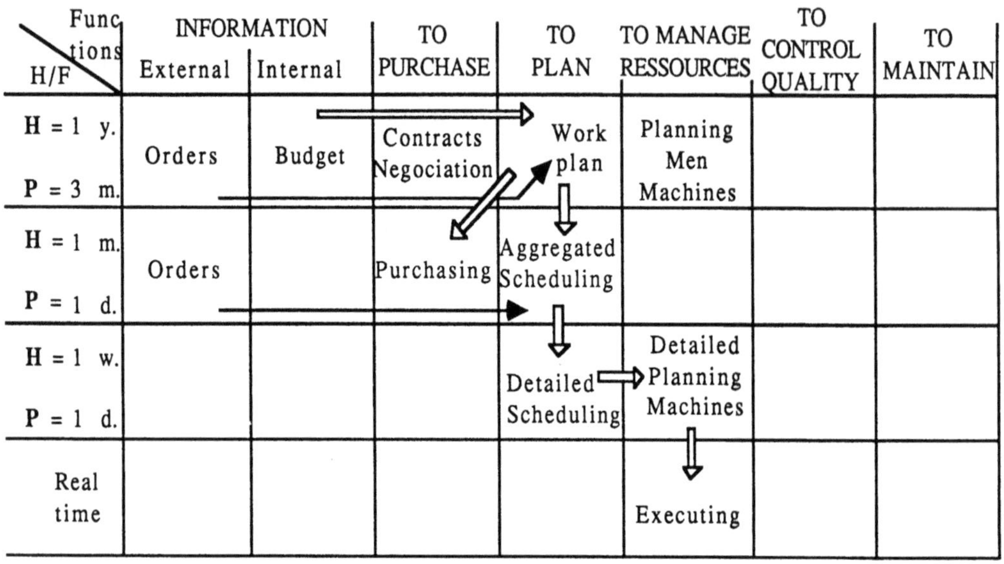

Figure 11 : GRAI Grid

GRAI Grid gives a representation of Macro Conceptual Model with the relations between decision centers. On the X-axis we have the various functions concerned by the manufacturing system and on the Y-axis we have the various level based on the decomposition criteria Horizon/Period. A double arrow indicates a frame betwen two Decision-Centers, the single arrows indicate the main information links between DC.

GRAI Nets give a detailed representation of Micro Conceptual Model.

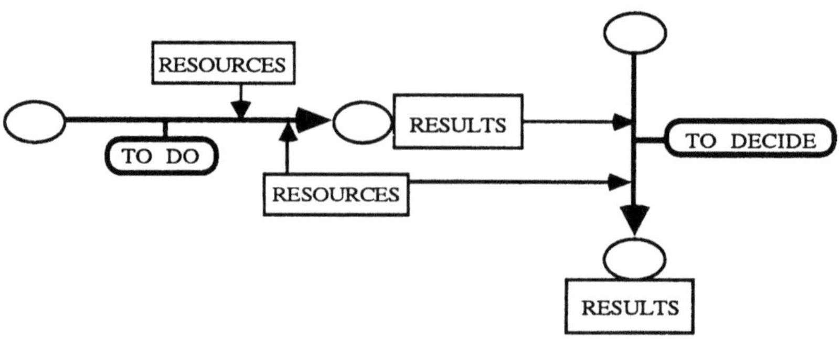

Figure 12 : GRAI Nets

In order to describe the various activities of each decision center we use a graphical representation through GRAI tool.

Through GRAI tool, an activity is represented by an horizontal arrow starting and ending with a circle. A circle represents a state. The activity transforms an initial state into a final state. The name given to this activity is a verb in the infinitive. To perform this activity, it is necessary to know information and means, as both may be the result of the preceding activities. Every information is written in frames linked to the activity with an arrow.

The results (one or several) of the activity are written in a rectangle close to the circle showing the end of the activity.

Remark : Decision activities are drawn the same way ; but a decision making is represented by a vertical arrow instead of an horizontal one. Besides, in that case, the decisional variables to be taken into account by the decision maker are mentioned in a frame.

We can thus represent the set of decisions of each decision center.

As this is done the same way for each decision center of the studied structure, the various links between these decision centers and also between main functions of the structure are thus determined. This representation of the activities of decision centers through GRAI tool is shown in Figure 12.

2.3 - Structured approach (Figure 13)

The application of GRAI method must be structured and abide strictly by good procedures. Its use requires :

- a synthesis group composed of the main future users who will have responsabilities ; this group has to follow the progress of the study, to check the results of the various steps and to guide the design thanks to their technical abilities and their suggestions for the new system ;

- an analyst group (with one or several analysts), whose job is, in particular, to collect all data needed for the study ;

- an interviewed group : people who are interviewed in the bottom up approach.

Some experts on GRAI method, or production management, or techniques needed for the study could be involved.

The GRAI method consists mainly in two phases :

- the analysis phase : to analyze the current system, to collect all data necessary for designing the new system ;

- the design phase : to design the system from data collected during the previous phase, by analyzing the inconsistencies between the current system and the ideal system.

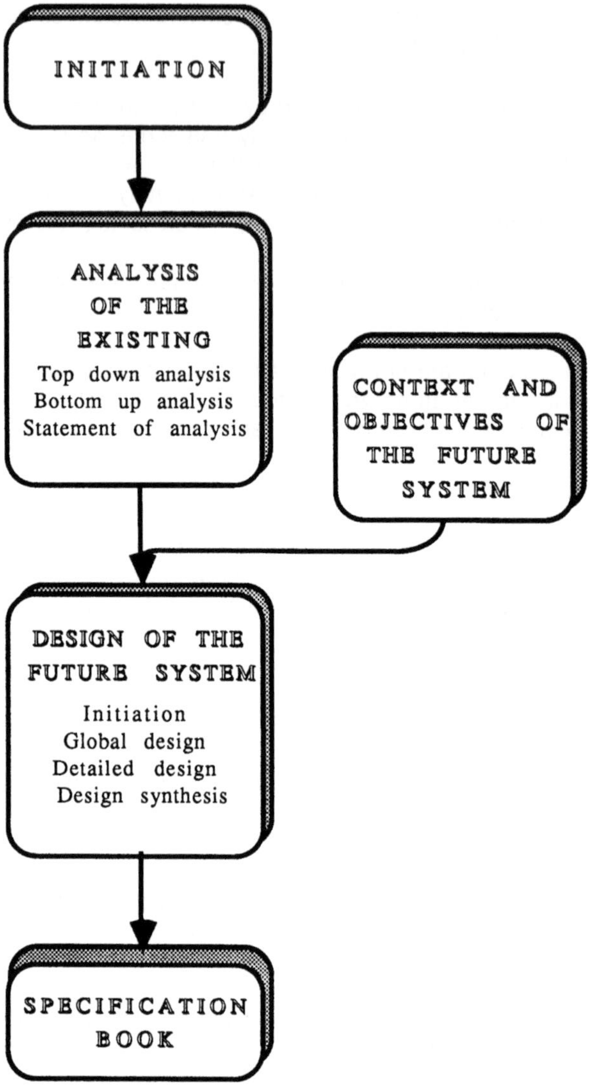

Figure 13 : GRAI structured approach

2.3.1 - Top down analysis

The first step of the analysis, top down analysis, consists in drawing up a frame, using the grid tool, showing the structure of the decision centers of the current system (Figure 13). The various levels are meant for the horizons/periods of decision making specified in the presentation of the conceptual model. The columns represent the various functions of Production Management. These functions are generally : to plan, to design, to purchase, to manage resources, to control, to deliver. It happens that some of them are irrelevant in a particular example as we will see in the example where we have extended the number of functions.

So each decision center is located in the hierarchy and its links with others are defined by two symbols :

====> transmission of a decision frame
----> transmission of information. This symbol is only used for external or very important flows.

Structural defaults may be detected from this grid, concerning wrong coordination between decision centers.

2.3.2 - Bottom up analysis

The second phase on bottom up analysis consists, for each decision center, in analyzing in details :

- decision made in the decision center,
- variables used in making decisions,
- frame of decision (constraints on variables, criteria...),
- constraints issued of controlled decision centers,
- information to make decisions,
- rules between inputs and outputs,
- dynamic aspects of production.

The analyst, a true sensor to collect information in manufacturing system, uses the techniques of interviews for this phase.

This analysis is supported by GRAI nets which represent both activities (decision or process) and information required by the activities (Figure 12).

2.3.3 - Analysing results

The aim of the analysis is :

- first, to take a picture of current system or current technique when we have to design a new system (it was the case for our study at SNECMA) ;

- second, to find out defaults or inconsistencies in Production Management System.

There are three types of inconsistencies :

- transmission of information :
  . no updating of lead time,
  . no knowledge of big orders,
  . very short transmission of information,
  . redundancies of data,
  . inconsistencies between measurement parameters.

- specific to a decision center, examples :
  . no influence of delays on purchasing negociations,
  . inventory level not enough detailed,
  . no "follow up" of stock outs.

- coordination between decision centers, examples :
  . between frequency of general planning and orders,
  . decisions in manufacturing induce perturbations in inventory control.

These inconsistencies may be solved in the design phase in which a new structure of production control system is defined.

2.3.4 - Designing process

The design process starts from two points :

- the objectives given by the management : economic objectives, technical objectives (for example: use J.I.T. techniques, ...) ;
- the list of inconsistencies.

The design phase includes two steps :

- drawing up of a frame representing the new structure of Production Management System ;
- drawing up of the GRAI net which shows, for each decision center, the main specifications of information and decision.

This design method is based on conceptual model. Rules may be defined to guide design but they take into account the external constraints applied to the internal constraints,related to the manufacturing of products (i.e. cycles).

2.3.5 - Specification book

In order to choose software packages or to develop some software for the production management system, a specification book is written in which the specifications of the system to be implemented are described.

This book may be used to make an invitation to tenders to engineering companies or hardware suppliers.

3 - APPLICATION OF GRAI METHOD

This specification deals with the design of an AMS (producing turbine disks) for SNECMA, an important jet engine company and the use of the GRAI method to design the integrated production management system.

In 1982, SNECMA company started a project for a flexible manufacturing system (FMS), in order to manufacture the turbine disks of its engines in the best technical, economic and social conditions.

This project was first intended for Corbeil factory, but as the project was developing, the company decided to build an advanced factory in which the node would be the FMS. The area of Le Creusot (France) was chosen.

The task of designing the integrated production management of this unit appeared very important. The company looked for an approach methodology and selected the GRAI method. The application was carried out by engineers of the company in close cooperation with reseachers of GRAI laboratory and engineers from an engineering firm : GRAI Productique S.A.

This AMS is certainly one of the most ambitious French project with an investment of 60 millions dollars.

After presenting the technical specifications of turbine disks, the production figures and the objectives of the AMS, we will briefly describe the theoretical foundation of the method, and the practical application of its process.

We will finally give the main outcomes of the study, and conclude on the user's opinion about the method.

3.1 - Overall context

3.1.1 - Manufacturing of turbine disks

The turbine disks of SNECMA company which have been manufactured so far by the Corbeil factory but will be made at Le Creusot in the future.

These turbine disks are major parts in the plane engines. They are located at the exit of jet engines and therefore undergo important constraints under high temperatures, their operating imposes very accurate machining dimensions. Their manufacturing requires much care and many quality tests all along the cycle.

These disks belong to an homogeneous family of products, their definitions are constant over the long term. Besides they represent (through their dimensions and machining) a lot of other elements included in the engine ; this may be an opening for the Creusot factory towards the manufacturing of other parts.

3.1.2 - Objectives of the future factory

In this project, SNECMA company has three general aims :

- to ensure its competitivity over the long term,

- to master an advanced technology : design and operating of an advanced manufacturing system,

- to generate an effect of impetus and innovation for the whole factory manufacture.

At LE CREUSOT, this AMS will enable the company to control the costs better (particularly by cutting down the value of work-in-process) and to decrease the cycles. The manufacturing cycle of a turbine disk will decrease from 5 months in Corbeil factory down to 1 month in the future factory. The overall manufacturing features of the Creusot factory are as follows :

- staff: 162 people
- manufacturing output: 150 parts/a month (during 4,5 days a week on a 3/8 shift basis)
- number of part references : 14
- cycle time : 1 month

3.1.3 - Outlines of the study

The analysis and the design of such a project required a global approach. That is why the Creusot factory has been considered as a system, each part of the factory has been carefully defined and analyzed, the links between each part has been specified (in order to point out the integration of the various sub-systems). This approach has necessitated the participation in the project of each department of the company.

The first studies has been carried out on the configuration of the physical manufacturing system, especially on the definition of the flexible manufacturing system.

It was then necessary to specify the whole production management system in order to define the computer system of control.

Therefore the GRAI method has been used. We illustrate in figure 14 the overall approach which has been followed by SNECMA.

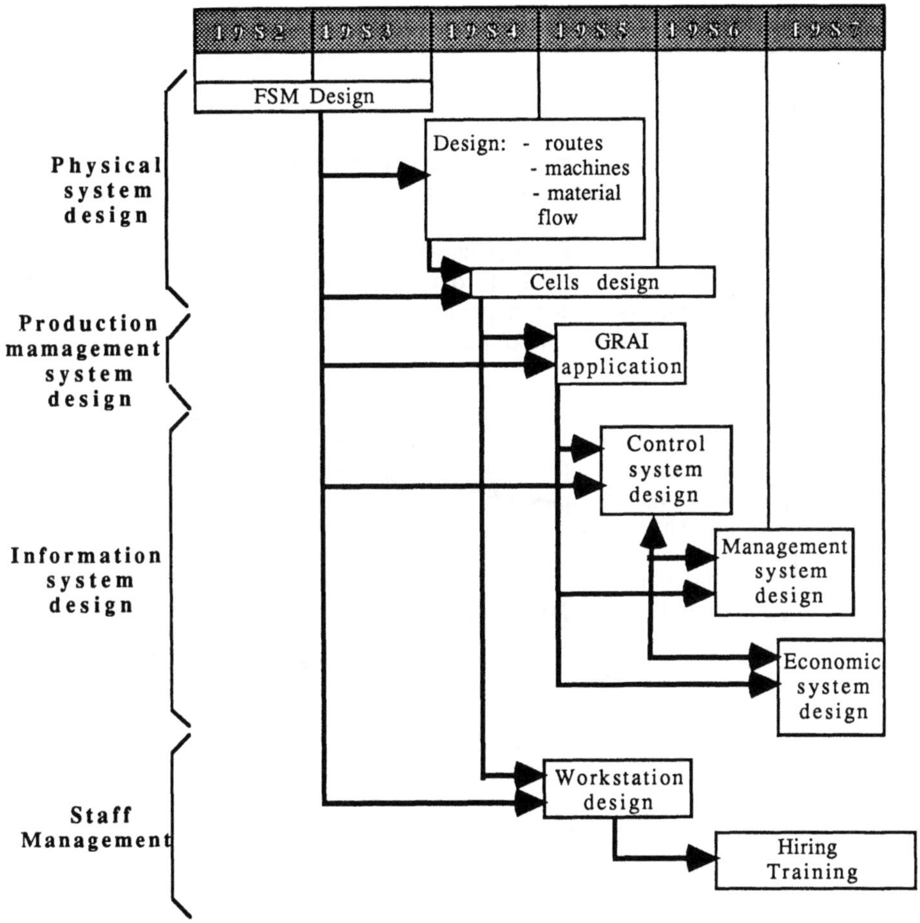

Figure 14 : LE CREUSOT plant design process

3.2 - Application of the GRAI method in SNECMA company

The aim of the study carried out by SNECMA company was to define the specifications of the integrated production management system for the future factory of Le Creusot.

Theses systems are related to the following functions :

- Production management,
- Manufacturing Engineering and process planning,
- Maintenance,
- Quality control
- Cost management,
- Staff management.

We will present successively :

- the analysis phase and the conditions of realization,

- the design phase,
- the use of the results.

3.2.1 - Analysis phase

Usually, the basic principle of the method is to start from current system. This approach allows to take into account the initial constraints and to discuss about their validity.

In our case, we have to design a new manufacturing system. In such situation, we start from a "pseudo-existing system" which is specified from the present techniques utilized by the firm and the knowledge of engineers. In this experiment, we have a good reference: the present manufacturing system to manufacture the disks at CORBEIL plant.

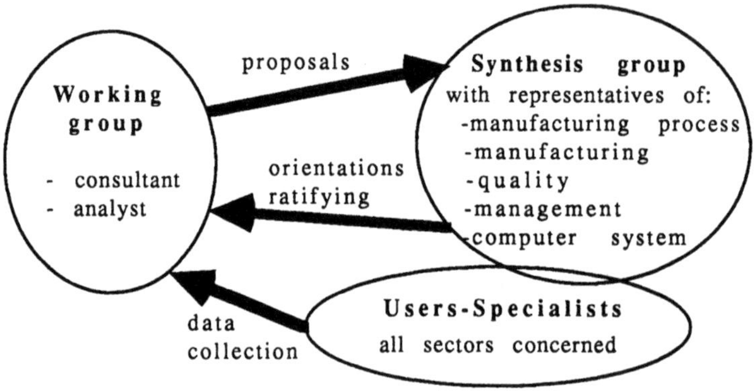

Figure 15 : Study groups organization

So, the synthesis group (Figure 15) was composed of the managers of each function mentioned above plus some future heads of LE CREUSOT plant. We think this composition of the group was ideal for experiences and continuity.

The analysis group was composed of a SNECMA analyst, one representative of the engineering company GRAI PRODUCTIQUE S.A. and one from the laboratory GRAI.

The results of SNECMA analysis are illustrated by several grids drawn by the synthesis group: a grid for the analysis of the whole production management system of Corbeil factory (pseudo-existing system, Figure 16) and 5 grids for the related functions. This "explosion" of the overall grid (Figure 17) into detailed grids is rather original and shows very clearly the necessary specifications, especially as far as coordinating and integrating aspects are concerned).

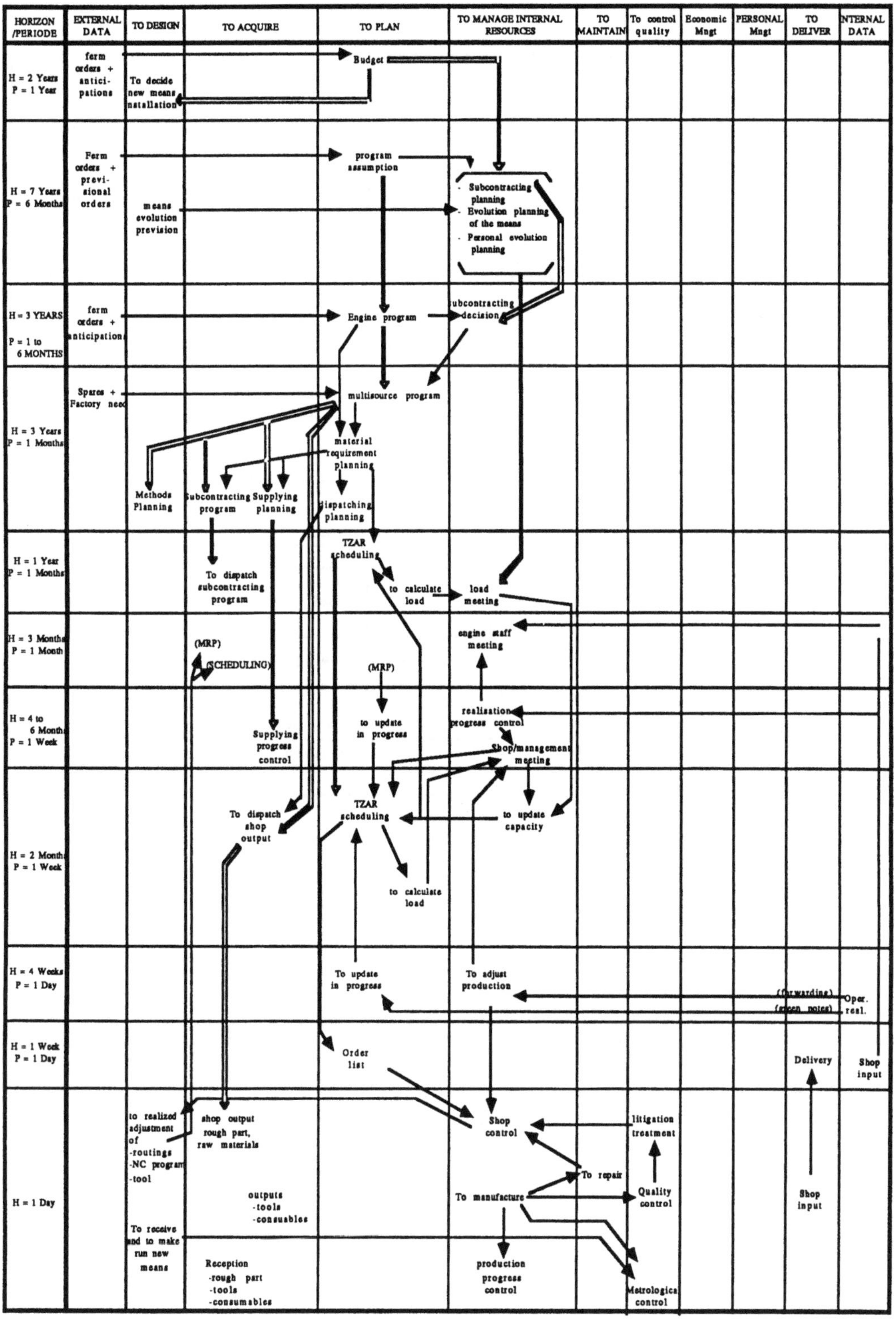

Figure 16 : Top down analysis global grid

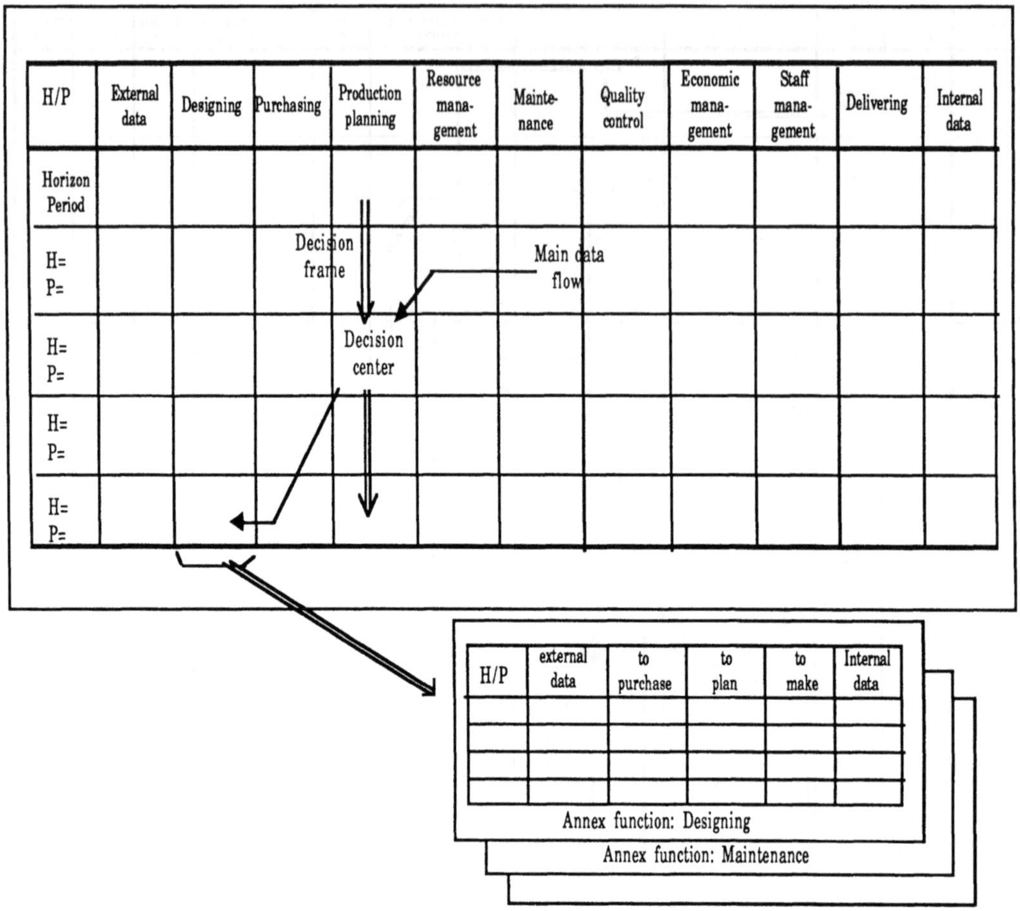

Figure 17 : "Explosion" of the overall grid into detailed grids

The studied related functions are :

- design (process planning aspect),
- quality control,
- maintenance,
- cost management,
- staff management.

We have one example for the function maintenance (Figure 18).

The main decision centers are identified (on the overall grid and on the detailed grids as well), analyzed according to a top-down approach and their activities are illustrated by graphs which point out the present running of the production management system through a bottom-up approach.

The strong implication of managers of the main functions allows to get a lot of significant information to prepare the design phase. We can emphasize also the ability to point out, on the grid, the constraints of integration between functions.

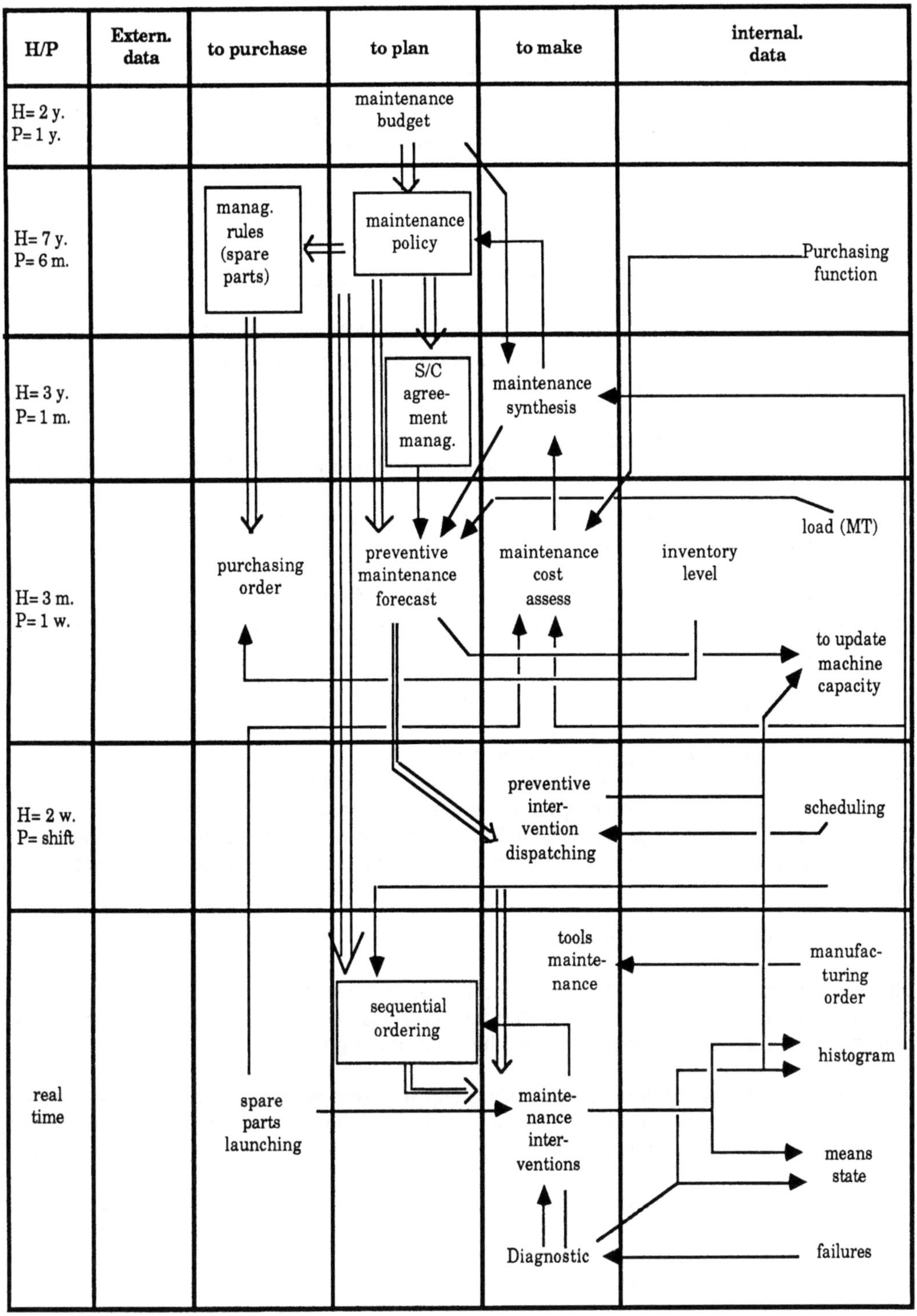

Figure 18 : Grid of the function maintenance

3.2.2 - Design phase

The design phase was specific to LE CREUSOT plant.

The first work was to delimit the functions of this plant, with respect to other company departements. The first level of decision to integrate at the lower levels was the FMS already specified in a previous study.

The design phase starts from :

- the result of the analysis phase with a list of constraints and inconsistencies ;

- the objectives given by the management of the firm: objectives of LE CREUSOT plant (see point 2), integration of FMS, constraints specific to the running of LE CREUSOT, incoherence with the running of the firm.

To study the integration of LE CREUSOT in the firm particularly on the point of view of purchasing , accounting, we start from the analysis grid and we indicate the interface level.

We can see on the design grid (Figure 19) this limit : between level H = 3 years, P = 1 month and level H = 2 years, P = 1 month.

So, the first work of the synthesis group was to build the new grid. If we compare with the analysis grid, we can remark some changes in the value of horizon/period, some additional decision center.

The result of the design phase was :

- general design grid (Figure 19) ;
- particular design grids for the previous mentioned functions ;
- detailed GRAI graphs for each decisional center allowing to check the coherence of the grid and giving the detailed information on the integration (Figure 20). The activities included in each graph were commented and the ones liable to be computerized were indicated.

3.2.3 - Main results

This study started in March 1985 and ended in November 1985. For constraints given by the need to order some equipments we divided the study into two parts, first part to get the global design (end in July) and second part to get the detailed design.

The frequency of meeting for the synthesis group was one meeting every two weeks with a duration of three hours. The load of the analyst was, for two of the three, 50 % of working time, the third one participated only in the meeting and some works with the analysis group.

The result of the study was a Specification Book which gives the structure of the plant system and the detailed specifications with GRAI tools. This Specification Book was send to three selected engineering firms, in order to recieve proposals. The proposals were evaluated in comparaison with the Specification Book.

From the viewpoint of the users who have participated in the study, we can make out the following comments about the GRAI method.

The GRAI method looks quite appropriate to :

- analyse the whole Production Management System and the related functions,
- give a synthetic representation of the studied system,
- point out the hierarchical organisation of the functions,
- enable the future users to validate the specifications,
- allow in the future an easy implementation and use of the system.

The following points must be improved :

- computer support for graphic tools,
- link with the next phase which consist in elaborating the specifications of the information system based on information model,
- systematic study of altered modes.

For the last point, the obligation of finishing the study on time does not allow to perform the analysis in perturbated situation, but this task is included in the method.
For the other point, the on-going work on the computerization of the GRAI method to be studied later will answer in the next future and certainly will improve the use of the method.

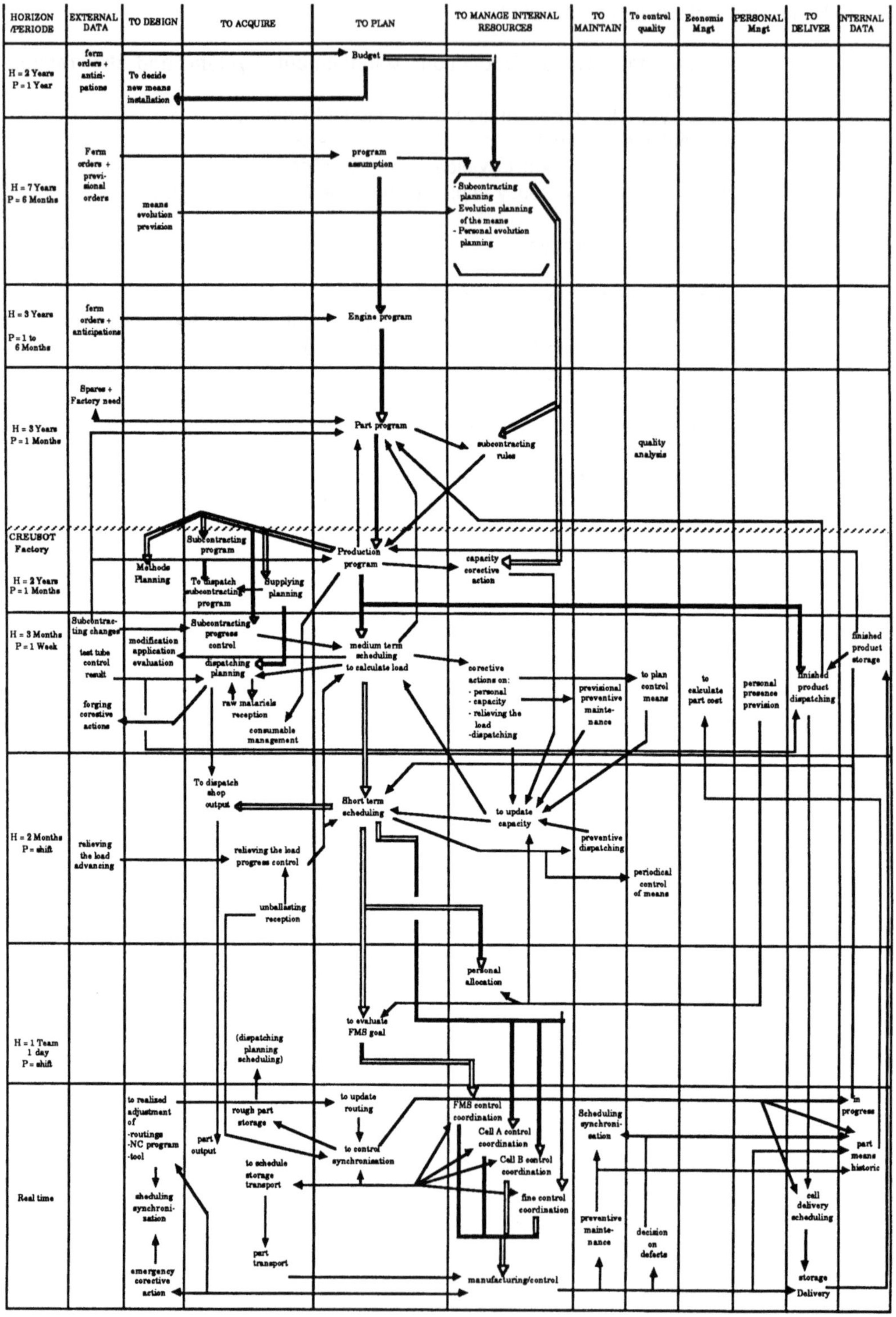

HORIZON /PERIODE
EXTERNAL DATA
TO DESIGN
TO ACQUIRE
TO PLAN
TO MANAGE INTERNAL RESOURCES
TO MAINTAIN
To control quality
Economic Mngt
PERSONAL Mngt
TO DELIVER
INTERNAL DATA
H = 2 Years P = 1 Year
ferm orders + antici-pations
Budget
To decide new means installation
H = 7 Years P = 6 Months
Ferm orders + previ-sional orders
program assumption
means evolution prevision
- Subcontracting planning
- Evolution planning of the means
- Personal evolution planning
H = 3 Years P = 1 to 6 Months
ferm orders + anticipations
Engine program
H = 3 Years P = 1 Months
Spares + Factory need
Part program
subcontracting rules
quality analysis
CREUSOT Factory
H = 2 Years P = 1 Months
Methode Planning
Subcontracting program
To dispatch subcontracting program
Supplying planning
Production program
capacity corective action
H = 3 Months P = 1 Week
Subcontracting changes
modification application evaluation
test tube control result
forging corective actions
Subcontracting progress control
dispatching planning
raw materials reception
consumable management
medium term scheduling to calculate load
corective actions on:
- personal
- capacity
- relieving the load
-dispatching
previsional preventive mainte-nance
to plan control means
to calculate part cost
personal presence prevision
finished product storage
finished product dispatching
H = 2 Months P = shift
relieving the load advancing
To dispatch shop output
relieving the load progress control
unballasting reception
Short term scheduling
to update capacity
preventive dispatching
periodical control of means
personal allocation
H = 1 Team 1 day P = shift
(dispatching planning scheduling)
to evaluate FMS goal
Real time
to realized adjustment of
-routings
-NC program
-tool
sheduling synchroni-sation
emergency corective action
part output
rough part storage
to update routing
to control synchronisation
to schedule storage transport
part transport
FMS control coordination
Cell A control coordination
Cell B control coordination
fine control coordination
manufacturing/control
Scheduling synchroni-sation
preventive mainte-nance
decision on defects
in progress
part means historic
cell delivery scheduling
storage Delivery

Figure 19

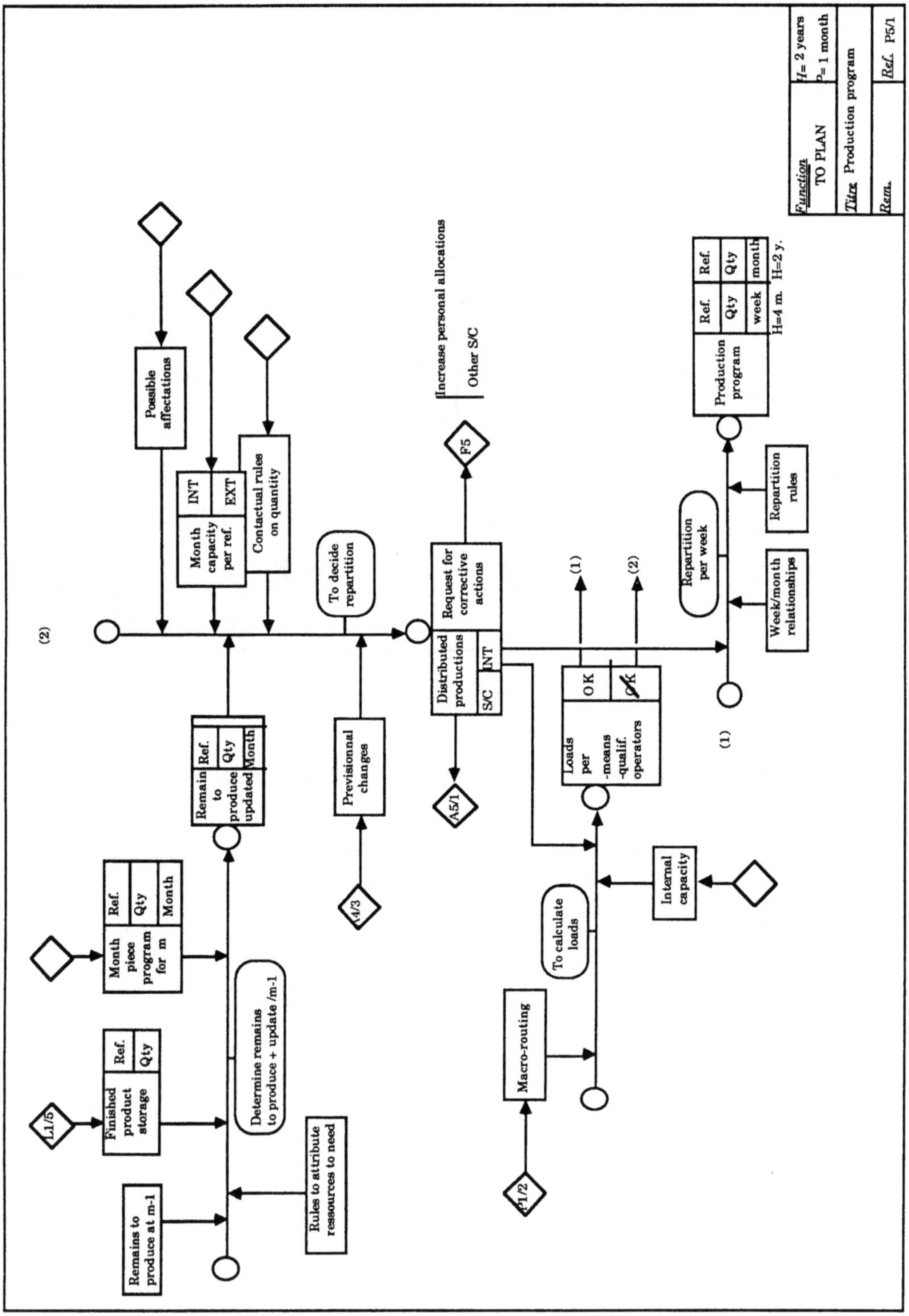

Remains to produce at m-1
Rules to attribute ressources to need
L1/5
Finished product storage
Ref.
Qty
Determine remains to produce + update /m-1
Month piece program for m
Ref.
Qty
Month
Remain to produce updated
Ref.
Qty
Month
(2)
Possible affectations
Month capacity per ref.
INT
EXT
Contactual rules on quantity
To decide repartition
A4/3
Previsionnal changes
Distributed productions
S/C
INT
Request for corrective actions
F5
Increase personal allocations
Other S/C
A5/1
P1/2
Macro-routing
To calculate loads
Internal capacity
Loads per -means -qualif. operators
O K
OK
(1)
(2)
(1)
Repartition per week
Week/month relationships
Repartition rules
Production program
Ref.
Qty
week
Ref.
Qty
month
H=4 m.
H=2 y.
Function
TO PLAN
H= 2 years
P= 1 month
Titre Production program
Rem.
Ref. P5/1

Figure 20

3.3 - Today's CREUSOT factory

The Creusot factory is being set up. The implantation of the physical system (Flexible Manufacturing System and cells) is over (Figure 21).

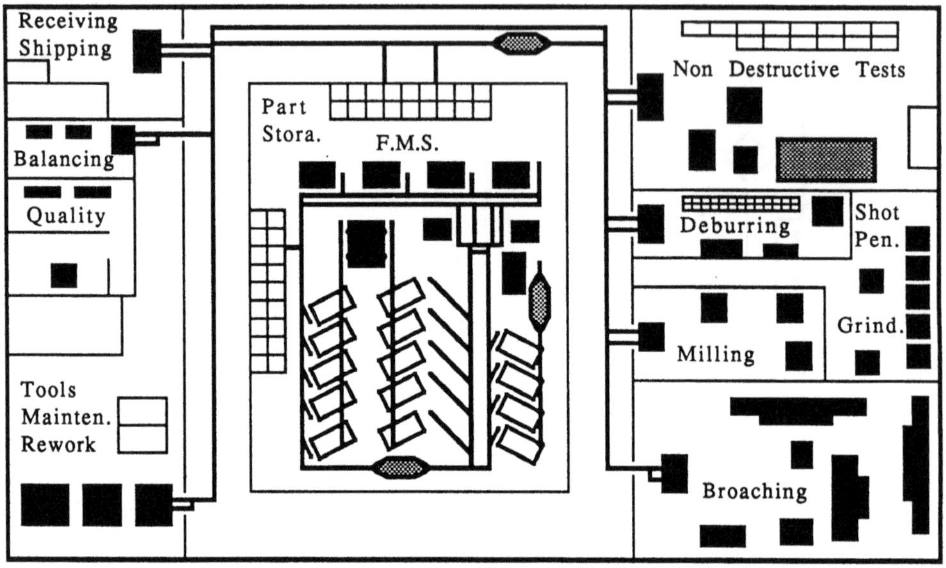

Figure 21 : Physical system implantation

Today the system is operated manually to start the manufacturing, and the company hires and trains its employees.

Meanwhile the computerized production management systems are being developed according to the specification book stemming from the GRAI study.

The first turbine disks manufactured at the Creusot factory are to come out in the last quarter of 1987.

The manufacturing should be fully automated and running around mid 1988.

4 - NEW TRENDS BASED ON GRAI METHOD

The applications of GRAI Method in industry permit us to discover several domains in which some concepts or some approaches developed in GRAI Method can be reused with various adaptations. We will summarize in the following pages some of these new developements mainly :

- the GRAICO Method
- the GRAIFLEX Method
- the generalized GRAI Method
- the computerization of GRAI Method
- the use of GRAI Method for economic evaluation

4.1 - GRAICO Method

The GRAICO Method aims to design the operating system of manufacturing system. At present GRAICO Method deals with continuous manufacturing such as iron and steel industry, chemistry, power plants, etc. but this method will be extented to the whole manufacturing in the future.

The "operating systems" take into account the following functions : supervision of process, control of physical system, technical management, management of information at physical level. These functions are rarely formalized for two reasons : usually production management works in terms of Decisional Activities Planning, so it is difficult to take into account Decisions and Information Processing controlled by events which represent the features of these functions. The second reason is a consequence of the lack of Integrated Design at the level of physical system control: the considered functions are processed step by step and not globally. We show on Figure 22 the position of operating system.

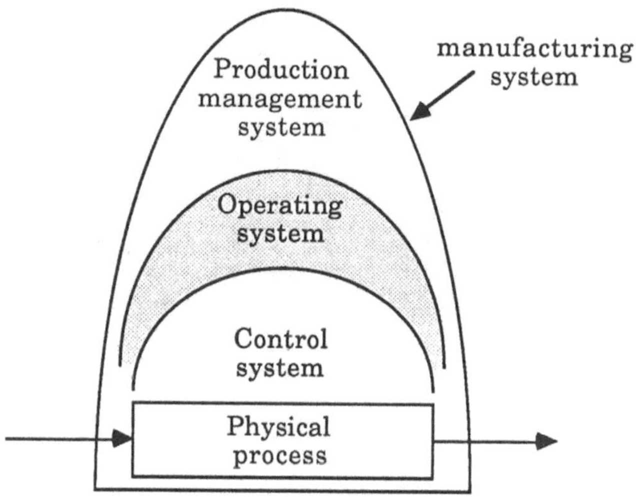

Figure 22 : Layers in manufacturing system

In reference to the GRAI Conceptual Model, we include the operating system inside the Physical System (Figure 23 a), but we can also split up the Physical System into two parts : Operating System and Physical System.

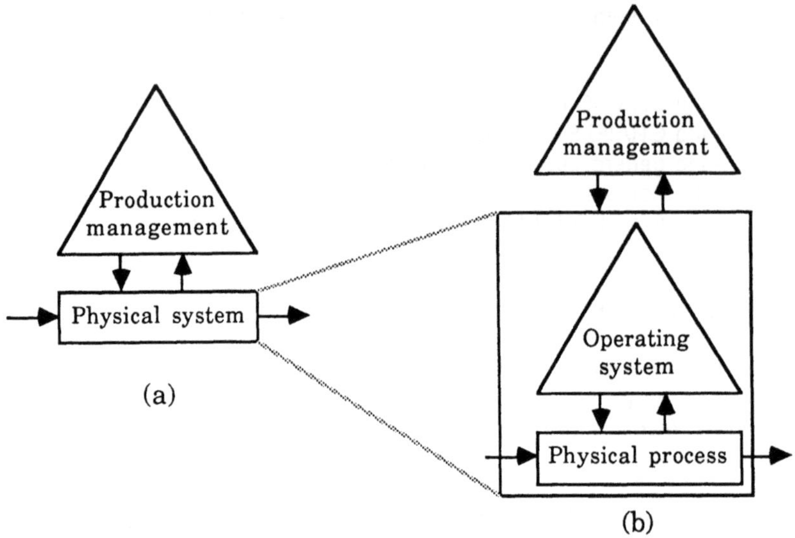

Figure 23 : GRAI Conceptual Model and Operating System

The operating system corresponds to the lowest level of decision system (Figure 24).

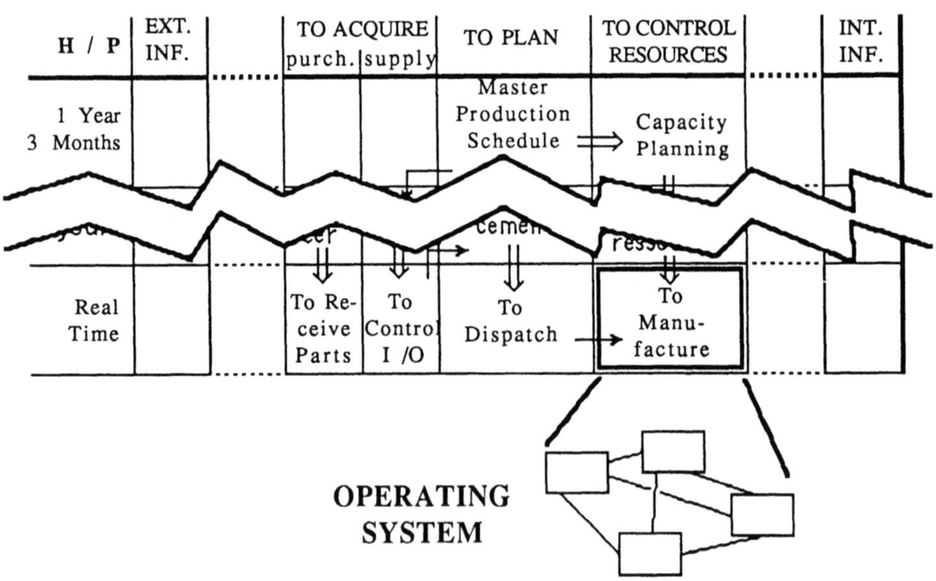

Figure 24 : Link between GRAI grid and operating system

The general structure of an operating system can be divided into two sub sets :

- information on :
  . products,
  . process,
  . models,
  . miscellaneous data.

- decisional and informational processing ; we distinguish two kinds of processing :
  . Internal processing :
    * feedback actions : tool set up, adjustment of manufacturing parameters,
    * coordination : constraint control, diagnosis, adaptative control,
    * self organization : parameters up to date, structure modification,
    * miscellaneous processing : supervision, product progressing, optimization, maintenance, simulation.

  . Interfacing processing : data acquisition on products and process, elementary control, operator dialogue, external exchange.

The figure 25 shows the operating system structure.

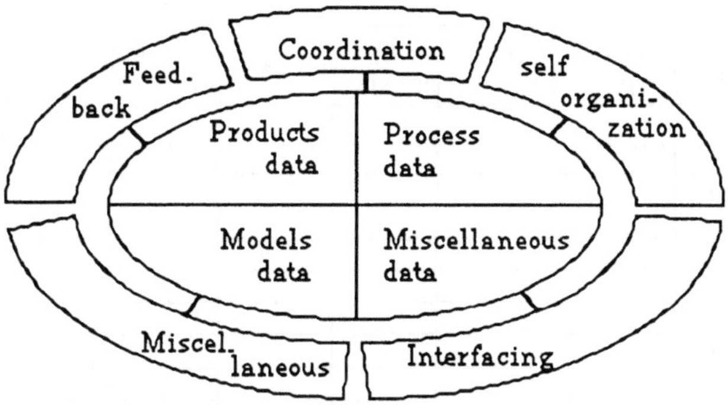

Figure 25 : Operating Systems Structure

Based on this model, a method was developped to design the operating system.

The GRAICO method owns five main steps :

- *Initialization* : preparation of the study, set up of the project management;

- *Process analysis* : the GRAICO method begins by process analysis. This analysis leads to a hierarchical model of the process. Two models are obtained: the existing process in a first use of the step and the process with some adaptations to the operating system at the end of the study ;

- *Operating function definition* : the use of a grid allows to link all the levels of the process analysis (step 1) to the set of operating functional types. The proposed set of operating functional types is very broad. Then, in a particular project, the usable set must be studied ;

- *Detailed analysis of operating functions* : the functions determined in the second step are analysed one after the other. This hierarchical analysis may be very detailed. This analysis particulary models the information processing, data flows between processing and linking (synchronisation) of these processings ;

- *Data structuration* : this last step builds a hierarchical data structure. These data are all the data present in the analysis of all the operating functions. They are grouped in a data dictionnary.

The GRAICO method mainly produces the "functional analysis". This report will be used to software design.

Today, a new advancement of the method lays stress on "security" and "production" goals in the control of the process. This allows an enhanced study of debased situations.

The Figure 26 introduces the grid with new sets of operating functional types following this advancement.

FUNCTIONS / PROCESSUS	CONTROL			PROGRESSING			SECURITY	
	OBJEC-TIVES	COORDI NATION	INDIS-PENSIAL	PROCESS	EQUIP-MENT	ME-MORIES	ALARMS	PROTEC-TIONS
PO								
P1								
P11								
P12								
P13								
P14								
P15								
P16								
P2								
P21								
P22								
P23								
P3								
P31								
P32								
P33								
P34								
P4								
P41								
P42								

Figure 26

4.2 - The GRAIFLEX Method

The GRAI method deals with the decisional sub-system, but a method for designing Flexible Manufacturing System has also been developed : GRAIFLEX. This method allows the design of the physical system in connection with the decisional one.

To describe the decisional system we use GRAI tool and for the physical one we use $IDEF_0$ tool. According to the analyse of the interface between these two sub-systems regarding the control of each activities of the physical system we can deduce the control system and which part of the production system it is necessary to automatize. The design process is very complex : no theory can guide us. We can illustrate this complexity in Figure 27 which represents the design process for an FMS .

In the GRAIFLEX Method we distinguish two parts : an analysis and a design step :

- The first step is dedicated mainly to the analysis of the process plan, the analysis of the current production control system in which the FMS will be implemented and the analysis of the quality procedure. During the analysis phase we define the architecture of the FMS, determining the features of new machines and new control procedures, the performances in accordance with the objectives and constraints.

-The final solution is obtained by :

. structuring in order to determine several possible architectures,
. simulating in order to assess the performances and then choose between potential solutions basing the decision on a dynamic point of view,
. making an economic evaluation in order to validate the hypotheses.

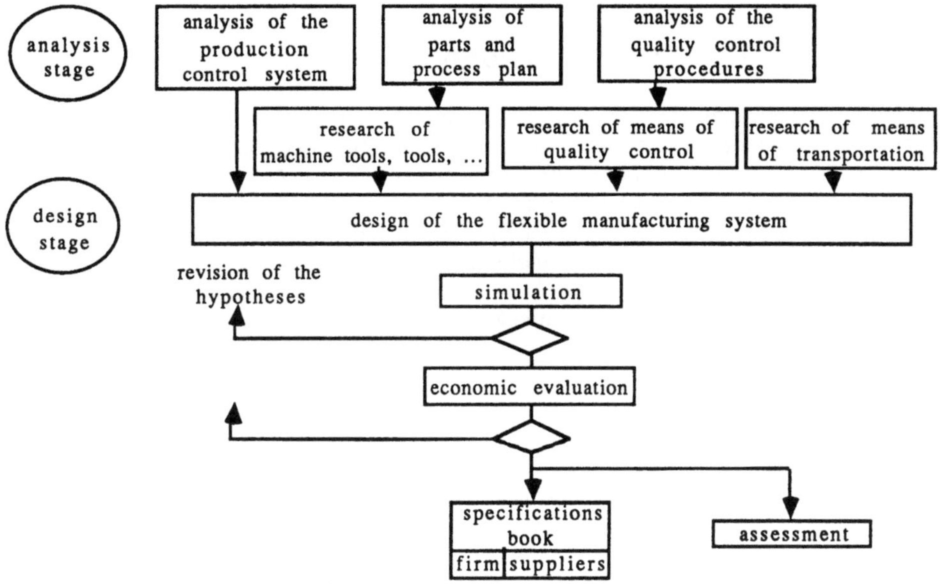

Figure 27 : GRAIFLEX Method

The initial data are :

- a set of parts given by the company,
- the current machine tools,
- the objectives of the company : to cut down inventory, lead time, direct labor, to increase the use of the equipment and its flexibility.

The answer will be :

- the parts which can be machined in the FMS (certainly a subset of the initial set) ;

- the features of new machine-tools, transfer systems, tooling, fixtures ;

- the quantitative specifications of the components of the FMS : number of machine tools, tools, transfer equipment... ;

- the specifications of the FMS management : scheduling rules, monitoring procedures, structures ;

- interactions between FMS and the other manufacturing functions.

As in GRAI method the users are involved in the design process.

This example shows the integration of several tools for the design : structuring of physical and decisional systems, simulation, economic evaluation.

Remark on simulation : we have developed our own simulation tool : SIMGRAI, in order to have an homogenous model: in the structuration phase we use GRAI Nets, for the simulation we derive directly the model from them.

4.3 - The GRAI Integrated Method (GIM)

Through the various research works we have performed and according to the complexity of an Advanced Manufacturing System we have developed an extended method to design AMS. This work was mainly supported by the ESPRIT Project 418 "Open CAM Systems" in which we are associated with PROCOS (DR) leader of the project, LOGICA (GB), CIG (B), WZL from University of AACHEN (FRG) and lately CRIF (B), and by the own research works of GRAI Laboratory with the support of Conseil Régional d'Aquitaine and French Ministry of Research.

The specification methods of AMS we propose have three different components based on GRAI approach (Figure 28) :

- a *conceptual model :* abstract model giving an invariant representation of a system for a problem-solving class ;

- *representation tools* which facilitate the dialog between participants and implementation ;

- a *structured approach* which deals with practical use in guiding step by step the study progress of the development, in order to have a project management.

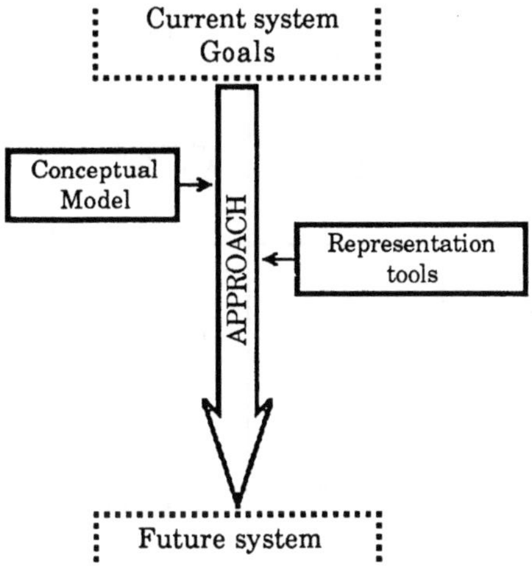

Figure 28 : Components of GRAI Integrated Method

For the GRAI Integrated Method (GIM), according the GRAI Manufacturing model which is divided into three sub-systems : decisional, informational, and physical, we proposed to integrate three methods well defined: GRAI, MERISE and IDEFo [ROBOAM 88.07].

We have already described GRAI Method. We will present now the two others.

4.3.1 - MERISE method (for data and treatment modelling)

MERISE method [TARDIEU 83] uses modelling of framework, activities and goals of organizationin a systemic way.

MERISE is supported by three cycles :

- the life-cycle which consists of :
  . design phase,
  . completion phase,
  . maintenance phase;

- the abstraction cycle which comes out onto a three steps hierarchy :
  . for Data (D): conceptual, logic, physical levels,
  . for processing (P): conceptual, organizational, operating levels ;

- decision cycle which introduces the design choices linked to existing constraints and criteria.

From this decomposition MERISE method build 6 different models : 3 for processing, 3 for data (one per abstraction levels) (Figure 29).

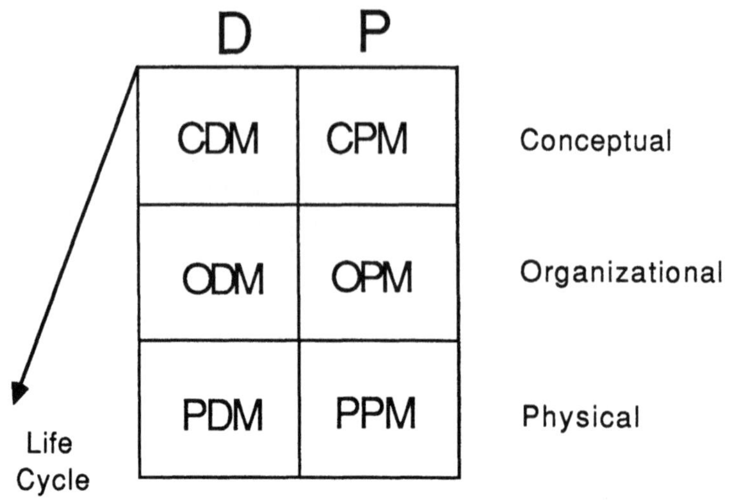

Figure 29

MERISE is split up into six steps :

- The master plan aims to link organisation strategic goals and information needs ;

- The preliminary study allows to get a compromise between conflicting constraints. During this step, information needed to prepare fair solutions to decision makers are collected ;

- The detailed study go from the choice made during the preliminary study to the achievement of the "functional specification book" (general to details) ;

- The completion is split up into two phases :
  . technical study describing data organization, data stores, processing framework and software
  . programming

- The implementation concerns :
  . preparation
  . organization
  . starting up

- The maintenance concerns system adaptation according to user needs, environment and technical progresses.

MERISE uses two types of modelling tools, one for dynamic modelling (processing), another for static modelling of data.

Dynamic modelling tool (Figure 29) derived from PETRI nets. This tool uses three concepts :

- the event : anything occuring,

- the process (operation) : any action following an event,
- the synchronization : starting up process condition.

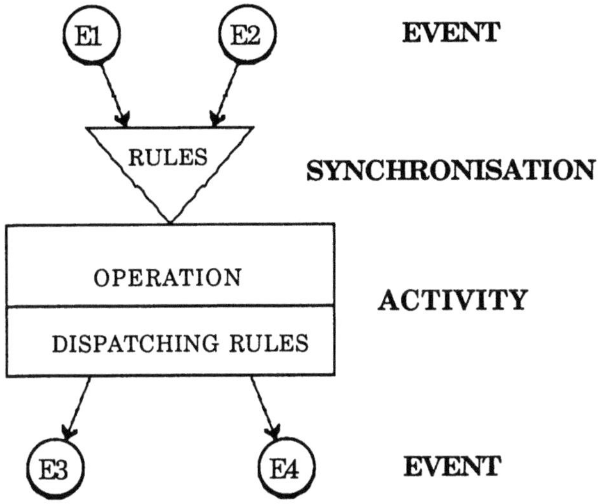

Figure 30 : MERISE CPM

It exists one model per abstraction level and per life cycle step. For "processing" MERISE has determined 3 different models corresponding to the level of details :

- CPM : Conceptual Processing Model (Figure 30)

- OPM : Organizational Processing Model which is a decomposition of each operation of the CPM into tasks, ... (Figure 31) and a correspondance with the physical organization, i.e. the station where the task is done (Figure 32). Besides, we add the degree of automation of the processing (A : automatic, M : manual, I : immediate, D: deferred)

- PPM : Physical Processing Model, which correspond to the realization of programs.

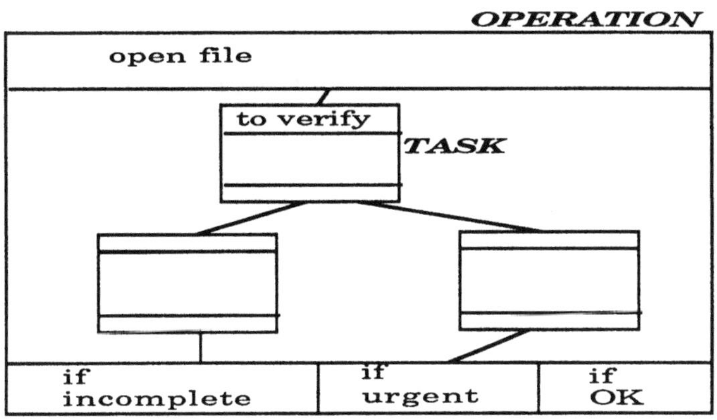

Figure 31 : MERISE OPM

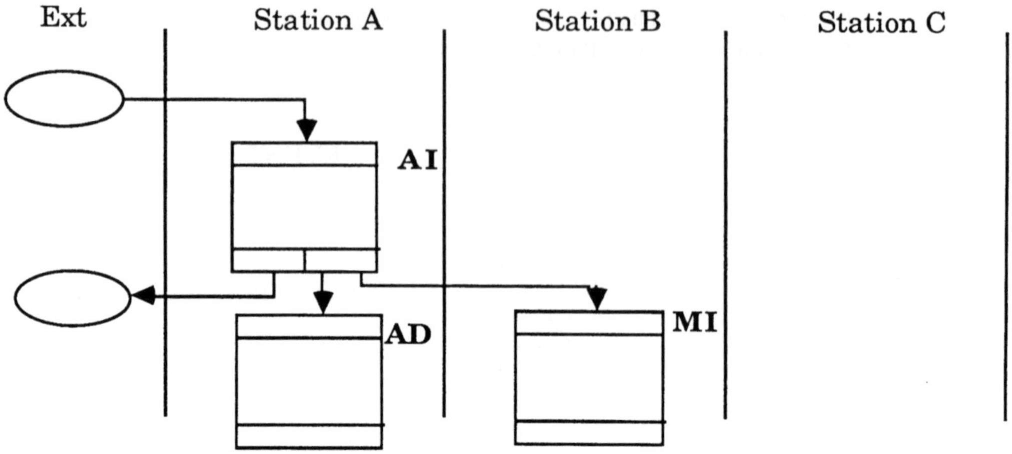

Figure 32 : MERISE PPM

MERISE uses entity/relationship model for data modelling. Data analysis is a methodology which links together the analysis of functions and data into an integrated and structured approach.

We will present now the various kinds of entity types, relationship types and entity modelling.

• **Entities and entity types** :

The building block upon which all of entity analysis is based, is called an entity, and an entity is "anything relevant to the entreprise about which information could be or is kept". An entity is represented by data, it is not itself data. For instance, a drilling machine exists as a machine but its capability, number of tools, availability and so on are just qualities which may be represented as data about her.

A second term used in entity analysis is entity type (Figure 33). An entity covers all entities relevant to the enterprise, which is a given definition.

We can determine several kinds of entity type :

- real entity types : these are tangible objects or things, such as machines, people, buildings...

- activity entity types : these are activities of interest to the enterprise, about which data could be kept, for instance : accident, inquiries...

- conceptual entity types : a business can create or use purely conceptual entity types, both intangible and in some cases unique to the business, which might be : employment, cost center, shop order...

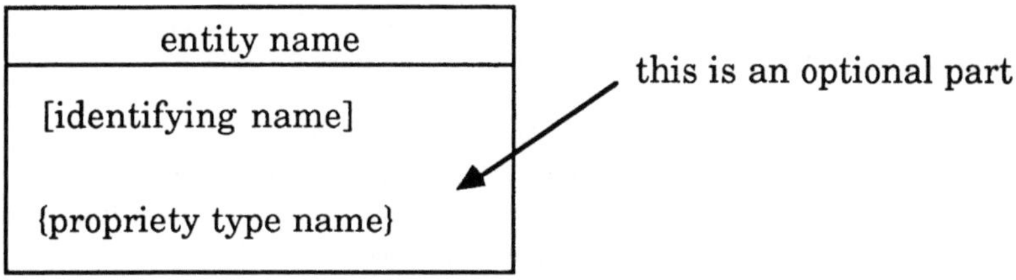

Figure 33 : Entity type

• **Relationship** :

A relationship is "an association between two or more entities which is of interest to the enterprise". Anything that shows or sharpens a connection between two or more entities may be thought of as a relationship.

The associated entities may be of one or two types, but not more than two. A relationship type comprises "all the relationship occurences which fit a given definition" (Figure 34).

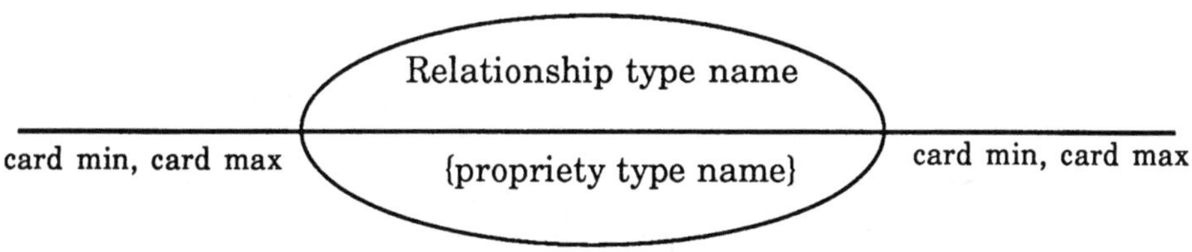

Figure 34 : Relationship

A relationship type does not denote direction. If one were to draw a parallel between relationship types and language the relationship type would be the verb and the two entity types the subject and predicate nominative noun. In language these are reversible using a different verb construction (active and passive). In other words we could just as easily have reversed the relationship type to read and meant exactly the same thing.

We can introduce the concept of degree in relationship. It exists several way to describe this concept. We present there the three main encountered :

- One to one : one entity of one entity type may have that relation type with one entity of another or the same entity type

- One to many : one entity of one entity type may have that relation type with one or more entities of another or the same entity type

- Many to many : many entities of one entity type may have that relation type with one or more entity of another or the same entity type.

It exists one model per abstraction level and per life cycle step. For "data" MERISE has determined 3 different models corresponding to the level of details :

- CDM : Conceptual Data Model (Figure 35)
- LDM : Logical Data Model, which is a modification of the CDM according to the technological constraints on data base or files. At this level we make the choice for the future structure of the data system. We have several choice: data base or files or ... If we want to implement a data base we have several possibilities : hierarchical or relational.

If we select for example a CODASYL Data base we have to modify the LDM according to these rules :

- binary relations 1,1 - 0,n or 1,1 - 1,n become SET type

- binary relations 0,n - 0,n ou 1,n - 1,n or 0,n - 1,n or 1,n - 0,n se become RECORD and 2 SETS types

- n-aire relation types become 1 RECORD and all entity types involved in the relationship type become SET, ... (Figure 36)

- PDM : Physical Data Model, which correspond to the implementation of data base.

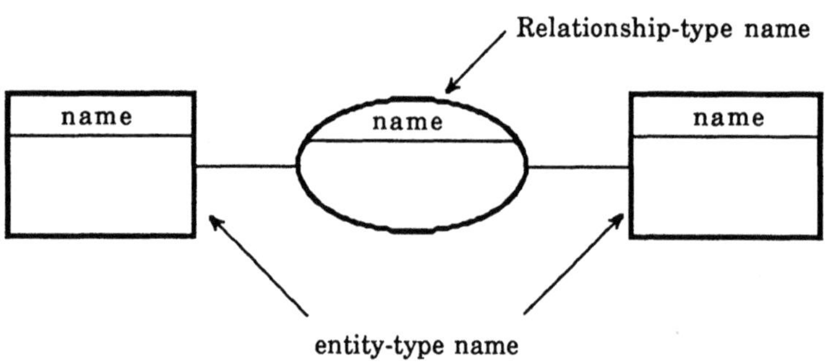

Figure 35 : CDM

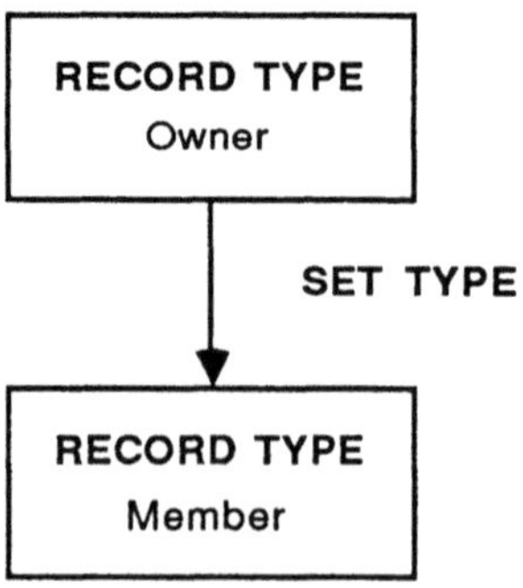

Figure 36 : Example of LDM

4.3.2 - IDEF0 method (for physical system modelling)

IDEF0 method models the whole system through a hierarchical structure (Figure 37).

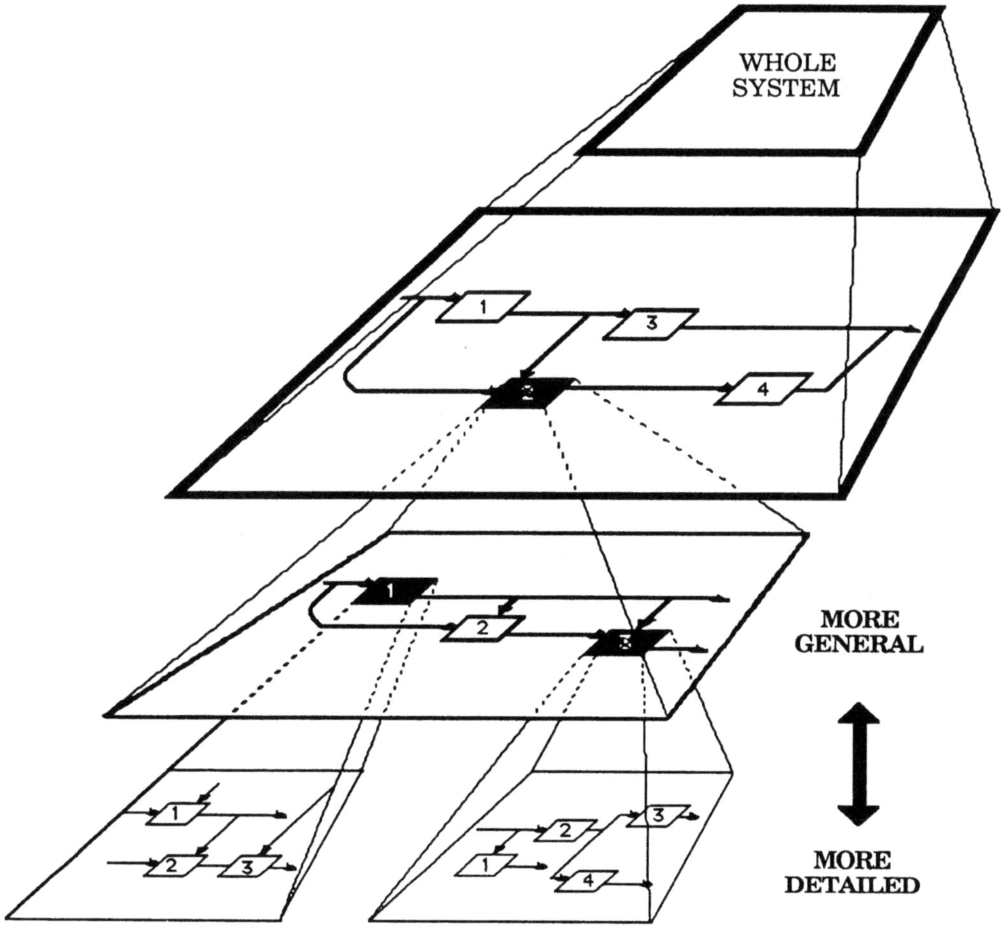

Figure 37 : IDEFo Hierarchical Structure

Besides, IDEF0 method models the whole communication aspect with actigrams. In actigrams, the boxes (Figure 38) represent activities or functions and the arrows represent data (either information or object). There is a breaking down procedure which enables us to perform the analysis according to a top down approach: the diagrams at the high levels encompass a wide range of details so that the words in the boxes and on the arrows indicate an aggregation of concepts. Successive diagrams at the lower levels disclose the details with more specific terms.

In the integrated methodology we use IDEF0 tool to describe the physical aspects. Accordind to this fact, each boxes represent a physical activity such as "to drill", "to mill", ... We have one box per machine at the lower level of decomposition.

The inputs and outputs of the description represent the part flow. Each flow corresponds to a routing. We identify specific output which represent the follow up used by the decisional sub-system. The mechanism represents the machines, people and tools. The control represent the technical data, the orders, ... (Figure 39).

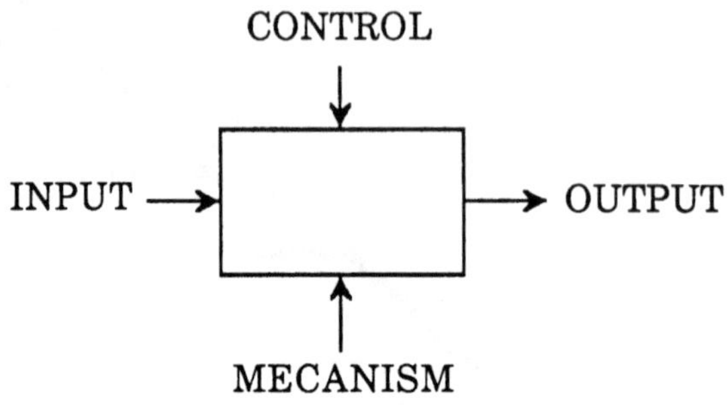

Figure 38 : IDEFo Actigram

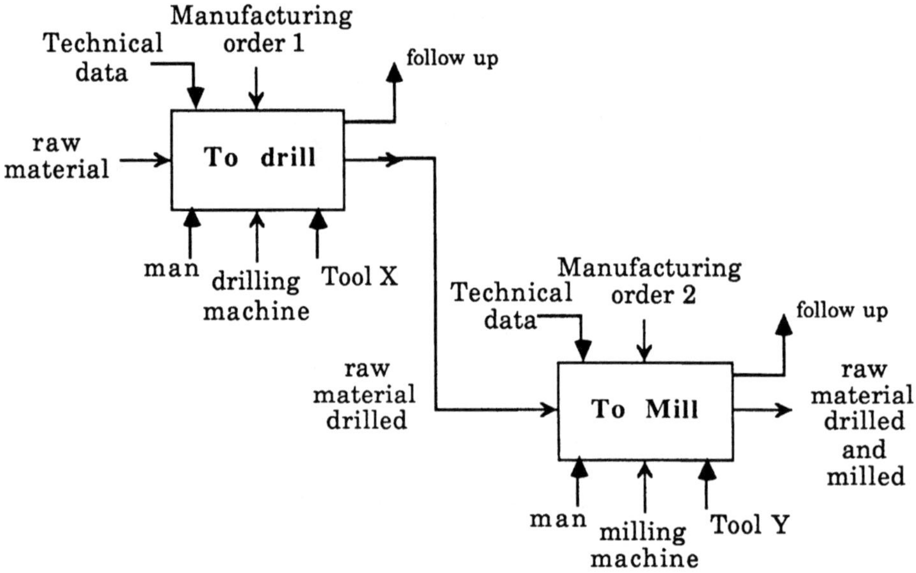

Figure 39 : Example of IDEFo Model

The decomposition of the physical system into several levels allow to identify the structuration of it in factory level, shop level, cell level, workstation level (Figure 40).

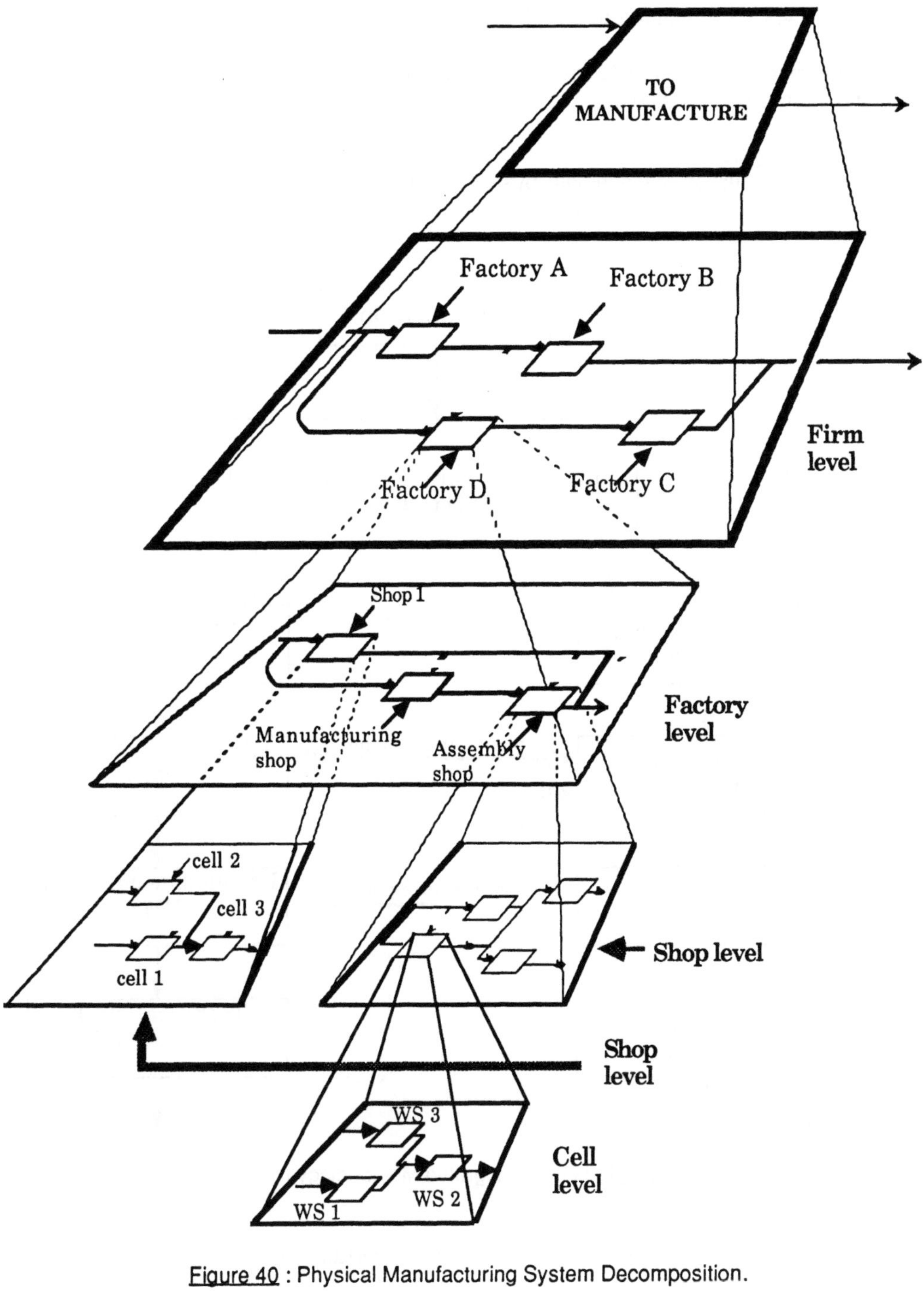

Figure 40 : Physical Manufacturing System Decomposition.

4.3.3 - Structured approach of the integrated methodology

We can summarize the integrated methodology on both Figures 41, 42. We see the different steps and the two parallel studies :

-physical system
-Production Management System (data and processing).

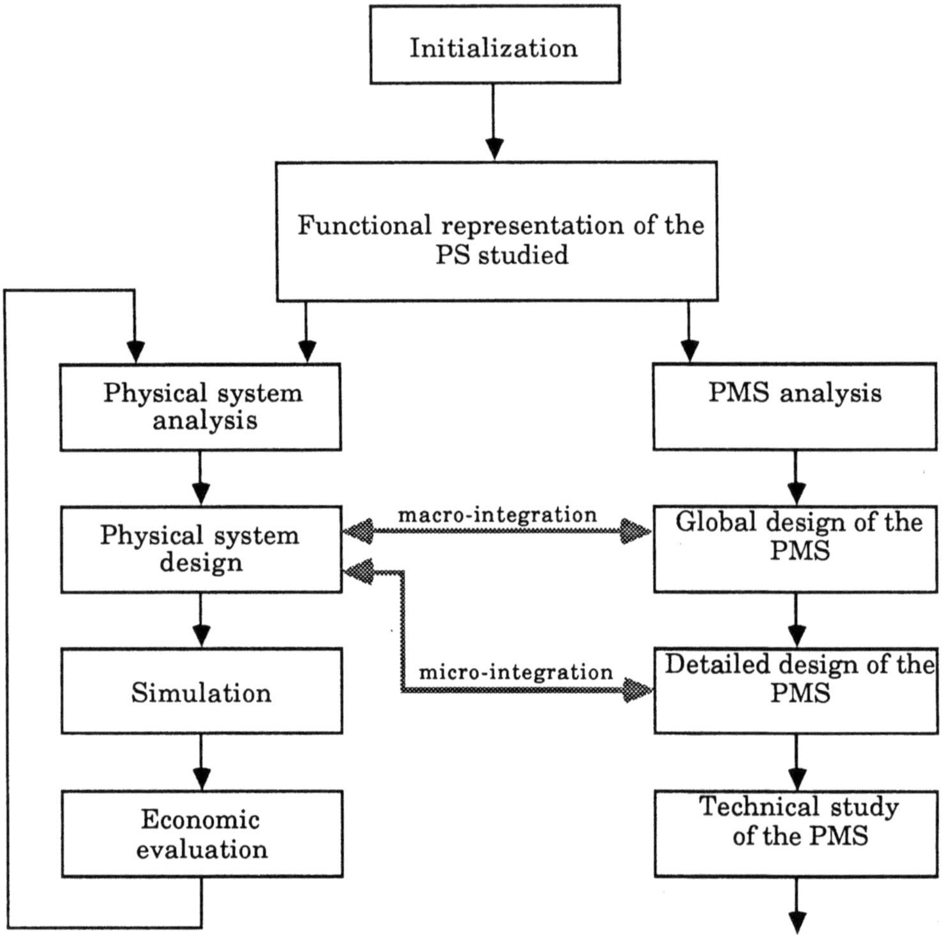

Figure 41 : GRAI Integrated Methodology (GIM)

The first step of our methodology consists in the realization of a functional model which describes (IDEF0 model) the whole system. We only describe on this model the systems and the different flows between these systems. We decompose after the Manufacturing System (MS) into another IDEF0 model and we make a separation between the PMS, described by GRAI-MERISE method, and the Physical system described by IDEF0 models. We can complete this study by the utilization of reference models on the various sub-systems to ensure an "a priori" integration instead of an "a posteriori" integration.

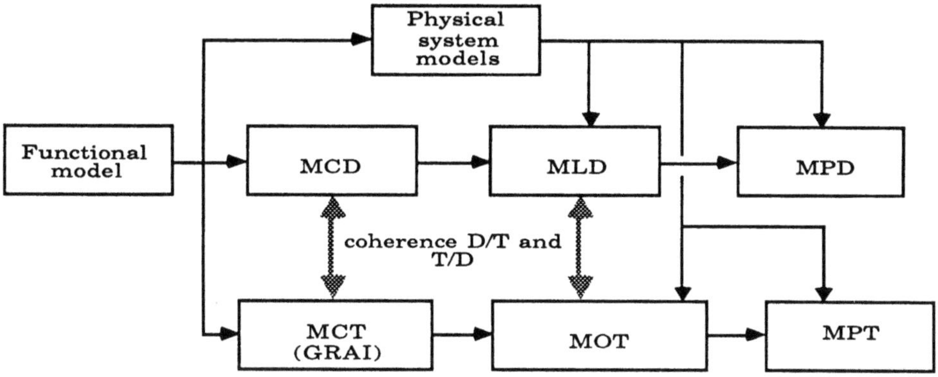

Figure 42 : Structuration of GIM

Once this global study is performed, we have to analyse in details the PMS and the physical system. For that purpose we build the various models determined on Figure 42. To ensure a coherence between the different sub-systems, we have to control and to integrate the models. We use for that some graphical tools: "D/T coherence tools", "T/D coherence tools", "Physical/T coherence" and T/physical coherence". The integration is done with several levels of details. We make at first a global integration called macro-integration and then we make a micro-integration.

• Main steps of the GRAI Integrated Methodology :

Initialisation phase

. first contact with the company management,
. information and training,
. definition of goals and study domain,
. planning of the study.

Analysis phase

Top down analysis :

. realization of the functional model of the global system with IDEF0 tool,
. realization of the GRAI grid (of the existing situation),
. realization of the CDM (of the existing situation),
. realization of the physical system model with IDEF0 tool (of the existing situation),
. planning of the interviews.

Bottom up analysis :

. realization of the interviews with GRAI net tool (of the existing situation),

. adjustment and validation of the GRAI grid (of the existing situation),
. realization of ODM on a limited domain (of the existing situation),
. adjustment and validation of the CDM (of the existing situation),
. adjustment and validation of the physical system model with IDEF0 tool (of the existing situation).

Check-up of the analysis :

. detection of inconsistencies on the GRAI grid and nets using formal rules [ROBOAM 88] and reference decisional models,
. detection of inconsistencies using coherence tools,
. detection of inconsistencies using coherence tools.

Design phase

Global design :

. identification of objectives and constraints,
. inconsistencies analysis,
. proposal for new architecture using GRAI grid tool (of the futur situation) using reference model,
. CDM proposal (of the futur situation),
. proposal for physical system model with IDEF0 tool (of the future situation),
. validation and adjustment of the different models using coherence tools,
. realization of GRAI nets (of the future situation),
. elaboration of the corresponding CPM (using MERISE tool),
. elaboration of the OPM,
. elaboration of the ODM,
. simulation of the physical system using computer tools,
. adjustment and validation of the models.

Detailed design :

. improvement of the OPM,
. improvement of the ODM,
. improvement of the physical model (organizational),
. validation of the models using coherence tools,
. detailed specification of the processing,
. detailed specification of the data,
. technical choice adjustment,
. planning for development and implementation.

Technical study

Software production

Implementation and utilisation of developed computer tools

The three last steps are directly issued from MERISE method [TARDIEU 85].

4.4 - Development of GRAI computerized tools

We have pointed out the difficulties to use design method particularly in the design phase. This situation explains the development of computerized tool to aid the designer in design process.

In that paper, we propose to present our latest research work performed in this domain through the various projects we have already presented in the GIM paragraph (4.3.3).

At first, we will study the needs and constraints met in the specification of computerized tools for GRAI method. Then, we talked about a particular point to be solved : reference models. We will show their implementation using AI techniques and the aid which could be brought in the design process. At the end of this paper, we will give the architecture of the computer tool used to support GRAI method.

4.4.1 - Needs and constraints in the specification of computerized tool for GRAI method

The support of computer techniques to GRAI method can be situated at two levels :

- the first one concerns "utilisation aid" : graphical interface, flexible and convivial graphical editor, project management. At this level, it is particularly a man/machine interface problem to be solved ;

- the second step concerns "decision aid" : the design of such computer tool is more complex and need the utilization of reference models. It consists in the formalization of the "know-how" of expert in production management for two specific phases of the GRAI method : detection of inconsistencies and design. The computer tool, to support these phases, must contain some reference to be able to make proposal to the user.

The computerized tools used must be particularly evolutionary and they must have an efficient man/machine interface. We could make the following remarks for our choice of implementation :

- *Object oriented representation* : we have choosen an object oriented representation because it is possible to manipulate the objects independantly from one another. Besides, this representation allows to create models and to modify them by using heritage models ;

- *Man / machine interface* : in design-aid system, we have two actors: the designer and the computer. They have to exchange easily information about concerned objects. So, the design-aid system must use graphics, mouse, multi-windows, ... ;

- *Link between man / machine interface and object oriented approach* : the structure of the computer model must be identical to the decision maker's mental model. According to the definition proposed by Zeigler, it must be structurally valid [ZEIGLER 76], [BERARD 87]. The object oriented approach allows to build such models in which data produced by a computer model are generated as the real system generates the true data. So, we propose to integrate object oriented model, man / machine interface and the procedures to use the system.

4.4.2 - Reference models

All design process activities are decisional activities. Each decisional activity, without taking into account the application domain, is performed exactly in the same way, using the same process. For each decision, the designer try to use reference which allow us to understand the reality to study. This understanding is done, generally, by a simplification of the real system to analyse and by comparison with reference models. These models represent the "know-how" of experts on the studied domain. If we arrive to build these models, it will be possible to use them for the design process because, at the present time, all design activities are performed by unconscious utilization of reference models.

GRAI method computerization needs the formalisation and utilisation of reference models. These reference models are used to support the utilization rules of the GRAI method. We can distinguish two kind of rules :

- the formal rules which are well defined and have been built according to the work on conceptual models and on concrete applications of the methodology ;

- the specific rules are not so well defined. These rules are, at the present time, strongly connected to the "know-how" of the analyst which make the study.

The classical steps of the GRAI method can be summarize on the Figure 43. It will be possible to use the reference models for two specific phases : detection of inconsistencies and design phase.

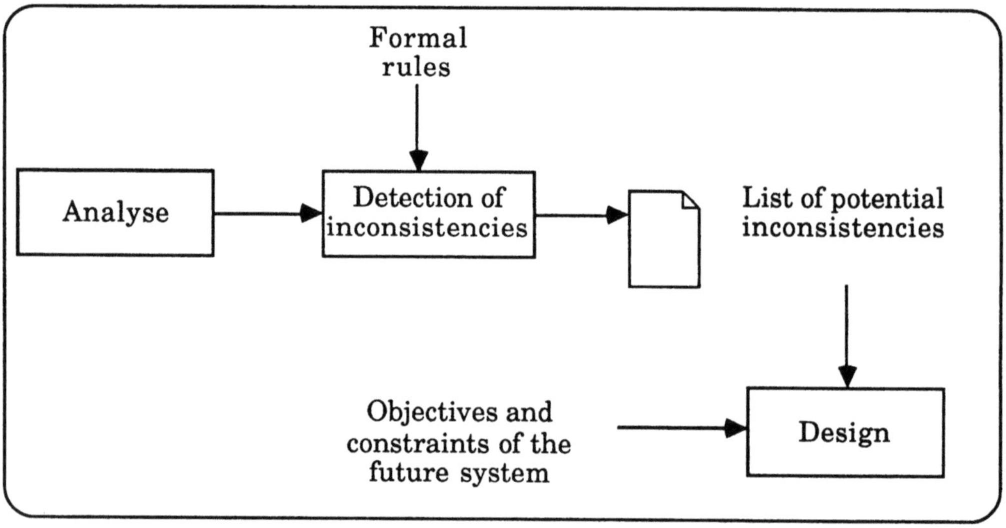

Figure 43 : Classical steps of the GRAI method

We have developed several types of reference models (Figure 44) according to their use in the decision aid procedures [ROBOAM 88.07] :

- Conceptual Global Decisional Reference Model (CGDRM) : it corresponds to a general vision of Production Management Systems (PMS). This model can be identified to conceive and describe the characteristic elements of all PMS. It is composed essentially of rules related to systemic problem approach ;

- Conceptual Particular Decisional Reference Model (CPDRM) : it allows to particularize the CGDRM to a specific PMS class. In [MII 86], a classification of PMS is proposed. Six classes of PMS have been identified and for each of them, we can built a CPDRM. These models have a character dedicated, because it describes a particular class of PMS and also a generic aspect because it is representative of all PMS included in the class. The CPDRM must be open, it means extensible and interconnectable on a specific class of problem. For that reason, it must contain parameters independant of a specific problem (specific factory). For example, the concept of H/P level can't be value because this is a specificity of a dedicated PMS. The utilisation of such a model on a specify PMS of a factory need an adjustment of the parameters. We have to particularize the CPDRM to a specific PMS. We have to derive from the CPDRM a specific model we call SDRM : Specific Decisional Reference Model ;

- Specific Decisional Reference Model : this model has the same structure than the CPDRM but we have valued the parameters of it.

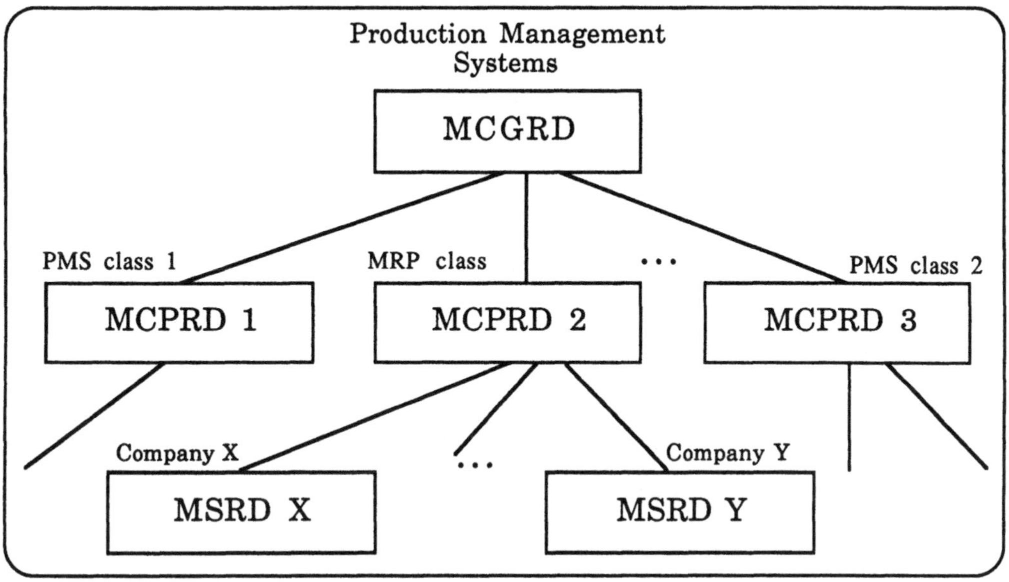

Figure 44 : several types of reference models

We are going to present the way to use the reference models into the two concerned steps of the structured approach of the GRAI method.

• Detection of inconsistencies phase

The Figure 45 provides a summary of the way to use reference models at this step. We have at first to identify the class of the studied PMS. For that purpose, we can use existing typology and derive from that the corresponding CPDRM in the CPDRM base. We have attached to each CPDRM a set of rules describing the specificities of the described PMS. With these rules and by a comparison between the structure of the selected CPDRM and the analysis model of the studied PMS, we can issue a set of inconsistencies. The comparison between these two models is easy because they are formalized and described with the same tools : GRAI grid and nets.

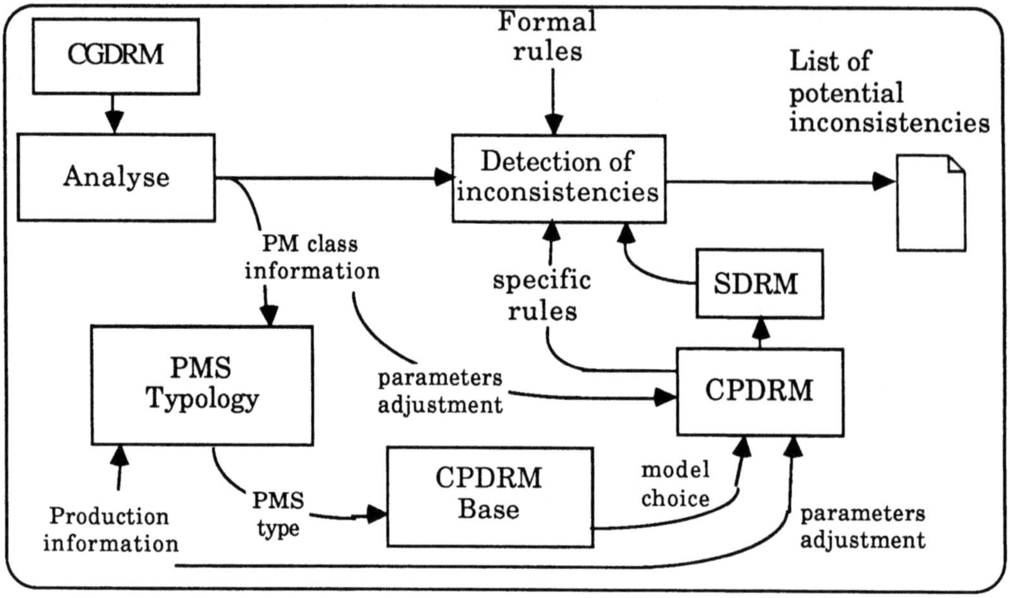

Figure 45

• Design phase

This step is the most difficult of the GRAI method. It exist no methodology which allows to define the evolution of the existing structure towards the design architecture.

The utilization of the reference models will be the same than in the previous step. We will apply an iterative approach which consists in a comparaison between the existing situation and a selected CPDRM. According to the list of inconsistencies, the objectives and constraints of the futur PMS we will guide the evolution of the existing situation towards the structure of the CPDRM which represents a "ideal" situation of the PMS.

4.4.3 - Computer implementation

Within the framework of the french AIP Project [AIP 88.05], the GRAI Laboratory has worked on the specifications of a computerized graphical editor for the utilization of the GRAI Method. Besides this computer tool will allow to apply inconsistencies rules.

To perform this work we have established a classification of the rules needed for the application of the GRAI Method. We have determined two different types :

- The rules for the representation analysis : it means rules to ensure syntactic verifications for several representations (grid, nets). This is necessary when an object can be examined on several points of view ;

- The rules for running mode analysis: it means rules to verify the running mode according to reference models. For example, the conceptual model of GRAI Method provides a semantic frame to some inconsistencies rules.

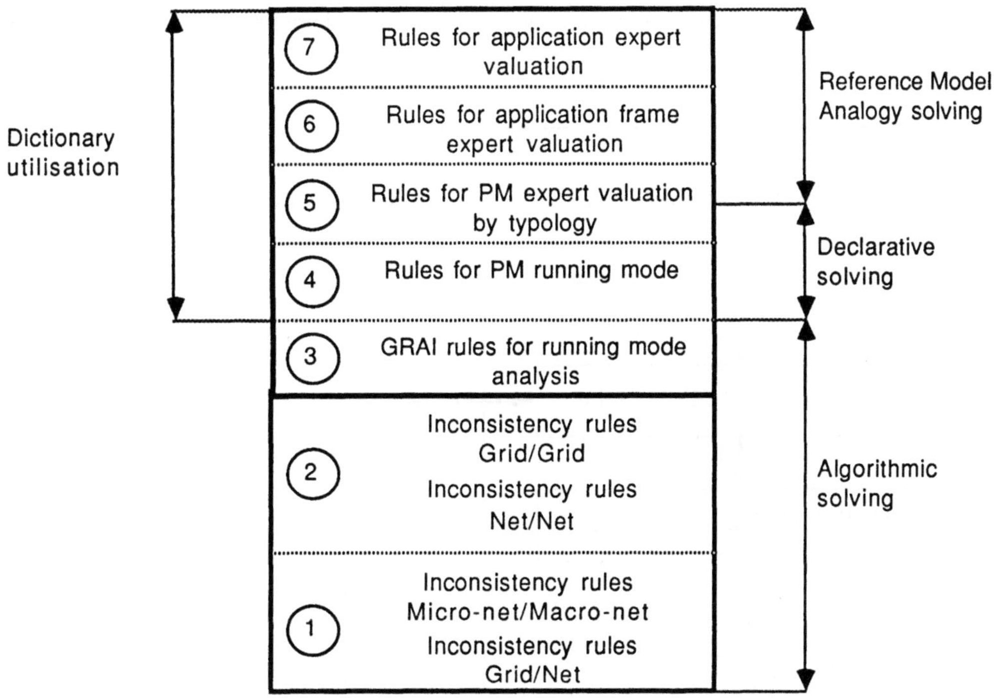

Figure 46

The Figure 46 presents this typology. We can remark the link with reference models. Indeed, the three first rule types correspond to formal rules of GRAI method. The specific rules of the classification (level four) related to Production Management are an important part of the CPDRM.

The rule classification is used for the computerization of the GRAI method. With these rules we intend to have an automatic detection of inconsistencies. At the present time, only the four first levels have been studied in details.

A convivial programming language allows the programming and implementation of rules.

But, this language has to take into account several solving mode :

- the four first levels concern algorithmic solving,

- the rules on PM and representing "know-how" of experts, use a declarative mode. For this second mode it will be necessary to use an inference engine,

- the next rules represent the reference models and may be used by analogy.

We have also a computer tool which allows the detection of inconsistencies. These inconsistencies are deduced from an ideal and global production management model (i.e. the CGDRM). Besides, it will support more specific reference models (SDRM) which allow a complete analysis with less global parameters. Our choice of implementation is given in the next lines.

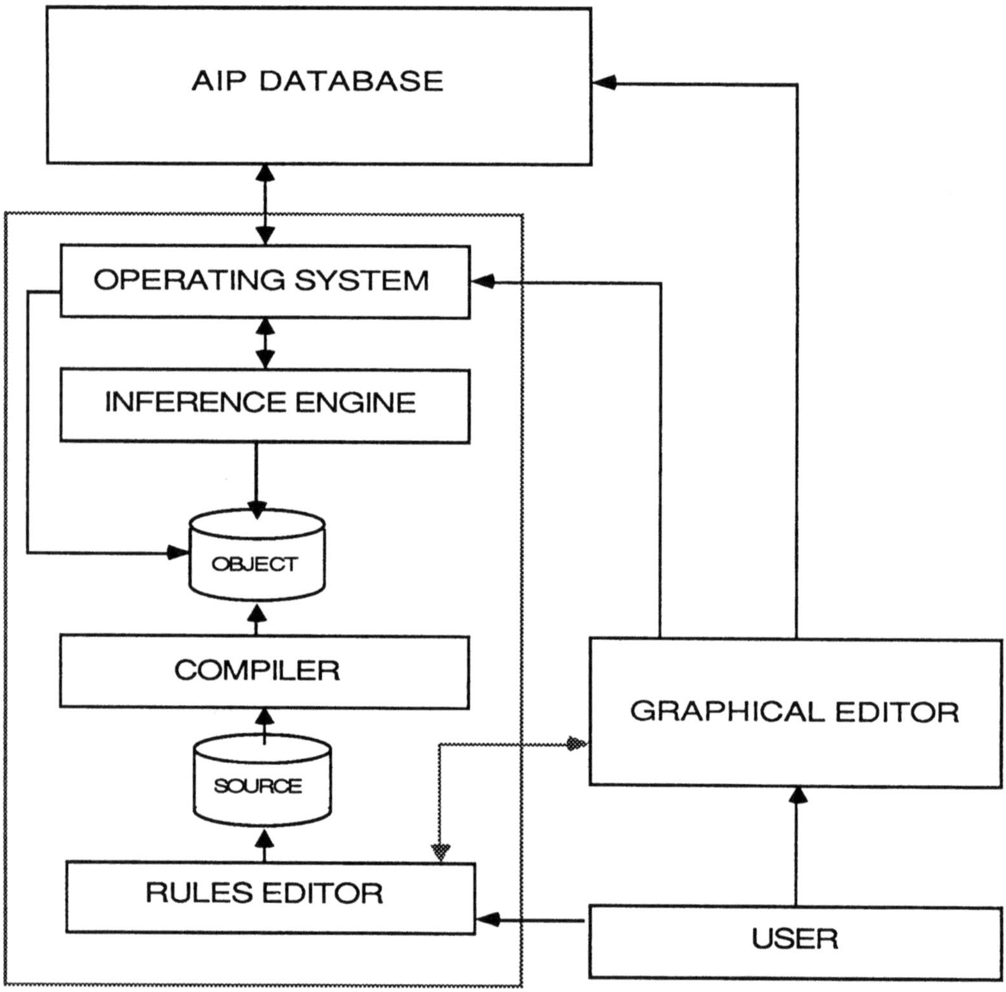

Figure 47 : Project architecture

The project architecture (Figure 47) contains the following software components :

- an oriented object database,
- a graphical editor,
- a detection of inconsistencies module.

The design and the development of such tools are realised with an oriented object philosophy. This is relatively easy to associate attributes and methods (for example, the control of each object) to a methodology. However, we need a module to establish the syntactic and semantic control of a designed modelization. The detection of inconsistencies module provides it. The prototype is going to be developed with computer languages LISP and PROLOG.

The detection of inconsistencies module contains :

- The operating system : it provides the "command" part of the different components of the module. It allows the design of rules, meta-rules for the control, the control at all, ... ;

- The inference engine : it allows the inference of declarative rules. The type 5 rules and some type 6 rules are inference rules which are generated in PROLOG language. The low level rules (type 1 to 4) are translated in LISP language ;

- The compiler : provides the lexical and syntactic consistency of each program describing one or several rules. It will be implemented by using PROLOG language ;

- The rules editor : It supports the rule programming. It must be very convivial and very easy to use. It also provides some syntactic and semantic control during the rule capture.

4.4.5 - Conclusion

The development of computer tools to support GRAI method is now in active phase. But more and more we carry on the task, we discover the complexity to go from the use of GRAI method by a human being to the use with computer aid.

However, we envisage some very interesting applications with this computer assisted design: AMS design diagnosis by evaluations (economic evaluation but also evaluation with others more or less formal criteria), design simulation with graphical visualisations, improvement of the relations between the designers and the users of designed system [WAGNER 88.09].

The prototype on computerized support tool for GRAI method will be available in 1989.

4.5 - Use of GRAI Method for Economic Evaluation

The strong involvement of the GRAI with the industrial world has shown the great interest of companies for new tools for economic evaluation.

The rapid evolution of Production Systems has not been followed by a similar one in the field of Economic Management Systems. This problem is now well studied, since the Harvard Business School (with M. R. KAPLAN) tried to formalize it and to communicate it towards academic and industrial community.

In 1982, the laboratory began research in the field of economics.

The general idea was not of course to become a laboratory of economics. Our objectives was :

- from the laboratory point of view: to developp inter-disciplinarity which is one of the characteristic of the system approach,

- from the scientific point of view: to bring a contribution with the help of the structured analysis tools used by the GRAI, and not so often used by in the field of management research.

The economic approach of such systems is not very easy, because their introduction into a company has many consequencies for the physical system and the management system. Hence, some of the costs and benefits will not be identified by the analysis of the system, but by analysing that company's environment which is the firm itself, in which the new system is to be implemented .

This new situation completely changes the methods for evaluating costs and benefits. Some years ago, this evaluation was based only on the performance of the physical system. The attention was then focused on the reduction of direct manufacturing costs.

Today, the economic approach must be global and integrated at each step of the design of the system.

Several works established the basis of our problem [KAPLAN 83.10], [GIARD 87.04], [BITTON 86.09], [BRIMSON 87.10]. We shall now summarize the different aspects.

For this purpose, we can refer to the manufacturing system life cycle. Traditionally, we distinguish three phases: design, exploitation and death.

Our needs can thus be seen on two levels, which correspond with two goals of our economic evaluation :

- investment analysis,
- performance measurement.

Investment analysis and performance measurement are necessarely supplied in the form of economic data by the cost analysis system.

The costs analysis constitutes the third aspect of the problem.

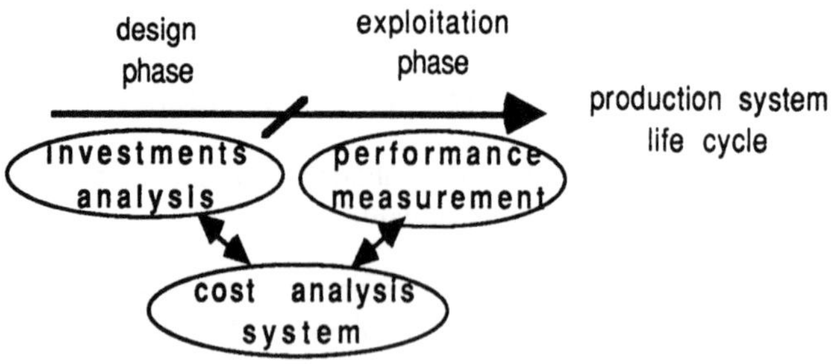

Figure 48 : The various aspects of the problem

Each aspect is concerned with the evolution of C.I.M. :

- Can we apply conventional studies for investment justification in systems such as flexible manufacturing systems [KAPLAN 86.03], [TARONDEAU 82]?

- How to build a performance measurement system, which is consistent with the objectives of the company, which is closely related to the manufacturing system, and which makes the most of the data resources which are offered to us by automation and computerization [KAPLAN 83.10], [TROXEL 86.12]?

- How to design costs analysis systems which do not distort the true picture by using arbitrary repartition rules of overhead costs [ANCELIN 87.04], [SHERIDAN 86.12]?

GRAI Laboratory has been working essentially in two domains :

- economic evaluation of investment in the design phase of CIM System,
- design of performance measurement systems.

We will describe the main results we get in both domains.

4.5.1 - Economic evaluation of CIM System

• Investments and CIM

The Identification of Costs and Benefit Sources

For many people, investment analysis is entirely related to the financial function of the company.

This vision can have negative effects. Indeed, if financial techniques such as ROI, Discounted Cash-flows, Payback Period, represent a certain need, we must nonetheless realize that our investment analysis problems are not solved simply by applying them [KAPLAN 86.03]. These techniques are efficient only if they are backed up by good technical data. In fact :

- they are not a "guide" for determinating these data,
- the precision of the financial evaluation is entirely dependant upon the precision of these data.

So, the fundamental problem, before we can make any cost calculation, consists in the identification (the most thorough kind) of all the costs and benefit sources, which follow the introduction of the new system.

To adopt such a global vision, we must consider two levels of identification.

The Levels of Identification

Generally, the costs identification is very close to the technical aspect of the system. As mentioned in the introduction, this point of view is now proving inadequate, as in the case of a C.I.M. system, at the point where it becomes relatively major. Indeed, integration, richness of information, strategic aspect are the strong characteristics of that kind of system, which combine to make us adopt a global approach : it is as important to study the system itself than to study the relationship of the system to its environment.

So, there are two main levels of identification (Figure 49) :

- C.I.M. system analysis and its place in the whole physical production system,

- Management system analysis and relationship to the company's one.

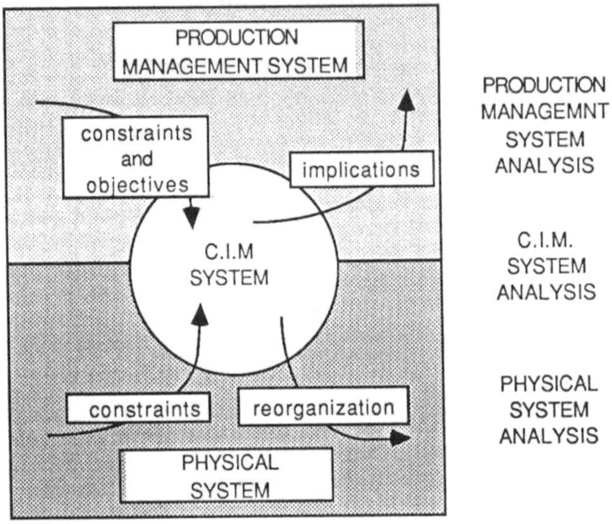

Figure 49 : The different levels of identification

Obvious Costs and Hidden Costs

The previous distinction permits us to introduce another one, concerning the costs: we will talk about "obvious costs" and "hidden costs".

By obvious costs, we mean all costs which are directly linked to the system: i.e. equipment purchased, hired personnel, and those which are at its near periphery : maintenance costs, staff training costs, etc...

Hidden costs have two sources : the influence exercised on the management system by the new system introduction : reorganisation, computerization... and the new standing of the company among its competitors, due to the improvement of quality products, of delays, of production management.

The obvious costs must be analysed in technical terms. This analysis is oriented towards the physical system.

On the contrary, hidden costs must be analysed in terms of organisation. This analysis is oriented towards the decisional system.
Nowadays, the studies are oriented towards the identification of obvious costs only, because of the lack of tools for evaluating hidden costs.

• The Proposed Approach

The previous sections have shown the need to lead the analysis of a new C.I.M. system on a double aspect :

- analysis of the physical system (static and dynamic),
- analysis of the decisional system

This double-edged analysis, which is absolutely necessary during the design phase of the C.I.M. system, allows us an (almost) complete identification of costs and benefits.

The method comprises four steps, which are represented in Figure 50.

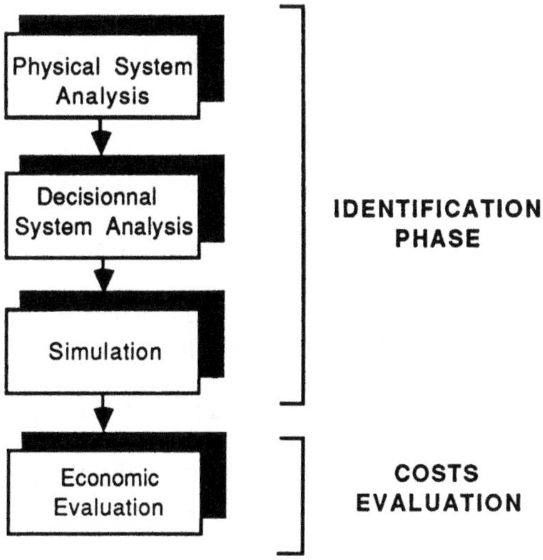

Figure 50 : The technical and economic approach

This approach seems logical, nearly ...evident (although many projects have been lead without using this form of logic).

In fact, the interest of this approach is, above all, the use, at each step of the approach, of a structured analysis method.

The choice of these methods is not of primary importance. What is important is to ensure their appropriateness to the steps of the approach.

Nevertheless, we can say that the important characteristics of these methods are :

- a clear modelling, with a graphic support which can be understood by every person taking part in the study (including of course non-specialists in the method) ;
- an interactive approach which installs a real dialog between all participants.

We generally use the following methods :

- PHASE 1 : IDEFo method,
- PHASE 2 : GRAI method,
- PHASE 3 : ECSL method.

We use, throughout the study, the same groups of persons that we define in the GRAI approach.

The results of each step mean we can :

- design the system,
- identify and calculate the costs.

In fact, there is of course feedback between design and economic evaluation.

In order to show the interest of linking an analysis step to a structured analysis method, we shall present in the following sections the GRAI method and its use for determining part of the hidden costs mentioned above.

• Use of GRAI Method for the Hidden Costs Analysis

Grai method helps to identify hidden costs on its two levels :

- "macro-level": with the drawing-up of the grid,
- "micro-level": with the drawing-up of the nets.

The grid enables us to understand the constraints (in the management field) which will be imposed by the new system on the rest of the company. For instance, the introduction of a new FMS may require the reorganisation of the purchasing department. A consequence of this is a better knowledge of the investment, in its totality, and consequently, a better evaluation of costs.

Another interest of this superior knowledge is the prevention of inadequate transition period.

Indeed, any new system which is introduced to a company requires a transition period. Some omissions in the design phase (i.e an inadequate specification of management tasks to be computerized) can extend this transition and can be the source of unexpected costs.

The nets allow us to evaluate the influence of the automation on management activities. We know that the richness of the information provided by integrated automation leads to greater efficiency in management departments. Nevertheless, this notion remains very qualitative, and by the way, not taken in account for the investment justification. However, as R. Kaplan says, it is better to determine a benefit which is "vaguely true" than to expect no benefit, which is "surely wrong"!

The study of each GRAI net allows us to determine which tasks are influenced by automation. So that we can determine, first qualitatively, then quantitively the costs and benefits provided by the automation of the physical system.

• Application

We have used the complete approach in an industrial case: the context was the design of a new workshop in which it was necessary to use a robot.

Each step of the method had given results :

- concerning the design,
- concerning the identification of costs and benefits.

Particulary, the use of the GRAI method allowed us to see that :

- It was imperative to computerize the Quality Control function, if the company wanted to produce the projected quantities ;

- the additional costs for having machines with automatic quality controls whose results took the form of listings, were balanced against benefits, not only at operational level, but also at the level of function manager in his office (this last benefit had not been taken in account before the GRAI nets had been realized).

The key-point is that these hidden sources of costs and benefits had not been noticed by the company because the Quality Control Department was common to the whole firm. They thought that the investment analysis was finished as soon as the analysis of the shop was finished. They did not examine the whole interface with the internal environment.

The approach, and the associated tools enable us to have a more global view.

• Conclusion

The more integrated a system is, the more difficult are its design and its evaluation to realize.

That is the reason why the analysis, the design and the evaluation must be the fruit of an approach which combines technical, organisational and economic aspects.

To be efficient, this kind of approach must be based on a systematic use of structured analysis methods. That is not to say that there is no other possible approach. What it does mean is that the use of the modelling offered by this kind of methods (on physical <u>and</u> decisional aspects) constitutes a very efficient guide for achieving the best identification of costs.

4.5.2 - Design of performances measurement systems

The research work we will describe now is the result of a cooperation between the GRAI and SNECMA Company (French national aircraft engines builder), which has recently set up a new plant with a very high level of automation and integration (see example described in chapter 3).

In the following sections, we present the method which has been designed, and we try to show the links between this method and the GRAI method.

• ECOGRAI

ECOGRAI uses the decisional approach for modelling the production management system for which we want to design the performance measurement system. This approach is naturally supported by the GRAI model.

The main reason of this choice is that the PMS is designed for the controlling of the production system in a decision aid way.

The problem for us will be to give "control board" for each actor of the production management system. A "Control Panel" will be a set of indicators determined relatively to the nature of the decision center which is studied.

ECOGRAI contains in its today's version, four main phases. This is represented in the Figure 51.

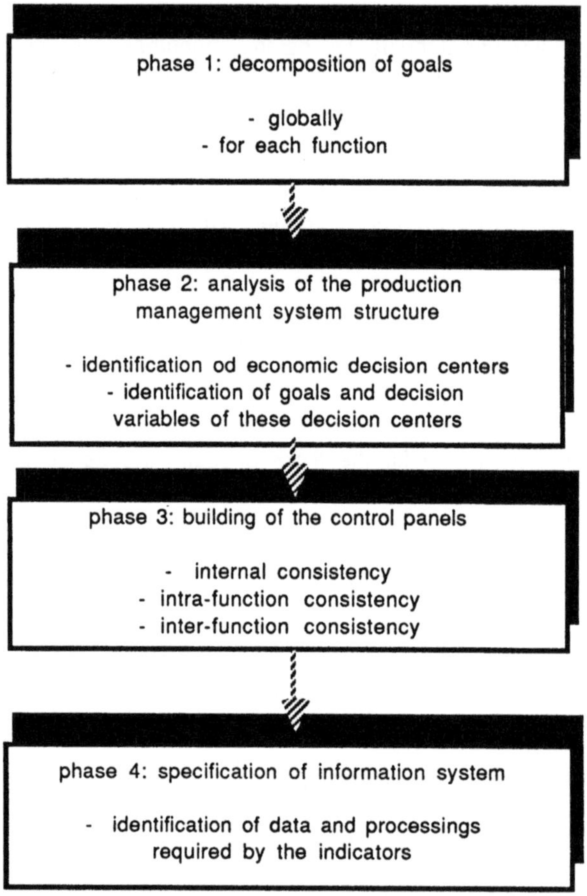

Figure 51 : The phases of the ECOGRAI method

We can say that this method has three strong characteristics :

- First, it adopts a top-down approach, since the starting point of the method is a top-down decomposition of objectives (from global level up to decision center level) ;

- Secondly, all the analysis is supported by the GRAI method, of which we use the graphical tools, and especially, the grids ;

- Thirdly, it tries to formalize the notion of "economic decision center", since one aspect of the problem is to identify what are the decisions which can increase or decrease the company's performances.

Phase 1 / Décomposition of objectives

One important aspect of the PMS is that indicators must be consistent with company's strategy and objectives.

So, as soon as we begin the ECOGRAI analysis, we try to build the system of objectives. We make a decomposition of objectives, starting from those given by the company, and determining these of the different functions of the production management system, so as they can contribute to the superior ones. This decomposition is supported by GRAI Grid.

Phase 2 / Analysis of production management system structure

* Notion of economic decision center :

The object of this phase is to "scan" the grid which represents the function, and to identify the economic decision centers. They are simply defined as "the decision center which have a strong impact on costs events".

These decision centers objectives are obtained by decomposition of those of the function in which they are.

The knowledge of these objectives is necessary to build indicators (following phase), but it is not sufficient : the phenomenon described by these indicators must be in the field of action of the considered decision center. This is then determined by the identification of the decision variables of each economic decision center.

Phase 3 / Building of the "control board"

The two previous phases allow us to propose a "Control Panel" for each economic decision center. The indicators are build, relating to the objectives, the technical features of the decision center, and the actions it can do.

They are build regarding three levels of consistency :

- "internal" consistency : adequation between objective, indicator and decision variable of a given decision center ;

- "intra-function" consistency : adequation between the different "control board" of a same function (problem of decomposition and aggregation) ;
- "inter-function" consistency : adequation of "Control Panel".

Phase 4 / Specification of the Information System

This is the key-point in our approoach, as indicators are in fact data and processings.

This phaseallows to check the operating feature of the approach.

ECOGRAI Approach

Besides the method itself, an approach, to apply it in the firme has been made.

The three characteristics of the approach are as follow :

- active participation of members of the company,
- definition of the indicators by the end-users,
- validation by the high decisional level, for respecting the company strategy.

Three groups are thus defined (Figure 52) : the analysis team, constitued by the external consultant and one corresponding member of the company. This is the "leader" group for the application of the method. The synthesis group is constitued by the persons in charge of the studied management system. This group validates the different phases of the study and gives the principal orientations. The users working groups are constitued by the members of each studied function (and especially, the persons in charge of the economic decisions centers).

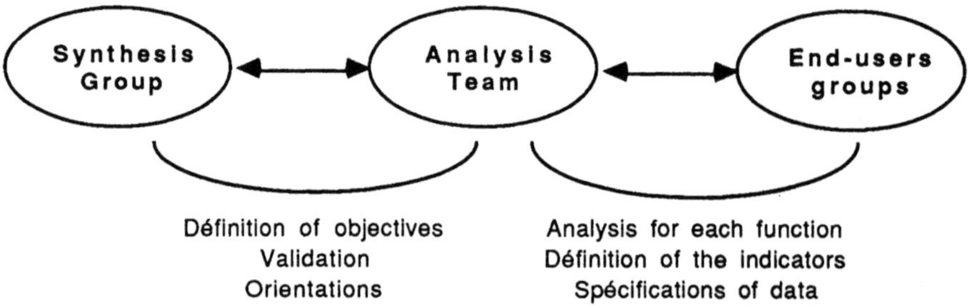

Figure 52 : Definition of the groups for the application

Conclusion

The proposed approach is lying on a global vision of of the production management system. This vision is supported by the GRAI model, which lead us to the "internal architecture" of the control system.

This systemic approach allows us :

- to ensure the adequation between the performance measurement system and the management system. This adequation is formalised by the multileve consistency study ;

- to give the specifications of the information system, real "vector" of the performance measurement system (especially, to determine the need in economic data for the technical decision centers) ;

- to give the expression of needs for decision support systems ;

- to make the link between techniques and economics, with the definition of economic decision centers ;

- to develop a participative approach, which tries to ensure the adequation reality of company / method principles.

The Figure 53 is a possible representation of what brings to us the GRAI methodological frame.

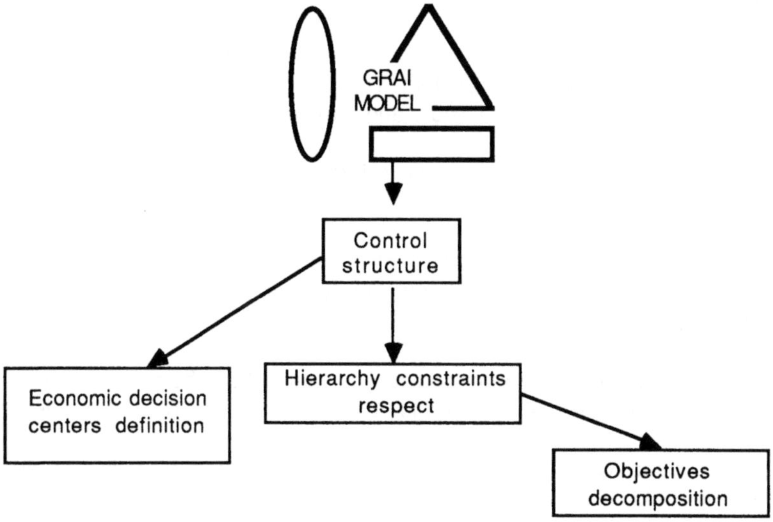

Figure 53 : The interest of the use of GRAI frame, for ECOGRAI

An application of ECOGRAI begun at the beginining of 1988, and should be ending in November. This application concerns the SNECMA plant, in Le Creusot, which is a very high level automated production system. The design of the production management system of this new plant had been made with the help of the GRAI method.

This application is already giving very interesting results, from practical and theoretical point of view. The results will be presented in October, in Grenoble, at the international congress organised by AFCET : AUTOMATIQUE 88.

5 - CONCLUSION

This paper gives a good overview about the work developed by GRAI Laboratory under the responsability of Professor Lucas PUN and ourself.

Through the research work we can point out the guiding idea : we start by the need of method to design Production Management System. To elaborate Method we have to set up Conceptual Model, graphical and computerized Tools and structured approach.

Starting from GRAI Method which is related with Decisional features of Production Management we have :

- extended the method to physical and informational systems to be able to design Computer Integrated Manufacturing. In such way we were obliged to develop adapted simulation tool,

- developed computerized tool to support method. This activity shows first the difficulty of the task and second the need to develop Conceptual Models and References Models,

- used GRAI Method for Economic Evaluation which is now an important subject in this field,

- we have not described two others works which are on going development but which are supported also by the GRAI method : the development of Decision Aid Tool to control manufacturing unit and on the other hand, by GRAI Method, the knowledge acquisition process for technical diagnosis.

In these various research activities, more we have progressed, more the need of basic research and theoretical tools increased, particularly for Manufacturing System. It is the reason why we try to set up recently with other colleagues from European Universities a research program on theory for manufacturing system.

BIBLIOGRAPHY

[ASHBY 52] W. ASHBY. "Design for a brain. The origin of adaptative behavior". Chapman & Hall (1952).

[SIMON 69] H.A. SIMON. "The science of the artificial" - MIT Process (1969).

[MESAROVIC 70] M.D. MESAROVIC, D. MACKO, TAKAHARAY. "Theory of hierarchical, multilevel systems". Academic press, New York, Etats-Unis (1970).

[MELESE 72] J. MELESE. "L'analyse modulaire des systèmes de gestion". Editions Hommes et Techniques (1972).

[ZEIGLER 76] B.P. ZEIGLER. "Theory of modelling and simulation", John Wiley & Sons Publisher, New-York, Etats-Unis (1976).

[YOSHIKAWA 81.09] H. YOSHIKAWA. "General design theory and CAD System" - IFIP working conference "Man-Machine communication in CAD/CAM", North Holland Publishing (September 1981).

[TARONDEAU 82] J.L. TARONDEAU. "Technologies flexibles et rentabilité", Harvard l'Expansion (Eté 1982).

[DOUMEINGTS 83.06] G. DOUMEINGTS, D. BREUIL, L. PUN. "La gestion de production assistée par ordinateur", Hermes Publishing, 183 p.. Paris, France (Juin 1983),

[DOUMEINGTS 83.09] G.DOUMEINGTS. "Methodology to Design Computer Integrated Manufacturing and Control of Manufacturing Unit"- Advanced Course on Computer Aided Manufacturing - CAM 83, Karlsruhe, FRG (September 1983).

[KAPLAN 83.10] R. KAPLAN."Measuring Manufacturing Performances: A New Challenge for Management Accounting Research". The Accounting Review, vol LVIII n°4 (October 1983).

[DOUMEINGTS 84.11] G. DOUMEINGTS. "Méthode GRAI. Méthode de conception des systèmes en Productique" - Higher doctoral thesis - University of Bordeaux I (November 13, 1984).

[DOUMEINGTS 85.03] G. DOUMEINGTS. "How to decentralize decisions through GRAI Model in Production Management" (Invited Paper), IFIP W.G. 5.7 International Working Conference, Munich, RFA (March 1985)

[DUMORA 86] E. DUMORA, J.F. HUET. "Spécifications du système de pilotage de l'usine de Le Creusot", Rapport interne (1986).

[MII 86] "Etude prospective GPAO pour le Ministère de l'Industrie", Société MII (1986).

[KAPLAN 86.03] R. KAPLAN. "Must CIM Be Justified by Faith Alone?". Harvard Business Review (March-April 1986).

[BITTON 86.09] M. BITTON. "Problématique de l'évaluation technico-économique des projets en productique". GRAI Laboratory, University of Bordeaux 1 (September 1986).

[DOUMEINGTS 86.11] G. DOUMEINGTS, E. POUMEYROL. "Computer aided design for advanced manufacturing systems". Working conference on New Tehnologies for Production Management Systems, Tokyo, Japan (November 1986).

[SHERIDAN 86.12] T. SHERIDAN. "Getting the Right Numbers for Manufacturing Management", Proceedings of the Symposium "Cost Management System Project", CAM-I, Nice, France (8-9 December 1986)

[TROXEL 86.12] R. TROXEL. "Performance Measurement". Proceedings of the Symposium "Cost Management System Project", CAM-I, Nice, France (8-9 December1986).

[BERARD 87] C. BERARD. "Le pilotage des systèmes discrets de production", Thèse d'Etat, Université de Bordeaux 1 (1987). A PARAITRE

[ANCELIN 87.04] B. ANCELIN, C. FRAY. "La Décision en Production : limites et besoin d'une instrumentation économique", 2nd International Conference on Production Systems, INRIA, Paris, France (Avril 1987).

[ERKES 87.04] K. ERKES (from WZL Aachen), M. CLARK (from LOGICA): "Public Domain report n°1", ESPRIT Project 418 "Open CAM System", (April 1987).

[GIARD 87.04] V. GIARD. "Evaluation economique et prise de decision en gestion de production". 2nd International Conference on Production Systems, INRIA , Paris, (April 1987).

[BRIMSON 87.10] J. BRIMSON."Cost Management for the Aerospace Industries". SITEF (October 1987).

[ROBOAM 87.10] M. ROBOAM, G. DOUMEINGTS (from GRAI Laboratory), K. DITTMAN and M. CLARK (from LOGICA): "Public Domain report n°2", ESPRIT Project 418 "Open CAM Sytem" (October 1987).

[AIP 88.05] JL WAGNER. "Spécifications détaillées du module de règles", Rapport interne du Projet AIP, (Mai 1988).

[GRABOT 88] B. GRABOT, B. VALLESPIR. "Failure modes analysis to design operating systems in industrial process" (1988).

[ROBOAM 88.07] M. ROBOAM. "Contribution à l'extension de la méthode GRAI : Intégration des méthodes d'analyse, élaboration de modèles de référence, de coordination et de synchronisation". Thèse de l'Université de Bordeaux I (Juillet 1988).

[WAGNER 88.09] JL WAGNER. "Utilisation des modèles de référence dans l'activité de conception des systèmes de Gestion de Production. Synthèse et perspectives d'évolution", mémoire de DEA, Laboratoire GRAI, Université de Bordeaux 1, (Septembre 1988).

LLAMA:
A LAYERED LOGICAL ARCHITECTURE FOR MATERIAL ADMINISTRATION

H.V.D. Parunak and R. Judd
Industrial Technology Institute
P.O. Box 1485
Ann Arbor, MI 48106
USA

Abstract

Material handling systems are an information bridge between high-level shop floor control and the sensors and actuators at the factory floor. When such systems are engineered as a monolith, each must be custom-built, and the cost of modification and maintenance is high. Drawing on parallels between communications (moving data) and material handling (moving material), we propose a layered architecture analogous to the ISO OSI reference model for communications. This article defines the problem, discusses layered architectures in general and previous layered systems in particular, and describes the functions and interfaces of the layers in our system. This is a working paper describing work in progress, not a finished specification.

1. The Material Handling Problem

We define the material handling problem in two steps. First, we outline an abstract model that distinguishes material handling from processing, thus delimiting the domain. Second, we exhibit the range of alternatives that a reference model for material handling must cover by describing some of the technologies and strategies in common use today.

1.1. An Abstract View of Material Handling

It is convenient to view the factory floor in a discrete parts manufacturing operation as a series of operators that change the state of in-process inventory. We group these operators into two classes, *material handling* and *processing.*

NATO ASI Series, Vol. F53
Advanced Information Technologies for Industrial Material Flow Systems
Edited by Shimon Y. Nof, Colin L. Moodie

Algebraically, the manufacture of discrete goods consists of applying a series of state-changing operators to inventory. For concreteness, we assume that a bit string represents the state of a part, with groups of bits representing such state information as shape, hardness, color, components, part number, location, and age. Then each operator defines a function on a part's bit string. With each operator is associated not only its function on the state of an input part, but also the set of equipment in the factory that it uses to perform the physical change represented by the function. This model is not practical for direct implementation, but offers a useful abstraction for thinking about what happens in a factory.

Different operators manipulate different fields of the state bit string. Operators corresponding to a lathe or a mill manipulate the shape field. A painting robot has operators changing the color field. Some fields can represent unknown values, and an inspection operator can resolve these values. A warehousing operator adjusts the age field of a part's state. Transport operators vary the location field. These examples all map a single state string to a single state string. Assembly operators map several state strings, representing components, into a single state string that represents the assembly and has the component state strings embedded in it in the appropriate fields. Similarly, disassembly operators map a single state string into multiple state strings.

Complex operators can be built up by composition of primitive operators, each manipulating only one field of the state string. Most operators are complex, since most factory operations require non-zero time, thus ageing their parts and requiring a component warehousing operator to increment the age field of the parts' states accordingly. A drying tunnel corresponds to a complex operator including at least three primitive components, one to record that the wet paint is now dry, one to increase the part's age, and one to record the change in its location.

Not all of the components of a part's state are equally important. Some state components, such as shape, hardness, and color, reflect the part's specification, and make up its *functional state.* Other components of state, such as a part's location in the plant or the length of time it has been there, are irrelevant to its function, and make up its *non-functional state.* "Functional" and "non-functional" in this context refer to the function for which a part is manufactured, not the functions of the factory as it makes the part. The manufacturing function depends on the fact that a part is located at a certain milling machine at a certain time, but this information is irrelevant to whether the part functions correctly in its intended application. The purpose of the factory is to make changes in a part's functional state. The non-functional state changes only to enable changes in its functional state, or as coincidental consequences of such changes.

In this framework, we call the set of all primitive operators that change a part's functional state, "processing." "Material handling" is the set of all primitive operators that change a part's non-functional state. Material handling thus includes the traditional functions of moving material from one workstation to another, and storing it in a warehouse. It also includes transportation and holding that occur as components of a complex operator, such as the drying operation described above. Thus material handling occurs in almost every machine in a plant, as parts and requests for parts move back and forth.

Our model of material handling implies that material handling pervades the entire factory, including machining workstations as well as AGV's, conveyors, and ASRS's. As a result, our architecture for material handling covers non-functional state changes within workstations as well as between workstations.

1.2. Technologies and Strategies for Material Handling

The abstraction developed in the last section gives us a theoretical perspective. Practically, any reference model for material handling must accommodate current practice and permit incremental growth if it is to be accepted in industry. Exhibit 1 identifies a number of material handling technologies in common use today and some common strategies that make use of them, and shows which technologies are useful for which strategies. In this section we briefly describe each technology and each strategy. Later in the paper, we will demonstrate the adequacy of LLAMA by showing how it models these current paradigms.

```
                          Strategies:
                Sequential         Non-Sequential
Tech-          Synch  Asynch    Push   Pull   CASCADE
nologies:                             (Kanban)

Dial           xxxxx
Machine

Pwr/Free       xxxxxxxxxxxxx

Free                  xxxxxx
Conveyor

Manual                xxxxxxxxxxxxxxxxxxxxxxxxxxxxxxx
(Fork
Lift)

Belt                  xxxxxxxxxxxxxxxxxxxxxxxxxxxxxxx
(Bosch)

AGV                             xxxxxxxxxxxxxxxxxxxxx
```

Exhibit 1

1.2.1. Technologies

The technologies in Exhibit 1 represent some of the hardware available for material handling today.

Dial machines move a set of parts in lock-step past a series of workstations. In one configuration, workstations are situated at intervals around the edge of a turntable on which the parts ride. All workstations have the same length of time to complete their tasks. When all are completed, the turntable indexes the parts by one workstation position, so that each workstation now has the part that was to the left (or right) of it, and the cycle repeats.

Power and free conveyors combine a chain drive for pallets with a release mechanism that allows them to queue up before a workstation.

Free conveyors are roller mechanisms that convey pallets from one point to another by gravity or as a result of a manual shove.

Manual methods include fork lifts, hand trucks, and other human-directed conveyances.

Belt conveyors include systems like the Weldun-Bosch switched conveyor that permit dynamic route decisions.

AGV's are individual carts that can steer themselves without direct human intervention.

1.2.2. Strategies

We divide strategies into *sequential*, where each part visits workstations in the same order, and *non-sequential*, where each part can follow its own path through the shop.

A *synchronous sequential* strategy, typified by a dial machine, not only moves all parts along the same path, but moves them all at the same time.

An *asynchronous sequential* strategy moves all parts along the same path, but the workstations can differ from one another in how long they require to perform their tasks. As a result, parts move at different times, and queues can build up behind the slower machines. This strategy is typical of assembly lines.

In a *non-sequential push* system, each part moves from one workstation to the next as soon as it is finished. This system can be viewed as multiple *asynchronous sequential* systems in the same shop, sharing some workstations.

A *non-sequential pull* system is like a random push system except that parts move when the next workstation calls for them, not when the previous workstation completes them. Thus it adds output queues to each workstation, where parts wait until they are requested. This system, combined with an upper limit on the number of parts that a workstation can hold in its output queue, is the classical Kanban system. [1]

CASCADE is a system developed at ITI that combines push and pull, and uses a neural model to permit a material handling system to be self-routing. [2, 3, 4, 5]

As Exhibit 1 suggests, some technologies are better suited to some of these strategies than to others. The strategies impose different constraints on a control architecture. Each of the strategies can solve problems in some environments that the others cannot, and so each is an important and viable alternative in the system designer's bag of tricks. The challenge for LLAMA is to devise a model that constrains the problem space usefully while still accommodating the different technologies and strategies that have been found necessary.

2. Layered Architectures

In this section we discuss existing examples of layered architectures, for communications, operating systems, and material handling. [6] Then we highlight some general features of layered architectures.

2.1. The Seven Layer ISO-OSI Model

Perhaps the most common example of a layered architecture is the OSI reference model for telecommunications. [7] The Manufacturing Automation Protocol (MAP) for shop floor communications, and the Technical and Office Protocol (TOP) for administrative communications, are perhaps the best known examples of architectures based on the OSI model. This model establishes a set of seven layers that extend across all communicating systems.

The *application layer* contains the processes that deal with the user's problem domain. These processes may incorporate many operations besides communications, but the OSI environment includes only those parts of these processes that deal with interprocess communications. The application layer also includes a number of application service elements (ASE's) that provide higher-level abstractions of communications for specific purposes. Typical ASE's are the Manufacturing Messaging Specification (MMS) for messages commonly needed in operating manufacturing equipment, the File Transfer Access Method (FTAM) for file transfers, and the Virtual Terminal Protocol (VTP) for interfacing with video display terminals.

The *presentation layer* isolates application processes from differences in message syntax. For example, it is responsible for providing an application that expects ASCII data with data in that form. In general, it permits different stacks to negotiate a common context in which to communicate.

The *session layer* organizes and structures the interactions between application processes. It provides services to deal with such issues as synchronization and simultaneous vs. alternating operation.

The *transport layer* manages the quality and cost of moving data from one end system to another, taking care of issues such as flow control, error handling and retransmission.

The *network layer* encapsulates the data transfer technology, including decisions concerning routing and relaying information. It provides unique addressing of individual systems in the global domain.

The *data link layer* provides for the movement of data between network entities. It detects and may correct errors that originate in the physical layer.

The *physical layer* provides standard mechanisms to access the physical medium. It is not the same as the medium, which is not usually considered a layer within the OSI model.

2.2. OS Structure

The general structure of computer systems follows a layered architecture.

Again, application programs, written by users to address problems in their domains, occupy the top layer.

Applications are built on support packages such as databases and communications packages, which offer standard services. This paradigm is especially strong in systems such as UNIX, which offers a wide range of standard functional modules on which user programs can call.

The support packages in turn call on the Operating System for more basic services, such as managing processes, file storage, and other resources.

The operating system often draws on a lower level of software device drivers that provide an interface to the physical hardware.

2.3. The Philips Model for Material Handling

At least one other group, working at Philips, has proposed a layered architecture for material handling. [6]

The application layer of the Philips architecture is the *workstation commander* (WSC), the communication and planning component of a manufacturing workstation. The lower layers handle material handling between workstations. This architecture thus ignores the common material handling functions that occur not only between but also within workstations.

The *route layer* provides reliable transportation of material between workstations.

The *product link layer* models the transport system as a series of tracks (which follow fixed paths) and nodes (where material can move in more than one direction). It provides reliable transportation of material between adjacent nodes. Material may have to traverse several such nodes in order to move from one workstation to another.

The *sensor/actuator layer* encapsulates the "standard displacements," the motion primitives of the transport system.

In the Philips model, unlike the OSI model, the *Medium* is an explicit layer.

2.4. Theory of Layered Architectures

All of these layered architectures show certain common features. They are *implementation hierarchies* rather than control hierarchies; they impose a common pattern on error handling; and they obey a "law of the weakest link."

2.4.1. Implementation Hierarchies

The structures we have discussed are all examples of "implementation hierarchies," [8] which begin with hardware at the bottom and add functionality incrementally in successive layers. The word "implementation" in the phrase "implementation hierarchy" indicates that the hierarchy organizes successive layers of an implementation. The hierarchy itself is an abstract model, not an implementation.

Implementation hierarchies offer two kinds of standard interfaces between modules. Each layer offers a number of standard functions, or *services*, to the layer above it, which can access these services through *service elements* consisting of requests and responses (directed downwards) and indications and confirmations (directed upwards). Communications between peer layers in separate systems are governed by *protocols* and take the form of *protocol data units*. Both service elements and protocol data units are abstractions, specifying the kinds of information that are exchanged, but not constraining the precise form of messages in an actual implementation.

It is important to distinguish between implementation hierarchies, which add functionality, and control hierarchies, which offer time-phased decomposition of tasks at successive layers with shifts in the granularity of tasks.

For example, as we step from the cell into LLAMA, which is at the workstation level, process plans shift from being lists of machines to visit, and become process steps within a workstation. In general, at factory level X, an X process plan consists of a partially ordered list of operations to be carried out at the layer below X. Thus a cell process plan consists of a list of operations to be carried out at the workstation level, while a workstation process plan consists of a list of operations to be carried out by the devices within the workstation. Similarly, resources to be scheduled at the cell level are workstations, while within a workstation they are parts and tools.

These shifts in granularity are a consequence of moving down the control hierarchy, not the implementation hierarchy. Sometimes the two hierarchies do become merged. The granularity shift between our presentation and routing layers reflects a control-type hierarchy embedded within the implementation hierarchy of material handling.

2.4.2. Error handling

The most that a layer in an implementation hierarchy can do for errors that originate above itself is to detect them. Sometimes it cannot even do that. Its ability to correct errors is limited to those that originate below itself.

2.4.3. Law of the Weakest Link

An actual implementation may provide fewer than the complete set of services at some layers. If this is done, higher layers may not be able to offer full functionality, even if they themselves have code to support their full set of services, since the lower-level functions needed will not be available. The chain is only as strong as its weakest link.

3. The LLAMA Architecture

In this section we outline our proposal for a layered model for material handling. After defining the entities and layers of LLAMA, we discuss the message interfaces between successive layers in one stack (services) and between corresponding layers in different stacks (protocols). Then we examine the options available at each layer, and finally show how the different material handling strategies described above can be constructed within LLAMA.

3.1. Overview of LLAMA

LLAMA carries the parallel between material transport and data communications further than does the Philips model, using the OSI layers to suggest services that might be needed in material handling. This section surveys the elements in LLAMA and the layers in its stack.

3.1.1. Elements

The elements of LLAMA are systems, layers, messages, and Material Transport Units (MTU's). Including the MTU's as elements of the model differs from OSI's exclusion of physical concerns such as electrons. We include the physical MTU's to emphasize that in our view, material flow is just as much a part of the information flow of the system as is the flow of PDU's. [9]

System

A *system* is a portion of the factory that is logically controlled as a unit from the perspective of material handling. Typical LLAMA models represent each workstation, each ASRS, and each material handling controller as a single system. Each system has its own stack of layers.

Layer

A *layer* is a functional module in an implementation hierarchy that adds value to services provided by lower layers in order to offer services to the layer above it. It can communicate both vertically (to the layers on either side) and horizontally (to peer layers in other systems).

Message

A *message* is the unit of communication between layers. Messages are of two types. *Service elements* pass between successive layers in a single system, while *protocol data units* pass between peer layers in different systems.

MTU

A *Material Transport Unit (MTU)* is the unit of material movement between systems. In many systems it is a pallet.

Conceptually, a part's data is attached to the MTU and travels with it as a unit. This data includes ID, history, state, next task, and process plan within the cell (giving only enough detail to determine what workstations to visit, but not including operations within a workstation). Possible implementations include a RAM chip attached to the MTU or a bar code keyed to an external database. Whatever the implementation, data is manipulated by a software object corresponding to whatever manipulates the MTU. In particular, part data does not go through the cell scheduler. If the implementation involves a central database, the layers of LLAMA communicate directly with the database.

The control needed to keep the MTU on the track (for example, the wire or paint stripe that guides an AGV, or the physical constraint provided by a conveyor to its pallets) is local to the MTU, not part of the central control system. A MTU will automatically stop if another MTU is in front of it. An MTU must also be able to start, stop, and change directions, either passively in response to actuators (as with pallets on a conveyor), or actively (as with an AGV).

3.1.2. Layers

LLAMA's layers follow the general functional pattern of the OSI model. In this section we outline a layer of LLAMA corresponding to each layer of the OSI model, though some of these are not actually developed in our system.

Application

Our *application layer* includes the cell controller and several application service elements (ASE's).

The cell controller is responsible for a collection of two or more workstations integrated by material handling, data communications, and database services. It establishes process plans (and thus product and tool routing) and operating parameters for the material handling system.

In the OSI model, the ASE's present an abstraction of communications tailored to a particular problem. LLAMA's ASE's similarly determine the abstract structure of the material handling system as seen by the cell controller. Particular examples of LLAMA ASE's are a synchronous transfer line, an asynchronous transfer line, point-to-point push transport, point-to-point pull transport (KANBAN), or the CASCADE neural architecture.

In the ITI FMS architecture, this layer on a workstation includes the Workstation Scheduler, [10] which shares many of the functions of the Philips Workstation Commander. However, where the Philips model has the Workstation Commander as the top layer of a protocol stack corresponding to the inter-workstation material handling system, in LLAMA each workstation has its own protocol stack, separate from and parallel to the inter-workstation material handling system.

Presentation

If different systems support different sets of ASE's, the *presentation layer* permits them to negotiate a common context so that they can exchange material. We have not defined LLAMA's presentation layer in detail, and its full exposition is a topic for future research.

Session

The *session layer* in OSI handles synchronization between systems. Its counterpart in LLAMA would handle such synchronization issues as coordinating the movement of multiple MTU's as part of a single operation and synchronizing if the transfer aborted after only some of the MTU's moved. Since LLAMA deals with single MTU's, we do not anticipate a need for this layer, and do not describe it further in this document.

Route

The *route layer* fills the role of the transport and network layers in the OSI model. Its main task is mapping destinations specified at higher layers into meaningful entities at lower layers.

LLAMA typically models each workstation, ASRS, and track system as a separate layered stack. The route layer routes products among the entities covered by one such stack. When material needs to move from one stack to another, the route layer is responsible for determining which adjacent system should receive the material, but does not reason directly about destinations several systems away.

OSI supports source routing (in which a message carries along a list of the systems through which it must pass) and route recording (in which a message records the systems that it traverses). The current LLAMA specification does not support either of these, but their usefulness is under investigation.

Material Link

Corresponding to the OSI data link layer is LLAMA's *material link layer*. We extend the title from the term "product link" used in the Philips model, since this layer can control the manipulation of tooling as well as product. Within a workstation, this layer corresponds to the accumulators. [10]

The material link layer consists of many communicating elements. Each control element at this layer corresponds on a one-to-one basis with a Material Transport Element (MTE) at the physical layer. The material link layer addresses the following control issues:

- *Divergence* - Control of the direction of the MTU at each point that has more than one exit.

- *Convergence* - Regulation of access to a point with multiple entrance points. Convergence is basically a resource sharing problem.

- *Storage* - Insuring that the number of MTU's in any section of track does not exceed the physical capacity of the track. Elements can have zero capacity, in which case material is allowed to flow through the element but cannot be stopped in the element.

Physical Layer

LLAMA's *physical layer* corresponds to the OSI physical layer and the Philips sensor-actuator layer. It encapsulates the physical hardware and models it to the material link layer as a set of material transport elements, or MTE's. Convergent switches, divergent switches, switches with both convergent and divergent characteristics, track sections with capacity, terminal track sections, and bidirectional track sections with zero capacity can all be represented as MTE's.

An MTE has the components indicated in Exhibit 2. The MTE has three components: a convergent switch, a divergent switch, and storage between them. The arrows above the diagram indicate the messages that travel between the material link layer and the physical layer, and suggest the portion of the MTE that those messages concern.

```
| Convergent |        Storage              | Divergent |
   MTUIn       Halt       NextMTU            Go    MTUOut
     ^          |           ^                |       ^
     |          |           |                |       |
     |          v           |                v       |


      \                                                ^
       \                                              /
     P*\                                            /P*
         v                                         /
  ----->O-------------------------|--O------>
  P*     ^   C*      ID*       P+    S+  \     P*
     P*/                                  \P*
      /                                     \
     /                                       v
```

Key:
```
 P  =  presence sensor
 S  =  stop
ID  =  ID sensor
 C  =  capacity sensor
 *  =  optional
 +  =  this presence sensor and stop are needed only with
       MTU's having nonzero capacity.
```

Exhibit 2

The presence sensors on the convergent and divergent sections are optional. However, at least one set of these sensors should be present to indicate positively when a pallet moves from one element to another. The protocols will function with no sensors, but material movement will be inefficient since the system will need to use timeouts to guarantee part arrival, and there will be no way to detect an external intervention that removes or adds an MTU unexpectedly.

If the MTE has nonzero capacity, then the sensor P+ initiates the negotiation for reserving space on the next MTE. The stop must be positioned less than one full MTU away from sensor P+, to insure that the MTU at the stop is the one for which the negotiation is taking place.

An optional capacity sensor at the beginning of the storage portion of the track can be used instead of a software queue to indicate when the storage capacity of a track has been exceeded. This sensor can reduce communications dependencies between the two ends of a long MTE by using the physical chain of MTU's to keep track of the queued material.

We consider the movement of material as much an element of communications as the movement of messages. [9] Thus we include the *hardware layer* as an explicit element of our model.

3.2. LLAMA Services and Protocols

This section describes LLAMA's interfaces between successive layers in one system (supporting layer services) and between similar layers in different systems (protocols). These interfaces are restricted to those involved in material handling. Other domains that would be included in a complete interface specification but are omitted here include quality, diagnostics, maintenance, tool maintenance, and negotiated task assignment.

Each section below describes the service elements (that it offers to the higher layers) and PDU's for one layer of LLAMA. Some service elements have two interpretations, depending on whether we are looking at a workstation (WS) or a material transport system (MTS). Also, among service elements we distinguish between a layer's requests (which the next higher layer sends to it) and its indications (which it sends to the next higher layer). LLAMA does not currently use the message types known in OSI as responses and confirmations.

These messages are logical. Actual implementations will include details such as immediate handshaking responses to guarantee that the message arrived, or verification of parameters, but these are omitted here to emphasize the overall pattern.

Some messages appear at more than one layer. In this case, they are defined the first time they appear, and only named in subsequent interfaces, unless there are layer-specific details.

3.2.1. Application

Service Elements: Requests

The application layer offers a set of common service elements to provide a material handling interface to the non-material-handling modules of an application.

- **SetSystemParameters <parameter-list>**.--"Parameters" in this case are more than just numbers, and include (for instance) workstation process plans. In an MTS, this message conveys such information as TOMP limits for CASCADE; the number of stations on a transfer line; or the connectivity matrix for a point-to-point system. In a workstation, it conveys buffer capacities, task assignments (including part type, number to produce, and time constraints), and workstation process plans (detailed sequences of operations and the tooling needed to carry them out, as well as the classes of input and output parts).

- **Start**.--Turns the system on (e.g., at the beginning of the shift).

- **Stop**.--Turns the system off (e.g., at the end of the shift).

- **Destination <MTU_id> <destination>**.--Sets a destination for the specified MTU.

Service Elements: Indications

- **Arrived <MTU_id>**.--Reports that the MTU has reached its destination. This report appears only if the destination is within the same system. If the presentation layer sets a destination outside of its own system, **Give** or **Take** will ultimately be issued between presentation layers, and the completion of the transfer will be marked by the **Received** PDU.

- **Exception <msg>**.--Asynchronous notification of an exception condition.

Protocol Data Units

- **Find <MTU_type> <key>**.--Requests recipient to locate and acquire an MTU of the designated type. This message command has three effects.

  - It leads to a mapping of MTU type to MTU id, which may require waiting indefinitely if an MTU of the desired type is not avilable in the system. Thus an immediate mapping cannot be guaranteed.

  - It reserves the identified MTU for the sender of **Find**, so other entities will not **Find** it. This function must be associated with the mapping, otherwise between the time a requestor learns the mapped id and asks to reserve it, someone else might have reserved it. The need to unify the mapping and reservation functions is analogous to the need for a unified "test and set" instruction for concurrency control in an operating system.

  - It begins to move the MTU toward the requestor. Since one only reserves an MTU in order to use it, it is natural to unify this with the reserve function.

  Find is non-blocking. The sender is notified asynchronously with **Found** when the required MTU has been found, reserved, and brought "near" to the requestor, where "near" is an implementation-dependent condition characterized by the fact that a requestor can be expected to block on a request to deliver the MTU (**Give**).

- **Found <key> <MTU_id>**.--Announces successful completion of **Find.** There is no reply if **Find** is unsuccessful. The original sender can impose a time-out if desired.

- **Give <MTU_id>**.--Asks recipient to hand over the specified MTU. The requestor blocks until the MTU is given. Applications can implement a give that requests a type rather than an id (for example, a transfer line) as a subroutine that

  - Calls **Find** with the type;

  - Waits for the **Found**;

  - Issues **Give** on the <MTU_id> returned by the **Found**;

  - Returns to the caller.

- **Take <MTU_id>**.--Asks recipient to remove the specified MTU.

- **Received <MTU_id>**.--Reports successful MTU transfer as a result of a **Give** or **Take. Received** travels between stacks in the opposite direction to the movement of the MTU. We view the movement of the MTU as an information exchange in one direction, and this message as the response to that movement in the other direction.

- **Lose <key>**.--Cancels a find. If the mapping has not yet been done, **Lose** stops the system from waiting for it. If an MTU has been reserved, **Lose** unreserves it. If an MTU is moving toward requestor, or has reached it, **Lose** allows it to move away (a direct consequence of removing the reservation).

3.2.2. Routing

In our current implementations, this layer is null, but we envision its usefulness to deal, for example, with differing implementations of loops in CASCADE (e.g., physical loops vs. parking lots). In a workstation, this layer consults the workstation process plan and determines the sequencing of accumulators.

Service Elements: Requests

- **SetSystemParameters <parameter-list>.**

- **Start.**

- **Stop.**

- **Destination <MTU_id> <destination>.**

Service Elements: Indications

- **Exception <msg>.**

- **Arrived <MTU_id>.**

3.2.3. Material Link

Service Elements: Requests

- **Destination <MTU_id> <destination>**.--<destination> here is typically at a finer granularity than that modeled by the route layer in its service to higher layers, but no finer than the MTE. This difference reflects a control hierarchy embedded within the implementation hierarchy that is the basic structure of LLAMA.

- **SetSystemParameters <parameter-list>**.--An important parameter at this interface is the sequence of elements that constitute a route. In a workstation, the parameters at this level specify accumulator parameters.

- **Start.**

- **Stop.**

Service Elements: Indications

- **Exception <msg>.**

- **Arrived <MTU_id>.**

Protocol Data Units

- **SpaceRequest <destination>**.--Asks peer whether space is available for one MTU.

- **SpaceAvailable.**--Response to **SpaceRequest.**

- **Coming <MTU_id>**.--Announces intent to send an MTU. Differs from **SpaceRequest** in that this message causes the recipient to configure its hardware for the MTU, not just reserve space. We pass the MTU_id to accommodate MTE's that do not have ID sensors.

- **Ready.**--Response to **Coming.**

- **Gone.**--Announces that MTU has been sent. On MTE's with exit sensors, results from MTU passing exit sensor. On MTE's with no exit sensors, this message is sent immediately before the MTU is released.

- **Got.**--Response to **Gone.** On MTE's with entry sensors, results from MTU passing entry sensor. On MTE's with no entry sensors, this message is sent automatically as soon as the **Gone** is received. This message permits the source MTE to clear the sent MTU from its internal data structures.

3.2.4. Physical

Service Elements: Requests

- **Go <MTE_id> <direction>**.--Sets MTU at the head of the MTE moving in the specified direction. (In one common implementation, drops the stop on the appropriate exit track.)

- **Halt <MTE_id>**.--Stops the next MTU to arrive at the head of the MTE. In one common implementation, raises the stop.

- **SetSystemParameters <parameter-list>.**

- **Start.**

- **Stop.**

Service Elements: Indications

- **MTUIn <branch_id>**.--An MTU has passed the entry sensor on the specified branch.
- **MTUOut <branch_id>**.--An MTU has passed the exit sensor on the specified branch.
- **NextMTU <MTU_id>**.--The designated MTU has arrived at the MTE presence sensor. The MTU_id comes from the ID sensor if there is one, and otherwise is null.

3.2.5. Hardware

We represent these messages in a keyword-argument fashion for consistency. In fact, in most implementations they will actually constitute signal levels on wires. The physical layer is responsible for filtering upcoming sensor information and performing sensor fusion to support the more abstract information delivered to higher layers.

Service Elements: Requests

- **Actuate <device_id> <parameters>**.--Send the indicated parameters to the specified device.

Service Elements: Indications

- **Value <sensor_id> <value>**.--The specified sensor has indicated the specified value.

Protocol Data Units

The only PDU's at the hardware level are the physical MTU's moving between systems.

3.2.6. An Example

Here we trace the roles of presentation and material link layers in a **Give**, to initiate movement of an MTU between two systems. Typically, a workstation issues **Give** to a neighboring transport mechanism to acquire parts or tooling. The two systems are sOurce and sInk, and their layers are labeled with their letters, thus: "I:Presentation." Mnemonically, we think of the MTU as moving Out of O and Into I. Two destinations are involved in the protocol. I sets a destination for the MTU at its load point, but O sets a pseudodestination that lies outside of itself and thus initiates the actual transfer of the MTU. The MTU never reaches this destination, so no **Arrived** is generated in O. The pseudodestination simply determines what neighboring MTS receives the MTU.

A **Take** sequence is exactly the same as a **Give**, except that the initial message travels from O to I instead of from I to O.

`I:Ap --> O:Ap:` **Give**.
`Both sides set destination:`
    `I:Ap --> I:Ro --> I:ML:` **Destination <load point>**.
    `O:Ap --> O:Ro --> O:ML:` **Destination <pseudodestination>**.
`O:ML --> I:ML:` **SpaceRequest**.
`I:ML --> O:ML:` **SpaceAvailable**.[1]
`O:ML --> I:ML:` **Coming**.
`I:ML --> O:ML:` **Ready**.

Exchange Sequence:[2]

---

[1] If I has no space available, it is deadlocked, but this is its fault, since it should have reserved space before initiating **Give**.

[2] The next sequence varies depending on whether O has an exit sensor or not. The protocol works for any combination of exit and entry sensors, including the case where neither MTE has either one. However, from an engineering perspective, every interface between MTE's should be required to have at least one sensor, either an exit sensor on the source or an entry sensor on the sink.

```
    Sequence if O has exit sensor:
        O:ML --> O:Ph: Go.
        O:Ph --> O:HW: Actuate.
        O:HW --> I:HW: MTU moves.
        O:HW --> O:Ph: Value <exit sensor>.
        O:Ph --> O:ML: MTUOut.
        O:ML --> I:ML: Gone.
        These steps only if I has entry sensor:
            I:HW --> I:Ph: Value <entry sensor>.
            I:Ph --> I:ML: MTUIn.
        I:ML --> O:ML: Got.

    Sequence if O does not have exit sensor:
        O:ML --> I:ML: Gone.
        O:ML --> O:Ph: Go.
        O:Ph --> O:HW: Actuate.
        O:HW --> I:HW: MTU moves.
        These steps only if I has entry sensor:
            I:HW --> I:Ph: Value <entry sensor>.
            I:Ph --> I:ML: MTUIn.
        I:ML --> O:ML: Got.

I:HW --> I:Ph: Value <presence sensor>.
I:Ph --> I:ML: NextMTU <MTU_id>.
I:ML --> I:Ro --> I:Ap: Arrived.
I:Ap --> O:Ap: Received.
```

3.3. A Closer Look at the Layers

In this section, we examine each layer of the LLAMA architecture and specify several alternatives for each. The alternatives are presented in increasing order of complexity. The next section shows how each of the strategies presented in section 1.2 can be created using these alternatives. When partial functionality is used at a given layer, the observation given in section 2.4.3 about the weakest leak should be considered.

3.3.1. Application

There are four different alternatives for this layer: sequential, sequential with MTS storage, Non-sequential with WS storage, and Non-sequential with MTS storage.

Sequential

For this case, this layer is null and simply passes the requests and indications through without modification.

Sequential with MTS Storage

Only the **Start** and **Stop** requests need to be supported.

The workstation application layer will request a part from the MTS using the blocking **Give** PDU. A blocking **Give** is appropriate in this scenario since the source of all parts for the workstation is fixed and the workstation cannot continue its work until this request has been satisfied.

Non-sequential with WS Storage

We assume that in this scenario the workstation does all the local buffering of parts. Since the MTS cannot store material except for the duration of the transfer, **Find**, **Found**, and **Lose** PDUs are not necessary.

When the workstation needs a part and has space to store the requested part it issues a **Give** to the MTS. The application then finds an appropriate WS that has the requested part in its output buffer and issues a **Give** on that part. The transfer of material is complete when these two **Give** PDU's are satisfied.

Non-sequential with MTS Storage

This application has the most capability of the application discussed so far. Not only can the MTS route parts between any two WS, but it also acts as the local buffering agent. In this scenario, the WS are conceptually simpler since they do not need to buffer material. However, since the MTS now stores parts, the entire set of PDU's outlined in section 3.2.1 must be supported. The **Give** provides a pull mechanism, the **Take** provides a push system, and the **Find**, **Found**, and **Lose** PDU's provide advance knowledge to the MTS that a WS needs a MTU so that the MTS can retrieve it from storage and deliver it to the WS in a timely manner.

3.3.2. Presentation

This layer is needed if two different types of application are connected to one another. It is not yet defined in LLAMA, but will be explored in the future.

3.3.3. Session

Since LLAMA deals with the transfer of a single MTU at all times, there seems to be no need for this layer.

3.3.4. Route

There are two alternatives for the route layer: sequential and random routing.

Sequential Routing

WS's are placed sequentially. The route between adjacent WS's is fixed in the hardware configuration of the MTS. In this case, the route layer is null and just passes requests and indications through without change.

Random Routing

The routing among the MTE's to deliver the MTU to is destination is accomplish by this layer. Also the mapping from the logical layout as represented in the application to the physical elements is done at this level. The **Destination** request sent from the application layer is converted to a series of **Destination** requests for the material link layer. The route layer may reroute MTU's based on congestion or exception indications received from the layer below.

3.3.5. Material Link

There are three alternatives for the material link layer: sequential processing, non-sequential processing, and non-sequential processing with random routing.

Sequential Processing

This is the first non-null layer in a synchronous sequential architecture, and follows a three step cycle of receiving parts from workstations, indexing the MTS, and giving parts to following workstations.

1. To guarantee that all workstations are done with their current part, this layer waits for the completion of the PDU interchange outlined above at each WS.

2. When all WS's have delivered their parts, the layer will issue a single **Go** request to the physical layer.

3. When this layer receives the **NextMTU** indication from the physical layer, it initiates the PDU sequence to deliver each part to the next workstation.

Non-sequential Processing

If the material link layer must handle nonsequential processing of arrivals and departures of MTU's to and from MTE's, this layer must supply two processes for each MTE. One process handles the transfers of pallets into the MTE and the other handles the outgoing pallets. The exchange sequence for the PDU has been outlined in section 3.2.6.

Non-sequential Processing, Random Routing

In addition to the capabilities of non-sequential processing mentioned above, the **Destination** request must be converted to a **Go** request for the physical layer. This conversion is based on a routing table generated by a **SetSystemParameter** request from the route layer.

3.3.6. Physical

There are three alternatives for the physical layer: sequential, non-sequential, and non-sequential with random routing.

Sequential

The model of the MTE is a track connecting two fixed WS's. The track accepts only a **Go** request, which transfers a MTU between the two WS's. No parameters need to be associated with the request. Normally, all such **Go** requests are issued simultaneously.

Sequential with Storage

The model of the MTE is still a track connecting two fixed WS's, but now the track queues MTU's. It must have a sensor at the head of the queue. This sensor triggers the material link PDU sequence. A sensor must record the transition from the WS to the MTS at the entrance of each MTU. This sensor can belong to either the WS or the MTE, as explained in section 3.2.3.

Random with Storage

In addition to all the functionality described above, the layer must be able to direct the pallet physically to different MTE's at the exit, and accept pallets coming from different MTEs at the entrance. The model for this MTE was given in section 3.1.2.

Random

An MTE as described above but without storage capability. It must respond to the same requests and protocols. The major difference is that it completes processing of a MTU before it accepts another MTU.

3.3.7. Hardware

Current material handling hardware includes dial machines, power and free conveyors, free conveyors, belt conveyors with switching elements, manual methods, and automatic guided vehicles. See section 1.2.1 for a description of each of these technologies.

3.4. System Design in LLAMA

Exhibit 3 indicates how each of the example stategies presented in 1.2 is implemented in LLAMA, using the different alternatives mentioned above. The alternative listed in the table is the alternative with minimal functionality needed to provide the desired services. Another alternative with additional services could be used in place of the indicated one. When designing a MTS the law of the weakest leak should be kept in mind, when deciding which alternative is appropriate at each layer.

	STRATEGIES:				
	Sequential		Non-Sequential		
	Synch	Asynch	Push	Pull (Kanban)	CASCADE
LAYERS:					
Application	Seq.	Seq. MTS store	Non-Seq. WS store	Non-Seq. WS store	Non-Seq. MTS store
Presentation	Null	Null	Null	Null	Null
Session	Null	Null	Null	Null	Null
Route	Seq.	Seq.	Random	Random	Random
Material Link	Seq.	NonSeq.	Nonseq. Random	Nonseq. Random	Nonseq. Random
Physical	Seq.	Seq. Store	Random	Random	Random Store
Hardware	SEE EXHIBIT 1				

Exhibit 3

4. Conclusions

LLAMA demonstrates that the techniques of layered decomposition in an implementation hierarchy that have served the communications community so effectively are equally promising in the arena of material handling. It offers the prospect of standardization in this field, with significant lifecycle economies resulting from vendor interoperability and reusability of software, and illustrates the value of a single reference model in serving as the framework for different specialized architectures.

We acknowledge with gratitude the careful review and helpful suggestions of our colleague K.H. Muralidhar. The programs of the Industrial Technology Institute are partially supported by the W.K. Kellogg Foundation.

References

[1] Y. Sugimori; K. Kusunoki; F. Cho; and S. Uchikawa, "Toyota production system and Kanban system: Materialization of just-in-time and respect-for-human system," *International Journal of Production Research*, vol. 15, no. 6, pp. 553-564, 1977.

[2] H.V.D. Parunak; P.W. Lozo; R. Judd; B.W. Irish; and J. Kindrick, "A Distributed Heuristic Strategy for Material Transportation," in *Proceedings of the 1986 Conference on Intelligent Systems and Machines*, Oakland University, 1986.

[3] H.V.D. Parunak; J. Kindrick; and B.W. Irish, "A Connectionist Model for Material Handling," in *Proceedings of the International Conference on Manufacturing Science and Technology of the Future*, MIT, 1987, pp. 183-187.

[4] H.V.D. Parunak; J. Kindrick; and B.W. Irish, "Material Handling: A Conservative Domain for Neural Connectivity and Propagation," in *Proceedings of the Sixth National Conference on Artificial Intelligence (AAAI'87)*, American Association for Artificial Intelligence, 1987, pp. 307-311.

[5] H.V.D. Parunak; J. Kindrick; and B.W. Irish, "A Connectionist Model for Material Handling," *Robotics and Computer Integrated Manufacturing*, vol. tbd, no. tbd, pp. tbd, tbd 1988.

[6] F.P.M. Biemans, "The Design of Distributed Transport Systems as a Major Standard Interface in Computer Integrated Manufacturing," *Computers in Industry*, vol. 7, pp. 319-331, 1986.

[7] J. Day and H. Zimmermann, "The OSI Reference Model," *Proceedings of the IEEE*, vol. 71, no. 12, pp. 1334-1340, December 1983.

[8] H.V.D. Parunak and J.F. White, "A Synthesis of Factory Reference Models," in *Proceedings of the IEEE Workshop on Languages for Automation, August 1987, Vienna*, IEEE, 1987, pp. 109-112.

[9] H.V.D. Parunak, "Manufacturing Experience with the Contract Net," in *Distributed Artificial Intelligence*, Huhns, M., ed. Pitman, 1987, pp. 285-310, originally presented at the 1985 Workshop on Distributed Artificial Intelligence, Sea Ranch, CA.

[10] H.V.D. Parunak; J.F. White; P.W. Lozo; R. Judd; B.W. Irish; and J. Kindrick, "An Architecture for Heuristic Factory Control," in *Proceedings of the American Control Conference*, American Automatic Control Council, 1986, pp. 548-558.

Simulation Technology for the Design and Scheduling of Material Handling and Storage Systems

F. Hank Grant, Ph.D.

FACTROL, Inc.
P.O. Box 2569
West Lafayette, IN 47906

ABSTRACT

The primary objective in acquiring material handling systems is usually the same: find the best system to meet objectives at the least cost. The design and operation of material handling systems is a complex and difficult task. Material handling is also not exactly a prestigeous focus for a career in U.S. industry. Our best talent is rarely assigned to manufacturing, much less material handling.

In Europe and Japan, material handling has long been recognized as a critical link in the manufacturing chain. In fact, it is the key element of integration in modern manufacturing systems.

This article will explore the use of simulation technology as one tool in the design and scheduling of material handling components of production systems. Material handling is examined as it is perceived in industry vis-a-vis its critical role in integrated manufacturing. Simulation and Scheduling is reviewed from a historical perspective and simulation-based tools for designing and scheduling production systems discussed. An overview of some important applications is provided followed by an overview of research needs regarding simulation and material handling.

NATO ASI Series, Vol. F53
Advanced Information Technologies for
Industrial Material Flow Systems
Edited by Shimon Y. Nof, Colin L. Moodie

INTRODUCTION

Manufacturing companies have a pressing need to understand new technology and the potential for its use in a rapidly changing industrial environment. Although simulation has been used traditionally to help explore the ramifications of new technology for manufacturing systems, many companies now require that simulation studies of proposed manufacturing facilities be performed before a final decision on implementation of either new or current technology. Simulation can provide insight into issues that are not apparent or are counterintuitive. Many software tools are now available to provide engineering support for system design projects for creating material handling systems. [Musselman, 1984 Haider and Banks, 1986, Pritsker, 1986]

Manufacturing companies in the U.S. face many difficult problems. Some of those problems stem from the lack of prestige given to manufacturing as a career. Further, within manufacturing, material handling is not highly regarded. The best personnel are rarely given manufacturing assignments and material handling is almost never included.

Attitudes are changing, however. It's now being recognized that material handling is a substantial part of what is key to making manufacturing successful. Also, manufacturing is beginning to understand the importance of material handling in integrated manufacturing. Material handling is viewed as the integrating element in CIM systems.

The use of simulation for scheduling and control of material handling systems is a natural outgrowth of its application in design. The objective in designing a manufacturing facility is to obtain knowledge sufficient to make capital commitment decisions that satisfy production objectives, and these same characteristics are also important in controlling production. Simulation, when used for production scheduling, provides a vehicle for achieving the discipline necessary for effective production management on the factory floor, thereby helping to achieve productivity goals through efficient, effective use of new, integrated technology. New computer technology, such as powerful workstations, coupled with advances in easy-to-use software systems, make simulation a useful tool for the production scheduler. [Standridge and Pritsker, 1987]

Simulation is an analysis tool that can be applied effectively to a variety of shop floor design and real-time shop floor control problems. Simulation can support longer-term system design evaluations of resource requirements, equipment needs, inventory buffer sizes and availabilities, and sensitivity analysis of a variety of product demand and equipment performance possibilities. Simulation can also support shorter-term decisions involving equipment scheduling, shop order release, and work order scheduling.

Historically, simulation techniques have been highly successful and used extensively for the planning and analysis of current operations or proposed designs. Although on-line simulation analysis is feasible and can be cost-effective, applications of simulation in this mode have been few. On-line applications of simulation for real-time shop floor control purposes can be applied effectively only if the data supplied to the simulation models are accurate, organized, and timely. In the past these constraints have proved difficult to overcome.

More recently, major advances have been made in data-base technologies. These improvements make the access and manipulation of data for real-time factory control a reality. A modeler can now construct a simulation model of an operating unit and supply this model with accurate and timely data describing the performance of machines and the operators and the expected demands on the manufacturing system.

Other data processing improvements that facilitate the use of simulation for on-line analysis relate to computational capability and the graphic display of results. Managers have historically disliked having to wait a long time for analysis. With the computing capabilities and graphic constructs that are now available, managers not only can get quick response for analysis of work order scheduling or work order release, they can also receive the information in a more easily understood form.

This paper discusses how these latest developments will support the use of simulation for designing and controlling material handling systems. It also identifies and discusses the important research areas and the challenges in effectively applying this technology.

MATERIAL HANDLING ISSUES

Material handling has long suffered from neglect in manufacturing systems. Part of the reason for that neglect is the development of a true understanding of the function of material handling in manufacturing and its importance in the overall scheme of production. Attention has been focused on material handling through the development of automated systems. In automated systems, it is clear that material handling will be the life's blood of production. The system will simply not work if the material handling components are not well designed and effective.

A more complete definition of the role of material handling in manufacturing and in integrated systems has been articulated as "using the right method to provide the right amount of the right material at the right place, at the right time, in the right sequence, in the right position, in the right condition, and at the right cost". [Tompkins and White, 1984] p. 166 Since material handling systems cut across the boundary of so many components of manufacturing, material handling can be considered as the system integrator for production systems. [White, 1984-1986]

A more traditional view of the role of material handling in manufacturing is shown in Figure 1. In this view, material handling is a secondary system, subservient to the production process, which is looked upon as the heart of manufacturing. Material handling has been traditionally seen as being a component of manufacturing which is difficult or impossible to manage, or even unnecessary to manage. The attitude has been that material handling management focused on the availability of material and its delivery to the production facility. Production personnel were then left to their own means to assure arrival of material to the appropriate production operation.

Automated systems have changed this view. And this has influenced the operation of manual systems as well. Figure 2 provides a more current view of the role of material handling in industry. Production cells are linked to each other and other parts of the production process by the material handling system. Material arrives at off-line storage from suppliers. From there, the delivery of the material to the production cells must be coordinated by material handling operations. Further, material handling must coordinate the movement of material between production cells and to on-line storage between the cells.

Thus, material handling is crucial in a successful manufacturing system. While production continues to be a focus, material handling is the component that provides the integration element in manufacturing.

If integration is provided to a large extent by carefully designed, effective material handling systems, what are the barriers to effective implementation? The following reasons are representative [White, 1988].

1. It is not easy to design and implement integrated systems. A detailed understanding is required and nothing can be left chance.
2. They are a radical departure from tradition. In some cases, it may be impossible to create an integrated system using a traditional manufacturing view.
3. There are few apparent rewards. The motivations for individuals to get involved are minimal.
4. Insurance is the conservative, preferred approach in manufacturing. This takes the form of inventories, excess people, etc.
5. Organizational barriers must be overcome. Effective integration projects cut across many disciplines.
6. The concept of an integrated system is not well understood. And mean different things to different people. Integration of information is different than integration of hardware which is different than integration of product design.
7. A champion is required for effective integration. A leader must be present and committed.
8. Few success stories exist, but numerous horror stories are available. Top management tends to remove itself from manufacturing and the belief continues that integration is expensive, risky, and complex.

9. Resources are limited. These include both capital and people.
10. Some individuals are threatened by the use of integration. Direct labor may be reduced. Indirect labor is the true target, since they provide the integrating link in many manufacturing systems.

Even with these barriers, the evidence is overwhelmingly on the side of the advantages of integrated systems. Simulation can provide the technology to address a number of the barriers cited above. In support of design, simulation provides a laboratory for the investigation of system designs and verification of operation prior to implementation. While the new system may be a radical departure from tradition, simulation can provide counterintuitive information about the behavior of the system and provide the confidence to proceed.

While the rewards may not increase, the probability of success can through the use of simulation. This should motivate more individuals to adopt integration. Insurance can be provided through the application of simulation based scheduling and sequencing. Since the prediction of system performance will be much more precise, inventories need not be as large, or can be eliminated entirely in some cases.

Organizational barriers will always be present as long as there are organizations. However, simulation for design and scheduling can provide a basis for communication that can help overcome those barriers. Simulation can also help by providing a common ground for understanding integration in the context of a manufacturing company's production operation. Various components and systems can be included and the individuals responsible can understand the operation of the system and the interaction of their part with others.

Next, the element of a champion is required in any success story. Simulation can provide that champion with a tool to support his cause and reaching the goals of integrated manufacturing. Success stories will take time in any departure from tradition that is poorly understood. Simulation, and other computer technology will be a part of that.

Resources are typically a problem in any manufacturing operation. Simulation can help to understand and justify those resources and the role they will play in the production process. While some individuals may feel threatened by integration, it is primarily the indirect labor component which will be reduced through integration. Hopefully, those individuals can move into more productive activities in the operation.

Simulation is not the ultimate panacea for manufacturing integration. But it is a major contributor to integration activities in designing production facilities and in scheduling and controlling them once they have been completed. The next section will provide a historical perspective on simulation for design and scheduling of material handling systems, and discuss some tools which will be particularly useful.

HISTORICAL PERSPECTIVE AND CURRENT TOOLS

Simulation has been available for many years and a variety of tools are now available to support its use. It is recognized as an important tool in making the multi-million dollar capital expenditure decisions in manufacturing operations.

Simulation has traditionally been used to design industrial facilities. However, there is also interest in using the same technology that predicts behavior in a newly designed system to predict the performance of production on a day-to-day basis for existing systems.

The use of simulation for the scheduling and control of material handling systems is now available through new technology which is based on simulation. The growing use of computers on the factory floor has made this possible.

The following discussion provides an overview of the use of simulation for design, the traditional methods of scheduling, and the use of simulation for scheduling and control.

Simulation Technology For Design

Simulation is the process of creating a representation, or model of the operation of a system on a digital computer. This model is created by providing to the computer with a description of the physical components of the system and the logic associated with the operation of the system. The model of the system under study is put together using a simulation language. This language gives structure to the model-building process by providing special constructs that relate to the system under study. For example, many languages provide constructs to represent components like queues and servers. Other more system-specific languages provide constructs to represent machines, parts, and process plans.

Many different simulation languages are available. These include general-purpose languages, such as SLAM II, SIMSCRIPT II.5, and GPSS, and more specialized languages oriented toward manufacturing, such as MAP/1, SPEED, and MAST [Haider and Banks, 1986; Miner and Rolston, 1983; Pritsker, 1986].

Once built, the simulation model serves as a laboratory in which various design alternatives can be tested and compared. By running the model on a computer, the actions of the proposed system are represented in detail, permitting inferences to be drawn about overall system performance. These inferences are made on the basis of numerous performance measures provided by the simulation, such as machine utilization, in-process inventories, part waiting times, and throughput. Through this process of experimentation, the best overall system design is selected.

Several features of simulation make it particularly advantageous for the design of manufacturing systems [Grant, 1988]. These features include physical and control system balance, system flexibility, random system behavior, and animation.

Physical and Control System Balance

Proper balance among the physical and the logical and control components of a manufacturing system must be maintained during system design. Should the design of one component be emphasized disproportionately, overall system performance could suffer. Designing a system with proportional importance given to each component is possible through the total systems perspective that simulation provides.

For example, in modeling a flexible manufacturing system (FMS), the system can be viewed as being composed of two interrelated subsystems: a physical subsystem and a control subsystem. The physical subsystem, which transports, stores, and processes parts, consists of programmable machines, material handling equipment, and in-process storage facilities. The control subsystem, which selects, sets priorities, routes parts, and controls traffic flow, consists of situation-dependent logic to coordinate part interchange in the physical subsystem. These two subsystems must work in concert. It is important, therefore, to consider the interaction between these two subsystems during the design. Isolating, for analysis purposes, one subsystem from the other or giving one preferential treatment will most likely result in misleading conclusions.

A more complete evaluation, with appropriate consideration being given to the interplay between these two subsystems, is possible with simulation. In contrast to strictly analytical approaches, simulation includes the system's operating and control strategies as an integral part of the model. As parts move through the various operations, the model processes them according to the system's situation-dependent control logic. Simulation provides a rich environment for the design of physical as well as logical control systems.

System Flexibility

The increasing popularity of simulation is due, in large, to its ability to represent various levels of detail. Analytical formulations, while offering closed-form solutions, tend to be restrictive, since details must often be neglected in order to accommodate the formulation. Simulation, on the other hand, can provide as much or as little detail as the analyst wants. All relevant system characteristics can be taken into account, such as processing time variability, equipment reliability, fixture restrictions, in-process storage requirements, complex routing decisions, operating policies, and scheduling constraints. Simplifying assumptions, such as reduced decision models, are not required. The result is a flexible experimental setting in which to test alternative design strategies where the analyst has control over the details and assumptions that are included.

Random System Behavior

The operational behavior of manufacturing systems can be quite dynamic over time. That is, the interaction of various system components can trigger unexpected behavior that may occur only infrequently. The system can exhibit, for example, significant variation in demand for resources, depending on the particular interactions of its components. Although this dynamic behavior tends to increase the complexity of the design, and often the model, knowledge and understanding of this behavior can lead to significant design efficiencies. Simulation provides a means of understanding system dynamics and learning about components of the system that may interact and behave in a counterintuitive manner.

Proper accounting of random variability of system parameters helps ensure that a proposed design will exhibit stable performance in a variety of configurations. Ironically, instability can actually be promoted when certain sections of a system are selectively overdesigned to guard against processing variations. As a result, there can be unforseen weak links in the system. Through systematic experimentation with a model, weak links can be revealed and actions taken to correct them.

An example of this use of simulation is in the study of fixture load stations. To examine the effectiveness of a particular configuration for a fixture load station, the part arrival sequences and their demand on the station must be understood. Unfortunately, given the variety of part types that could be simultaneously processed and the processing variations that could occur for each part type, the number of different part arrival sequences is extremely large. Examining all possibilities in great depth is unreasonable. However, rerunning a representative, randomly selected set of part mixtures in the simulation model can provide insight into the station's performance. If the design of this station were found to be deficient, it could be modified and then tested against a broader range of part arrival sequences. This ability to deal with random behavior with the models and to create representative samples that stress the system's capability is a major benefit of simulation.

Animation

One aspect of simulation technology that has made considerable progress in recent years is graphical animation. Besides the usual plots of system performance, it is now possible to animate a manufacturing system operation in great detail [Grant and Weiner, 1986]. By bringing the model of the physical subsystem to life graphically, the operation of the control subsystem can be studied in action. This provides a detailed understanding of the implications associated with the system's control policies. Subtle errors can be identified and corrected before the system becomes operational.

Simulation has been quite successful in the design of material handling systems. Many tools are available in simulation languages to represent and analyze systems including autoguided vehicles, conveyors, and storage and retrieval systems. These are typically special features imbedded in the languages to make the modeling efforts straightforward.

Scheduling material handling systems represents another set of challenges. Before discussing how simulation can be used, a review of traditional scheduling techniques is in order.

Traditional Scheduling Methods And Their Limitations

Shop floor scheduling is an important task in managing a production system. It entails complex decisions that affect such objectives as meeting delivery due dates and maintaining a desired level of inventory. Although it is not possible to consider all of the variables that determine the effectiveness of a particular schedule, major productivity improvements can be realized by making the production scheduling process more effective. Effective scheduling and control of the material handling system make these productivity improvements possible.

The quality of a production schedule involves many, sometimes conflicting, objectives. Whereas maximizing throughput is certainly one important consideration, an ideal schedule will also have the following characteristics:

- o Delivery due dates will be met.
- o Inventory costs will be maintained at acceptable levels.
- o Equipment, personnel, and other limited resources will be well utilized and have balanced workloads.
- o Adaptations can be made quickly in the case of an unexpected event, such as equipment failure or raw material shortage.

Since it is difficult to "optimize" a schedule over all these characteristics, one or two characteristics are often chosen, depending on current production objectives. Generally, trade-offs must be made to reach a balance between these more limited objectives.

Production scheduling is done in many ways in industry. Probably the most common methods of scheduling are purely manual techniques. In the most straightforward form, an expert such as the department foreman or the machine operator selects the next job from those waiting in front of the machine. The criteria used in this circumstance often reflect the measures by which the scheduler is evaluated and may not reflect overall business objectives. Job status control boards are also used to lay out schedules visually.

A more analytical approach to scheduling is sequencing by dispatching rules. This method uses rules that set priorities for the jobs waiting for processing. Research has demonstrated that rules such as the weighted shortest processing time rule can generate reasonable schedules. The effectiveness of the schedule may vary widely, depending on the particular rule selected, the type of production facility, and the mix of jobs to be produced. It is impossible to predict which dispatching rules will work best in most manufacturing systems by traditional methods. They are also limited in the scope of what they consider and are often hard to implement on the shop floor in a cost-effective manner.

Material requirements planning (MRP) was one of the earliest computerized techniques for factory management. In its earliest form it generated unconstrained production schedules, which were based on a bill of materials and estimated production time requirements. Manufacturing resource planning (MRP-II) expanded the scope of MRP to consider many other facets of production facility management. In addition to sophisticated factory accounting capabilities, modules for capacity planning and shop floor data collection were also provided. These techniques are effective for longer-term scheduling and order launching, but they lack the detail necessary for effective day-to-day production scheduling.

Recently, there has been emphasis on more sophisticated definitive capacity planning tools as a means of generating production schedules. These methods determine the expected critical resource and then schedule forward and backward around that resource. At present, these techniques do well on a global level but are inadequate when highly interactive components are present and frequent changes are needed. The computational time required to generate schedules is usually large, with interactive execution being impractical. Also, the critical, or bottleneck, resources tend to change, based on production demand.

As an alternative to these approaches, simulation-based scheduling can provide an effective tool for shop floor scheduling while requiring few assumptions. The schedules generated are based on an accurate, realistic model of the production facility.

Scheduling of material handling systems has unique requirements. Scheduling is difficult in that many types of material handling resources are typically not well controlled. (i.e., fork lifts and industrial trucks) But there are application of material handling where these types of devices have been scheduled, and very effectively.

Automated systems also command effective material handling before they will work at all. These systems require frequent scheduling and rescheduling as production executed as well as an effective information system to drive the material handling operations.

Simulation can be an effective scheduling tool. The following discusses simulation's role, and tools for its application in the scheduling environment.

Simulation-Based Scheduling

Simulation practitioners are familiar with the ability of simulation models to predict system behavior in great detail. It is a natural extension to attempt to apply simulation on a day-to-day basis to predict schedule performance and resolve problems before they occur. Simulation is well suited to this type of analysis and, with proper support, can be used successfully to create and evaluate production schedules. The simulation model provides a computer replica of the department in the factory. The model plays through the schedule and provides performance information. Event trace information can provide the details of a feasible schedule, given the constraints specified in the model. The detailed interaction of various production limitations can be included at any level of detail.

The dynamic interactions between resources can be captured and analyzed with simulation. For example, a material-handling vehicle may deliver product to several machining stations. Even though this vehicle may have an expected total workload requiring only 50 percent of its available time, two stations that need material at the same time will cause the system to perform differently than it would if there were no conflict for the vehicle's services. Simulation can be used to evaluate the effect of these conditions on schedules and to resolve the conflict.

In another case, the scheduler may have many options for selecting the transfer batch size in a production order. The entire order may be processed at each operation, or the order may be split into two or more separate loads and allowed to move independently through the operation. Simulation would allow the scheduler to contrast alternatives and select the one that provides the best performance.

Because this analysis is performed on a computer, many alternatives can be explored with relatively little expense or risk. Simulation models can represent a large variety of the factors critical to a manufacturing system's performance in great detail. The influence on tooling, personnel, and other resources can be evaluated. Capacity changes, such as a machine breakdown or scheduled maintenance, may also be included in the model.

Most simulation languages have the modeling features necessary to represent production systems. But there are many additional needs in scheduling that are not typically provided. To be effectively used for scheduling, simulation models must provide reports that can be readily understood and used by the production scheduler. Preferably, reports should be provided that serve as actual job performance schedules for equipment and personnel and are consistent in format. Also, a strong user interface is needed to support the scheduler and to allow schedules and other reports to be created easily. The scheduler should not have to interact with the details of a simulation model but still should be able to use it easily. Most simulation languages do not provide these features.

Traditionally, large simulation models of production facilities have been too expensive to build and too cumbersome for use on a daily basis. The modeling process has required a highly trained and experienced analyst with a solid understanding of both the simulation language and system under study. Even after the model is completed, execution of detailed models could take too long to be useful in production scheduling. This problem has been addressed in part by the advance of computer technology. Workstations that provide mainframe-like capabilities are moving rapidly to the factory floor.

One Available Scheduling Tool

Currently there are few commercially available products that provide tools for applying simulation to production scheduling. One product that addresses scheduling is called FACTOR [Grant, 1986; Grant, 1987, Pritsker et al., 1986]. It is designed to be highly interactive and used by shop floor scheduling supervisors. The architecture for applying FACTOR in production systems is shown in Figure 3.

FACTOR supports two types of users in providing a shop floor scheduling system. The primary user is the factory floor production supervisor, who uses it to generate production schedules. The second user is the individual responsible for modeling the factory and developing the data that characterize the manufacturing system and the products that are produced in it. The model development activity typically occurs only once in the installation of FACTOR. Modification of the FACTOR model is easily accomplished to incorporate changes in the production department after it has been installed.

FACTOR can be integrated with the production management software and plant data to provide a useful tool for the production scheduler. One important feature that FACTOR provides is an interface to the MRP system to extract information related to a master schedule. It also interfaces with the shop floor status system to acquire production data. In FACTOR, this is typically a straightforward interface to the particular MRP system that is in use. This interface collects the current list of orders to be processed at the production department and automatically loads them into the FACTOR system. A similar interface is provided to collect information regarding the current location and status of released orders from the shop floor data collection system. This information is also loaded into FACTOR to initialize the scheduling system automatically to start at current shop floor status. These data are stored in a local data base that may be accessed and edited by the production supervisor.

Once the scheduling supervisor has gathered the current production requirements, FACTOR is used to simulate production and identify potential problems. The supervisor can select from a large variety of ranking priority codes to sequence jobs through the production system. These codes include familiar priority measures such a due date and shortest processing time. In addition, more complex ranking priorities are provided, such as estimated remaining processing time based on past performance as well as critical or tactical ratio. The critical ratio measure is the ratio of the scheduled remaining production time to the estimated production time. The tactical ratio is the inverse of the critical ratio. Additional priority schemes are available based on production costs and other system status variables.

Tools such as FACTOR make many contributions, including implementation of technology necessary to schedule the factory floor. Another is the development of the framework necessary for further developments in scheduling technology.

FACTOR provides a number of tools to address the needs of scheduling material handling systems. Extended modeling options are provided to address automated material handling components, including autoguided vehicles, conveyor systems. Also, other types of material handling devices such as forklifts and industrial trucks can also be addressed. FACTOR also provides tools to support storage and retrieval systems and the generation of schedules which reflect their use.

APPLICATIONS

Simulation has had much success in applications to material handling system environments. Its success in designing industrial facilities is well known. The following example applications will focus on success in design as well as in scheduling and control. [Davies, 1988; Grant, 1987]

CTS Corporation

Shop floor control is the focus of the application at CTS. This application is part of an overall CIM project to address major issues in operating their business in a highly competitive environment. These issues focus on cost control, high quantity levels, and the requirement for shorter lead, cycle, and delivery times.

The CTS plant, in Berne, Indiana, produces resistor networks for the data processing, telecommunications, automotive and assorted electronic markets. Some customers order as many as 2000 part numbers, so scheduling for timely delivery at the same time as for effective manufacturing is a complex process.

Simulation was used as part of the design process in creating the facility, which is a highly automated production environment. Once product is loaded, it can progress through the system with little human intervention.

The scheduling system (FACTOR) is integrated with the MRP-II system to provide order entry, capacity requirements planning, and master scheduling. This higher level planning data is then routed to the minicomputer where FACTOR resides to assign the work to specific machines and resources. A detailed sequence for production is produced and then executed by the system controller.

The use of simulation for shop floor scheduling has enhanced the flow of material through the shop. With the focus on just-in-time delivery, efficient simulation based scheduling provides shop-floor control via quicker production, less inventory, and better control over quality.

Garrett Turbine Engine Company

Mainframe based scheduling was already available when Garrett implemented simulation based scheduling in their production system in Phoenix, Arizona. It provides a high level prioritization of production. Simulation based scheduling provides the detailed view to schedule and sequence at the machine level or to cope with rescheduling due to machine failures.

Garrett tested the concepts an theoretical problems for a variety of production environments. These varied from a general manufacturing department to a dedicated cell. Variations at the dedicated cell include a direct interface to the cell controller to update the simulation information relative to machine status.

The approach worked well, particularly in the automated environment. Garrett is now implementing the technology in scheduling a high volume production line for turbine fan blades as well as in a more flexible cell to produce gears.

Digital Equipment Corporation

DEC is a long time user of simulation for design for production systems. Their Hudson, Massachusetts facility produces simiconductor wafers for internal use in DEC's line of CPUs. In investigating the use of simulation for shop floor scheduling, their primary needs were to reduce work-in-process and increase throughput, while providing a tool that the production people could effectively use.

One particular step in the production process was especially critical. This was the photo area. Product cycles throughout this area many times in the production process and it was critical that they schedule it carefully to avoid excess buildup of material prior to entry as well as delays in downstream steps.

Their goals were also to provide a tool which could be used by production personnel through a friendly easy-to-use interface that would shield their users from simulation technicalities. Material handling management requirements are also satisfied through the systems modeling capabilities.

Productivity goals are to cut cycle time in the department by approximately 50%. Current cycle time is approximately one day and each production order goes through critical departments several times.

CONCLUSIONS

This paper has discussed the effective use of simulation for both the design and scheduling of material handling systems. Simulation as a technology is intuitively appealing to a wide audience and can provide much insight to reduce or remove the barriers to effective integration, which is the real focus of material handling systems in modern manufacturing.

The future growth of this technology will depend upon the effective use of both hardware and software technology for successful integration projects. Cheaper and more powerful computers will make this technology more prevalent on the factory floor.

There are many research opportunities in the use of simulation for material handling design and control. More effective, easy-to-use simulation tools are needed to build and analyze material handling systems. The link needs to be formalized between control systems and simulation models so that the simulation modeling components can generate the control system design. Finally, work needs to be done to better integrate the control system operation with short term decision making as provided by simulation tools.

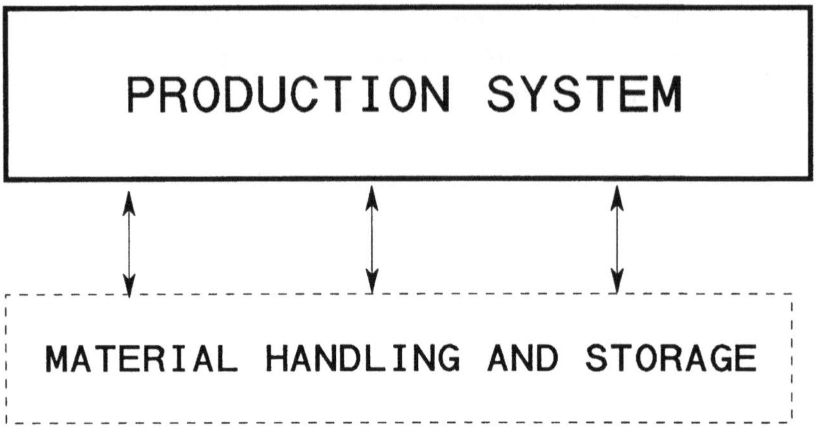

Figure 1. Traditional Material Handling View

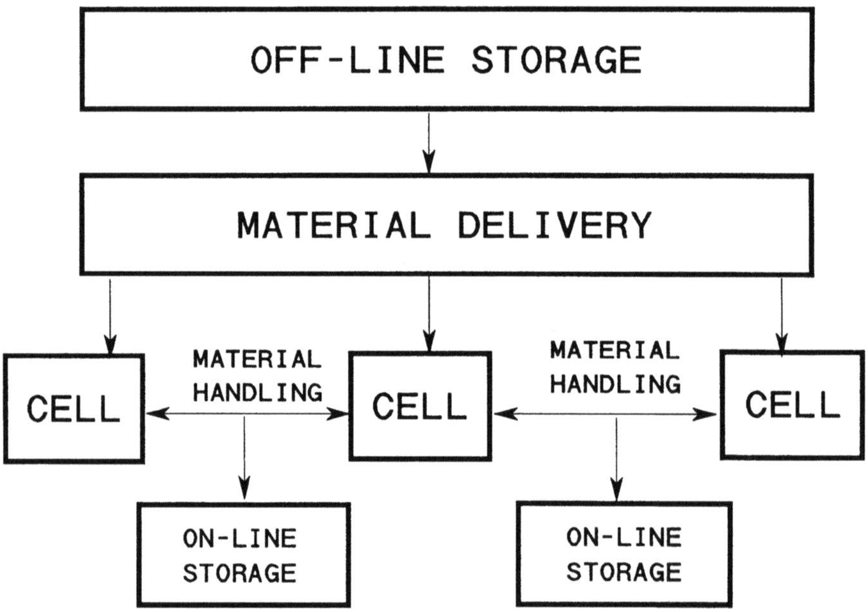

Figure 2. Integrated Material Handling

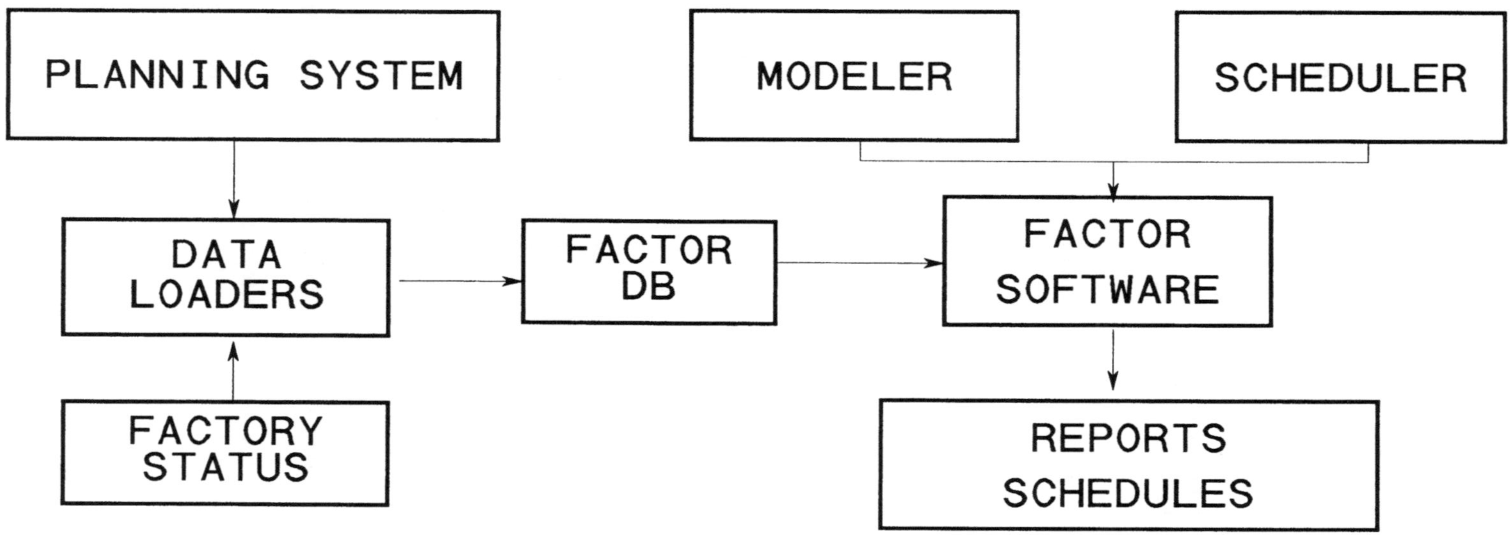

Figure 3. FACTOR Architecture

REFERENCES

Davies, O. Gaining flexibility in scheduling. Managing Automation. February, 1988.

Grant, F. H. 1986. Production scheduling using simulation technology. Pp. 129-138 in Proceedings of the Second International Conference on Simulation and Manufacturing. Bedford, England: IFS Conferences, L'.d.

Grant, F. H. 1987. Scheduling and loading techniques. In Production and Inventory Control Handbook, second edition, F. H. Grant and J. H. Green, Eds. Falls Church, VA: American Production and Inventory Control Society.

Grant, F. H. 1987 Simulation in Designing and Scheduling Manufacturing Systems. Design and Analysis of Integrated Manufacturing Systems. National Academy Press. 1988.

Grant, J. W., and S. A. Weiner. 1986. Simulation Series, Part 4: Factors to consider in choosing a graphically animated simulation system. Ind. Eng. 18(8):37-38 and 65-68.

Haider, S. W., and J. Banks. 1986. Simulation Series, Part 3: Simulation software products for analyzing manufacturing systems. Ind. Eng. Errata 18(9):87.

Miner, R. J., and L. J. Rolston. 1983. MAP/1 User's Manual. West Lafayette, Ind.: Pritsker & Associates, Inc.

Musselman, K. J. 1984. Simulation: A design tool for FMS. Manuf. Eng. 93(3):117-120.

Pritsker, A. A. B. , F. H. Grant, and S. D. Duket. 1986. Simulation in real time factory control. Presented at a conference on Real Time Factory Control, May 13 and 14, 1986. Dearborn, MI: Society of Manufacturing Engineers.

Standridge, C. R., and A. A. B. Pritsker. 1987. TESS: The Extended Simulation Support System. West Lafayette, IN.: Pritsker & Associates, Inc.

Tompkins, J. A., and J. A. White. 1984. Facilities Planning. New York: Wiley.

White, J. A. 1985. Selling integrated systems. Modern Material Handling 40(3):29.

White, J. A. 1986a. Becoming the systems integrator by the year 2020. P. 372 in Proceedings of Fall Industrial Engineering Conference. Norcross, GA.: Institute of Industrial Engineers.

White, J. A. 1986b. Impediments to system integration. Modern Material Handling 41(10):23.

KNOWLEDGE BASED APPROACHES FOR PRODUCTION CONTROL AN ASSESSMENT

Amitava Dutta
Management Science Department
College of Business
University of Iowa
Iowa City, IA 52242

ABSTRACT

With increased automation in the manufacturing process, planning and control of material flow in a factory is being performed by computer based systems. Control of the material flow involves responding to unanticipated events. In this paper, we review the major activities associated with the control process, and then evaluate current efforts to apply knowledge based techniques in performing them. Our assessment is that, while such efforts have met with some success in generating corrective responses, there still remains considerable room for the application of such techniques in the detection of exception conditions. Furthermore, under a hierarchical view of factory operations, knowledge based methods appear to provide a basis for aggregation/disaggregation of information necessary to detect exceptions and take corrective actions.

KEYWORDS: Automated Manufacturing, Production Control, Knowledge Based Systems, Scheduling.

NATO ASI Series, Vol. F53
Advanced Information Technologies for Industrial Material Flow Systems
Edited by Shimon Y. Nof, Colin L. Moodie

1. INTRODUCTION

There are two basic flows in any manufacturing environment. One is the physical flow of material, the other, flow of information. Obviously, the two flows are related and affect each other. Automation of the manufacturing activity has involved the increasing use of computers in managing these two flows, so as to achieve increased flexibility. This is seen in the use of CNC machines and computerized material handling systems, as well as in the increasing use of computer based techniques to perform planning, scheduling and control at almost all levels of operation. Our concern here is mainly with the flow of information, and the decision processes associated with that flow. For this, it is first necessary to adopt a systematic view of factory operations. The predominant view is a hierarchical one [BlM83] [Spu87] [Wec87] [McL81], such as that shown in Figure 1. It is clear that some operations in the hierarchy are time-critical, while others are not. This is reflected in the information flows and decision processes at each level. Conceptually, a control function executes corrective actions in response to detected deviations from a prespecified plan of activity. Such deviations arise inevitably due to a variety of unanticipated events on the factory floor. Under the hierarchical view just mentioned, this implies that corrective actions must be undertaken at each level, in order to achieve total factory control. This control activity is the focus of attention in this paper. More specifically, the characteristics of control activity at each level are reviewed, and the applicability of knowledge based methods is evaluated. Current efforts at using such methods are also reviewed.

The interest in using knowledge based methods (KBMs) for control, stems from several reasons. Among them is the absence of good control models that can evaluate the global effects of corrective actions. The complexity of the problem is one of the reasons why control actions are often very heuristic in nature. KBMs are specially suited to representing such fragmented and evolving heuristic knowledge on computers.

The structure of the paper is as follows. We first analyze the characteristics of control activity at different levels of the operations hierarchy. This provides the basis for identifying potential advantages of applying KBMs. Next, special emphasis is placed on the control, using knowledge based methods, of activities at and below the scheduling level. This is followed by an identification of those issues which, in this authors view, appear to have been neglected in the study of automated factory control, and which could benefit from the application of knowledge based techniques.

10Yrs___ Plant Design & Prototyping, Machine Selection

1Yr ___ System Simulation, Product Design (CAD), Process Design, Fixture Design,Production Planning.

1Mo ___ NC Programming

1Day ___ Factory Operation, Part Scheduling Part Shipping & Receiving, Part Transport, Part Warehousing, Part Inspection

1Hr ___

1Min ___ Part Handling

Loading/Unloading
Assembly

1Sec ___

Machine control, Intelligent Machines, Process Control

1mSec___

Figure 1. A Hierarchy of Factory Operations

2.0 CHARACTERISTICS OF HIERARCHICAL CONTROL

As mentioned earlier, the need for control stems from a desire to follow prespecified plans as far as possible, recognizing that it is usually not possible to plan for every eventuality. The term planning is being used here in a somewhat general sense, to mean a predetermined scheme or blueprint for activities. Under the hierarchical view, plans at higher levels impose constraints on those at lower levels. In order to administer control, two generic activities must be performed at every level. One is that of detection of deviations from plan, the other being execution of appropriate corrective actions. Note that corrective actions at one level may propagate down to lower levels. The reverse propagation is also possible. For instance, if a machine fails for an extended period, considerable rescheduling may be required. Alternatively, if excessive congestion is observed, or deadlines are being repeatedly violated, this may be an indication that a capacity expansion is warranted. We will have further comments about the propagation of control actions towards the end of this paper. We now proceed to investigate the nature of control activity at each level, paying special attention to the type of knowledge required and the time constraints involved in taking corrective action, as well as the process of detecting exception conditions (deviations from plan). This will help identify those levels that can most benefit from KBMs.

Clearly, the first layer is an "offline" operation in that it is not time-critical. Changes in plant design and/or machine types are very infrequent. The knowledge required for this activity largely consists of models using queuing theory, mathematical programming and simulation analysis, that can help in determining plant design and machine selection, see [VGV86] for a review and reference list. "Control" activity, i.e.

response to changes, usually consists of complete execution of the relevant models, and a redetermination of the complete plan at this level. Changes at this level generally have serious repercussions at lower ones. The deviations or exceptions at this level usually consists of fairly gross changes in the operating environment or production objectives. Thus, some exceptions are deliberate, while others are due to uncontrollable factors. In any event, the process of "detection" usually occurs over a prolonged period, is quite poorly structured, and not amenable to automation. It is best carried out by humans, with computer based aids as necessary.

Activities at the next two layers are not time critical either. Product, process and fixture design, usually evolve very gradually. The knowledge used for this purpose involve engineering design models, design databases, and sophisticated user interfaces, indicating both algorithmic and heuristic components. "Corrective" measures at these levels are again evolutionary in nature, requiring repetition of many of the planning activities themselves. The process of detecting deviations usually consists of recognizing shortcomings in product or process designs. In addition, deliberate changes may be made to products or manufacturing processes for a variety of reasons. As with level one, detection of exception at this level is rather gradual, and it is not clear that the detection process can or needs to be automated. Current practice leaves this detection and correction process largely in the hands of humans, who ofcourse, make extensive use of computer based aids. For instance, design tools can help determine effects of suggested changes in product characteristics etc..

The remaining layers can be characterized as being time critical, with the severity increasing as one descends the hierarchy. The knowledge required to generate static schedules for factory operations is largely algorithmic in nature

[CMM67],[Gra81], although there have been efforts under way for some time to apply KBMs for this purpose. In principle ofcourse, control actions should be proposed and evaluated based on their effects on some performance measure(s) of interest, examples of which include machine utilization, average flow time, makespan, etc. It would thus be natural to consider the use of mathematical models of scheduling as the basis for evaluating and choosing corrective actions. Theoretically, these models ought to be able to determine the global effects of corrective actions, and thus should provide the best choice. However, since at these levels, control responses must be made within fairly short time frames, these decisions are often taken with limited local information and with limited knowledge of their global consequences. Furthermore, there do not exist good models of the global effects of control actions. Thus at these lower levels, the control actions do not usually consist of complete replanning. Rather, continual adjustments are made in a fairly heuristic manner.

The process of recognizing exceptions at these lower levels consists of analyzing numerous system status parameters that are monitored continually. The task of identifying exception conditions from monitored status values is quite nontrivial in general. While catastrophic exceptions, such as machine breakdowns, are easy to identify, considerable sophistication is usually required to identify more subtle exceptions. For instance, combinations of status parameter values may signify an exception, while each independently seems harmless enough. Further, small deviations at the present time may be indicative of serious exceptions at a future point in time. In manual factory environments, most of this detection process has typically been done by humans, through training and experience. Unfortunately, this process often cannot be articulated in a way that makes it amenable to computer representation. It is clear however, that in order to automate the factory operations, there is a critical need to perform this detection task as well as that of generating corrective responses, using computer based approaches. The tighter time constraints on operations in the automated factory, further complicate the detection and correction process at lower layers.

In short, we have seen that while corrective actions at higher levels are best characterized as "replanning", those at lower levels are more incremental in nature, employ heuristic knowledge and are often hard to characterize algorithmically. Further, the process of detecting exceptions is not easily formalized either.

3.0 APPLICABILITY OF KNOWLEDGE BASED METHODS

By the term knowledge based methods, we include those techniques developed in the AI literature, to represent and manipulate different types of knowledge. The common representation frameworks are state-space search, semantic nets, rule based systems, logic based systems and frames. While other areas of AI such as machine vision, speech understanding etc. have made considerable contributions to automated manufacturing, KBMs provide the techniques that would be most useful in automating the decision processes associated with control.

Knowledge based techniques can be used at the higher levels of factory operations, to facilitate the reactive process. This usually takes the form of intelligent user interfaces that can access data conveniently and present results in appropriate forms. However, the decision process is still essentially being carried out by humans. We will instead, give more attention to lower level operations, where time constraints dictate that the decision process be itself computerized. Thus we are focussing on parts scheduling, part handling and part transport levels. Henceforth, control at this level will be referred to as s_control.

Ideally, s_control actions should be taken based on some global model of their effects. Thus it would seem natural to consider mathematical scheduling models as the basis for such decisions. After all, these models are global in the sense that they show the effect of scheduling decisions on some global objective function. However, two observations about the s_control function argue against this approach. First, s_control is a dynamic operation. Any mathematical model must be continually updated to reflect the current state of the environment. More importantly, this updating may involve making structural changes to the model. For instance, if a task is just completed, it is not enough to simply record that fact. Appropriate structural changes must be made to the model to ensure that this particular task is not repeated as a result of any subsequent s_control action. In the absence of such updates to the model, it is conceivable that its solution may, not so occasionally, suggest nonsensical corrective actions. It is not clear how such models could be updated in an automatic manner. The second observation is that the NP-completeness of mathematical scheduling models is very well documented [GaJ79, pp236-244], underscoring the computational difficulties of obtaining good solutions to them.

In the absence of good s_control models, automation of this function requires that the heuristic knowledge referred to in section 2, be somehow represented on the computer. KBMs offer distinct advantages over more procedural representation techniques in this regard. Examples of the latter are representations using procedural languages or simulation representations. Knowledge representation techniques strive to achieve several objectives. Those of special importance for s_control are:

Extensibility: The data structures and manipulation methods should be able to incorporate extensions and changes to the stored knowledge "gracefully" without requiring extensive alterations. While this objective is met to varying degrees, its importance in view of the heuristic nature of s_control knowledge is obvious.

Explicitness: If the storage structures are chosen judiciously, the important elements of the environment are made explicit. For instance, the use of frames may highlight important structures in the environment. Logic based systems are good at capturing consistency conditions etc. Explicitness helps in the process of codifying heuristic knowledge, by providing constructs that are particularly powerful in capturing the essential characteristics of this knowledge in a compact and usable form.

The manipulation techniques associated with KBMs allow one to draw conclusions from a body of knowledge, using symbolic reasoning, as well as numerical computations as necessary. These manipulation techniques allow us to mimic, at least to some degree, the human decision processes that underlie s_control actions. Procedural representations require a complete description of s_control actions in algorithmic form. This is hard to obtain as mentioned earlier. We now proceed to review some knowledge based approaches to s_control.

4.0 SYSTEM ARCHITECTURES

Figure 2. shows a blackboard architecture to perform S_control functions. This architecture provides a convenient reference structure to review different knowledge based approaches to s_control. It consists of functionally distributed components communicating through a blackboard. Since the components perform specialized and dissimilar tasks, their structure and representation vary considerably. The blackboard mechanism allows such disparate representations to coexist and cooperate. In particular, it allows algorithmic knowledge to be integrated with heuristic knowledge in the control task.

4.1 Generating Corrective Responses

There has been more emphasis placed on this aspect of s_control than on detection. Some examples of proposed schemes are [Dut87],[Dut88],[Smi87], [Ben86]. We are excluding from

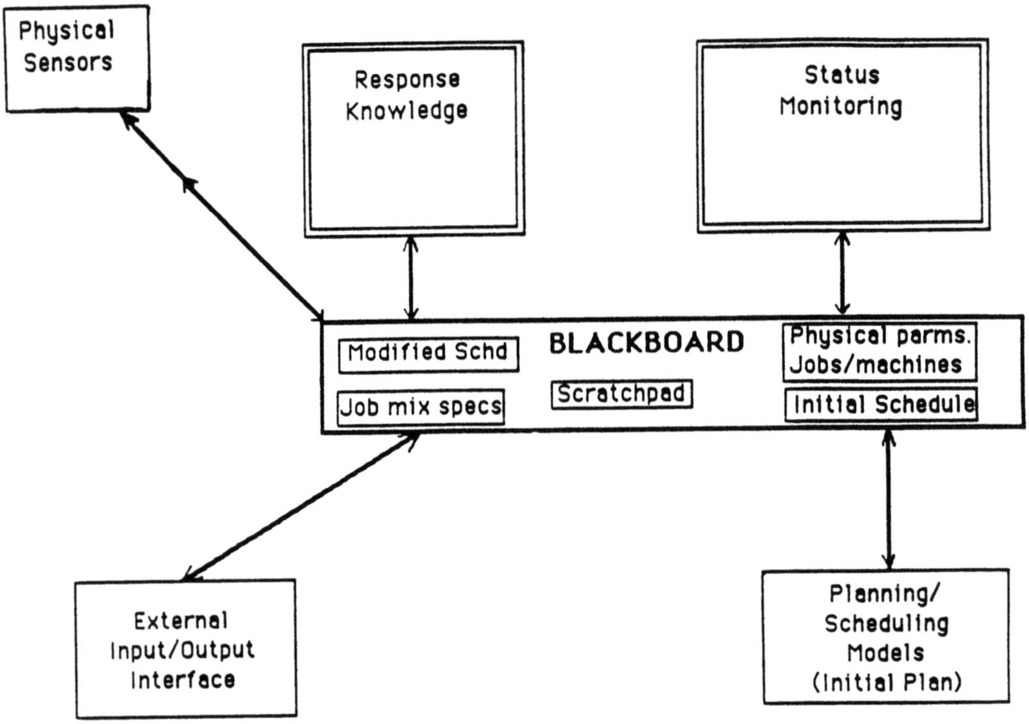

Figure 2. A Blackboard Architecture for Control

consideration, knowledge based factory simulation models, which essentially study scheduling and control policies in an offline manner [KEE84]. In [DuS87], [Dut88], we have used structured rule bases to represent the knowledge required to make correct-

ive responses. The justification is that rule based techniques have been widely recognized as an effective way of representing knowledge about actions [Hay85]. Structure is imposed by meta-rules that focus attention on a portion of the knowledge base relevant to the exception conditions being corrected. At the time of writing, we have developed responses to machine failures, dynamic introduction of new jobs, and dynamic changes in job priorities. Development of responses has been guided by specific strategies. Among them are the following:
(i) Control attempts to be as myopic as possible and procrastination is allowed. It was found that the computational costs of making farsighted adjustments is not too worthwhile, specially in environments characterized by high exception rates. Note that these exceptions need not only be failures of some kind.
(ii) Control attempts to be opportunistic. While many exceptions have detrimental effects on existing schedules, this is not always so. Opportunities frequently arise due to exceptions, and these should be availed.
(iii) Control attempts to be optimistic in its corrective actions. It does not assume worst case conditions. For instance, when a machine fails, its expected failure duration affects the corrective action, but control is prepared for the situation where the machine becomes operational earlier than expected.

The control mechanism always strives to maintain the objective of the initial schedule as far as possible. The current implementation address minimization of average completion times, with additional objectives under study. Experimental tests indicate that the mechanism is quite robust, in that degradation of performance is quite gradual.

An alternate approach is that represented by the work at CMU, stemming from the ISIS framework for scheduling [Fox84]. Here, the control activity stems from "constraint propagation".

The factory scheduling problem is viewed as one of coordinating sequences of manufacturing operations so as to obey the temporal restrictions of production processes and capacity limitations of a set of shared resources, and balance a diverse and conflicting set of preferential concerns. The initial schedule represents the "best" compromise given an initial set of operational constraints. Unanticipated deviations are considered to alter the constraint set. Therefore, the bulk of the control process consists of "propagating" the consequences of change in the constraint set. This propagation results in some inconsistencies and some opportunities. The former have to be resolved through appropriate conflict resolution rules while attempts are made to take advantage of the latter. Three basic types of exceptions are recognized, time conflicts, capacity conflicts, and time vs. capacity conflicts. This detection is made possible through an explicit representation of the current state of the schedule. In the authors' view, this constraint based approach provides a unifying view of scheduling and control [Smi87].

4.2 Detecting Exception Conditions

The process of detecting exception conditions, unfortunately, has not in our view, received quite the attention it deserves. Most of the work on generating corrective responses implicitly assume that exception conditions have already been identified. This is not surprising since the exception conditions most frequently considered in the literature are what might be called "catastrophic" ones; machine failure, dynamic introduction of new jobs, change in a job's deadline, other changes in resources etc.. There is not much decision making that has to be performed in identifying such exceptions. However, more complex and subtle exceptions can occur, which may warrant corrective action. For instance, small drifts in duration of some operations may be indicative of an impending machine malfunction. In general, exceptions can occur spatially or temporally. A major reason for the complexity of the detection process is that fragmented spatial and temporal observations must usually be combined in order to recognize several types of exceptions. Complete automation of the detection process is still very far from being achieved. The reality at this time is the use of human supervisors for exception detection and reactive management. It has been proposed that these human operators be aided in their tasks by computer based techniques to create a hybrid system.

A notable effort along these lines is the "supervisory control paradigm" [AGM87]. This paradigm proposes an integration of humans, analytic models, and knowledge based systems into a philosophy for system design and control, where the roles of both control automation and human operators are explicitly engineered and designed into the system. These concepts are being tested on the GT-FMS(imulator), a research tool based on this paradigm. Under this paradigm, the detection and corrective actions are considered to depend on higher level cognitive and perceptual functions, that the human possesses. He is responsible for establishing set points, system reconfiguration and exception handling which involves failure detection, fault isolation etc. The underlying computer based techniques, provide the necessary information, in a way conducive to his decision making capabilities. For this purpose, it is necessary to address human operator modelling issues [Rou81],[MGA84]. At this time [KMT87], however, it appears to this author that the emphasis is still on generating appropriate corrective responses. Presentation formats are discussed in [KMT87], but it is not clear if any knowledge based techniques are being used in processing the data displayed to the human for detection purposes.

We suggest here, however, alternate knowledge based approaches that have special appeal in addressing the issue of detection. Essentially, the task is one of diagnosis; given some observations about the system, to attempt to identify probable "illnesses" indicated by them. Diagnostics has been a favorite application area for knowledge based methods [BaF82a], [BaF82b] ,[CoF82], although their success has varied. Application domains which consist mainly of mechanical/physical systems, have tended to enjoy more success than biological or social systems. The predominant representation technique used for diagnosis is rule based, while some have attempted the use of frames. We conjecture that for s_control applications, frames can be appropriate when the factory operations are relatively stable; as are part mixes. These conditions would favor the identification of some structure in the failure patterns, which frames can exploit. On the other hand, when such structure is absent, and only isolated fragmented observations are available, rule based approaches can be more appropriate. This is more likely to be the case in random FMSs where part mixes and characteristics tend to change relatively rapidly.

Regardless of which approach is used, it would be necessary for the exception detection process to be hierarchical in nature. This would parallel the hierarchical view of factory operations mentioned earlier. A hierarchical detection mechanism provides for aggregation of stimuli and prevents control from being overly jittery. Knowledge representation methods, rules and frames among them, provide means to capture such a hierarchical structure. For instance, frames can be refined to different levels, while rule sets can be aggregated through the use of meta rules to focus attention during the detection process. This technique has, for instance been used in medical diagnosis knowledge bases.

5.0 PROPAGATION OF EXCEPTIONS AND CONTROL ACTIONS

Under the hierarchical view of factory operations, it is

important to understand the propagation of exception conditions and effects of corrective actions, across levels. The granularity of control becomes finer as one descends the hierarchy. In short, there is information aggregation in travelling up the hierarchy, and disaggregation downward. Hierarchical knowledge representation techniques can be helpful in performing such aggregation/disaggregation of information. We now study the information flows associated with the propagation of exception conditions across the hierarchy, followed by the propagation of control actions. The implications for automation of these functions is also noted in so doing.

The propagation of exception conditions can proceed in both directions in the hierarchy. Typically, exceptions originating at the lowest levels will not necessarily imply an exception at the upper layers. However, each layer has to decide whether or not exception conditions need to be passed on to higher layers. The information passed on is usually more aggregate in form. For instance, individual machine controllers have to detect aberrations in machine operations, and decide if this warrants higher level correction. Similarly, if a task is taking longer than its scheduled time, a decision will have to be made as to whether or not resequencing is in order.

Propagation down the hierarchy often occurs due to "deliberate" exceptions, as opposed to chance happenings. For instance, if the due date of a job is moved forward, this can cause fairly major disruptions at lower levels. The same holds true if a new job is introduced dynamically, if a product design changes etc. The downward propagation of exceptions calls for disaggregation in information. In our view, while there is an awareness of this issue of propagation of effects [Smi87pp15], it calls for further investigation. This is specially important for the time-critical lower level actions which are the primary focus of automated control.

6.0 CONCLUSION

Our assessment of the use of knowledge based methods for factory control, is that the emphasis has largely been on the generation of corrective responses for a given set of exception conditions. Relatively less emphasis has been placed on the exception detection function. We believe knowledge based approaches can be helpful here as well. Also, the review indicates that KBMs have been used largely at the scheduling level. There is a need to address the issue of integrating control at this level with other levels of factory operation. Again, we believe that KBMs can be helpful in achieving this integration because they can facilitate the aggregation and disaggregation of information necessary for this purpose. Also, since the type of knowledge required for detection and correction is substantially different, the representation techniques would be different for them. We have seen a need for these disparate representations to communicate and coordinate their activities. We feel that a blackboard type architecture facilitates co-operation between such disparate agents. We would also suggest explicitly separating the detection and response functions, instead of having them closely coupled in one control mechanism. The complexity of the control function suggests that comprehensive models of control will be hard to develop. It is more likely to evolve in a fragmented manner. KBMs have the flexibility to assimilate updated knowledge without having to make monumental changes in representation.

It is also our assessment that "control" at the highest two or three levels is still best performed by humans. It is unlikely that KBMs will provide much help in automating those functions. Instead, KBMs can help by providing intelligent decision support to facilitate the human decision makers' activities. In short, we believe that the greatest potential for KBMs in automating the control function, lies at and below the scheduling level. In this context, there is one important issue that we have excluded from the scope of this paper. It concerns hardware developments. Most practical sized knowledge bases are notoriously slow for time critical environments. We will only mention in ending, that improvements in hardware environments will be necessary in order to take full advantage of the capabilities of KBMs.

REFERENCES

[AGM87] Ammons J.C., Govindraj T., and Mitchell C.M., "Human Aided Scheduling for FMS: A Paradigm for Human-Computer Interaction in Real Time Scheduling and Control," Proc. 2nd. ORSA/TIMS Conf. on Flexible Manufacturing Systems, 1986, pp443-454.

[BaF82a] Barr A., Fiegenbaum E.A., "The Handbook of Artificial Intelligence," Vol. 1, Kaufmann, Los Altos, California, 1982.

[BaF82b] Barr A., Fiegenbaum E.A., "The Handbook of Artificial Intelligence," Vol. 2, Kaufmann, Los Altos, California, 1982.

[Ben86] Ben-Arieh D., "Knowledge Based Control System for Automated Production and Assembly," in Modelling and Design of Flexible Manufacturing Systems, Andrew Kusiak (Ed.), Elsevier, 1986, pp347-368.

[BlM83] Bloom H. and McLean C.R., "Standardization Suggested by The AMRF-A Research Testbed for The Factory of The Future," Presented at the First International Symposium on Automated Integrated Manufacturing, April 5-6, 1983, San Diego, CA.

[CMM67] Conway R.W., Maxwell W.L. and Miller L.W., Theory of Scheduling, Addison-Wesley Publishing Co., Reading, Mass.,1967.

[CoF82] Cohen P., Fiegenbaum E.A., "The Handbook of Artificial Intelligence," Vol. 3, Kaufmann, Los Altos, California, 1982.

[DuS87] Dutta A., and Sen A.,"Production Control in Automated Manufacturing Environments," Proc. IEEE Conf. on Systems Man and Cybernetics, 1987,pp517-521.

[Dut87] Dutta A., "Knowledge Based Control of Production Schedules," working paper College of Business, University of Iowa, Iowa City, IA 52242, 1987.

[Fik85] Fikes R., and Kehler T., "The Role of Frame Based

Representation in Reasoning," Comm. ACM, Vol. 28, No. 9, Sept., 1985, pp904-920.

[Fox84] Fox M.S. and S.F. Smith, "ISIS: A Knowledge Based System for Factory Scheduling", Expert Systems Vol. 1, No. 1, July 1984, pp25-49.

[GaJ79] Garey M. and Johnson D.S., Computers and Intractibility: A Guide to The theory of NP-Completeness, W.H. Freeman and Co., San Francisco, 1979.

[Gra81] Graves S.C.,"A Review of Production Scheduling," Operations Research, 29,4, July-Aug 1981, pp646-675.

[Hay85] Hayes-Roth F., "Rule-based Systems", Comm. ACM, Vol. 28, No. 9, Sept. 1985, pp921-932.

[KMT87] Krosner S.P., Mitchell C.M., and Tipton B.C., "The Role of Human Operators in Control of an FMS with Random Operation Times," Proc. IEEE Conf. on Systems Man and Cybernetics, 1987, pp671-674.

[MGA84] Mitchell C.M., Govindaraj T., Ammons J.C.,"Human Machine Interface in The control of Flexible Manufacturing Systems," Proc. IEEE Conf. on systems Man and Cybernetics, October 1984,pp59-62.

[McL81] McLean C.R., Bloom H.M. and Hopp T.H.,"The Virtual Manufacturing Cell," NBS Report.

[Rou81] Rouse W.B., "Human-Computer Interaction in The Control of Dynamic Systems," ACM Computing Surveys, Vol. 13, No. 1, March 1981, pp71-100.

[Smi87] Smith S.F., "A Constraint Based Framework for Reactive Management of Factory Schedules," Proc. International Conf. on Expert Systems and The Leading Edge in Production Planning and Control, Charleston, SC, May 1987.

[Spu87] Spur G., and Specht D., "Computer Integrated Manufacturing in Future Factories," Robotics and Computer Integrated Manufacturing, Vol. 3, No. 2, 1987, pp147-155.

[VGV86] Van Looveren A.J., Gelders L.F., and Van Wassenhove L.N., "A Review of FMS Planning Models," in Modelling and Design of Flexible Manufacturing Systems, A. Kusiak (Ed.), Elsevier, 1986, pp3-31.

[Wec87] Weck M. and Kiratli G., "Applicability of Expert Systems to Flexible Manufacturing," Robotics and Computer Integrated Manufacturing Vol. 3, No. 1, 1987, pp97-103.

Implementation and Management Issues

INFORMATION PROCESSING IN MATERIAL FLOW SYSTEM MANAGEMENT

Agostino VILLA

Dipartimento di Tecnologia e Sistemi di Produzione
Politecnico di Torino
corso Duca degli Abruzzi, 24
10129 Torino
Italy

ABSTRACT. Production Management in a manufacturing plant requires to handle flows of two types of "entities":

- product orders, which must be propagated backward from a work stage to the upstream ones, to assign the production orders to each stage;
- material & components, which must be transferred forward, from a stage to the downstream ones, to assure availability of items to be processed at each stage.

The aim of any Production Management (PM) System is that of synchronizing production order flows with item flows.
Two alternative approaches can be adopted in order to solve the PM Problem:

a) planning of production orders for all work stages in the plant, and "pushing" of items to the input buffers of all stages, according to planned loading conditions;
b) planning the throughput of the final stages, and "pulling" items from upstream stages, according to item consumption.

Depending on the adopted PM approach, different PM Architectures can be designed. Obviously, each one of them should be tuned on the market situation of the factory to be managed. To this aim, the question asked to the PM Architecture designer by a Production Manager is always of the following type:

"is it possible to compare PM Architectures, in front of a given evolution of the product orders in the future, and of a given structure of the plant?" .

The problem underlying this question is particularly important in case of factories composed by a number of plants, often located in different towns, i.e. distributed industrial environments where the management of item flows assumes a relevant importance. This paper will try to offer a theoretical solution to the above question. First, to allow a comparison of existing PM Procedures, a common formulation of both PM approaches will be stated, in case of a single-plant factory. SADT representations of the necessary INFO Systems (for collecting and addressing necessary information and planning/control decisions), for each one of the resulting PM Architectures, will be derived. Then, the multi-plant situation will be analyzed, to discuss the necessary generalizations (and, possibly, integration) of PM Architectures assuring a reasonable efficiency in front of variable demands.

NATO ASI Series, Vol. F53
Advanced Information Technologies for Industrial Material Flow Systems
Edited by Shimon Y. Nof, Colin L. Moodie

INTRODUCTION

The problem of Production Management in a large-scale manufacturing environment can be schematically represented in terms of four sub-problems:

1. Product Demands Forecasting, i.e. estimation of the future demands of products which should be required by the market, according to information available from present and past orders portfolio;

2. Workload Scheduling, i.e. optimal ordering of workloads, conditioned on the production capacity of the plant;

3. Production Capacity Planning, i.e. definition of the best patterns of item flows inside the plant, and of the service rates of the component work stages;

4. Production Control, i.e. adaptation of the work stage service capacity to the occurrence of unpredictable failure events.

Intuitively, the Product Demand Forecasting requires a top-layer module, able to smooth past data and propose scenarios of product demands.
The crucial PM Sub-Problems appear to be the other three ones, practically corresponding to the so-called PRODUCTION PLANNING & CONTROL PROBLEM.

At present, a formal solution of this problem is practically unusable. The simplified solution that is generally, and frequently unconsciously, adopted is a type of "open-loop-feedback" control, accounting for the two steps of production planning and production control [1,2]:

a) a deterministic open-loop planning problem, based on the current state pattern, is solved over a given horizon, by optimizing a suitable performance index summarizing the plant efficiency;

b) the result of the planning step plays the role of a reference input for a decentralized feedback control, trying to counteract perturbation (uncertainty) effects at each stage;

c) an updating of the reference plan is required each time the feedback control strategy becomes unsufficient to assure an accurate tracking of the plan itself, due to the occurrence of a large-scale event, either endogenous (hard failure), or exogenous (a new product order).

By its "large-scale" nature, the Production Planning and Control (PPC) Problem must be decomposed into a network of sub-problems. From a practical point of view, two "decomposition modes" can be adopted, based on the exploitation either of the product structure, or of the process structure. The former decomposes the PPC Problem into a number of sub-problems, each one associated to a work stage in the plant. The latter tries to solve a PPC

sub-problem for each type of products to be processed.
Depending on the decomposition assumption, different production planning procedures can be developed. A well known product-structure-based procedure is the so-called Manufacturing Resource Planning (MRP II) [6]. Different process-structure-based procedures have been developed, depending on the type of planning phylosophy adopted, namely "push" or "pull". OPT appears to be a push-type proces-structure-based planning procedure [8], while JIT is a pull-type process-structure-based control procedure [9]. The PPC Hierarchical Decentralized Architecture (HDA) developed in Torino essentially is a push-and-pull-type process-structure-based one [2, 7, 10].

However, if one accounts for a possible application of one of above mentioned Production Planning and/or Control Procedures in an industrial environment, even if composed by a single plant, some immediate questions arise:

> "is is possible to evaluate the efficiency of a Production Planning and/or Control Procedure, in front of a product market characterized by a fast variability?",

and also

> "is it possible to compare the structures of the different Production Planning and/or Control Procedures, such to recognize what procedure could be more adaptable to a particular industrial environment and to its technological constraints?".

The practical relevance of these two questions is evident; an answer for both of them is not immediate.
A possible answer could be derived by developing a general simulator of a manufacturing plant, and then applying the different existing PPC Procedures to it, once the simulator parameter values have been calibrated for the particular case to be analyzed. This solution is apparently easy to implement, but cannot surely cover a wide range of real plants.
An alternative approach can be stated by summarizing both above questions into the following:

> "is it possible to give a common formalization to all existing PPC architectures ? ".

Apparently, this question could appear only a theoretical one, but it is not true. If a positive answer could be found, existing PPC Procedures could be theoretically compared, thus assuring an "absolute" evaluation, parametrized on the factory estimated market of products and conditioned on the structure of the plant to be analyzed.
On the other hand, this unified formulation of PM Architectures could allow to simplify the Information Processing (INFO) System that must be designed for supplying necessary data and planning/ /control decisions to the architecture modules. Evidently, in case of a unique representation, a standardized design of the INFO System could be envisaged, thus reducing the main obstacle to the effective implementation of a PM Architecture, mainly in

medium/small-dimension environments.
Finally, a unified representation could suggest the best modes to integrate together PM Architectures of different types. Note that this last specification is extremely important in multi-firm industries, often composed by a number of medium/small-scale plants, each one oriented to serve a particular market.

By these reasons, the present paper is organized as follows:

i) a first part is devoted to state a unified formulation of the production planning & control problem, and to analyze its solution approaches, based either on the product structure, or on the process structure;

ii) in the second part, the resulting architectures will be represented in terms of SADT diagrams [12], to exploit required information flows; based on these representations, a comparison of their features will be proposed;

iii) the third part will discuss the necessity of generalizing the PM Architecture design in case of multi-plant firms.

PART I. THE PRODUCTION PLANNING & CONTROL PROBLEM

I.1. PPC Problem Statement

Formally, the PPC problem can be stated as a control optimization problem for a large-scale dynamic system, whose evolution is influenced by the occurrences of some discrete events at random times [3,11].
Admissible events can be of two different types :

A. "events to be planned" (and accomodated), when new orders have to be accepted, and then their processing times have to be scheduled (Exogenous Events);
B. "events to be counteracted" (and possibly avoided), when random failures will occur at some work stage, and their maintenance have to be decided (Endogenous Events).

Each time an event will occur, the operation modes of the manufacturing plant will require some modification:

A. in case of an occurrence of an exogenous event (a new order has been accepted), the plant production capacities must be modified in order to assure the completion of the updated pattern of product orders, while verifying the order due dates;
B. in case of the occurrence of a failure at a work stage, the production capacities of the others must be updated to assure the processing of orders by the partially-failed plant.

It must be noted, however, that any occurred event can normally be detected and recognized in a very short time (often, almost

instantaneously), by the monitoring devices mounted on any machining centre with some automated functions.
Then, from a practical point of view, the Production Planning & Control (PPC) Problem in any manufacturing plant with some degree of automation is going to be implemented as a Problem of Updating the Plant Production Plan in front of the Occurrence of Events which could significantly affect the plant response to market (i.e., either new product orders, or hard failures).
From a theoretical point of view, this PPC specification reflects into the necessity of stating an Event-Driven Open-Loop-Updated EDOLU Problem for optimizing, at each event occurring time, the future production plan.
Since the occurred event can be assumed instantaneously recognized, the EDOLU Problem can be solved at each event time, once stated as a deterministic dynamic optimization problem [2,3,11].
At each event occurring time t_0, a large-scale optimization problem must be solved, by extremizing the expected value of a suitable performance index over an assumed horizon H, conditioned on the measured state of the system at time t_0, and on the event-modified values of the plant model parameters.

For problem formalization purpose, let us consider a plant composed by a forward chain of work shops, each shop being defined by a network of machining centers (or work stations) [2].
Each work shop is then able to share its service time among the different part types which it can process. Depending on the order scheduling and the part processing at upstream shops, each work shop must serve a mix of parts variable in time [7].

Let j (j = 1,...,N) and i (i = 1,...,M) be the indexes associated to work shops and part types, respectively.
According to the serial structure of the whole manufacturing plant, this one can be partitioned into "elemental blocks", each one composed by a set of inter-shop parallel buffers (one for each part type), together with the upstream and downstream work shops (see Fig. 1).

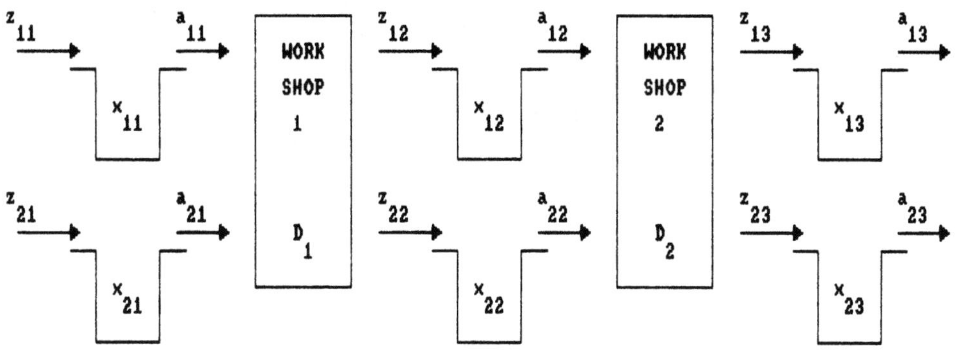

Fig. 1. Skecth of a multi-stage multi-product manufacturing plant.

More precisely, to each buffer set B(j), j = 1,...,N+1, either located at the input of the work shop j (for j = 1,...,N), or located at the final output of the work shop series (for j=N+1), a state describing the inter-shop local inventory levels can be associated.
For each buffer set B(j), j = 1,...,N+1, and each part type i, the following notations are introduced :

$a_{ij}(t)$ flow rate at which parts of type i are loaded on the work shop j (<u>loading rate</u>);

$x_{ij}(t)$ storage level at the shop input buffer (<u>shop state</u>), for part type i;

$z_{ij}(t)$ flow rate at which parts of type i are released at the output of the work shop j-1 (<u>processing rate</u>);

For each work shop j (j = 1,...,N), it must be accounted for:

$D_j(t)$ <u>internal control variables</u> of shop j (for part routing and station service control).

Then, <u>for the whole manufacturing plant</u>, the set of acquired product orders can be represented by introducing the following:

τ_m is a future instant at which an order of products of any type should be delivered (set of the orders due dates); then,

$$\tau_m \ \varepsilon \ \{ \tau_1 = t_0 , \ldots, \tau_m , \ldots, \tau_{NO} \}$$

being NO the total number of product orders acquired for all product types at time t_0;

$NO_i(t_0)$ is the number of orders of product of type i, acquired at time t_0;

$$Q_{mi}(\tau_m) \begin{cases} = \text{the dimension of the m-th order of products of type i to be delivred at time } \tau_m ; \\ = 0 , \text{ if no order of product type i must be delivered at time } \tau_m ; \end{cases}$$

$\Theta_i(t_0)$ is the ordered index-set of required product orders for part type i; this NO-dimensional vector will be organized as follows:

$$\Theta_{im}(t_0) = \begin{cases} = k , \text{ if the k-th order of product type i must be delivered at time } \tau_m ; \\ = 0 , \text{ if no order of product type i must be delivered at time } \tau_m . \end{cases}$$

REMARK 1. Note that, in above proposed notations, no account has been made to endogenous events. The motivation is that endogenous events, when occurring, will modify the production capacities of the plant work stages. These effects will reflect on modifications of the values of some parameters of the plant model which will be stated in the following. Due to the assumption of immediate recognition of the event effects, the model parameters can be assumed already known after each event occurrence, and before the related updating of the production plan.

REMARK 2. Each exogenous event (new product order) will reflect on a variation of the plant loading condition (more products to be delivered). Then, these events effects will be described by accounting for new orders in the constraints stating the amount of product to be processed, assumed known immediately after the exogenous event occurrence.

REMARK 3. Note that whenever a sub-script or super-script is missing, we will refer to the whole vector of variables originated by letting it vary over its all range.

The state dynamics at each buffer set B(j) can be described by a "balance equation" of input and processed part flows, formally stated by the following <u>State Equation Model</u>:

$$\dot{x}_{ij}(t) = z_{ij}(t) - a_{ij}(t) \quad ; \quad x_{ij}(t_0) = x^{\circ}_{ij} \qquad (1)$$

for $j = 1,\ldots,N+1$; $i = 1,\ldots,M$.

where

t_0 denotes the <u>current time instant</u>, i.e. the time where an updating of the planning and control strategies is required, due to the occurrence of a discrete event, either exogenous or endogenous.

Note that both loading rates and processing rates, defining the input/output flows at each input buffer of the set B(j), will depend on the production control strategies $D_{j-1}(t)$ and $D_j(t)$, adopted at the work shop j and at the upstream work shop j-i. Then, at each work shop j, it must be verified the following <u>Interaction Variable (or Consistency) Constraints</u>:

$$a_{ij}(t) = z_{i,j+1}(t) \quad , \text{ for } j = 1,\ldots,N \; ; \; i = 1,\ldots,M \; ; \qquad (2)$$

where

$$z_{ij}(t) = W_i(\, D_{j-1}(t)\,) \; , \text{ for } j = 2,\ldots,N \; ; \; i = 1,\ldots,M; \qquad (2\text{-a})$$

$$a_{ij}(t) = S_i(\, D_j(t)\,) \; , \text{ for } j = 1,\ldots,N \; ; \; i = 1,\ldots,M; \qquad (2\text{-b})$$

where $W_i(.)$, $S_i(.)$ denote the functional dependence of the processing rates and loading rates for a part type i, respectively, on the internal production planning strategies of the work shops located at the input and output of the buffer j, i.e. work shops j-1 and j.
Note that these functions will depend on the capacity parameters of the j-th work stage: the parameter values, possibly affected by a failure, will be updated just after the failure occurrence.

In case of the first buffer in the chain of work shops, i.e. for j = 1, the input flow rate must satisfy to the following Supply Constraints:

$$z_{i1}(t) \leq r_i(t) \quad , \quad i = 1,\ldots,M \; ; \tag{3}$$

where

$r_i(t)$ is the maximum value of the raw material release rate.

At the last buffer set of the line (for j = N+1), the output flow rates must be related to the set of product orders by the following Order Servicing Constraints:

$$\int_{t_0}^{\tau_m} a_{i,N+1}(t)\,dt = \sum_{m \varepsilon \Theta_i(t_0)} Q_{mi}(\tau_m)\; h(\Theta_{im}(t_0)) \tag{4}$$

for each $m = 1,\ldots, NO_i(t_0)$

where

$$h(\Theta_{im}(t_0)) = \begin{cases} 1 & \text{if } \Theta_{im}(t_0) \text{ not equal to } 0 \; ; \\ 0 & \text{if } \Theta_{im}(t_0) = 0 \; . \end{cases}$$

The planning/control strategies to be applied to all work shops in the plant, in order to assure the complete service of all required product orders at the updating time considered, must be optimized by minimizing some performance index associated to the manufacturing plant.
The case here considered is based on the following choice of performance index: the Plant Work In Process (WIP), defined over a future horizon H a-priori unknown:

$$WIP = G(x) = \sum_{i=1}^{M} \{ \sum_{j=1}^{N+1} \{ \int_{t_0}^{H} x_{ij}(t)\;dt \} \} \tag{5}$$

where

H is the a-priori unknown planning horizon adopted at the planning updating time, i.e. the future horizon where all product orders will be completely served.

It must be noted that the "clearance horizon" H will correspond to the period required for processing all product orders available at the considered updating time. This period should be such that all orders' due dates are satisfied, i.e.

$$H \le \max \{ \tau_k , \quad 0 \le k \le NO \}$$

However, no a-priori condition can assure that all "required due dates" can always be respected.
Then, <u>it is here assumed that</u>:

a) the clearance horizon H (also named "planning horizon") is a-priori unknown;

b) the above mentioned due dates $\{\tau_1, \ldots, \tau_k, \ldots \tau_{NO}\}$ are initial tentative due dates for the corresponding product orders, which can be modified by any planning procedure.

In order to give a compact formulation of the resulting <u>dynamic optimization PPC Problem</u>, let us rewrite the above introduced dynamic model and constraints into a matrix form, as follows:

* State Equation Model, from (1):

$$\dot{x}(t) = z(t) - a(t) \quad ; \quad x(t_0) = x^\circ \tag{6}$$

* Consistency Constraints, from (2):

$$a(t) = C\, z(t) \tag{7}$$

where C is a (N+1)-dimension square matrix

$$C = \left[\begin{array}{c|c} 0 & \begin{array}{c} I \\ (NxN)\mathrm{dim} \end{array} \\ \hline \multicolumn{2}{c}{0} \end{array} \right]$$

* Supply and Order Servicing Constraints (3) and (4):

$$\Omega\, [\, x(t), r(t), \{\Theta(t_0)\,] \le 0 \tag{8}$$

To above re-written equations, one must add the following

Physical & Storage Capacity Constraints

$$0 \leq x(t) \leq x^* \ , \quad t_0 \leq t \leq H. \qquad (9)$$

being x* the vector of buffer storage capacity bounds.

By using these last formulations, the problem of defining the production planning and control strategies to be assigned to the work shops in the plant can be stated as follows:

PPC PROBLEM: minimize $G(x)$ w.r.t. $H, D(t), \{\Theta(t_0)\}, a_{N+1}(t)$,

for $t_0 \leq t \leq H$,

subject to:
* State Equation Model (6);
* Consistency Constraints (7);
* Supply and Order Servicing Constraints (8);
* Physical & Storage Capacity Constraints (9).

I.2. Solution Approaches to the Production Planning Problem

In order to have coherent formalization of the PP Problem, either based on the product-structure decomposition or on the process-structure one, let us assume that the product tree of each part type i is composed by N levels, each level j denoting a working phase to be accomplished on the shop j (see Fig. 2).

Taking into account the process-structure of the multi-stage multi-product manufacturing system , the PP Problem can be rewritten in the following way, by introducing, as interaction variables at each work shop j, the processing rates of upstream work shop j-1 (i.e., the incoming flows of all workable part types).

PPC PROBLEM - PROCESS-STRUCTURE-BASED FORMULATION: minimize

$$WIP = \sum_{j=1}^{N+1} G_j(x_j) = \sum_{j=1}^{N+1} \{ \sum_{i=1}^{M} \int_{t_0}^{H} x_{ij}(t) \ dt \}$$

w.r.t.

$$H, \{D_j(t), j = 1,\ldots,N\}, \{\Theta(t_0)\}, a_{N+1}(t), t_0 \leq t \leq H,$$

subject to

* State Equation Model:

$$\dot{x}_j(t) = z_j(t) - \begin{cases} S\,(\,D_j(t)\,)\,, & \text{for } j = 1,\ldots,N\,; \\ a_{N+1}(t)\,, & \text{for } j = N+1\,; \end{cases}$$

$$x_j(t_0) = x^{\circ}_j\,, \quad \text{for } j = 1,\ldots,N+1;$$

* Consistency Constraints:

$$z_j(t) = W\,(\,D_{j-1}(t)\,)\,, \quad \text{for } j = 2,\ldots,N+1\,;$$

* Supply and Order Servicing Constraints:

$$\Omega_j\,[\,x_j(t),\, r(t),\, \{\Theta(t_0)\}\,] \le 0$$

and to $0 \le x_j(t) \le x^*_j\,, \quad t_0 \le t \le H.$

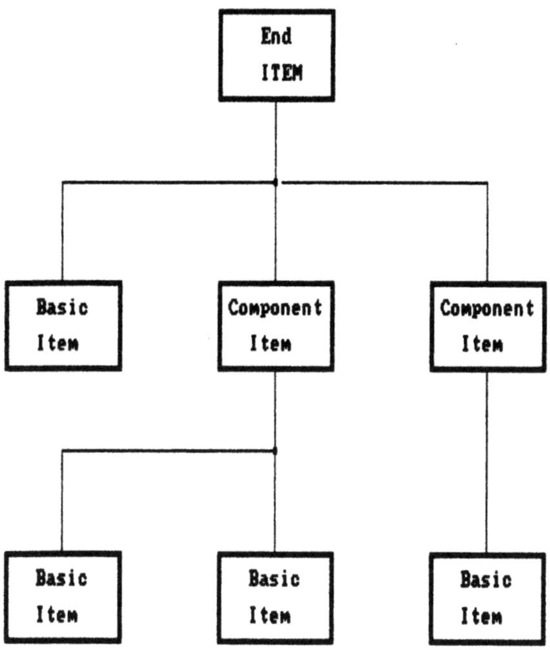

Fig. 2. Sketch of a "product tree".

Three different approaches, all based an above stated optimization problem formulation, and derived from Large Scale System (LSS) theory, can be outlined [4].

I.2.1. The Interaction/Prediction Approach

By this approach, the interaction variables $z_j(t)$ are replaced by a vector of functions $b_j(t)$, a-priori assumed, so that the j-th local controller treats the interaction variables as given.

By said position, the overall Process-structure-based Form can be splitted into a family of N+1 parametrized (on b_j) sub-problems:

SUB-PROBLEM $P_j(b_j)$: Minimize

$$WIP(j) = G_j(x_j) = \sum_{i=1}^{M} \{ \int_{t_0}^{H} x_{ij}(t)\, dt \}$$

$$\text{w.r.t.} \begin{cases} H,\ D_j(t),\ \text{for } t_0 \le t \le H,\ \text{when } j = 1,\dots,N\ ; \\ H,\ z_j(t),\ D_j(t),\ \text{for } t \le t \le H,\ \text{when } j = 1\ ; \\ H,\ \{\Theta(t_0)\},\ a_{N+1}(t),\ \text{for } t_0 \le t \le H,\ \text{when } j = N+1\ ; \end{cases}$$

subject to:

* State Equation Model:

$$\dot{x}_j(t) = z_j(t) - \begin{cases} S(D_j(t)),\ \text{for } j = 1,\dots,N\ ; \\ a_{N+1}(t),\quad \text{for } j = N+1\ ; \end{cases}$$

$$x_j(t) = x^{\circ}_j, \qquad \text{for } j = 1,\dots,N+1;$$

* "Imposed" functions describing Interactions between work stages:

$$z_j(t) = b_j(t),\quad \text{for } j = 2,\dots,N+1\ ;$$

* either Supply or Order Servicing Constraints (see eqs. (3) and (4)), in case of the first (j = 1) or the last (j = N+1) buffer set in the plant, but lacking in case of internal buffer set B(j) for j = 2,...,N :

$$\Omega_j [x_j(t), r(t), \{\Theta(t_0)\}] = \begin{cases} \text{equation (3), for } j = 1 ; \\ \text{equation (4), for } j = N+1 ; \end{cases}$$

and to $0 \le x_j(t) \le x^*_j \quad , \quad t_0 \le t \le H.$

The solution of the overall PP Problem can now be envisaged as follows :

STEP A. The j-th local controller receives the parameters $b^k_j(.)$, being k the index of the current solution iteration, from the coordinator, and it solves Sub-problem $P_j(b^k_j)$.

Let $x^k_j(t)$ be the obtained optimal state trajectory, and

$$e^k_j(t) = b^k_j(t) - z^k_j(t) \quad . \qquad (10)$$

STEP B. According to a relaxation procedure, a new parameter value $b^{k+1}_j(.)$ is computed, starting from $b^k_j(.)$ and and $e^k_j(.)$.

Iteration will stop as soon as no relevant variations of the interaction functions will be found.

This approach formalizes the so-called "Interaction/Prediction Principle" [5], and, in mathematical programming terms, it is known as "Projection/Feasible Directions" method.
The above two steps perform as follows :

- a higher-level phase is devoted to the prediction of the interaction variables, tacking into account past differences between predicted variables and the ones computed on the basis of the local optimizations' results ;
- a lower-level decentralized phase implements decoupled local optimizations, each one referred to a buffer set B(j), and then to the corresponding shop j.

It must be noted that the higher-level phase does not perform any active coordination, but only verifies if the iteration is directed towards a reduction of (10).
Said feature gives reason of the fact that an Interaction/Predic-

tion approach cannot be used if a PP strategy - based on a process-structure decomposition of the overall PP problem - is desired.

On the contrary, the Interaction/Prediction approach - <u>based on a product-structure decomposition</u> - gives rise to the following set of Sub-problems

<u>BASIC SUB-PROBLEM IN A PRODUCT-STRUCTURE-BASED DECOMPOSITION</u>

SUB-PROBLEM $P_i(b_i)$: Minimize

$$WIP(j) = G_i(x_i) = \sum_{j=1}^{N+1} \{ \int_{t_0}^{H} x_{ij}(t)\, dt \}$$

w.r.t. H, $\{D_{ij}(t),\ j = 1,\ldots,N\}$, $\{\Theta_i(t_0)\}$, $a_{i,N+1}(t)$,

for $t_0 \le t \le H$,

subject to:
* State Equation Model:

$$\dot{x}_{ij}(t) = W_i(D_{i,j-1}(t)) - S_i(D_{ij}(t)) \;;$$

$$x_{ij}(t_0) = x^{\circ}_{ij}, \qquad \text{for } j = 1,\ldots,N \;;$$

* "Imposed" functions describing the Shop-Internal Control Variables:

$$D_{ij}(t) = b_{ij}(t), \quad \text{for } j = 1,\ldots,N \;; \tag{11}$$

* Supply and Order Servicing Constraints (see eqs. (3) and (4)), in case of the first (j = 1) or the last (j = N+1) buffer set in the plant, but lacking in case of internal buffer set B(j) for j = 2,...,N ; <u>note</u> that these constraints are referred only to the part type i :

$$\Omega_i[\, x_i(t),\ r_i(t),\ \{\Theta_i(t_0)\}\,] \le 0$$

$$\text{and to} \quad 0 \le x_{ij}(t) \le x^*_{ij}, \quad t_0 \le t \le H.$$

In this case, the interaction variables among sub-problems P_i

represent the dependence of the Internal Control Strategy applied to each work shop j, on the requirements of additional service rates for processing the product orders of all other types.
In other words, the vector of interaction functions

$$b_i(t), \text{ for } t_0 \leq t \leq H ,$$

which are assumed known, according to the Interaction/Prediction Approach, will define the <u>"service rate margins"</u> that can be assigned to part type i at each work shop j, in front of the necessity of processing also the other part types.
Each local sub-problem $P_i(b_i)$ will then correspond to the problem of optimizing the service of only one part type at a work shop, i.e. of controlling the "efficiency margin" at each work shop, conditioned on the planning patterns for the other products.
Practically, this is the case of the production capacities of all shops, that clearly depend on the patterns of products to be processed.
Such interactions, as in the case of Sub-problems P_j, can be a-posteriori evaluated, and a solution procedure according to the two above outlined steps can be again implemented.
The Manufacturing Resource Planning (MRP II) appears to be a procedure just implementing such two mentioned steps, under an approximated solution of each Sub-problem P_i [6].

I.2.2. The Interaction/Balance Approach

By this second approach, the interaction variables $z_j(.)$ are included in the j-th control optimization sub-problem, while a comon Goal Coordination Problem is introduced, to assure verification of consistency constraints [2,7,11].
The complete PPC Problem can then be decomposed into a

<u>Two-Level Hierarchy of Sub-Problems</u>:

i) <u>a set of Lower-Level Local PPC Sub-Problems</u>, for each j=1,...,N+1, stated as follows

LOCAL PPC SUB-PROBLEM $P_j(\beta_j)$: Given the Local Augmented Cost:

$$G_j(x_j, \beta_j) = \{ \sum_{i=1}^{M} \int_{t_0}^{H} [x_{ij}(t) +$$

$$+ \alpha_{ij}(t) (z_j(t)) - S_i(D_j(t)) - \dot{x}_{ij}(t)) + \qquad (12)$$

$$+ \beta_{ij}(t) (z_{ij}(t) - W_i (D_{i,j-1}(t)))] dt \}$$

find the MAXIMUM w.r.t. $\{\alpha_{ij}(t), i = 1,\ldots,M\}$, of the MINIMUM

$$\text{w.r.t.} \begin{cases} H, D_j(t), \text{ for } t_0 \le t \le H, \text{ when } j = 1,\ldots,N ; \\ H, z_j(t), D_j(t), \text{ for } t \le t \le H, \text{ when } j = 1 ; \\ H, a_{N+1}(t), \text{ for } t_0 \le t \le H, \text{ when } j = N+1 ; \end{cases}$$

for given functions $\beta_{ij}(t)$, $D_{i,j-1}(t)$, $t_0 \le t \le H$

and for given sequences $\{\Theta_i(t), i = 1,\ldots,M\}$,

subject to $0 \le x_{ij}(t) \le x^*_{ij}$, $t_0 \le t \le H$.

ii) a common <u>Goal</u> <u>Coordination</u> <u>Sub-Problem</u>, stated as follows

GOAL COORDINATION SUB-PROBLEM GCP:

$$\max_{\beta_j} \min_{\{\Theta(t_0)\}} \sum_{j=1}^{N+1} G^*_j(\bar{x}_j, \beta_j) \tag{13}$$

subject to constraints (3) and (4), for $t_0 \le t \le H$.

Note that, in above formulation, the symbol "$\bar{x}_j$" means the locally optimized state trajectories at the buffer set B(j), i.e. the ones resulting from the solution of Sub-Problem $P_j(\beta_j)$.

In this case, the solution of the overall PPC Problem can be obtained according to the following steps :

STEP A. The j-th local controller receives the Lagrangian variables $\beta^k_j(t)$ from the central coordinator and it solves its local sub-problem $P_j(\beta_j)$;

STEP B. The coordination procedure must compute the

Lagrangian variables $\beta_j^{k+1}(t)$ by solving the Goal Coordination Sub-Problem, conditioned on the locally optimized strategies, and evaluate the best sequencing of the product orders $\{\Theta(t_0)\}$, based on the locally optimized production capacities $a_j(t)$.

This approach formalizes the so-called Interaction/Balance Principle [5], and, in mathematical programming terms, it is known as "Dualization / Feasible Directions" method.

Such a decomposition gives rise to a hierarchical control architecture where the higher-level phase now plays the role of an effective coordinator.
Here the Lagrangian variables

- can be computed by standard mathematical procedures, and mainly
- can assume an interesting physical meaning, that of sensitivity variables of the problem solution with respect to the verification of interaction constraints.

The PP Hierarchical Decentralized Architecture, developed at the Politecnico di Torino, is a procedure just implementing said solution [2,7,10].

I.2.3. The Decomposition/coordination Approach

By this third approach, the interaction variables $z_j(.)$ are taken into account directly in the j-th PPC sub-problem, that can be stated as

SUB-PROBLEM P°_j : Minimize

$$WIP(j) = G_j(x_j) = \sum_{i=1}^{M} \{ \int_{t_0}^{H} x_{ij}(t)\, dt \}$$

$$\text{w.r.t.} \begin{cases} H,\ D_j(t),\ \text{for } t_0 \le t \le H, \text{ when } j = 2,\dots,N\ ; \\ H,\ z_j(t),\ D_j(t),\ \text{for } t_0 \le t \le H, \text{ when } j = 1\ ; \\ H,\ \{\Theta(t_0)\},\ a_{N+1}(t),\ \text{for } t_0 \le t \le H, \text{ when } j = N+1; \end{cases}$$

subject to:

* State Equation Model:

$$\dot{x}_j(t) = W(D_{j-1}(t)) - S(D_j(t)) \;;$$

$$x_j(t_0) = x^\circ_j \;;$$

* either Supply or Order Servicing Constraints (see eqs. (3) and (4)), in case of the first (j = 1) or the last (j = N+1) buffer set in the plant, but lacking in case of internal buffer set B(j) for j = 2,...,N :

$$\Omega_j [\, x_j(t), r(t), \{\Theta(t_0)\} \,] = \begin{cases} \text{equation (3), for } j = 1 \;; \\ \text{equation (4), for } j = N+1 \end{cases}$$

and to $0 \le x_j(t) \le x^*_j \;, \quad t_0 \le t \le H.$

Obviously, the local control strategy $D_j(t)$, that can be obtained solving Sub-problem P^o_j, depends not only on the j-th shop state $x_j(.)$, but also on the remaining planning/control variables of the other shops.
Said dependence can be implicitly represented as follows [4]:

$$D_j(t) = K_j(\, x_j(t), \; \delta_j(t) \,) \tag{14-a}$$

where

$$\delta_j(t) = d\, \bar{G}_j(.) \,/\, d\, D_j(t) \,, \tag{14-b}$$

being $\bar{G}_j(.)$ the part of the overall performance index G(.) not retained into $G_j(.)$.
Then, being $\delta_j(.)$ the sensitivity of said part $\bar{G}_j$ with respect to the j-th control variables, that function represents the "coordination need" between the j-th local sub-problem and the other ones.
Such a coordination function cannot be easily computed by directly solving (14). Generally, it should be handled by some relaxation formula, when inserted into an iterative solution procedure for the overall PP Problem, as follows :

STEP A. Solve each local Sub-problem P°_j, depending on assumed values for the coordination functions $\delta_j(.)$;

STEP B. Modify the coordination functions $\delta_j(.)$ by means of some relaxation formula, and iterate.

This approach, in mathematical programming terms, is referred to as "Dualization / Relaxation" approach [5].

Referring to PP Problem solution, the following particular form of above stated approach has been adopted.
Given the multi-stage manufacturing system, <u>assume that shop j is the "bottleneck" for a given production pattern.</u>
In this case, all other shops n (n = j, 1 < n < N) appear to be strongly connected to shop j, either by the "blocking effect" (the upstream shops), or by the "starving effect" (the downstream ones).
Pratically, it can be seen that all coordination functions referred to shops n are very higly valued, thus implying that the solution of the related sub-problems P°_n just only depend on said coordination functions.
On the contrary, referring to shop j, this one does not depend on other shops' strategies; then, its own coordination functions are almost null.
By said considerations, an approximated solution of the overall PP Problem can be obtained by the following steps :

1st step : solve the local optimization sub-problem for shop j ;

2nd step : obtain control strategies for the other shops by only verifying consistency constraints (2).

The Optimized Production Technology (OPT) appears to be a procedure just implementing said two mentioned steps [8].

<u>REMARK 4.</u> Note that, as in the case of the Interaction/Prediction approach, also the outlined Decomposition/Coordination one - at least, in the formulation that we have proposed and that allows OPT interpretation - does not perform any effective coordination, but it directly verifies constistency constraints.

I.2.4. The Backward-Conditioned Local-Feedback Decomposition

A particular decomposition, that cannot be related to any existing result of mathematical programming theory, but that derives from Control Theory applied to series systems,is obtained if the control strategy of each shop j is defined by solving a local sub-problem, when

- the interaction variables linking shop j to the downstream shop j+1 are assumed as reference signals ;
- while the interaction variables linking shop j to the upstream one are computed as result of the local optimization.

By said position, the overall process-structure-based Problem PP can be splitted into a family of sub-problems P^d_j, each one conditioned on the solution of the downstream sub-problem P^d_{j+1} :

DOWNSTREAM-CONDITIONED SUB-PROBLEM P^d_j : Minimize

$$WIP(j) = G_j(x_j) = \sum_{i=1}^{M} \{ \int_{t_0}^{H} x_{ij}(t)\, dt \}$$

$$\text{w.r.t.} \begin{cases} H,\ D_{j-1}(t),\ \text{for } t_0 \le t \le H,\ \text{when } j = 2,\dots,N\ ; \\ H,\ z_j(t),\ \text{for } t_0 \le t \le H,\ \text{when } j = 1\ ; \\ H,\ \{\Theta(t_0)\},\ D_{j-1}(t),\ a_{N+1}(t),\ \text{for } t_0 \le t \le H, \\ \qquad\qquad\qquad\qquad \text{when } j = N+1; \end{cases}$$

subject to:
* State Equation Model:

$$\dot{x}_j(t) = W(D_{j-1}(t)) - a_j(t)\ ;$$

$$x_j(t_0) = x^\circ_j\ ;$$

* Downstream Conditioning Constraints:

$$a(t) = S(D_j(t)),\ \text{for } j = 1,\dots,N\ ;$$

* Order Servicing Constraints (4), in case of the last (j = N+1) buffer set in the plant ;

and to $0 \le x_j(t) \le x^*_j$, $t_0 \le t \le H$.

REMARK 5. Note that, in this formulation, the Supply Constraints (3) have not been taken into account. The supply function r(t) will be obtained as the one that assures sufficient supply of material at the input buffer set, over the whole planning horizon (<u>Condition assuring existence of a Backward-Conditioned Local Production Control</u>).

In this case, each local strategy defines reference state trajectories for the upstream shop PP problem.
At the last shop (j=N), a reference trajectory for the product demand is defined by the set of product orders required by customers, once a product order scheduling problem has been solved.
At each intermediate shop j $(1<j<N)$, the interaction variables with the downstream shop are imposed, while the ones with the upstream shop can be computed after the solution of the local PP sub-problem.
Said approach does not emphasize any multi-level architecture, since it gives rise to a chain of conditioned optimizations.
The Just-In-Time (JIT) philosofy appears to be an effective implementation of this approach [9].

PART II. SADT REPRESENTATION OF THE PRODUCTION PLANNING & CONTROL ARCHITECTURES

II.1. Building SADT Diagrams of PM Architectures

From the point of view of the involved information flows, above formulated PP Problem approaches can be described by simplified SADT (Structured Analysis Design Techniques [12]) block diagrams. Irrespective of the different solution approaches, the PPC Problem can be represented in a common way , as in Fig. 3.

Such a representation shows that, in any case, the production planning aims at minimizing the cost of the production plan , closely related to holding costs of materials and estimated backlogging costs.
This portion of the planning activity is "open-loop", since it is based on the model, without verification by feedback data collected from the AMS.
After a while, since a production plan has been submitted to the plant, the feedback loop is closed, to trigger the forecast lot schedule with the system state.
Depending on the solution approach, the planning procedure (high layer block) requires a different representation.

The Interaction/Prediction Approach implies the definition of a two-level planning architecture, implementing the following steps :

STEP A : decentralized planning, at each work stage, according to a "push-type" approach, i.e. by independently satisfying due dates constraints required to each shop ; this procedure is the so called MRP - Material Requirement Planning, shown in Fig. 4;

STEP B : centralized prevision of interaction constraints, to be performed by
B 1. scheduling product orders (MPS - Master Production Scheduling) ;

B 2. verifying the required inter-stage inventories, to be compared with storage capacities;
if overload situations are recognized, messages are sent to MPS to modify input lots ordering (as shown in Fig. 5).

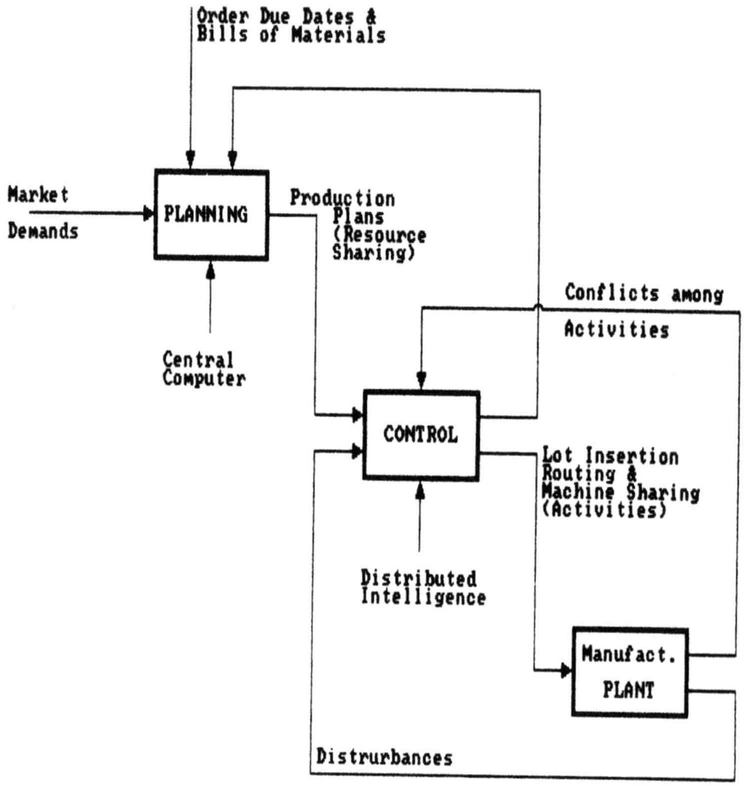

Fig. 3. SADT representation of the PPC Problem component blocks.

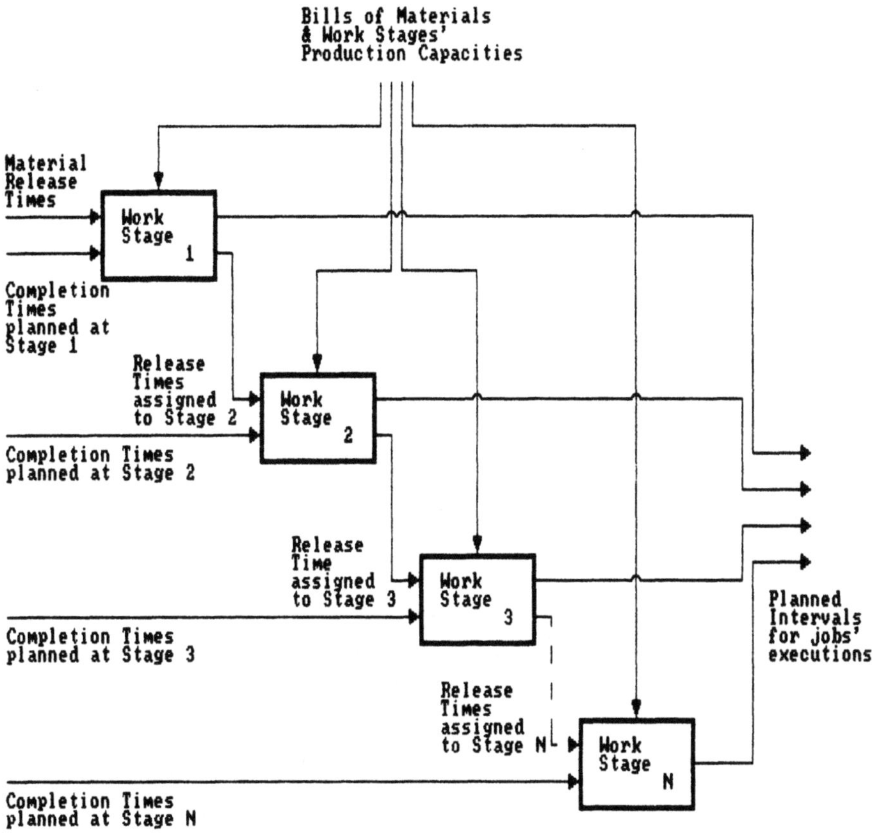

Fig. 4. SADT representation of the Material Requirement Planning (MRP) procedure.

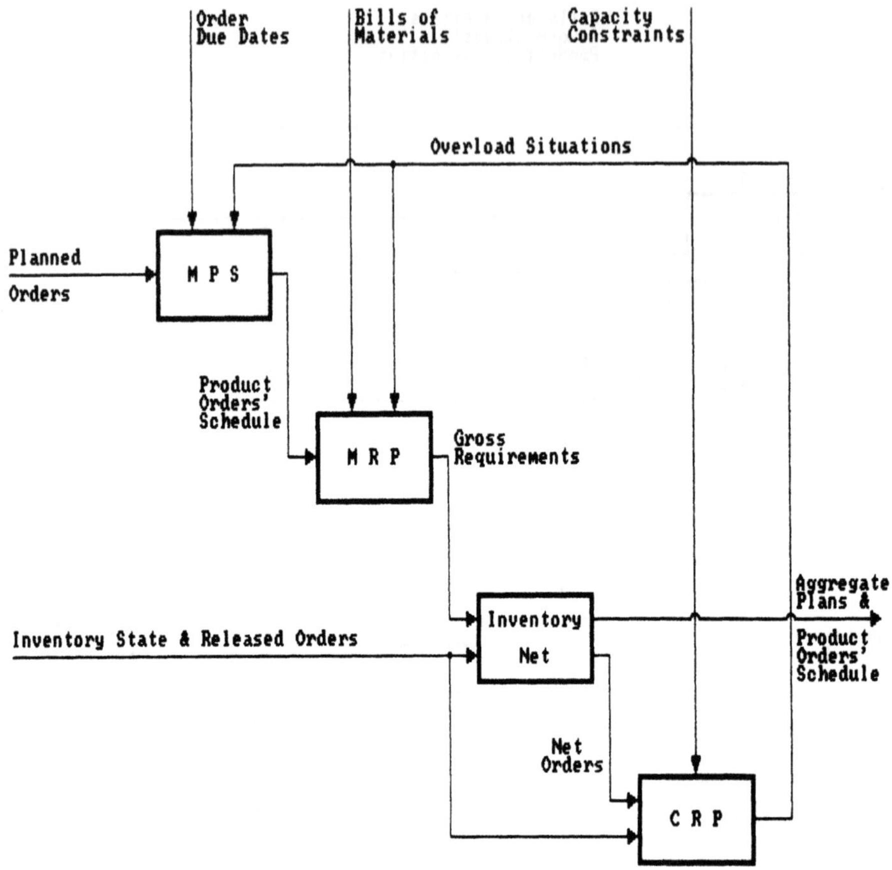

Fig. 5. SADT representation of the Manufacturing Resource Planning (MRP II) procedure.

The particular version of the Decomposition/Coordination Approach, giving rise to the OPT procedure, can be represented by a non-recursive scheme in which the fine schedule starts from the bottleneck shop and proceeds both backward and forward along the chain.
The role of the coordination block now involves the recognition of the bottleneck shop, as in Fig. 6.

If the work shop chain presents a single bottleneck, this strategy may lead to very good sub-optimal plans. On the other hand, if the system is alsmost balanced, non-bottleneck shops may become overloaded during the fine scheduling phase, and the resulting planning strategy could be not so good. The same applies if there are more than one bottleneck.

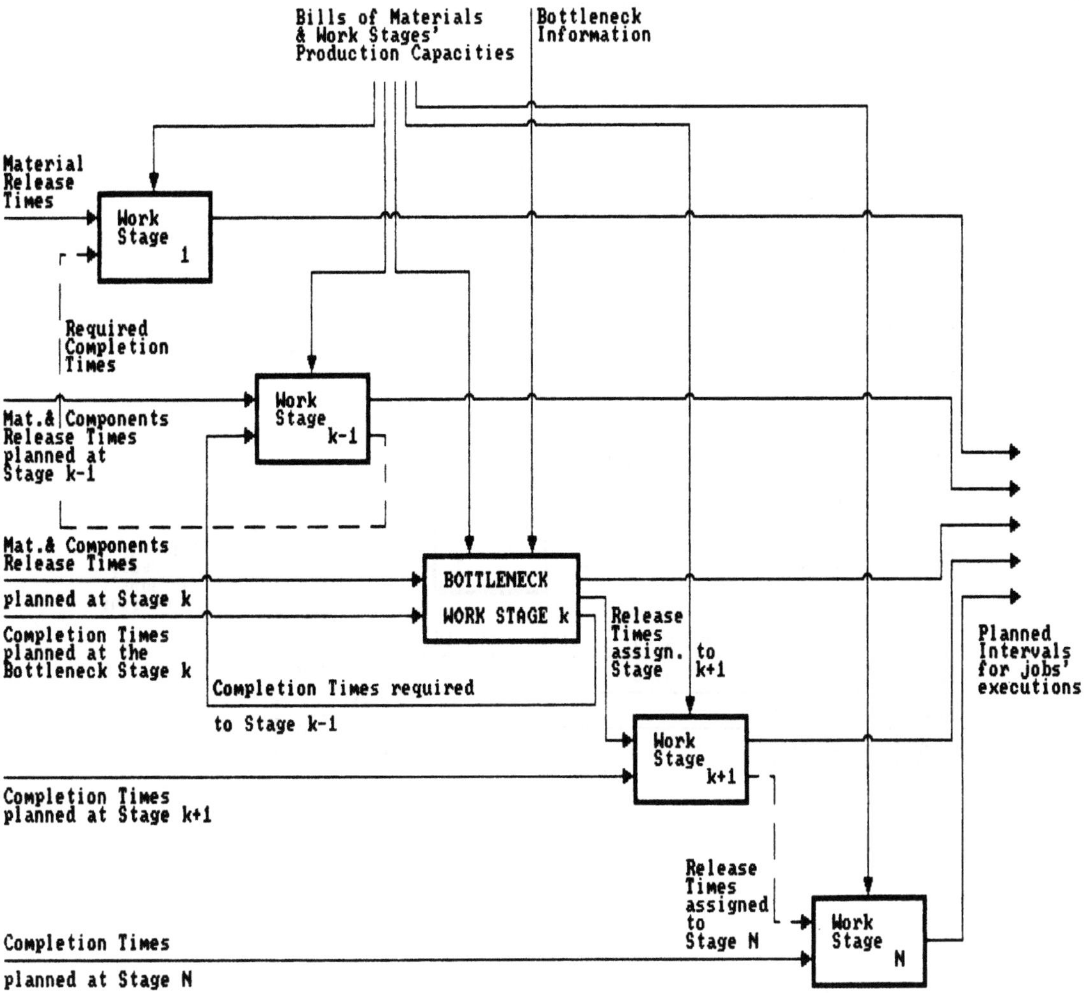

Fig. 6. SADT representation of the Optimized Production Technology (OPT) procedure.

The Backward-Conditioned Local-Feedback Decomposition implies the definition of a single-level planning architecture, implementing the following steps :

STEP A : centralized short-term prevision of the product orders to be satisfied (last shop desired throughput);

STEP B : decentralized planning, at each work stage, according to a "pull-type" approach, i.e. by verifying backward along the line due dates constraints required to each shop ; this procedure is dual with respect to the above mentioned MRP , as shown in Fig. 7.

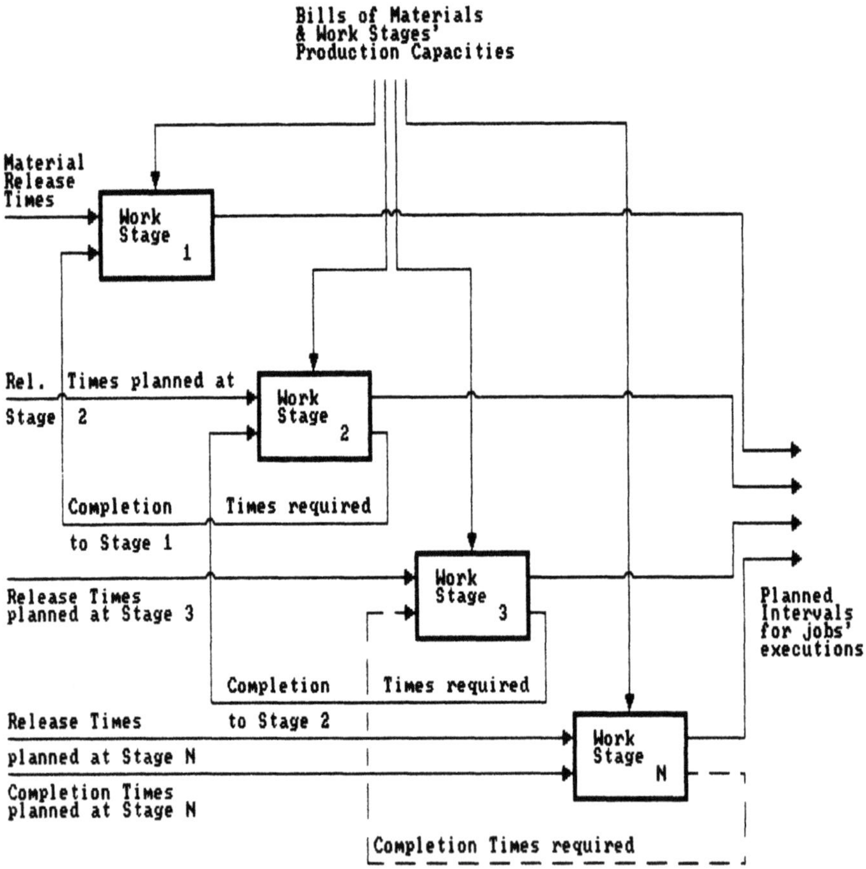

Fig. 7. SADT representation of the Just-In-Time (JIT) approach to production control.

The Interaction/Balance Approach implies the definition of a three-level planning architecture, implementing the following steps (see Fig. 8):

STEP 1 : an MRP II module provides the initialization of the planning procedure, based on an aggregated model of the AMS ; its results will be forecast due dates for lots to be processed ;

STEP 2 : the initial aggregate production plan from MRP II is submitted to local optimization procedures - one for each shop - by a coordination block ; this one interacts with local optimizations, by supplying them weighting factors (to account for interaction constraints), and by receiving from them locally optimized plans ;

STEP 3 : each local plan optimization is carried out to minimize a cost function dependent on the weighting factors supplied by coordinator ; in addition, each local optimization procedure communicates its own completion times to downstream shops, and its own release times to upstream shops, to make known their interaction variables (see Fig. 9).

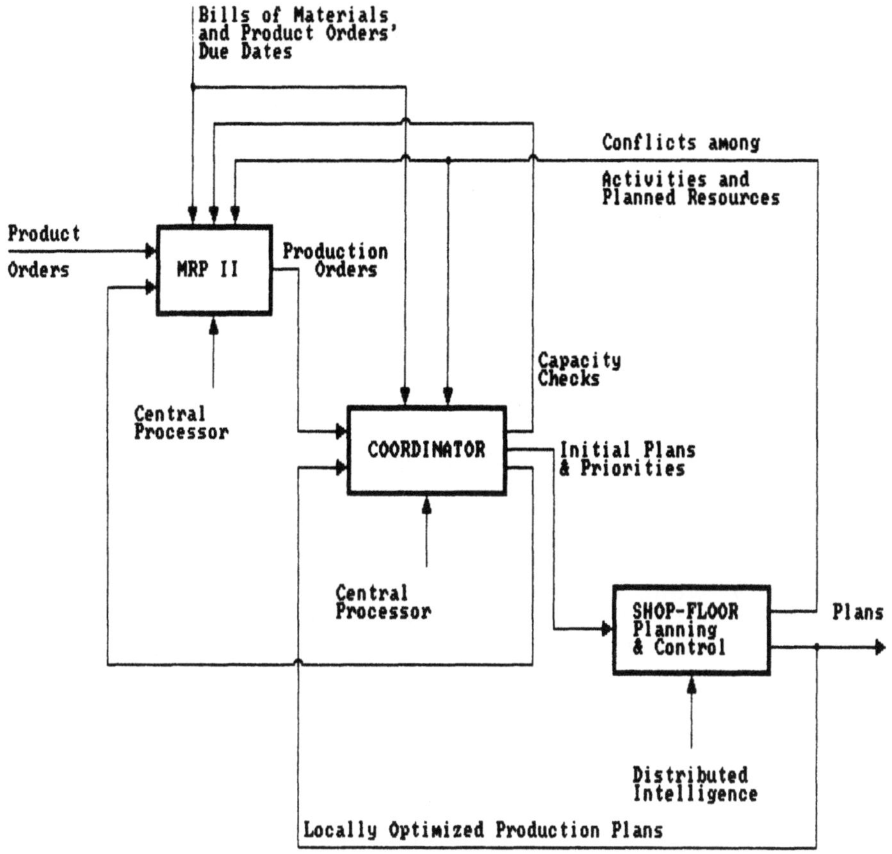

Fig. 8. SADT representation of the Hierarchical Decentralized Architecture (HDA) for PPC Problem solving.

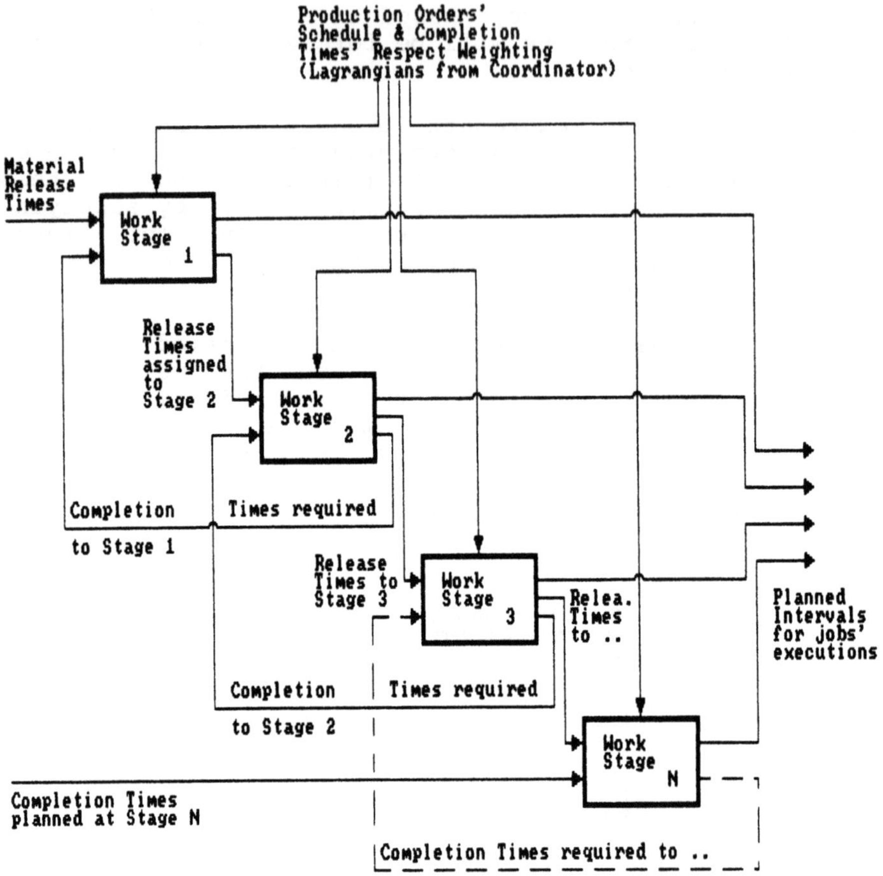

Fig. 9. Handling interactions among local PPC procedures in the HDA framework.

II.2. Some Considerations about PM Architectures Comparisons

The main result of above considerations appears to be the common statement of the production planning problem, in terms of a large-scale mathematical programming problem to be updated each time a new event occurs.
Such a common formalization, never developed before, allows to compare, in analytical terms, the different decompositions of the production planning problem for a chain of work shops, that can be classified as

- push-type procedures, if based on the interaction/prediction approach or on the decomposition/coordination one (as MRP II);

- pull-type procedures, if defined according to a backward-conditioned local-feedback structure (as JIT);

- mixed (both pull- and push-type) procedures, if based on an interaction/balance approach (as OPT and HDA).

The resulting problem statements clearly show the main features of all procedures.

The push-type ones present "apparent" coordination actions, i.e. mathematically forced by relaxation formulas. This fact reflects into a difficulty in efficiently handling large demand variations while assuring small inter-stage inventories, since large perturbations ask for strong coordination actions. This feature justifies the difficulty in efficiently handling the Capacity Requirement Planning (CRP) in any MRP II procedure.

The pull-type procedure still appears to be unable to handle demand peaks, but due to its inability to use forecasts in order to anticipate production. This feature is clearly shown by the JIT procedure representation in terms of a chain of backward-conditioned feedback loops.

On the contrary, the mixed approach allows to define a planning architecture with both a coordinating facility, and an interconnected sequence of local optimizations, thus balancing integration requirements and pull-type ability of reducing local inventories. The coordination efforts are weak in OPT procedures, based on the assumption that no improvement can be obtained by spending planning/control efforts at non-bottleneck stages. On the other hand, HDA assures an integrated management of all work stages, to accelerate as much as possible the stage servicing, thus reserving production capacity for incoming new orders.

In addition, it must be remarked that many practical production environments, inventory-driven product demand (that can be served by push-type strategies) and customer-driven one (requiring pull-type strategies) are generally present together. Proposed common formulations of the production planning & control problem can offer, in these situations, an analytical frame to compare different strategies that could be used.

PART III. GENERALIZING PM ARCHITECTURE DESIGN IN CASE OF MULTI-PLANT INDUSTRIAL ENVIRONMENTS

The third part of the paper will discuss the problem of Production Planning in an industrial environment composed by a number (generally small) of different factories, located in different regions. We will refer here to an industrial environment, at present under study, where the factories are devoted to process electrical parts, but each one of them is oriented to process a proper sub-set of part families. It means that all factories require a common management of material supply, while each one of them calls for a proper Production Planning, in front of proper demands either from final product inventory, or from local market.

For a better characterization of the considered production environment, we assume that the final products can be classified into a very large number of types, to be grouped into families. For each product sub-set, a-priori assigned to a plant, products can be classified into two groups:

- Group α, including products whose evolution in time can be estimated and then forecast, due to availability of a sufficient past history;
- Group β, including products whose demands are unpredictable.

Above outlined statement of the multi-plant manufacturing system allows to envisage an organization of the Production Management (PM) Architecture composed by two Main Decision Layers (MDL's):

* MDL-1, devoted to the common management of the material inventories, in order to supply materials and components to each plant at the right time and in required volumes; this (top) layer will contain three Modules, one for each plant, to solve the Material Supply Planning Problem, in terms of:
  a) the product-into-family aggregation/disaggregation;
  b) the demand forecasting for products of Group α;
  c) the Long-Range Production Programming (LRPP) for the whole set of plants, based on demands' forecasts (when avaliable), and on order portfolio;

* MDL-2, including a set of mid-term single-plant PM Architectures of the types outlined in previous parts, as modules defining the production plan at each individual plant.

Owing to the very large number of product types to be processed in each plant, solutions of the PM-Problems referred to MDL-1 are of crucial importance.

This last part of the paper will analyze the PM Problems referred to MDL-1, with the following objectives:
1. outline a Complete-PM (C-PM) Architecture for a multi-plant industry: this objective will require to verify integration modes among lower-layer PM Architectures assigned to individual plants;
2. investigate possible solution procedures, stated in a form easy to be integrated with the lower-layer decentralized PM Architectures.

III.1. The Market-Oriented Family Aggregation/Disaggregation

The problem of forecasting a large variety of product types surely appears the top-level tasks to be solved. Item flow assignment to each plant, as well as locally optimized production plans, hardly depend on product demands forecasts.
Due to the large number of product types, the use of product-into-family aggregations is mandatory.

Local PM Architecture, however, requires availability of two different types of product families, i.e. Market-Oriented and Production-Oriented ones. The former must be used for forecasting purpose of Group α product demands: the reason is to have at disposal, for each MO-Family, a sufficient past history even if in front of few data for each product type in the family. The latter is needed to optimize production plans over a reduced number of families, instead of the very large number of product types.
It is well known that PO-Families can be obtained by reasoning derived from Group Technology [13]. What is still a matter of investigation is the way of building MO-Families.

III.1.1. MO-Family Aggregation

The approach for MO-Family aggregation here proposed is based on the concept of "past evolution of sales of a product in the market", denoted as market trajectory of the product. Obviously, for each product, it will correspond to a "sampled signal", whose values have been collected over a past history of finite lenght.
Note that this formulation is suggested by the individual plant model adopted for designing local PM Architectures: this model has been based on the concepts of inventory states and of input and inter-stage transfer flows. Then, the market trajectory appears to be the "reference demand trajectory" which each local plan must track.
Denote by

p, $p=1,2,\ldots,NP$, the product index;

$y_p(t)$, $t \in [-L, t_0]$ the market trajectory of product p over the the past horizon of lenght L.

The best rule to aggregate products into an MO-Family can be derived by direct application of the Fourier concept of signal processing, i.e.

MO-Aggregation Rule:
Given two products, indexed by p and q, their market trajectories are similar

$$\text{if} \quad \sum_{f \in F(p,q)} [\, Y_p(f) - Y_q(f) \,] \le \sigma_M \, ,$$

$$\text{and if} \quad \sum_{f \in F(p,q)} [\, \Phi_p(f) - \Phi_q(f) \,] \le \sigma_\Phi \tag{15}$$

where the frequency bandwidth used to compare the two "signals" is

$$F(p,q) = \{\, f \text{ s.t. } Y_p(f),\ Y_q(f) \ge M \,\} \, ,$$

and

$Y_p(f)$, $Y_q(f)$	denote the Fourier spectra of the product p trajectoriy in the past horizon;
$\Phi_p(f)$, $\Phi_q(f)$	denote the relative phase functions;
σ_M, σ_Φ	are suitable upper-bounds of the difference between spectrum amplitudes and phases ;
M	is a lower-bound of the spectral amplitude.

REMARK 6. The MO-Family Aggregation Rule is a formal statement, in the frequency f domain, of the concept of similarity of two signals, normally used in frequency identification based on signal Fourier transform.

III.1.2. MO-Family Disaggregation

The problem of disaggregating an MO-Family into its component products is the dual problem with respect to the family aggregation one. In fact, the result of the MO-Family Aggregation Rule is that the mix of products, in each MO-Family, is almost constant over the past horizon.
Let k be an MO-Family, including NG(k) products, and p a product included. Denote by

$$y_k(t) = \{ \sum_{q=1}^{NG(k)} y_q(t) \} 100, \text{ for } t \in [-L, t_0]$$

the cumulative demand associated to the k-th MO-Family, and by

$$P_p(t), \quad t \in [-L, t_0]$$

the percentage of product p, with respect to the cumulative demand referred to the i-th MO-Family.
The Aggregation Rule above stated implies that:

$$P_p(t) = y_p(t) / y_k(t) \approx P_p = \text{const.} \qquad (16)$$

Then, the estimated future evolution of any product p demand can be obtained by propagating, for $t \geq t_0$ = "present time", the past dynamics of the MO-Family cumulative demand, weighted by P_p.

III.2. <u>The Long-Range Production Programming (LRPP) Procedure</u>

The LRPP-Procedure is here approached under the following <u>Practical Assumptions:</u>

A-1) Only Group α product demands can be forecast.

A-2) Production capacities of each plant, for the different products that can be processed, are obtained by averaging the capacity values resulting from the solution of the mid-term PPC Problem at the plant considered. In addition, production capacity is lowered to allow accomodation also of un-predictable Group ß product demands.

A-3) Being each plant oriented to serve its own market demands, the set of plants will be considered, by the Material Supply System, i.e. by the LRPP Problem solver, as <u>parallel services</u>, able to share their processing capacity among a sub-set of PO-Families.

A-4) Necessary materials can be obtained by suppliers at an incoming rate surely bounded.

A-5) The Long-Range Planning horizon T is related to the mid-term planning horizon H, common to all local PM Architectures driving individual plants, by the constraint $T = \mu H$, where $\mu \geq 10$.

Denote by

$n = 1,\dots,NF$, the PO-Family index (being PO-Family aggregation obtained by GT techniques [13]);

$m = 1,\dots,NI$, the index of the plant composing the considered "distributed firm";

$\{ D_n(t), t_0 \leq t \leq T \}$ the estimated demands of PO-Family n ;

$E(n)$ the set of plants where PO-Family n can be processed;

$G(m)$ the set of PO-Families that can be processed by the plant m ;

$x_{nm}(t)$ the storage of materials, at time t, in the input buffer of the plant m, which will be used for processing PO-Family n ;

$w_{nm}(t)$ the input flow of these same parts, loaded in the input buffer of plant m at time t (<u>loading rate</u>, to be planned);

$s_{nm}(t)$ the average processing capacity of plant m, for products belonging to PO-Family n (<u>production capacity</u> of plant m, to be planned in order to define long-range reference strategies on wich the local PM Architecture should be tuned).

By these notations, the model describing the <u>Dynamic Assignment</u> of material flows to the set of plants can be stated as follows:

$$\dot{x}_{nm}(t) = w_{nm}(t) - s_{nm}(t) \; , \; x_{nm}(t_0) = x^{\circ}_{nm} \qquad (17)$$

for $n = 1,\dots,NF;\ m = 1,\dots,NI$;

under the following constraints:

* Physical & Storage Capacity Constraints

$$0 \le x_{nm}(t) \le x^*_{nm} , \quad t_0 \le t \le T ; \qquad (18)$$

* Supply Saturation Constraints

$$\sum_{m \varepsilon L(n)} w_{nm}(t) \le w^*_n , \quad t_0 \le t \le T ; \qquad (19)$$

By these positions, it can be stated the following

<u>LRPP-Problem:</u> minimize

$$A = \int_{t_0}^{T} \{ \sum_{n=1}^{NF} [D_n(t) - \sum_{m \varepsilon E(n)} s_{nm}(t)_{pj}] \, dt \} \qquad (20)$$

w.r.t. $w_{nm}(t)$ and $s_{nm}(t)$, $t_0 \le t \le T$,

for $n = 1,\dots,NF$, $m = 1,\dots,NI$,

subject to constraints (17) to (19).

<u>REMARK-2</u>. The LRPP Problem essentially aims at assigning to each plant a proper work load: w(.) plays the role of splitting rates of the demand of each PO-Family amount n towards the plants able to process this same product type. By this reason, the LRPP Problem can be formally solved as a linear dynamic constrained optimization problem [14].

III.3. <u>Final Discussion about the Complete PM Architecture</u>

Above results allow to state an SADT representation of the hierarchical organization of the Decision Modules to be included in the C-PM Architecture, as illustrated in Fig. 10.
From a practical pint of view, the C-PM Architecture appears to be composed by:

a) a Module devoted to estimate the product demands, over a long-range future horizon (M-1), for products of Group α; forecasts may possibly be computed by heuristic rules, and should be updated in front of requests from LRPP, owing to inadequacy of the plants' capacities to satisfy estimated demands;
b) a Module M-2 defining the PO-Family aggregation of products of Group β, whose future demands cannot be forecast;

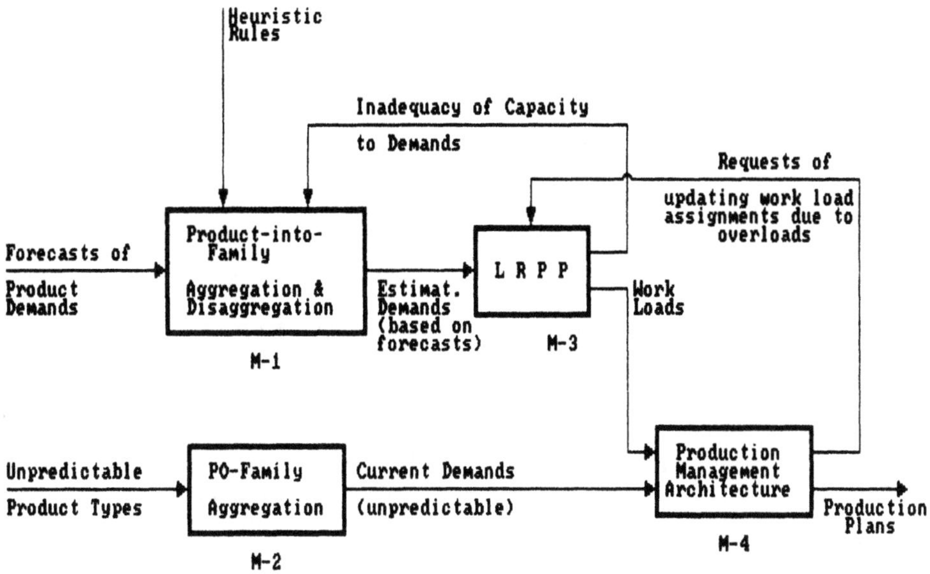

Fig. 10. SADT representation of the main Modules of the Complete C-PM Architecture.

c) an LRPP Module evaluating work loads for each plant, based on Group α estimated demands;
d) a Module including the set of PM Architectures, one for each plant, to locally define mid-term production plans.

In principle, production flows at each plant could be planned by a particular PM Architecture, among the ones analyzed in Part II. In fact, interactions between the lowest-layer Module M-4 and the LRPP Module consist only of:
- work loads, estimated by LRPP and sent to M-4;
- requests to update the work loads, in front of overloading situations, detected by an individual PM Architecture, and sent to LRPP.

According to this consideration, a PM Architecture could be designed by applying anyone of the above stated methods (MRP II, OPT, JIT, or HDA).
This conclusion, however, is not true: each PM Architecture must handle also unpredictable demands, and it must be noted that unpredictable demands require

a) to define production plans for Group α product orders, <u>and</u>
b) to control production orders execution by some JIT-based procedure.

The necessity of this integration, between MRP- or OPT-based planning, and JIT-based control, is the <u>effective</u> <u>crucial</u> <u>node</u> of

Production Flow Management.
HDA has been designed to satisfy this integration specification. However, also the other PM Procedures, if organized according to the proposed formulations, can satisfy the same specification.

REFERENCES

1. Villa. A. and Rossetto. S.: Towards a hierarchical structure for production planning and control in flexible manufacturing systems. in: Modelling and Design of Flexible Manufacturing Systems (A. Kusiak. ed.). pp. 209-228. Elsevier Science Pub., Amsterdam (1986)
2. Conterno. R., Fiorio. G., Menga. G., Villa. A.: A large scale systems approach to the production planning and control problem. in: Proc. 24th IEEE Conf. on Decision and Control. Fort Laurendale. pp. 2016-2021. 1985
3. Villa. A. and Rossetto. S.: On the joint problem of dynamic part routing and station service control in flexible manufacturing systems. Material Flow. 2. 97-110 (1985)
4. Sandell. N.R., Varaiya. P., Athans. M., Safonov. M.G.: Survey of decentralized control methods for large scale systems. in IEEE Trans. on Autom. Contr. AC-23. 108-128 (1978)
5. Geoffrion. A.M.: Elements of large scale mathematical programming. Management Science. 16. 652-691 (1970)
6. Adam. H., Cox. J.F.: Manufacturing Resource Planning: an information system model. in Long Range Planning. 18. 86-92 (1985)
7. Conterno. R., Menga. G., Villa. A.: Hierarchical and decentralized control for batch and repetitive manufacturing. in: Proc. 1987 IEEE Int. Conf. on Robotics and Automation, Raleigh, NC, pp.1855-1860. 1987
8. Fox. R.E.: OPT vs. MRP - Throughware vs. software. Inventories and Production Magazine, 4 (1984)
9. Hall. R.W.: Zero Inventories. Dow Jones-Irwin. Homewood. IL. (1983)
10. Kusiak. A., Villa. A.: Decision architectures for production planning in multi-stage multi-product manufacturing systems. submitted to Annals of Operations Research (1988)
11. Villa. A.: An hybrid knowledge-based/analytical control system for failure-prone manufacturing systems. in: Proc. of 1987 IEEE Int. Conf. on Robotics and Automation. Raleigh. NC. pp. 289-294 (1987)
12. Marca. D.A., McGowan. C.L.: SADT - Structured Analysis and Design Technique. McGraw-Hill. New York (1988)
13. Burbridge. J.L.: The Introduction of Group Technology. Wiley. New York (1975)
14. Singh. M.G., Titli. A.: Systems Decomposition, Optimization and Control. Pergamon Press. Oxford (1978)

CIM SYSTEM DEVELOPMENT FOR SMALL TO MEDIUM SIZE DISCRETE-PARTS MANUFACTURING COMPANIES+

Robert E. Young*
Institute of Production Management and Industrial Engineering
Technical University of Denmark
DK-2800 Lyngby
Denmark

ABSTRACT

This paper discusses analysis and design tools for use by small and medium sized discrete-parts manufacturing companies for CIM system development. The development lifecycle consists of eight primary technical tasks: Problem Analysis, Requirements Definition, Preliminary Design, Detail Design, Construct & Verify, Integrate & Validate, Implementation & User Acceptance, and Maintenance & Support. The discussion is based upon experiences with the CIM/GEMS project, a DKr 15 million (US$ 2.3 million) project to develop Danish industry's ability to *manufacture the manufacturing system*.

+ Some material from this paper was presented at the IFS 6th International Conference on Flexible Manufacturing Systems, Turino, Italy, November 1987.

This work was partially funded by the Council of Technology (Teknologirådet) in Denmark.

* Currently on leave from the Department of Industrial Engineering, North Carolina State University, Raleigh, NC, USA.

NATO ASI Series, Vol. F53
Advanced Information Technologies for Industrial Material Flow Systems
Edited by Shimon Y. Nof, Colin L. Moodie

INTRODUCTION

This paper discusses analysis and design tools for use by small and medium sized discrete-parts manufacturing companies in CIM development. It is based upon experiences with the technical tasks for the Danish CIM/GEMS project. CIM/GEMS is concerned with GEneral Methods for specific Solutions. It is a DKr 15 million (US$ 2.3 million) project jointly funded by the Danish Government, the Technical University of Denmark, and private industry to develop Danish industry's ability to *manufacture the manufacturing system* (see [30] for a detailed description of the project). It is targeted to provide methods and tools for small and medium sized companies, those companies that have 200 workers or less comprise 97% of the companies and employ 52% of the people in Danish discrete-parts manufacturing (incidentally, in the USA, companies that have 250 workers or less comprises 96% of the companies and employ 43% of the people in discrete-parts manufacturing).

The focus of CIM/GEMS is on engineering methods used to engineer CIM systems and not the application of the most advanced hardware technology. With this focus we have examined tools that support the CIM development process. This paper is a result of examining the tools available to support CIM development. Initially, we describe CIM and the development constraints for small to medium size companies. We then define the CIM development lifecycle and identify tools that support the main tasks in each lifecycle phase.

DEFINITION OF CIM

We define *manufacturing* as the services and activities necessary to support the production of tangible goods as well as the physical production itself. In this context then, *Computer-integrated Manufacturing* (CIM) is defined as the automation of the information flow in manufacturing through the application of computers. For small and medium size companies this most likely will involve little automation.

BACKGROUND AND RATIONAL FOR CIM

The focus of the CIM-development efforts is on the indirect-cost activities in companies, specifically, the information processing that supports, directs, and controls the physical manufacturing operations. In other words, CIM aims it efforts at the manufacture of the manufacturing system, rather than at the manufacture of the product. Thus, CIM is a business strategy whereby computers are efficiently integrated and effectively used to perform the information work in the manufacturing system.

THE RISE IN INDIRECT LABOR AND ITS COST

For many decades, companies have sought to improve their productivity in the direct "touch" labor of the shopfloor (e.g., mechanization, hard automation, methods engineering, change in work organization). However, these efforts generally require greater production preparation, which is done by higher-skilled and more costly non-production workers. Together with externally imposed changes (internationalization of markets, demand for customized products, etc.), requiring greater organization skill from management, the consequences for many companies have been that the "non-touch" labor has increased to over 60-70% of the total labor cost of producing batch-manufactured goods [37].

THE NATURE OF MANUFACTURING'S INDIRECT LABOR

The nature of this "non-touch" labor is intellectual rather than physical, and the tasks are primarily those of data manipulation and information processing. The modern manufacturing system is becoming dominated by this information-processing superstructure, which has evolved to provide service to the actual production operations. Seen from this viewpoint, the principal activity of most people engaged in manufacturing is creating, analyzing, transmitting and managing information, while the actual material transformation is secondary [36, 37].

However, because the amount of information in a manufacturing system is so great, managing it has become a runaway cost for many companies [36, 37]. The problem is compounded by the fact that no two companies are identical--either in their policies, their operations, or their products--so the information resident in and used by the manufacturing system varies from company to company, making a general solution impossible. If this information--the principal underpinning of and key to true system integration--is to be used to build more efficient manufacturing systems, then powerful and flexible tools are needed to manage and control this important resource.

SOFTWARE ENGINEERING BY PRODUCTION PEOPLE

Until now, the computers used for most technical applications in manufacturing have been mainframe or supermini computers running software designed by software houses or in-house computer specialists. Often, the computer applications and solutions produced by these specialists have been unsatisfactory. The problems have not gone away with vendor-dominated standalone computer systems for CAD, CAM, CAE, etc. These have tended to create non-integrated islands of manufacturing-support activity without substantially improving the manufacturing system as a whole, or reducing its indirect costs. However, computer technology has evolved sufficiently now to allow manufacturing and production people access to its capabilities. With the appropriate analysis and design tools, these people who know the manufacturing system best can rely on themselves to do the system requirements and specifications for their CIM systems.

THE NEED FOR SYSTEM-MODELING AND INTEGRATION TOOLS

The analysis tools and methodologies needed for specifying and designing CIM systems are not yet generally established in Denmark. A major task of the CIM/GEMS project is to investigate and try out these tools, importing those that it believes will benefit and advance CIM development in small to medium size manufacturing companies

When used by the people in the company who know its manufacturing operations best--usually the production people--these tools can harness in models the specific information that defines a company's manufacturing activities. The models can then be used by these same people to produce precise and defensible plans, specifications, requirements, and designs for a CIM system that meets the needs of the manufacturing system. With these tools, a company can follow a total-system design approach to CIM that emphasizes careful and detailed planning before implementation.

AN EXAMPLE OF THE INFORMATION PROBLEMS IN A SMALL MANUFACTURING SYSTEM

CIM activity is directed at investigating the information content of manufacturing system functions--both technical and administrative--and not at the material transformation associated with shopfloor automation. As an example, lets consider a small manual factory which has a sufficiently complex indirect cost structure with minimal process equipment. Although the factory produces only four part types in part lots of 15, it requires 64 documents to produce a single part lot. On a one-shift-per-day basis, the factory can produce 130,000 parts per year, requiring 12,000 documents for this production. However, if a lot size of one is implemented, the documentation requirement explodes to 463,000 documents per year. Clearly, the information-handling problems of this small manufacturing system mirror the larger ones of manufacturing companies.

THE NEED TO USE LOW-COST TECHNOLOGY

Small and medium size companies typically have low capitalization, they need methods, tools, and equipment that are low-cost. For this reason they are interested in CIM development on personal-computer (PC) technology, which has considerable capability without being exorbitantly expensive. Although immense computing power and effective software tools have been around for years, recent breakthroughs in the migration of this power to inexpensive PC's now provide the user with the power of a superminicomputer for the price of a PC [4, 16]. It is also possible now to expand and move PC-developed CIM systems to larger computers without modification.

THE CIM DEVELOPMENT LIFECYCLE

A major tool in CIM projects is the recognition and use of the CIM development lifecycle model. The need to define the CIM development lifecycle was first recognized by the U.S. Air Force's ICAM project and formalized in their System Development Methodology [14]. We used this development lifecycle as a basis for our project's development cycle. From [14], the CIM development lifecycle has eight primary technical tasks grouped as follows under four phases:

* **Phase 1** - Understand the Problem
  Problem Analysis

Requirements Definition

* **Phase 2** - Formulate & Justify Solution
  Preliminary Design
  Detail Design

* **Phase 3** - Construct & Integrate Solution
  Construct & Verify
  Integrate & Validate

* **Phase 4** - Implement & Maintain Solution
  Implementation & User Acceptance
  Maintenance & Support

This is a modified software development lifecycle as initially defined by Royce in 1970 [27] and subsequently discussed in all software engineering texts such as Sommerville [28], and is consistent with software development as a principle component within an CIM project. As a result, many of the tools have been borrowed from software engineering developments.

PHASE 1 - UNDERSTAND THE PROBLEM

PHASE DESCRIPTION

This phase consists of Problem Analysis and Requirements Definition tasks. In the Problem Analysis task the manual operation is analyzed to determine the way it currently exists. The analysis consists of building function models, information models and behavior models (simulation models) of the current manufacturing system. The act of constructing these models forces the CIM project team to bound the problem domain and understand the system within the bounds. With an understanding of the manufacturing system, we can identify problems in the structure of the existing system.

Within the Problem Analysis Task the "System Environment Specification" (SES) defines the current system. This specification consists of the following:

* System functional relationship model
* System information structure model
* Dynamic behavior model (simulation model)
* Physical layout
* System document library

In creating the system environment specification the CIM team moves through a discovery process that ensures all project personnel thoroughly understand the system and its problems. The SES is a consensus definition of the existing manufacturing system and its problems.

In the Requirements Definition task the Problem Analysis task results are evolved to produce models of what the CIM system should be. This includes the process and material handling flow, and computer hardware and software data flow. An example of data flow is the time between activities that would require file accesses on a network server. This information is used to determine network and file server capacity requirements.

Through these models a system is defined that corrects the problems identified in the Problem Analysis task. They form the basis for a set of requirements specifying the structure and behavior of the CIM system and are incorporated into the simulation model. The model's results are evaluated to determine the acceptability of possible ways to solve the problems identified in the Problem Analysis task. Requirements are altered, incorporated into the model, and the model's results evaluated until the requirements produce acceptable performance. The models define the system that must result from the other three phases.

The results from the simulation model can also be compared to the results from the problem analysis model to show system improvements. Additional activities in this task determine the testing requirements necessary to show that the completed system meets the requirements specifications.

The product of this task are formal CIM system requirements -- specification documents against which the system is designed, formal testing requirements specification documents specifying how the system testing is performed, and models of the CIM system. The requirement specifications are used in the remaining phases and tasks of the project as a reference against which to design, implement and test the CIM system.

APPLICABLE TOOLS

The principle tools used in this phase evolved from tools developed by the USAF ICAM program. They support the modeling activities for creating the System Environment Specification. Although initially developed by ICAM, they are now commercially available in Europe and America.

* IDEF0 Function Modeling, which captures those activities in the system that transform inputs under constraints into outputs, and shows the information flow between these activities.

* IDEF1 Information Modeling, which captures the rules, conventions, and laws that govern the structure and operation of information in a manufacturing system.

* IDEF2 Behavior Modeling, which captures the time-varying behavior of the system, which is commonly done in and known as computer simulation.

* SDM System Development Methodology, which plans, organizes, directs, controls, and analyzes major system-development efforts so that appropriate administrative and technical tools and methodologies can be chosen.

References to IDEF0 include the SADT papers from Ross et.al. [22, 23, 24], the ICAM reports [13], the IDEF0 course manual [6], and Wallace, Stockenberg, and Charette's recent text [32]. Automated tools include *IDEFine 0* from IDEFinite b.v. [15].

References to IDEF1 include Ruoff [25, 26], Wallace, Stockenberg, and Charette's text [32], Young [35, 36], and Appleton [2]. Although the Flavin monograph [8] is a good introduction to information modeling, perhaps the best text describing how to do information modeling is Howe's book [12]. Tsichritzis text [29] contains a highly technical discussion on data modeling. Automated tools include *The ER Package* from Chen & Associates [3], and *JANUS* from D. Appleton Co. [1].

IDEF2 is a simulation tool and is not used by our project. Instead we have used the fine commercial products available today. Examples are SIMAN [19] and SLAMII [21]. In addition, the use of simulation in the CIM development lifecycle will not be discussed further since it has been examined in detail by Young, et. al. [31].

SDM [14] may be the most important tool since it addresses both the administrative and technical views of a project. Although the technical aspects of SDM were adapted from software engineering and have been widely discussed elsewhere (see for example reference [32]), the administrative aspects of a CIM project are rarely discussed at more than a superficial level. It is the strength of the administrative guidelines which make the ICAM SDM unique.

The documentation system is rarely viewed as a powerful administrative and technical tool. Documentation captures a project's history and provides a metric for a project's progress and an archives for its technical accomplishments. It is particularly important for a CIM project because it represents the interim results and may be the only visible project output until the implementation phase. By providing access and control for all project information, it is the key to ensuring that the technical staff uses the most current information to perform their job. In addition, without detailed documentation, a project's lessons are not available to others, decision rationales can't be examined after the fact, and system maintenance and later expansion are often difficult and sometimes impossible. Horowitz and Williamson [11] provide a good description of a project documentation system.

PHASE 2 - FORMULATE & JUSTIFY SOLUTION

PHASE DESCRIPTION

This phase consists of the Preliminary Design and Detailed Design tasks. In the Preliminary Design task alternative approaches to meet the requirements specifications are conceptualized and models constructed. The models are used to select a specific alternative. The selected alternative determines the basic design approach to be taken in developing a solution and the models' detail is increased for use in the Detail Design task.

In the Detail Design task the structure and components of the alternative selected from the Preliminary Design task are designed in detail. In addition, component testing procedures that are consistent with the system testing procedures are designed. This phase results in the detailed system design used to implement the CIM system and procedures to test the system components.

APPLICABLE TOOLS

This phase has been the focus of activity in software engineering during the past decade and has the best developed support tools. Yau and Tsai [34] provide an excellent survey of the accepted techniques. The most widely used tools are data flow diagrams and structure charts, with the classical tools being DeMarco [5] and Gane and Sarson [10]. Excellent automated tools exist such as *Excellator* [7], *CASE 2000* [18], *PSL/PSA* [20]. For real time systems state transition diagrams are widely used and supported by *Excellator RT* [7].

PHASE 3 - CONSTRUCT & INTEGRATE SOLUTION

PHASE DESCRIPTION

The Construct and Integrate Solution phase consists of the Construct & Verify task and the Integrate & Validate task. In the Construct & Verify task the detailed logic associated with the detailed design is specified and used to build the system components. These components are verified and assembled into subsystems that are also verified. Verification is the process of ensuring that the software program logic matches the specified logic.

In the Integrate & Validate task the interfaces between components and subsystems are tested and integrated together to form the CIM system. This system is tested to validate that it meets the system requirements specifications.

System Testing Normal system operation is the way the system should operate when no errors or failures have occurred. Exception handling is the way the system should operate under error or failure conditions. Both operation modes are tested using several testing modes: deterministic step-through operation, deterministic continuous operation, and random-event continuous operation. In addition, the physical capacity of the computer hardware and software is determined through stress testing. The results of this task will be an operating CIM system that has been shown to meet the requirement specifications.

APPLICABLE TOOLS

The primary tools here are *code development environments* that support converting detailed logical designs into program code. Unfortunately, this is an emerging technology with few commercially available tools. Most work is performed manually with some support from windowing systems, interactive debuggers, and incremental compilers. An example of an advanced support tool is version 5.0 of Microsoft's C compiler [17].

PHASE 4 - IMPLEMENT & MAINTAIN SOLUTION

PHASE DESCRIPTION

Phase 4 is the last phase and consists of the Implementation & User Acceptance tasks, and the Maintenance & Support task. In the Implementation & User Acceptance task the CIM system is installed in a production environment and users are trained to operate the system. The system operation is monitored and evaluated to determine if the requirements specifications are met under production conditions. Invariably, problems arise and the system needs to be evolved into a fully functional production CIM system.

In the Maintenance & Support task system problems that arose while operating the CIM system are analyzed, and engineering changes to the hardware and software are specified to solve them. The changes are implemented and the system operations evaluated for correct performance. This task and the previous task, Implementation and User Acceptance, have a high degree of overlap and involve cycling until an acceptable performance under production conditions is achieved. When this occurs, the result is an operational CIM system.

APPLICABLE TOOLS

This area has been almost completely neglected in terms of support tools. The primary tool used is a system emulator that is designed and developed by the project group for the specific CIM system.

TECHNICAL REVIEWS

Technical reviews are one of the most powerful tools available to ensure the success of an CIM system. These are not management reviews and are not conducted by management. Their purpose is not to review project progress, but to ensure technical correctness. Management reviews are conducted to review progress and should be scheduled separately. Technical reviews are focused on the interim technical results and are conducted by skilled technical people. Management receives their results, but shouldn't participate in the activity.

As noted by Freedman and Weinberg [9], technical reviews are tools that support both the administrative tasks and the technical tasks throughout the entire development cycle. They allow management to track activities and correct problems while corrective action is still possible. They provide management with a better appreciation for staff competence, both individually and collectively. Within the project staff, they increase morale and professional attitude, reduce turnover, produce more reliable estimating and scheduling, and better appreciation of management's role in the project's success.

Projects with a full system of reviews report a ten times reduction in the number of errors reaching each stage of testing. The cost reduction for testing efforts runs between 50 and 80 percent, even when the cost of reviewing is added to testing costs. In addition, they note that typically a program that has been thoroughly reviewed during development and for which maintenance changes are also reviewed shows a five to one reduction in maintenance costs. For a detailed description of the "what, how and when" of walkthroughs, inspections, and technical reviews the reader is referred to [9] and [33].

SUMMARY

We have presented a brief description of CIM and the tools that comprise general methods for solving the specific solution of each manufacturing system. We defined CIM to set the context for describing tools and methods, defined the CIM development lifecycle and identified tools that support the main tasks in each lifecycle phase. This is by no means a complete list and could be enhanced in many ways. However, we hope that this paper will help direct other engineers in manufacturing to the techniques we have found useful in our own CIM project.

REFERENCES

1. D. Appleton Co. (DACOM), Los Angles, CA, USA.
2. Appleton, D. S., "Organizing and Managing Information Resources in the Data-Driven Enterprise," *Industrial Engineering*, Vol. 18, No. 6, (June 1986).
3. Chen & Associates, Inc., 4884 Constitution Ave., Suite 1-E, Baton Rouge, LA 70808, USA.
4. "Computers: The New Look," a special report, *Business Week*, (November 30, 1987).
5. DeMarco, T., *Structured Analysis and System Specification*, Prentice-Hall (Yourdon Press), Englewood Cliffs, New Jersey, (1979).
6. Deuermeyer, B., and Culliane, T., *IDEF0-Function Modeling Course Manual*, ICAM Technology Transfer under ICAM SEM Project 1701, ICAM Program Office, AFWAL/MLTC, WPAFB, OH 45433, (1983)

7. Intech Corporation, 18662 MacArthur Blvd. #200, Irvine, CA 92715, USA.
8. Flavin, Matt, *Fundamental Concepts of Information Modeling*, Yourdon Press Monograph, (1981).
9. Freedman, D. P. and Weinberg, G. M., *Handbook of Walkthroughs, Inspections, and Technical Reviews*, 3rd edition, Little, Brown Computer Systems Series, Boston, (1982).
10. Gane, C., and Sarson, T., *Structured Systems Analysis: Tools and Techniques*, Prentice-Hall, Englewood Cliffs, New Jersey, (1979).
11. Horowitz, E. and Williamson, R. C., "SODOS: A Software Documentation Support Environment - Its Definition," *IEEE Transactions on Software Engineering*, Vol. SE-12, No. 8, pp. 849-859, (August 1986).
12. Howe, D. R., *Data Analysis for Data Base Design*, Edward Arnold Press, London, (1983).
13. *Integrated Computer-Aided Manufacturing (ICAM) Architecture Manuals*, AFWAL-TR-81-4023, U.S. Air Force Materials Laboratory, 1981.
14. *System Development Methodology User's Manual*, UM 17013100, ICAM Program Office, AFWAL/MLTC, WPAFB, OH, (1983).
15. IDEFinite b.v., Rotterdamseweg 5, 2628 AH, Delft, The Netherlands.
16. "Compaq's Deskpro 386/20 Sizzles," *PC Week*, Vol. 4, No. 45, pp. 119 & 135, (November 10, 1987).
17. Microsoft Corporation, 10700 Northup Way, Box 97200, Bellevue, WA 98009, USA.
18. NASTEC Corporation, 24681 Northwestern Highway, Southfield, MI 48075, USA.
19. Pegden, C. D., *Introduction to SIMAN*, Systems Modeling Corporation, State College, PA, (1985).
20. Meta Systems Inc., 315 E. Eisenhower Parkway #200, Ann Arbor, MI 48104.
21. Pritsker, A. A. B., *Introduction to Simulation and SLAMII*, A Halsted Press Book, John Wiley & Sons, New York, (1984).
22. Ross, D., "Applications and Extensions of SADT," *IEEE Computer*, (1985).
23. Ross, D., "Structured Analysis (SA): A Language for Communicating Ideas," *IEEE Transactions on Software Engineering*, Vol. SE-3, No. 1, pp. 16-34, (1977).
24. Ross, D., and Schoman, K. E., "Structured Analysis for Requirements Definition," *IEEE Transactions on Software Engineering*, Vol. SE-3, No. 1, pp. 6-15, (1977).
25. Rouff, K. L., "Practical Application of IDEF1 as a Database Development Tool," *Proceedings of the IEEE International Conference on Data Engineering*, pp. 408-415, (April 1984).
26. Rouff, K. L., "CODES: A DATABASE DESIGN EXPERT SYSTEM PROTOTYPE," *Proceedings of the 1984 IEEE Computer Society Conference on Artificial Intelligence Applications*, pp. 439-445, (1984).
27. Royce, W. W., "Managing the Development of Large Software Systems: Concepts and Techniques," *Proceedings 1970 WESCON Conference*, (1970).
28. Sommerville, I., *Software Engineering*, 2nd edition, Addison-Wesley Inc., International Computer Science Series, Amsterdam, (1985).
29. Tsichritzis, D. C., and Lochovsky, F. H., *Data Models*, Prentice-Hall, Englewood Cliffs, New Jersey, (1982).
30. Vesterager, J., Young, R. E., Wichmann, K. E., and Heide, J., *CIM/GEMS: General Methods for Specific Solutions, A Description of the Danish CIM/GEMS Project*, technical report CG/TR:87-1, CIM/GEMS Project, Institute of Production Management and Industrial Engineering, Technical University of Denmark, (1987).
31. Young, R. E., Vesterager, J., Wichmann, K. E., and Heide, J., "Simulation Uses in CIM Development," *Int. J. Computer Integrated Manufacturing*, Vol. 1, No. 1, pp. 50-54, (1987).
32. Wallace, R. H., Stockenberg, J. E., & Charette R. N., *A Unified Methodology for Developing Systems*, Intertext McGraw-Hill, (1987).
33. Weinberg, G. M., and Freedman, D. P., "Reviews, Walkthroughs, and Inspections," *IEEE Transactions on Software Engineering*, Vol. SE-10, No. 1, pp. 68-72, (January 1984).

34. Yau, S. S., and Tsai, J. J. P., "A Survey of Software Design Techniques," *IEEE Transactions on Software Engineering*, Vol. SE-12, No. 6, pp. 713-721, (June 1986).
35. Young, R. E., *IDEF1-Information Modeling Course Manual*, ICAM Technology Transfer under ICAM SEM Project 1701, ICAM Program Office, AFWAL/MLTC, WPAFB, OH 45433, (1983).
36. Young, R. E., and Mayer, R., "The Information Dilemma: Conceptualizing Manufacturing as an Information Process," *Industrial Engineering*, Vol. 16, No. 9, pp. 28-34, (September 1984).
37. Young, R. E., "The Present and Future Impact of Automation on Productivity and the Future Focus of Engineering to Promote the Advancement of U.S. Industry," *Proceedings of the Hearing on the Role of Automation and Robotics in Advancing United States Competitiveness*, Testimony before the Subcommittee on Science, Research and Technology of the Committee on Science and Technology, U.S. House of Representatives, Ninety-ninth Congress, document no. 62, pp. 108-123, (October 7, 1985).

AN APPROACH TO SOLVE WORKSHOP REAL TIME SCHEDULING PROBLEMS

Jacques ERSCHLER and François ROUBELLAT
Laboratoire d'Automatique et d'Analyse des Systèmes du C.N.R.S.
7, avenue du Colonel Roche
31077 Toulouse Cédex
France

ABSTRACT

Real time production scheduling problems are considered, when the items to be performed have release times and due dates. The real time scheduling decision process is based on an original approach, aiming at the search for characteristics of schedules compatible with the main manufacturing constraints to be satisfied (limit times, item routings, limited amount of available resources). At any time when a decision has to be made in real time, these characteristics allow to select actions which are coherent with the constraints. Two procedures to search for these characteristics and to use then for real time scheduling are described

1. INTRODUCTION

Workshop real time scheduling is one of the key factors to improve production unit efficiency. If methods and tools are available for the upper levels of the multilevel management decision structure, very few efficient results have been obtained for the part flow control on the shop floor level. Obviously this is a very important problem, because all the best decisions made at the upper levels will be inefficient if the shop floor level is not able to execute them in an efficient way. One of the major problems at the shop floor level is the disturbed environment of the manufacturing process (machine failures, lack of operators, unavailability of raw materials or components...). To take account of these disturbances, the production plans proposed to the workshops need to include many degrees of freedom of various types (large flow times including time margins, under loading of the technical resources to take account of machine failures,...). Thus on the shop floor decision level, these degrees of freedom are available to react in real time to the various

NATO ASI Series, Vol. F53
Advanced Information Technologies for
Industrial Material Flow Systems
Edited by Shimon Y. Nof, Colin L. Moodie

disturbances, without modifying the proposed production plan. Thus there exists a direct connection between the amount of degrees of freedom necessary to react to disturbances and the efficiency of the real time scheduling procedures. The greater will be this efficiency, the lower will be the necessary amount of degrees of freedom. But it must be noticed that the flexibility associated with these degrees of freedom is costly (high level of in process inventories, under loading of the resources...). Thus it is very important to be able to reduce the requested flexibility by using efficient real time scheduling procedures on the shop floor level.

The purpose of this paper is to propose an approach for solving this real time scheduling problem. The general characteristics of the proposed approach are presented in section 2. In section 3, the considered scheduling problem is defined and the outline of the proposed methods to characterize sets of admissible schedules are briefly described. Two approaches to search for characteristics of admissible schedules are presented in section 4 and 5, with an emphasis on the use of this information for real time scheduling.

2. REAL TIME SCHEDULING PROBLEMS IN THE MANUFACTURING CONTEXT

2.1. General objectives of the scheduling function

To manage a production plant, a multilevel decision structure is generally used. Three main levels are considered [1] :

- a strategic decision level where long term decisions are made : investments, sale policy, new product definition...

- a tactical decision level where an overall production plan is defined to meet customers' requirements by using available resources in the best way.

- an operational decision level where the preceding production plan is carried out through a precise scheduling of the elementary operations of the planned items on the available resources.

Only this lowest level is considered in this paper. For small or medium size batch productions, the basic data for the scheduling problem are the following :

- the production plan issued from the tactical decision level defines the set of items to be performed, their release times and due dates;

- every item is characterized by its manufacturing process which defines the sequence of operations which have to be performed. Each operation is characterized by information about the resources needed for processing and the processing time.

Thus the objective of the operational decision level is to assign the operations to the resources and to have them scheduled on these resources, in order to satisfy the proposed production plan. Due to the various disturbances occuring on the shop floor, such a decision problem has to be solved in real time.

2.2. Proposed approach for real time scheduling

With the classical scheduling methods [2, 3], a unique schedule is defined. It corresponds to a plan for performing the operations. This plan will be usable on the shop floor until the cumulative amount of disturbances or the occurence of a constraint not included in the model will generate a state incompatible with it. At this time, a new schedule taking into account the actual state could be generated, defining a new plan to perform the operations, and so on. With such an approach, the flexibility concept which is needed to take account of unexpected situations is not explicitely introduced in the solving procedures.

It is the reason why we have proposed a new approach to solve real time scheduling problems [4, 5, 6, 7). As a starting point in this approach, the scheduling problem on the shop floor is considered as a real time decision making problem, the objective of which being to select the decisions needed to process the operations by using the available resources. In the disturbed environment of a manufacturing process, the objective of the decision making process is to select an action compatible with the considered scheduling problem constraints. As we are interested in scheduling problems for small or medium size batch industries, we assume that the production plan supplied to the workshops defines a set of items to be manufactured, each of them having a release time, a due date, and a specified manufacturing process. Then the main scheduling objective is to use the available resources in the most efficient way to satisfy this production plan. In order to coordinate perfectly the tactical decision level which has proposed this plan and the operational scheduling level which executes it, we have decided to consider these plan data as constraints to be satisfied. This implies that the item release times and due dates are considered as constraints for the scheduling problem. Thus, if the scheduling decision level cannot satisfy them, they will have to be modified by the tactical upper level which has proposed them, for this level is the only one to have all the information and data needed to make this decision correctly.

In such a context, a scheduling problem is fully characterized by a set of constraints to be satisfied and a decision will be acceptable as soon as it satisfies all these constraints. This is the reason why the approach which is described aims at defining a set of actions compatible with these constraints at any time when a decision has to be made : this makes the originality of the proposed approach. Consequently, the classical difficulties to model a real environment are reduced, because it is always possible to reject those of the proposed actions which are not compatible with unmodeled constraints. Furthermore, the proposed set of actions is very well suited with a decision aid approach because the selection of the suitable action at any time can be made by the decision maker, according to his specific decision criteria.

To achieve the proposed approach, a set of admissible actions has to be defined at any time when a decision has to be made. To solve this problem, we have defined a two step procedure :

i) in the first step, the characteristics of a set of schedules compatible with the main constraints of the problem (defining the basic scheduling model) are defined : this is an analysis step.

ii) in the second step, these characteristics are used in real time to define a set of possible actions at any time when a decision has to be made, by taking account of unmodeled characteristics and of the real state of the workshop. This is a decision aid step, for which some additional information can be attached to the suggested actions in order to assist the decision making process.

Thus during the first step, some degrees of freedom compatible with the proposed production plan are exhibited, which will be used during the second step at any time when a decision has to be made.

3. CONSTRAINTS BASED ANALYSIS IN SOME SCHEDULING PROBLEMS

According to the above proposed approach, it seems useful to characterize the set of the admissible schedules for a problem where objectives are considered as constraints. This makes clear the problem admissibility as well as the degrees of freedom which are available for scheduling decisions. This "constraints based analysis" process brings out information usable as a decision aid, on the basis of a basic scheduling model which takes account of the core of the problem.

3.1. The considered basic scheduling model

In practical situations, there exist many types of scheduling problems. Just now, we have considered only the scheduling problems with the following characteristics :

. the production plan issued from the tactical level defines a set of n items to be performed.

. every item i has the following characteristics :
  - a release time r_i,
  - a due date d_i,
  - a specified routing (technological constraints)

. every routing is made of an ordered set of operations (i,j), $1 \leq j \leq h_i$ which cannot be split.

. to perform one operation, one resource subset m_{ij}^k is needed. This subset is selected among a set $M_{ij} = \{m_{ij}^k\}$, $k=1,\ldots,\ell_{ij}$ associated with (i,j).
  . $q_{ij}^{k\ell}$ is the amount of resource ℓ in m_{ij}^k needed to perform (i,j) when m_{ij}^k is selected.
  . p_{ij}^k is the processing time of (i,j) when m_{ij}^k is selected.
  . Q^ℓ is the amount of resource ℓ available for processing the operations.

The above type of scheduling problem occurs very often in practical manufacturing situations. It will be the basic scheduling model used in this paper. In the next sections, we will describe the basic principles of the methods which have been developed :

- to search for the characteristics of a set of schedules compatible with the above defined constraints;
- to use the so defined characteristics as the basis for real time scheduling, in a disturbed environment.

3.2. Characterization of a set of admissible schedules

For the general case described in subsection 3.1, two problems have to be solved :

- an assignment problem, the purpose of which being to select the specific subset of resources m_{ij}^k which will process operation (i,j), inside the set M_{ij};
- a scheduling problem, the purpose of which being to define the starting time t_{ij} of every operation (i,j).

The two types of unknowns associated with these two problems characterize the schedules compatible with the considered scheduling model. Thus the first step of our approach described in 2.2. will aim at defining these two types of characteristics for a set of schedules compatible with the constraints described in 3.1.

A rigorous use of our approach would be to define these characteristics for the whole set of schedules compatible with the considered constraints. We have proposed a general method to find these characteristics [5, 8]. But for practical situations, with large sized problems, such an exhaustive characterization of the whole set of admissible schedules is not realistic, due to the combinatorial complexity. It is the reason why two approaches are developed in order to approximate the general constraints based analysis principles :

- a first approach consists in searching for only necessary admissibility conditions by cutting down the analysis procedure. Such an approach is presented in section 4.

- a second one consists in searching for only sufficient admissibility conditions by fully characterizing specific admissible schedule subsets. Such an approach is presented in section 5.

Moreover it should be noted that more restrictive hypothesis, like disjunctive resources, lead to specific results stronger than those which are obtained with the general case of cumulative resources. This point will be illustrated later in the paper.

4. RULES AND INFERENCE PROCESS TO CHARACTERIZE ADMISSIBLE SCHEDULES [9, 10, 11]

The characterization of admissible schedules (constraints based analysis) can be achieved through an inference process. This process uses rules which act on a base of facts. These facts consist of limit times and sequencing conditions which represent necessary admissibility conditions for the schedules.

In this section, the assignment problem is not considered. The specific characteristics of the scheduling problem and the notations are given firstly. Then, the basic principles and the main results which are used in the inference process are presented.

4.1. Problem statement

The basic scheduling problem which is considered is characterized by :

- a set of items (or jobs) J
- a set of resources M
- an item i is processed through a set of operations
- an operation O_i cannot be interrupted and it is characterized by :
  . the subset of resources it uses : $m_i \subset M$
  . the amount of each resource it needs : $q_i = \{q_i^\ell \mid \ell \in m_i\}$
  . its processing time p_i

The starting time t_i associated with each operation O_i is the scheduling variable. The set of starting times defines a schedule.

An admissible schedule has to satisfy three types of constraints :

i) the resource constraints : each resource ℓ available in a prescribed limited amount Q^ℓ

ii) the technological coherence constraints : sequencing conditions have to be satisfied between operations of a same item.

iii) the limit time constraints : the items have to be performed within time intervals which are defined by an earliest starting time and a latest finishing time.

By propagating the limit times defined in iii) along the technological coherence constraints, an earliest starting time est_i and a latest finishing time ℓft_i can be initially associated with each operation O_i.

4.2. Modeling the resource constraints

Resource constraints can be expressed by using the concept of critical set of operations. A critical set of operations Ic is a minimal set of operations using an amount of resources greater than the available one.

Consider, for example, the following set of 4 operations using a resource whose available amount is equal to 5. Let q_i be the resource amount that operation O_i uses :

	O_1	O_2	O_3	O_4
q_i	2	3	5	2

The critical sets are :

$$\{O_1, O_3\}, \{O_2, O_3\}, \{O_4, O_3\}, \{O_1, O_2, O_4\}$$

The critical sets of operations represent the dominant conflicts for resource utilization. Indeed to solve these conflicts it is necessary and sufficient to sequence two operations in each critical set. So, for the above example the following conditions can be derived :

(O_1 precedes O_3) OR (O_3 precedes O_1)

and

(O_2 precedes O_3) OR (O_3 precedes O_2)

and

(O_4 precedes O_3) OR (O_3 precedes O_4)

and

(O_1 precedes O_2) OR (O_2 precedes O_1) OR (O_1 precedes O_4) OR
(O_4 precedes O_1) OR (O_2 precedes O_4) OR (O_4 precedes O_2)

This shows that conflict resolution introduces a sequencing aspect in the scheduling problem.

Constraint based analysis lies in analyzing the interaction between time constraints and sequencing conditions resulting from the solution of conflicts for resource utilization. This leads to characterize the admissible schedules through time intervals and sequencing conditions.

A special case occurs when every critical set is a pair of operations. It is called the disjunctive case. This is the job shop scheduling case where each operation uses a specific machine. In this situation, every pair of operations which uses the same machine is a critical set.

In the following, three results are presented which can be associated with deductive rules :

i) an elementary result in the disjunctive case which concerns only pairs of operations

ii) a more general result in the disjunctive case which leads to compare an operation with a set of operations.

iii) a result which applies to the non disjunctive case.

4.3. Elementary sequencing and updating rules in the disjunctive case

Let $\{O_i, O_j\}$ be a critical set of operations (for example O_i and O_j have to be performed by the same machine).

The following sequencing rule can be used to solve the conflict associated with $\{O_i, O_j\}$:

Rule 1
If $(\ell ft_i - est_j) < p_i + p_j$
Then, (O_i precedes O_j)

Reciprocally, if an essential precedence constraint such as (O_i precedes O_j) has been obtained, the following updating rules can be used to update the time intervals of O_i and O_j;.

Rule 2a
If (O_i precedes O_j) and $(est_i + p_i > est_j)$
Then, est_j can be updated to the value $(est_i + d_i)$

Rule 2b
If (O_i precedes O_j) and $(\ell ft_j - p_j < \ell ft_i)$
Then ℓft_i can be updated to the value $(\ell ft_j - p_j)$.

Example

O_i	est_i	ℓft_i	p_i
O_1	3	9	3
O_2	0	7	4

$$(7 - 3 < 3 + 4) \xrightarrow{R1} O_2 \text{ precedes } O_1$$

$$\left.\begin{array}{c} O_2 \text{ precedes } O_1 \\ \text{and} \\ (0 + 4 > 3) \end{array}\right\} \xrightarrow{R2a} est_1 \text{ is updated to } 4$$

$$\left.\begin{array}{c} O_2 \text{ precedes } O_1 \\ \text{and} \\ (9 - 3 < 7) \end{array}\right\} \xrightarrow{R2b} \ell ft_2 \text{ is updated to } 6$$

4.4. Generalized sequencing and updating rules in the disjunctive case

Let Ω be a set of operations such that every pair of operations which belongs to Ω is a critical set of operations (for example every operation of Ω uses the same machine). Let 0_i be an operation of Ω.

The following sequencing rules can be used to solve partially the conflicts which arise in Ω /

Rule 3a

$$\text{If } \left(\max_{\substack{0_j \in \Omega \\ 0_j \neq 0_i}} (\ell ft_j) - est_i \right) < \sum_{0_j \in \Omega} p_j$$

$$\text{Then } \bigvee_{\substack{0_j \in \Omega \\ 0_j \neq 0_i}} (0_j \text{ precedes } 0_i)$$

Rule 3b

$$\text{If } \left(\ell ft_i - \min_{\substack{0_j \in \Omega \\ 0_j \neq 0_i}} (est_j) \right) < \sum_{0_j \in \Omega} p_j$$

$$\text{Then } \bigvee_{\substack{0_j \in \Omega \\ 0_j \neq 0_i}} (0_i \text{ precedes } 0_j)$$

where V is the logical OR.

The conditional precedence constraints which can be derived from rule 3a (rule 3b) mean that one operation at least in the set $\{\Omega - 0_i\}$ has to precede (follow) operation 0_i. To simplify, such a conditional precedence constraint is denoted by :

"$\{\Omega - 0_i\}$ precedes 0_i"

or

"0_i precedes $\{\Omega - 0_i\}$"

Reciprocally, if a conditional precedence constraint has been obtained, the following updating rules can be used to update the time interval of 0_i :

Rule 4a

If $(\{\Omega - O_i\}$ precedes $O_i)$ and $\left(\min_{O_j \in \{\Omega - O_i\}} (est_j + p_j) > est_i\right)$

Then est_i can be updated to the value $\left(\min_{O_j \in \{\Omega - O_i\}} (est_j + p_j)\right)$

Rule 4b

If $(O_i$ precedes $\{\Omega - O_i\})$ and $\left(\max_{O_j \in \{\Omega - O_i\}} (\ell ft_j - p_j) < \ell ft_i\right)$

Then ℓft_i can be updated to the value $\left(\max_{O_j \in \{\Omega - O_i\}} (\ell ft_j - p_j)\right)$

Example : $\Omega = \{O_1, O_2, O_3\}$

O_i	est_i	ℓft_i	p_i
O_1	2	12	2
O_2	1	9	4
O_3	0	8	3

$\mathrm{Max}(9,8) - 2 < 2 + 4 + 3 \overset{R3a}{\Longrightarrow}$ (O_2 precedes O_1) OR (O_3 precedes O_1)

(O_2 precedes O_1) OR (O_3 precedes O_1)
and $\min(1 + 4, 0 + 3) > 2$ $\Bigg\} \overset{R4a}{\Longrightarrow}$ est_1 is updated to 3.

Remark : The results in 4.3 can be considered as a special case of the above results, when the set Ω consists of only two operations.

4.5. Sequencing rules in the non disjunctive case

Let Γ be a critical set of operations consisting of more than two operations, and let O_i be an operation of Γ.

The following sequencing rules can be used to generate conditional precedence constraints.

Rule 5a

If $\forall (O_j, O_k) \in \{\Gamma - O_i\}^2$ with $O_j \neq O_k$,
$\ell st_k - est_j < p_k + p_j)$
and $\forall\, O_j \in \{\Gamma - O_i\}$, $\ell ft_j - est_i < p_j + p_i$
Then $(\{\Gamma - O_i\}$ precedes $O_i)$

Rule 5b

If $\forall(O_j, O_k) \in \{\Gamma - O_i\}^2$ with $O_j \neq O_k$,

$\ell st_k - est_j < p_k + p_j$

and $\forall\, O_j \in \{\Gamma - O_i\}$, $\ell ft_i - est_j < p_j + p_i$

Then (O_i precedes $\{\Gamma - O_i\}$)

When such a conditional precedence constraint is generated, the updating rules 4a and 4b can be used to update the time interval of O_i.

Example : $\Gamma = \{O_1, O_2, O_3\}$

O_i	est_i	ℓft_i	p_i
O_1	0	13	6
O_2	3	13	6
O_3	3	13	6

$$\left.\begin{array}{l} 13 - 3 > 6 + 6 \\ \quad\text{"}\qquad\quad\text{"} \\ \quad\text{"}\qquad\quad\text{"} \end{array}\right\} \xrightarrow{R5b} (O_1 \text{ precedes } O_2) \ \underline{OR}\ (O_1 \text{ precedes } O_3)$$

$$\left.\begin{array}{l} (O_1 \text{ precedes } O_2)\ \underline{OR}\ (O_1 \text{ precedes } O_3) \\ \text{and } \max(13 - 6,\ 13 - 6 < 13) \end{array}\right\} \xrightarrow{R4b} \ell ft_1 \text{ is updated to } 7$$

4.6. Propagation of limit time updating

The updated values of the earliest starting (resp : latest finishing) times which can be obtained by applying rules 2, 4a and 4b, can be propagated forward in time to operations following O_i (resp backward in time to operations preceding O_i). This propagation is achieved through the technological coherence constraints.

4.7. The inference process for constraints based analysis

The inference process consists in applying alternately the sequencing rules and the limit time updating rules. As soon as a new precedence constraint is found, a limit time updating is searched for. If a limit time is updated, the new value is propagated as far as possible, along the technological precedence constraints and the other precedence constraints which have been found already. When there is no more possible updating, a new precedence constraint is searched for by using the sequencing rules. Thus, the characteristics of the admissible schedules (time intervals and precedence constraints between conflicting operations) are progressively refined.

This inference process has been implemented by using PROLOG language. The associated module which is called M.A.S.C.O.T., brings out information about scheduling admissibility. This information may be used either in a static way in order to generate a plan or in a dynamic way in order to make real time decisions. In both cases, another knowledge type is needed to take into account the specificity of the application.

5. WORKSHOP REAL TIME SCHEDULING BY USING AN OPERATION GROUP SEQUENCE

In this section, the proposed approach described in subsection 2.2. is implemented through the full characterization of a subset of admissible schedules (sufficient admissibility conditions in subsection 3.2.). Then the degrees of freedom associated with this subset can be used to exhibit for the decision maker actions compatible with the considered constraints. If after disturbances, all the schedules included in this subset are no more admissible, a new subset of schedules will have to be characterized. If it is no more possible to find such a subset, the production objectives will have to be modified by the tactical level.

In this section, we will consider the disjunctive resource case only m^k_{ij} is made of only one machine with $q^k_{ij} = 1$; each machine can process only one operation at a time. M is the total number of machines in the workshop.

To clarify the presentation, two cases will be considered, according to the value of Card (M_{ij}) :

- Card $(M_{ij}) = 1$: each operation has to be performed by a specified machine : it is the classical job shop scheduling problem.
- Card $(M_{ij}) > 1$: we have to select the machine which will process one operation. Thus we have to solve simultaneously an assignment problem and a scheduling problem.

5.1. Card $(M_{ij}) = 1$

5.1.1. Scheduling characteristics and degrees of freedom

In such a case, only the starting times t_{ij} are unknown. Thus a schedule is defined by the set T such that

$$T = \{t_{ij} \ / \ i = 1,\dots,n; \quad j=1,\dots,h_i\}$$

A schedule is said to be admissible if it satisfies all the constraints defined in 3.1.. Thus this set T is a time solution for the problem, made only of time characteristics.

Another way to characterize a solution is to define for every machine k a permutation s_k of the operations it has to process. Thus the set $S = \{s_k \ / \ k=1,\dots,M\}$ corresponds to a sequential solution of the scheduling problem. A set of schedules can be associated with a sequence S : they correspond to an arbitrary choice of the starting times of the operations ordered according to s_k on every machine k. If at least one of the schedules in such a set is admissible then the sequential solution S is an admissible one.

With every sequential solution S, a set $\tau(S)$ of admissible time solutions T can be associated. It is based on the earliest starting times λ^S_{ij} and the latest finishing times λ'^S_{ij} for the operations of the sequential solution S. $\tau(S)$ is such that :

$$\forall T = \{t_{ij} \ / \ i=1,\dots,n; \quad j=1,\dots,h_i\} \in \tau(S)$$

$$\lambda^S_{ij} \leqslant t_{ij} \leqslant \lambda'^S_{ij}$$

Thus, time degrees of freedom associated with the operations ordered according to the sequential solution S are exhibited. This is the classical concept of slack time defined for the critical path methods.

Thus there exists two types of degrees of freedom in the problem : time degrees of freedom and sequence degrees of freedom.

5.1.2. Definition of an operation group sequence

To characterize a subset of admissible schedules, we need firstly to define what type of properties this subset has. To make easier the decision step, we have tried to define a subset of schedules inside which the sequential degrees of freedom will be easily exhibited [12, 13]. To achieve such an objective, a possible solution is to define for every machine an ordered sequence of groups of operations and to consider only the sequential solutions derived from an arbitrary choice for the sequence of operations inside any group. Therefore, sequential degrees of freedom are exhibited, since the sequence for processing the operations belonging to the same group is unrestricted and without any connection with the sequences chosen inside the other groups.

It is obvious that if an arbitrary sequence of groups is defined, some of the associated sequential solutions will not be admissible. Thus, a group sequence is said to be admissible if and only if every sequential solution issued from this group sequence is an admissible one.

If every sequential solution associated with a group sequence satisfies all the constraints of the scheduling problem, but the due dates, this group sequence will be said feasible only.

The following notation will be used for a group of operations : $g(\ell,k) = [u,v,w,\ldots]$ where ℓ is the machine number, k is the rank of this group on this machine and u,v,w... are the operations included in this group.

<u>Example</u>. Let us consider the following problem :

$$m_{ij} = \begin{bmatrix} 1 & 2 \\ 2 & 1 \\ 1 & 2 \end{bmatrix} \qquad p_{ij} = \begin{bmatrix} 1 & 1 \\ 1 & 1 \\ 1 & 1 \end{bmatrix}$$

$$r_1 = r_2 = r_3 = 0 \qquad d_1 = d_2 = 3 \qquad d_3 = 4$$

The group sequence G_1

$$\begin{cases} g(1,1) = [(2,2)],\ g(1,2) = [(1,1),\ (3,1)] \\ g(2,1) = [(1,2)],\ g(2,2) = [(2,1),\ (3,2)] \end{cases}$$

is neither admissible nor feasible.

The group sequence G_2

$$\begin{cases} g(1,1) = [(1,1),\ (3,1)],\ g(1,2) = [(2,2)] \\ g(2,1) = [(2,1)],\ g(2,2) = [(1,2),\ (3,2)] \end{cases}$$

is feasible, but it is not admissible.

The group sequence G_3

$$\begin{cases} g(1,1) = [(1,1),\ (3,1)],\ g(1,2) = [(2,2)] \\ g(2,1) = [(2,1],\ g(2,2) = [(1,2)],\ g(2,3) = [(3,2)] \end{cases}$$

is admissible, and thus feasible.

An ordering relationship can be defined over a set of sequential solutions : "A group sequence G is included in a group sequence G_1 if any sequential solution from G belongs also to G_1".

Thus a dominant group sequence can be defined : a feasible (r. admissible) group sequence is a dominant one if it does not exist a feasible (r. admissible) group sequence including it.

Example

Using the data of the preceding example, the following group sequence

$$\begin{cases} g(1,1) = [(1,1)],\ g(1,2) = [(3,1)],\ g(1,3) = [(2,2)] \\ g(2,2) = [(2,1)],\ g(2,2) = [(1,2)],\ g(2,3) = [(3,2)] \end{cases}$$

is not a dominant one, because it is included in the admissible group sequence

$$\begin{cases} g(1,1) = [(1,1),\ (3,1)],\ g(1,2) = [(2,2)] \\ g(2,1) = [(2,1)],\ g(2,2) = [(1,2)],\ g(2,3) = [(3,2)] \end{cases}$$

5.1.3. *Graphs associated with a group sequence [13, 14]*

First of all, we define a graph which allows to obtain the earliest starting times for the operations.

Let us consider a group sequence G. For every operation (i,j), a maximum earliest starting time μ_{ij} can be defined : it is the lowest starting time which is compatible with all the permutations of the operations to which (i,j) is linked. Three types of links or constraints have to be considered for an operation (i,j) :

i) a routing constraint which implies

$$\mu_{ij} \geqslant \mu_{ij-1} + p_{ij-1}$$

if overlapping between two successive operations is not permitted.

ii) if (i,j-1) belongs to a group g(a,b), thus every operation in this group is linked with (i,j) by a "group constraint"; this constraint takes account of all the possible permutations for the operations in g(a,b) :

$$\mu_{ij} \geqslant \max_{\substack{(u,v)\in g(a,b)\\ (u,v)\neq(i,j-1)}} \mu_{uv} + \sum_{(u,v)\in g(a,b)} p_{u,v}$$

iii) if (i,j) is in a group $g(\ell,k)$, the operations in the preceding group $g(\ell,k-1)$ on machine ℓ induce "machine constraints" for (i,j). This constraint takes account of all the possible permutations for the operations in $g(\ell,k-1)$:

$$\mu_{ij} \geqslant \max_{(u,v)\in g(\ell,k-1)} \mu_{uv} + \sum_{(u,v)\in g(\ell,k-1)} p_{uv}$$

Thus, by taking account of the three preceding types of constraints, the earliest starting time for (i,j) can be defined as :

$$\mu_{ij} = \max \left[\mu_{ij} + p_{ij-1};\ \max_{\substack{(u,v)\in g(a,b)\\ (u,v)\neq(i,j-1)}} \mu_{uv} + \sum_{(u,v)\in g(a,b)} p_{uv};\ \max_{(u,v)\in g(\ell,k-1)} \mu_{uv} + \sum_{(u,v)\in g(\ell,k-1)} p_{uv} \right].$$

For j=1, the routing constraint is equal to r_i and there is no group constraint.

It must be noticed that this earliest starting time is a pessimistic one, which takes account of all the possible permutations. It is obvious that for certain sequential solution S issued from this group sequence G, it is possible to obtain an effective earliest starting time λ^S_{ij} that could be lower than μ_{ij}.

When all these times μ_{ij} have been obtained, it is possible to define for every item its earliest finishing time. If a fictitious operation (i, h_i+1) of zero duration is added to every item, the earliest starting time of this operation will be the earliest finishing time of the item i. Only a routing constraint and a group constraint are associated with this operation.

Thus the quantity $(\mu_{i,h_i+1}-d_i)$ is the tardiness (if it is positive) or the earliness (if it is negative or equal to zero) of item i.

Let us define μ_H as :

$$\mu_H = \max_{i=1,\ldots,n} (\mu_{i,h_i+1} - d_i)$$

Then if $\mu_H \leqslant 0$ all the due dates are satisfied and the group sequence is admissible.

If $\mu_H > 0$, at least one of the due dates is not satisfied and the group sequence is not admissible.

All the previously defined earliest starting times can be obtained by applying the classical critical path method to a specific activity on node graph $G(X, U_G)$ associated with the group sequence G [14]. Thus the length R_G of the critical path in this graph is an admissibility index : if $R_G > 0$, at least one of the items is late, and the group sequence is not admissible. If $R_G \leqslant 0$, all the due dates are met and the group sequence is admissible.

Furthermore, this length R_G has the following property : if a group sequence G' is included in a group sequence G, then $R_{G'} \leqslant R_G$. This is obvious, because all the sequential solutions issued from G' are included in the set of sequential solutions issued from G. This property will be used in the next subsection.

It must be noticed that, in a similar way, a latest finishing time γ_{ij} can be defined for every operation. γ_{ij} can be obtained by using another graph, the definition of which is similar to the previous one [14].

5.1.4. <u>Method to search for an admissible group sequence</u>

Based on the preceding results, a heuristic procedure has been defined to search for an admissible group sequence [13, 14]. It is divided into three steps :

<u>Step 1</u> : initialization

An initial feasible group sequence G_0 is defined in this step. This initial group sequence is obtained by using a procedure similar to the generation procedure for active schedules [3]. Obviously, searching for an admissible schedule is equivalent to searching for a schedule minimizing the maximum tardiness, which is a regular criterion. The operation selection rule used in the generation procedure is the lowest margin. The considered margin is obtained by distributing the total slack time of the item over each operation of its routing proportionally to its duration. As we want to generate a group sequence, a processing interval is associated with each operation, by using the defined margin. This exhibit time degrees of freedom, which are used to generate sequential degrees of freedom, through the generation of appropriate groups of unordered operations. Generally, this initial group sequence is not admissible; it is only feasible. Thus it satisfies all the constraints, but the due dates.

Step 2 : search for an admissible group sequence

If a group sequence is not admissible, it includes at least one inadmissible sequential solution. To reach admissibility, inadmissible sequential solutions have to be rejected. Such a rejection can be obtained by splitting one group $g_0(\ell,k)$ into two new groups $g_1(\ell,k)$ and $g_1(\ell,k+1)$. This orders the operations of these two groups, which were initially unordered in $g_0(\ell,k)$. This splitting principle generates a new group sequence G_1 which is included in G_0. Thus, we have $R_{G_1} \leqslant R_{G_0}$. By iterating this splitting principle, it could be possible to obtain an admissible group sequence.

To have an efficient use of this splitting principle, it is necessary to find the groups the splitting of which will be the most efficient for reaching admissibility. The above defined graph for earliest starting times μ_{ij}, is the basis for the splitting principle. Indeed, it is obvious that a splitting has to be made only for groups including operations located on the critical path(s) of this graph. For every such group, rules based on the type of arcs issued from these operations have been defined, which allow to distribute operations of $g_0(\ell,k)$ among the two groups $g_1(\ell,k)$ and $g_1(\ell,k+1)$ in order to decrease lateness [14].

Thus the search for an admissible group sequence is based on an iterative use of the so defined splitting principle. This search is stopped when one of the two following situation occurs :

- even an admissible group sequence has been obtained;
- or no more splitting is feasible and no admissible group sequence can be found, based on the initial group sequence and the splitting principle. Thus, a new initial group sequence has to be found to try again, or the tactical level has to be informed of this failure; in this case, it will have to propose new tentative due dates or to agree with the last obtained earliest starting times μ_{i,h_i+1}, which correspond to admissible due dates for the generated group sequence.

Step 3 : search for a dominant group sequence

When an admissible group sequence has been obtained, it is possible to search for a dominant group sequence. Such a dominant group sequence can be obtained by trying to gather into one group the operations belonging to two adjacent groups. If such a gathering maintains admissibility, it is made. If not, it is cancelled. By iterating this gathering principle, a dominant and admissible group sequence is obtained.

5.2. <u>Card (M_{ij}) > 1 [15]</u>

In this case, we have to combine an assignment problem with a scheduling problem. Due to the difficulty of solving these two combinatorial problems simultaneously, we have adapted the preceding procedure to solve these two problems sequentially. The proposed procedure uses all the basic concepts presented in subsection 5.1. It is also made of three steps :

<u>Step 1</u> : initialization

A generation procedure similar to the previous one is used. To take account of the assignment problem, each operation (i,j) concerned by the selection procedure during the generation process is systematically assigned to the machine in M_{ij} which allows to complete it earliest. This assignment rule is coherent with the minimum lateness objective. It must be noticed that at each step of the initialization procedure, all the assignments of the unscheduled operations are updated, according to the preceding rule. Then the generation procedure described when card (M_{ij}) = 1 can be applied, because only a scheduling problem remains to be solved.

<u>Step 2</u> : search for an admissible group sequence

It is based on the graphs associated with a feasible group sequence. The splitting principle defined in subsection 5.1.4. can still be used to search for an admissible group sequence. But in the present case, we have an additional process that can be used to decrease the lateness of the group sequence. Indeed if in a group $g(\ell,k)$ including a critical operation, one arbitrary operation (i,j) is removed from the structure and the length of some arcs in the graph are modified and the length of the critical path will decrease, and consequently the lateness. As Card (M_{ij}) > 1 , this removed operation (i,j) can be assigned to another group g(u,v) of one of the remaining machines inside M_{ij}. If this new assignment for (i,j) does not modify the earliest starting times and latest finishing times of the operations in g(u,v), then the so defined transfer process will decrease the group sequence lateness, without using any splitting principle.

Thus, for Card (M_{ij}) > 1 , the search for an admissible group sequence is based on a combination of successive transfer and splitting processes. Starting with an initial group sequence, these allow to modify the operation assignments and group compositions in order to obtain an admissible group sequence if there exists. Thus in the proposed heuristic, the two available types of unknowns are used for solving the problem. As in the previous case, the search may fail to obtain an admissible group sequence.

__Step 3__ : search for a dominant group sequence

This step is similar to the one described in subsection 5.1.4.

__Example__ :

The following job shop scheduling problem is considered. The workshop is made of 5 machines. 6 items have to be performed, with $r_i = 0\ \forall i$ and $d_i = 100\ \forall i$. The routing process of each item is made of 3 operations and each operations has the following characteristics :

$$M_{ij} = \begin{bmatrix} 1 & 2 & 3 \\ 4,5 & 2 & 1 \\ 5 & 1 & 3 \\ 4,5 & 2 & 3 \\ 1 & 4,5 & 3 \\ 4,5 & 2 & 3 \end{bmatrix} \qquad P^k_{ij} = \begin{bmatrix} 10 & 15 & 5 \\ 10,10 & 20 & 15 \\ 5 & 10 & 20 \\ 10,10 & 15 & 10 \\ 15 & 10,10 & 20 \\ 10,10 & 5 & 20 \end{bmatrix}$$

The initial group sequence is presented in figure 1 by using a diagram similar to a bar chart. This sequence is not admissible : $\mu_H = 20$.

To improve the admissibility of this sequence, the transfer principle is applied to g(4,1) : (6,1) is transferred from machine 4 to machine 5. The lateness decreases : $\mu_H = 10$ (figure 2).

Then the splitting principle is applied to group g(2,1) : this group made of (6,2) and (2,2) is split into two consecutive groups made of only one operation : (6,2) in the first one, (2,2) in the second one. μ_H remains equal to 10, but the multiplicity of the critical path is reduced to 1 (figure 3). Then the splitting principle is applied again to g(3,1) and it generates two new groups made of only one operation each. At this step, the group sequence becomes admissible ($\mu_H = -5$) (figure 4).

In order to define a dominant and admissible group sequence, the gathering principle is applied to g(2,2) and g(2,3), to g(3,2) and g(3,3), and to g(3,4) and g(3,5). This final dominant and admissible group sequence is described in figure 5.

5.3. __Real time scheduling by using a dominant group sequence [15, 16]__

The group sequence obtained by using the preceding methods defines ordered groups of operations on every machine. Thus, to perform the items according to this group sequence, for every machine ℓ, the operations of the group $g(\ell,1)$ have to be processed firstly; then, when all the operations of $g(\ell,1)$ have been processed ($g(\ell,1)$ is empty) the operations of $g(\ell,2)$ have to be processed, and so on. At any time, the non empty group $g(\ell,k)$ such that all the operations of the groups $g(\ell,\alpha)$, $1 \leqslant a < k$ have been performed, is named "running group".

Thus a group sequence can be used as the basis for decision making or decision aid. It defines sequential information through the partial ordering of operations linked with the group ordering, and time information, through the earliest starting times μ_{ij} and the latest finishing times γ_{ij}. For an efficient use of these data for real time scheduling, the following additional information would be very useful :

- is it possible to define indexes to measure the flexibility that remains available when one specific operation is selected inside a running group?

- is it possible to define indexes to measure the impact of the disturbances occuring in the workshop on the characteristics of the used group sequence, especially on its admissibility?

To answer these two questions, the sequential degrees of freedom have been especially considered. A concept of "free sequential margin" has been introduced, in order to measure the amount of variations on the starting times of the operations that preserves this type of degrees of freedom.

5.3.1. Free sequential margin definition

For an operation (i,j) included in a running group $g(\ell,k)$, let $\bar{\mu}_{ij}$ be the latest starting time of (i,j) that preserves all the permutations for the operations of $g(\ell,k)$, and furthermore for all the other groups. Thus, a necessary condition for preserving all the permutations starting with (i,j), in $g(\ell,k)$ is :

$$t_{ij} \leqslant \bar{\mu}_{ij}$$

t_{ij} being the actual starting time of (i,j).

Then, the free sequential margin, denoted by $m(i,j)$, is given by :

$$m(i,j) = \bar{\mu}_{ij} - \mu_{ij}$$

In subsection 5.1.3., the latest finishing time γ_{uv} of an operation (u,v) has been defined as the limit for preserving all the permutations in all the groups following (u,v). Thus $\bar{\mu}_{ij}$ is defined by :

$$\bar{\mu}_{ij} = \min \left[\gamma_{ij} - p^{\ell}_{ij} ; \min_{\substack{(u,v)\in g(\ell,k) \\ (u,v)\neq(i,j)}} \gamma_{uv} - \sum_{(u,v)\in g(\ell,k),} p^{\ell}_{uv} \right]$$

By setting :

$$m_g(i,j) = \min_{\substack{(u,v)\in g(\ell,k) \\ (u,v)\neq(i,j)}} \gamma_{uv} - \sum_{(u,v)\in g(\ell,k)} p^{\ell}_{uv} - \mu_{ij}$$

the free sequential margin can be defined by

$$m(i,j) = \min \left[\gamma_{ij} - p^{\ell}_{ij} - \mu_{ij} ; m_g(i,j) \right]$$

The first term in this margin is due to overlapping prohibition between two successive operations of one item. $m_g(i,j)$ is directly linked with the existence of all the permutations in $g(\ell,k)$.

Thus, this free sequential margin m(i,j) will play a essential role for the real time use of a group sequence.

5.3.2. *Free sequential margin use for decision making [15, 16]*

In real time, the scheduling decision making objective is to select the operation of the running group $g_q(\ell,k)$ that will have to be performed by an empty machine ℓ. This decision making is equivalent to a splitting of $g_q(\ell,k)$ where $g_{q+1}(\ell,k)$ is made only of the selected operation and $g_{q+1}(\ell,k+1)$ is made of the remaining operations of $g_q(\ell,k)$. Thus, after a decision making, we need to update the margin m_g of the operations inside the new running group $g_{q+1}(\ell,k+1)$.

The following preliminary result has been proved [14] :

Let . $g_q(\ell,k)$ be the running group
. (r,t) be the selected operation
. m_g be the margin of an operation in $g_q(\ell,k)$ (before decision making)
. m'_g be the margin of the same operation in $g_{q+1}(\ell,k+1)$ (after decision making).

Thus we have the following results :

$$\begin{cases} . \ \forall (u,v) \in g_q(\ell,k) \text{ such that } m_g(u,v) \leqslant m_g(r,t) \\ \qquad m'_g(u,v) \geqslant m_g(u,v) \\ . \ \forall (u,v) \in g_q(\ell,k) \text{ such that } m_g(u,v) > m_g(r,t) \\ \qquad m'_g(u,v) \geqslant m_g(r,t) \end{cases}$$

5.3.2.1. Decision making in an undisturbed environment

In such a case, for all the running groups $g(\ell,k)$, $m(u,v) \geqslant 0$ $\forall(u,v) \in g(\ell,k)$. Due to the preceding result, the best choice for decision making on machine ℓ is to select the operation of $g(\ell,k)$ with the largest margin m_g : all the margins of the remaining operations will increase, which could be fruitfull to take account of disturbances in the future.

5.3.2.2. Decision making in a disturbed environment

i) <u>Considered types of disturbances</u>

The main disturbances occuring in a workshop can be classified into two types :

- **"item failure"**

(i,j) being completed, (i,j+1) cannot be performed even if the corresponding machine would be available : for instance, a needed tool is missing, or a component needed for this operation is missing... Such a failure will increase the earliest starting time of (i,j+1) and consequently of all the operations linked with (i,j+1).

- **machine failure**

This is the classical machine breakdown, which delays all the operations waiting for this machine, by increasing their earliest starting time.

These two types of disturbances cover the main part of the failures occuring on the shop floor. They induce an increase of the earliest starting time of certain operations. This will restrict the admissibility of the considered group sequence by reducing the number of admissible sequential solutions.

ii) <u>Consequences of disturbances for the group sequence admissibility</u>

The free sequential margin is a useful index for admissibility, easy to update when disturbances occur, because it is directly depending from the earliest starting times.

Thus, as a disturbance implies an increase of the earliest starting time μ_{ij} of at least one operation (i,j) from $g(\ell,k)$, we will have the following situations :

. if $\gamma_{ij} - p^{\ell}_{ij} - \mu_{ij} < 0$, thus the group sequence is no longer admissible, whatever may be the permutation issued from $g(\ell,k)$.

. if $\gamma_{ij} - p^{\ell}_{ij} - \mu_{ij} \geqslant 0 \ \forall(i,j) \in g(\ell,k)$ then
- even $m_g(i,j) \geqslant 0$: all the permutations from $g(\ell,k)$ starting with (i,j) are admissible.
- or $m_g(i,j) < 0$: at least one permutation from $g(\ell,k)$ starting with (i,j) is not admissible.

Then $m(i,j) \geqslant 0 \ \forall(i,j) \in g(\ell,k), \forall g(\ell,k)$ in the set of running groups is a necessary and sufficient condition for the group sequence admissibility.

iii) Decision making after a disturbance

Having updated the margins after a disturbance, two situations can occur :

- $m(u,v) \geqslant 0 \ \forall(u,v) \in g(\ell,k)$ in the set of running groups : the used group sequence remains admissible and can be used for decision making;

- there exists in a running group at least one operation (i,j) with $m(i,j) < 0$; thus two cases have to be considered :

. if $m(i,j) < 0 \Leftrightarrow \gamma_{ij} - p^{\ell}_{ij} - \mu_{ij} < 0$: the group sequence is no longer admissible and a new one has to be searched, by coming back to the analysis step.

. if $m(i,j) < 0 \Leftrightarrow \left[m_g(i,j) < 0,\ \gamma_{ij} - p^{\ell}_{ij} - \mu_{ij} \geqslant 0\right]$, the group sequence is no longer admissible. However, before coming back to the analysis step, it is possible to use some local procedures which try, by considering only the running groups, to define a new admissible group sequence. If such a procedure fails, the analysis step is used again.

5.3.3. *Practical implementation*

The practical implementation of the proposed approach has been based on a decision aid context. Thus an interactive real time scheduling software has been developed (ORABAID = ORdonnancement d'Atelier Basé sur une AIde à la Décision). It is made of three modules :

- an "ANALYSE" module, the purpose of which is to search for an admissible group sequence by using the proposed approach.

- an "ARBITRE" module, which is the core for real time scheduling. It uses the group sequence issued from the ANALYSE module to propose to the decision maker information about the running groups, the operation margins, the earliest starting times and so on. It is based on the results presented in subsection 5.3.. In order to update all the preceding information, especially when disturbances occur, this module needs to know at any time the real state of the workshop. Thus, it has to be linked with the real time monitoring system of the workshop.

- a user INTERFACE, which allows to have an interactive use of the ARBITRE module.

It must be noticed that in the above structure, only the ANALYSE module is concerned with the complexity of the overall scheduling problem. But this module is generally used off line and thus there are no crucial constraints on the computation time. For the real time part of the system, it is located in the ARBITRE module, which performs only simple computations that can be performed easily in real time with a small computer.

A package for workshop real time scheduling has been developed by using the ORABAID software as a core. It is now sold by a french software house. Since one and a half year, it has been implemented in 15 companies and it has produced in some cases a significant productivity increase.

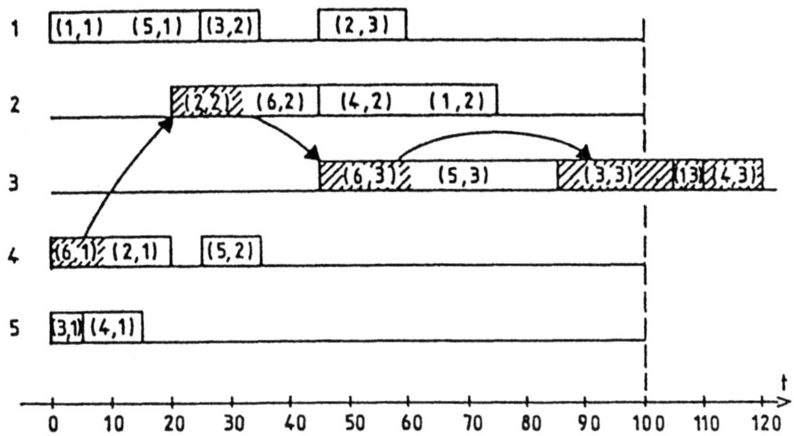

Figure 1

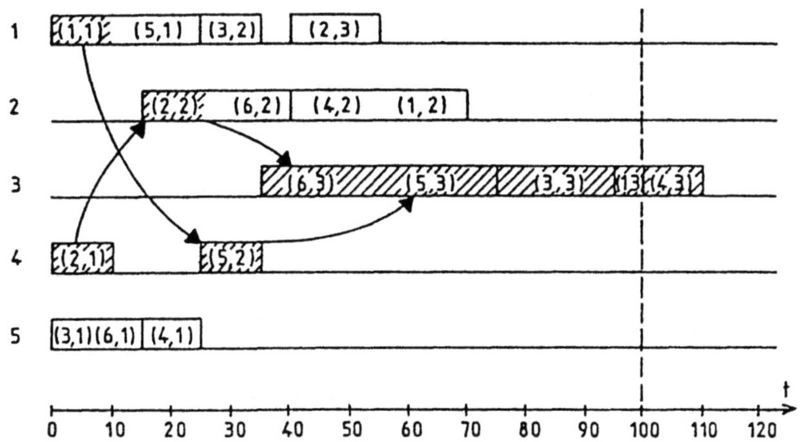

Figure 2

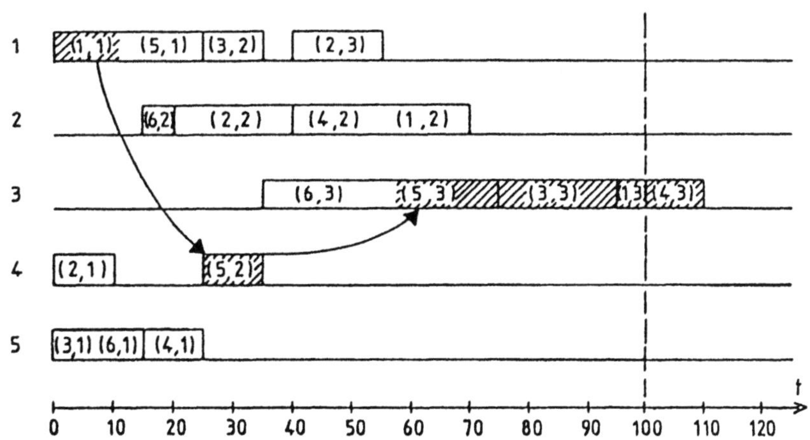

Figure 3

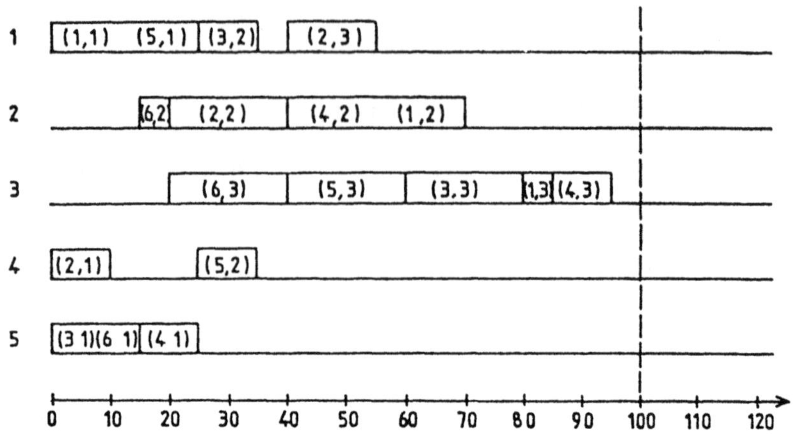

Figure 4

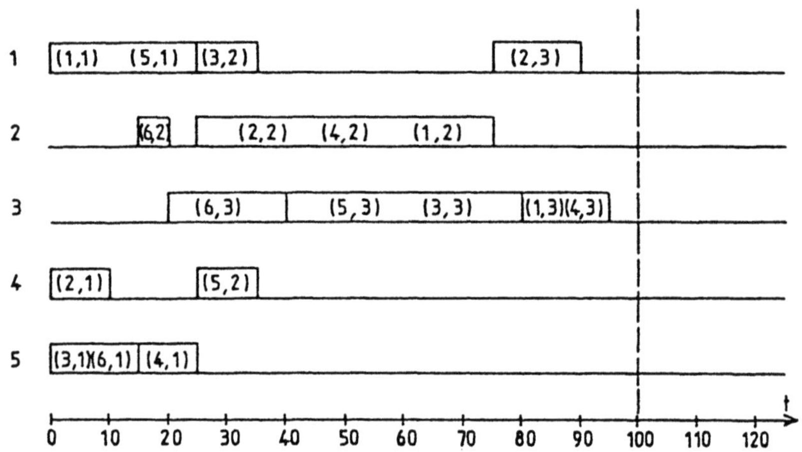

Figure 5

REFERENCES

1. Melese J.: L'analyse modulaire des systèmes de gestion. Ed. Hommes et Techniques, 1972.
2. Baker K.R.: Introduction to sequencing and scheduling. Wiley, 1974.
3. Conway R.W., Maxwell W.L., Miller L.W.: Theory of scheduling. Addison - Wesley 1967.
4. Erschler J., Roubellat F., Vernhes J.P.: A decision making process for the real time control of a production unit. Int. J. Prod. Res., vol. 14, n° 2, 1976.
5. Erschler J.: Analyse sous contraintes et aide à la décision pour certains problèmes d'ordonnancement. Doctorat ès-Sciences Thesis, Univ. Paul Sabatier Toulouse, France, 1976.
6. Erschler J., Fontan G., Merce C., Roubellat F.:Interaction between constraints and decision making in scheduling problems. 20th IEEE Conf. on Decision and Control, San Diego, USA, 1981.
7. Demmou R., Erschler J., Roubellat F.: Aide à l'ordonnancement en temps réel de la fabrication par la mise en évidence de groupes de tâches permutables. IFAC/IFORS Symposium, Toulouse, France, 1979.
8. Erschler J., Roubellat F., Thuriot C.: Scheduling with resource constraints. International EURO VII Conference, Bologne, Italy, 1985.
9. Erschler J., Esquirol P., Decision aid in job shop scheduling : a knowledge based approach. IEEE International Conference on Robotics and Automation, San Francisco, USA, April 1986.
10. Erschler J., Esquirol P., Règles et processus d'inférence pour l'aide à l'ordonnancement de tâches en présence de contraintes. 2ème Conf. Inter. Systèmes de Production, INRIA, Paris, France, Avril 1987.
11. Esquirol P.: Règles et processus d'inférence pour l'aide à l'ordonnancement de tâches en présence de contraintes. Doctorat Thesis Université Paul Sabatier Toulouse, France, 1987.
12. Demmou R.: Etude de familles remarquables d'ordonnancements en vue d'une aide à la décision. Docteur-Ingénieur Thesis, Univ. Paul Sabatier Toulouse, France, 1977.
13. Demmou R., Erschler J., Roubellat F.: Aide à l'ordonnancement en temps réel de la fabrication par la mise en évidence d'ensembles de tâches permutables. IFAC/IFORS Symposium, Toulouse, France, 1979.
14. Thomas V.: Aide à la décision pour l'ordonnancement en temps réel d'atelier. Docteur de 3ème Cycle thesis, Univ. Paul Sabatier Toulouse, France, 1980.
15. Roubellat F., Thomas V.: Une méthode et un logiciel pour l'ordonnancement en temps réel d'ateliers. 2ème Conf. Int. Systèmes de Production ,INRIA, Paris, France,1987.
16. Le Gall A., Rodriguez D., Roubellat F., Thomas V.: Aide à la décision pour l'ordonnancement d'atelier en temps réel. Technical Report LAAS n° 84056, 1984.

CYBERNETICS APPROACH TO MATERIAL FLOW CONTROL

Detlef F. Pape
Institute for Company Cybernetics
Zeppelinstraße 301
4330 Mülheim a.d. Ruhr
West-Germany

Introduction

The world of today is very often characterized as the beginning of the era of information. The value of information has never been so high in the past as nowadays. The problem we have to solve is not to have just any information, but the problem we need to tackle is to find the correct information at the right time out of the abundance of information which overflows us every day. The information networks we operate in companies, towns, states which function via telephones, mail or satellites enable us to be informed about incidents on the other side of the planet which have only happened a few seconds ago. - Thus information ties the elements of our world very close together. This entails that reciprocal relations of the world's elements do not only exist, but that their importance in the daily life increases heavily.

But this is also true for the daily life of companies' members. It is no longer possible with the contraction and shrinking of markets what means high competition that companies are organized in process-killing, functional ways. As long as companies' members understand their role only as a tayloristic carrying out of functions the dynamism and flexibility of such systems will tend to very low rates. This paper will try to

NATO ASI Series, Vol. F53
Advanced Information Technologies for
Industrial Material Flow Systems
Edited by Shimon Y. Nof, Colin L. Moodie

show that today's functional structures are not suitable for the evergrowing importance of interrelations between the elements of nearly every system.

This is especially significant for a complex task like material flow control. Ancient approaches were deterministic like the understanding of the production process was. Today's problems in the production area demand other views. The aim of this article is to show that a cybernetics approach could cope with these new requirements.

Organizations and Organisms

Do you know a company which did not declare bankruptcy or had to end its existence for more than a hundred years? Or two hundred? Or a lot more than thousands of years? - Well, you know one. It is the company of Mother Nature which keeps on going already for several million years. Although the surroundings of earth changed, different angles of the earth's axle provoked glacial periods, ice-ages or hot periods, Mother Nature always found a way how to adapt and to keep alive.

When we talk about artificial creations of man like organizations, companies and so on, we tended to say that their main target in our western life-style is to make profit. There is a growing part of people now who state that the main point of organizations is not just to optimize the profit but to ensure their existance. Then organizations are acting like organisms and the rules for their being might be applied for them. Herbert Marcuse (1) already said that one of the most eminent reasons for capitalism surviving for such a long time and equally probably its most important performance is that capitalism tends very strongly to integrate its opposition. So the system we all rely on and are satisfied with acts like a part of Mother Nature, like an organism that tries to cope with the changes of its surroundings. Adapting to those influences it also changes its own shapes and features.

But Man himself tends to be conservative and preserving. We all like our small habbits and nice customs. This might explain why companies present quite a statical structure. Nothing in today's life is more difficult to change than organizations. Very often companies need a profound crisis before they react on urgent requirements.

Eastern economic rules in western companies

But this is not yet the end. The way we organize our companies is extremely contrasted to the way we operate our economy. In business we have some guidelines set by laws. But as the reason for success of our system we preach the flexibility and initiative of the single subject. But what do we do inside our companies? - We try to organize everything, to tell everybody exactly what, how and when he has to do something.

Especially in production we leave very little room for personal initiatives. We have planning sheets, operating instructions, progress control, operation scheduling, operation descriptions and very sophisticated loading plans. But still we have a lot of problems with lead times, capacity loads, missing parts or quantities, work-in-process and so on. These problems do not exist in the model of production. There everything fits very well together, according to the master plan.

It is very striking that we operate our companies with a tool which is cited by so many economists as the main reason for the failure of the eastern economic systems. It is said that the master plan rends economies stiff and inflexible so that you always produce what you do not need exactly in that way at that time. Another aspect is that this is an initiative-killing-way of organization where the old plans are more important than personal thoughts. Nobody cares any more what happens, no enthusiasm appears any more. This picture drawn resembles a lot the situation we have in our companies or production sites. If we want to change this we need to change it profoundly. We need new concepts, not only some new machinery linked together to automatic systems.

In the future we have to invest a lot more into the most important riches we have in our companies: human capital or brains.

The empty promise of total automation

The answer to get rid of the incalculable and inconvenient factor "human being" in the production process was automation. It seemed to become possible to create fully automated systems where no one was no longer dependant on the human factor. Millions were spent on those ideas but the success never really appeared. The reason for that can be shown with a simple equation. If you have a system consisting of a hundred elements which each have a reliability factor of 99.5%, then the probability that the whole system works is $0.995^{100} = 0.61$.

A hundred elements is not very much for a FMS (Flexible Manufacturing System), a whole production site has a lot more. Okay - robots agree to work in three shifts, they do not strike for higher wages and their home rate is zero - but with a systems reliability of maybe a lot less than 60% it is hard to justify investments.

Cybernetics approach

When Norbert Wiener (2) and his research fellows tried to improve weapons during World War II he ascertained as the main difference between dead and lifely things the ability of the latter to adapt to changes by feed-back. So he started with a new branch of science he later called cybernetics. The main feature of this scientific approach is to model things as systems consisting of elements where the interactions of the elements, their feed-back to each other, play an important role. So why don't we try to structure our production areas the same way? - An aggregate cybernetics view of a company shows figure 1 (from 3).

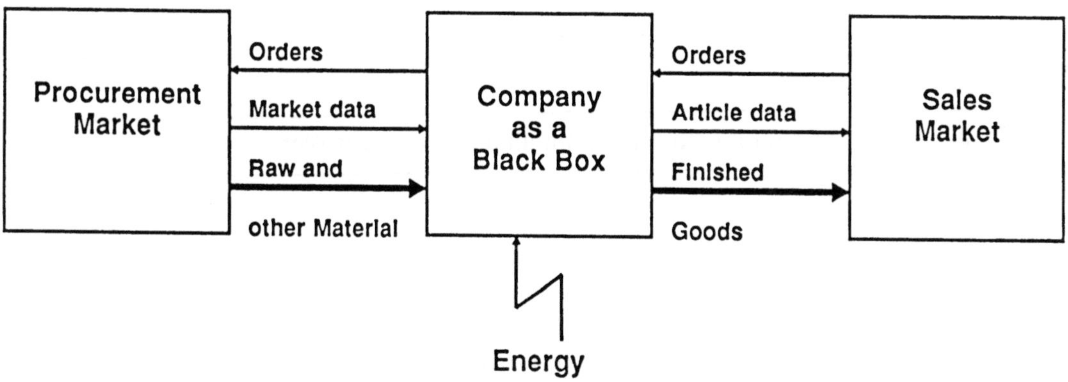

Fig. 1: Cybernetics view of a company

We could try to disentangle the now existing production systems and create new structures with hierarchical control circuits. Each level has one or more closed loops. The guidelines of the systems do not go straight top-down and bottom-up through the whole system. Now each level gives some ranges of guiding principles to the beneath system with enough allowance for the now responsible and competent human being to be able to react on suddenly appearing incidents. Those socio-technical control circuits (see an example in figure 2) might be better in performance as they combine the advantages of automatisms' accurateness and human beings' flexibility. The following sections give some ideas how such an approach could be transformed into reality.

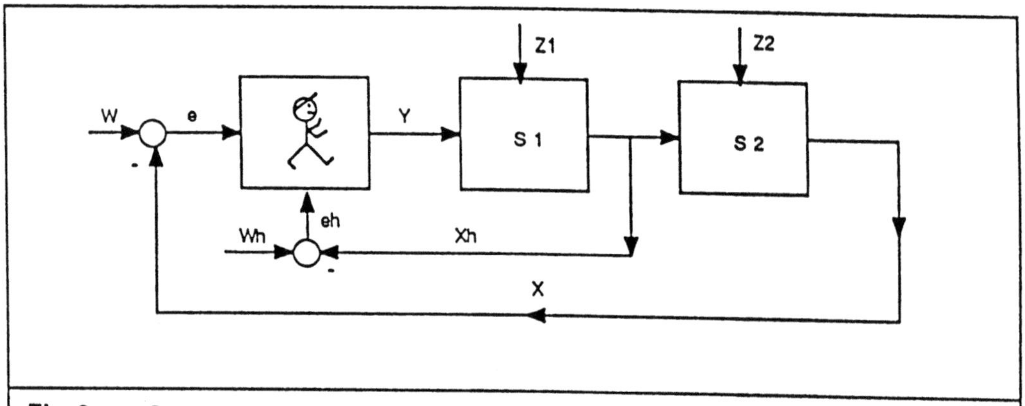

Fig. 2: Control circuit with human intelligence as controler

The system must model all important features

The view of production is, historically affected, machine-oriented. So are our production models and the relevant software packages. In fact only a part of reality is taken into consideration when orders are planned via scheduling algorhythms onto machines. Usually transport and waiting times are only considered by averages. The logistical capacity is calculated as infinite. Transfer time matrices were invented to make up for this. I think that this is not enough. We need models of production with elements corresponding to each important activity in production. Figure 3 gives an example.

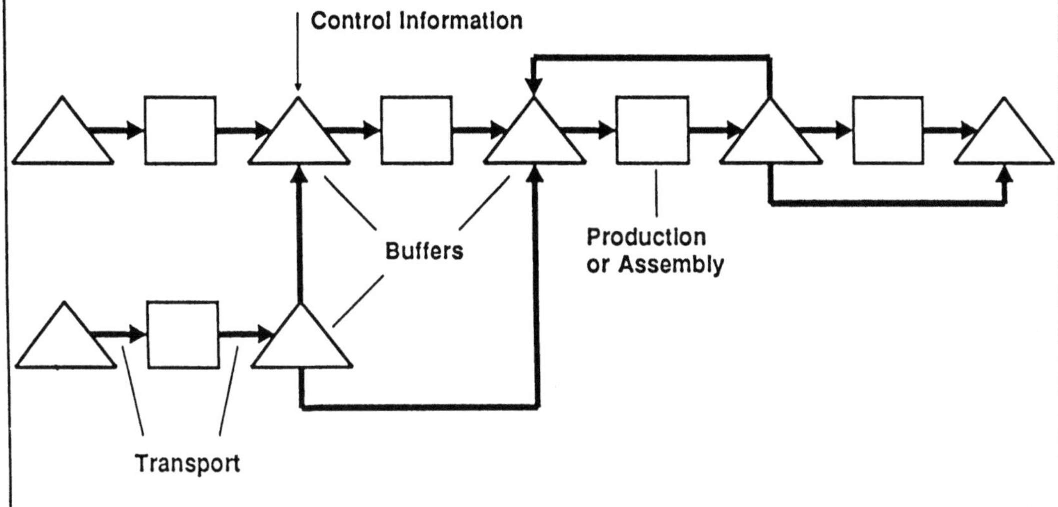

Fig. 3: Comprehensive model of production

The definition of control sections

The next step would be to define rules in order to create highly autonomous subsystems. Their external way of interaction with each other and their internal way of organization need to be created in that way that the human being is in its focus and that the now existing informal procedures of production control are formalized.

Without going more into detail the author dares to say that it is possible to create more effective production systems in that way.

But we need more than only new concepts.

The importance of education

I think the biggest problem we will face in the future will be the reeducation of people. Nowadays our education system is devided into a subset of subjects which are usually taught seperated from each other. Students learn something about money, psychology and economy. But very seldom they get to know the interactions of those three important factors. In school and at university we learn a lot to study the behaviour of system elements detached from the context.

So the politician speaks about politics, the broker about money and the priest about souls. Nobody would like to admit that his special subject has to deal with environmental polution. But nevertheless some interrelations are obvious.

Transferring this to the shop floor and higher levels: we need staff that does not only think and live in their own little piece of work. They have to understand that the whole company is a living organism. A little change here can provoke a big problem elsewhere. If we want to improve the competitiveness of our production sites we must give to our staff the chance to learn and understand the whole system as well as their own role in it. But this would not only improve the effectiveness of work. It would also improve working atmosphere and lead to a more worthwhile living.

Summary

The main target of this article is to underline that with the growing complexity of our life the importance of interactions cannot be neglected any more. This leads to new company and control models. The company models especially in the production area need to be more realistic: the production is not just a set of technical elements but a very complicated system. If we do not include processes like transport and buffering as well as disturbances into our models we do not copy reality. Thus the results of our planning do not work (4).

In order to cope with the handling of complex production systems we can apply natural laws. The definition of hierarchical control levels leads to a simplification of the whole. With the human being in its center as the main controlling element we can achieve two main advantages:

- the subsystems can at least partly control themselves and thus are more reliable
- the return of competence and responsibility to the staff creates an atmosphere which is a lot more fertile for companies' success.

The main problems we will face when we try to transform this into reality are to

- carry out comprehensive education and training and
- develop socio-technological control concepts and the relevant hard- and software structures.

References

1. Marcuse, H.: The One-Dimensional Man. Studies in the Ideology of Advanced Industrial Society. Beacon Press, Boston, Mass. 1964.
2. Wiener, N.: Kybernetik. Regelung und Nachrichtenübertragung in Lebewesen und Maschine, (Orig. amerik.) Rowohlt Verlag, Hamburg 1968.
3. Büser, M.; Pape, D.; Stewing, F.J.: Simulation of Integrated Information and Material Flow in Logistics Systems: an Application of the Modelling Tool, HIT. Müller, H.; Reddy, R.(ed.): Simulation in the Factory of the Future. Proceedings of the European Simulation Multiconference, Nice, France 1988.
4. Pape, D.: Channelized Information Flow within the Enterprise. An overall logistical Viewpoint integrating Production and Administration, in: Proceedings of the Material Handling Focus '87, Georgia Institute of Technology, Atlanta, USA.

RESEARCH PANEL REPORTS

Introduction

An important objective of the advanced research workshop is to stimulate research panel discussions by the participating scientists. The participants were divided to four panels. Each panel had an assigned chairman or co-chairmen, and a provacator, who had all prepared in advance to lead the panel discussions. The panels met in parallel and focused on four main themes, which are significant to the workshop topic:

- Material Flow Data and Knowledge
- Communications and Intelligent Systems
- Planning and Control Models
- Implementation and Economics

Since the panels met independently throughout the workshop, they produced four different summary reports. However, because of the intensive interactions during the workshop, there are some interesting overlapping ideas.

The panel members were asked to discuss:
(1) Traditional, new, and growing research problems in the panel theme;
(2) Traditional, emerging, and promising research tools and techniques.

The panel members were asked to generate:
(1) Priority list of research problems and projects;
(2) Methodological recommendations;
(3) Potential international cooperation on research plans.

The panel reports follow. (MF - Material Flow; MFS - Material Flow Systems)

RESEARCH PANEL No. 1: MATERIAL FLOW DATA AND KNOWLEDGE

Participants: A. Dutta, A. Hodgson, A. Kusiak (Co-chair), B. Montreuil, D. Weber, A. Whinston, T. Woo (Provocator), R. Young (Chair)

This research panel summarized its discussions in a list of 11 key research issues. Each issue is followed by a brief context and explanation of the specific need.

1. Issue: Graphic representation methods for presenting and manipulation of design and analysis models and results.

Context: Due to the mathematical and computer-based nature of material flow modeling techniques for design and analysis, it can be difficult for practicing engineers to comprehend. With the availability of low cost, high speed, and high resolution graphics on small computers it seems viable to try to develop methods to graphically represent complex ideas to aid the practicing engineer.

2. Issue: An architecture for MF control that could be used to identify specifications for generic software modules.

Context: The idea is to develop a framework in which research *and* development of usable non-proprietary CIM software can take place. The future availability of a coherent set of software systems for a major area of CIM could reduce one of the greatest barriers to CIM for small/medium (discrete, batch) manufacturers. Clearly, any software so developed should be usable to some degree in its basic form, but should also be amendable to extension/amendment or to plug-in-plug-out replacement.

3. Issue: The relationship between design and control knowledge and data.

Context: Increasingly faster life cycles for products and material flow systems, coupled with increasingly more modular and flexible facilities and equipment, it is becoming necessary to continuously examine potential design alterations and control strategies in a dynamic mode. This leads toward a need for integrating design and control decisions, knowledge and data. For example, a priori knowledge that a system is to be controlled using a given strategy may permit to generate a more adapted and efficient design (and vice-versa). This creates a whole new field of research.

4. Issue: Characteristics and architecture of an intelligent knowledge-based design environment for material flow, including:

- Knowledge base searches and inference strategies
- Domain knowledge
- Tandem vs. non-tandem architecture

Context: Intelligent knowledge-based design environments represent the next-generation research frontier in computer support for material flow systems design. such design environments will aim at integrating leading edge artificial intelligence techniques, optimization modelling, simulation modelling, interactive graphics, etc., in order to support the designer throughout the entire design process. In order for such design environments to mature, major research efforts have to be undertaken in domain knowledge representation, overall environment architecture, internal knowledge representation, knowledge-based searches and inference strategies, optimization and simulation modelling, interenvironment interfacing, etc.

5. Issue: Structures necessary to represent constraints and relationships to support planning, scheduling and control knowledge, both precise and imprecise, in MFS.

Context: In order to build general information systems to support material flow the philosophy of modularization should be adopted. In order to integrate the separate modules we propose a high level language to specify constraints and relations between the separate modules. A constraint language processor would be developed which would translate the high level statement of the constraints into executable code. An approach to the design of such a language involves the concept of "blackboard variables".

6. Issue: A standard benchmark(s) for evaluating MF methods and techniques.

Context: There is currently no standard way to evaluate results coming out of R&D groups. A standard benchmark would help R&D and allow industry to select results to focus their efforts on.

7. Issue The perspectives needed to adequately describe a MFS and the methodologies necessary to support modeling those perspectives for MF design. Examples would be: a scheduling perspective, spatial perspective, functional perspective, etc.

Context: Some of the applications of modern MF systems have not been successful. Analysis of these applications leads to the conclusion that at the design stage operation, spatial and functional issues have been ignored.

8. Issue: Investigation of issues dealing with applications of new architecture of knowledge-based systems, e.g., tandem architecture, parallel processing, multiagent collaboration, distributed systems in Material Flow.

Context: Design and management problems in modern MF systems characterize high computational complexity. In order to effectively solve the MF problems, the use of tandem architectures of knowledge-based systems, parallel processing, multiagent collaboration, and distributed system architectures should be considered.

9. Issue: Software support for exception handling.

Context: Exception handling and recovery procedures are important tools for managing a material flow system. Developing a uniform framework for allowing designers to specify these concepts in a high level language, possibly menu based, which then generates the exception handling software would be useful.

10. Issue: Interfaces between MF system and other components, such as software components, physical components, human components, etc.

Context: An MF system strongly interacts with the components of its environment. Therefore, the interface between the MF system and the components is of primary concern.

11. Issue: Information sharing, communication and coordination between heterogenous knowledge sources.

Context: Different types of knowledge are needed to perform the information processing function associated with material flow. Do we have adequate tools to represent such knowledge on computers for automation? What mechanisms are available, or need to be developed to get such heterogenous knowledge representation to coordinate their activities? Identification is needed of different types of knowledge, such as algorithm knowledge, heuristic knowledge, and subclasses within them.

The motivation for this topic is the increasing use of automation in factories. Since material flow, as well as information flow are needed to produce and deliver products, it is not sufficient to automate simply the physical processes associated with production. Automation of complementary information flow requires a thorough investigation of the type of knowledge involved and how such knowledge may be represented on a computer.

RESEARCH PANEL No. 2: COMMUNICATIONS AND INTELLIGENT SYSTEMS FOR MFS

Participants: A. Carrie (Chair), R. Dillmann, C. Fritsch (Provocator), J. Gascoigne, S. Nof, U. Rembold (Co-Chair), S. Schneider

Keywords in the development of manufacturing systems and material flow systems of the next 20 years will be *standardization, integration, configurability, error-tolerance, smart machines* and *information management.*

Basically, manufacturing operations consist of planning, scheduling, factory control, quality control and maintenance. These operations are typical for the manufacture of all discrete products. For this reason, many common manufacturing processes, computer structures, software and hardware interfaces, and software packages have been developed. In the factory of the future, an attempt will be made to develop configuration methods and systems to tailor these components to the need of the user according to a virtual manufacturing concept. With this concept, the manufacturing engineer can describe interactively to the computer his manufacturing requirements. He will do this with an easy to learn natural and/or graphical language. The decision of the computer to select the required system components will be supported by a hierarchically structured series of expert systems which have domain knowledge about specific manufacturing areas. A further trend in automation will be the integration of the manufacturing configuration system into the product design activity. Here, the basic manufacturing configuration will be derived from the three-dimensional product model and the required product functions. The computer has access to knowledge-based generative or variant product planning packages which will select the processes and manufacturing sequences. For scheduling, a database about the available manufacturing resources is available which contains an online status of all operations. To speed up processing of the scheduling algorithms, they will be cast into fact VSLI silicon chips or wired into special computer architectures. The flow of the manufacturing process is controlled by hierarchically structured control algorithms which can be configured to the prevailing factory situation. If, for example, a machine fails, the manufacturing run will be automatically reconfigured. For the support of the routing and manufacture of the product, interactive communication terminals and intelligent sensors will be available which can be readily connected to the communication system via standard interfaces and protocols.

In the following sections, these concepts are expanded within the context of communication and intelligent systems under three main headings:

1. The need for human-machine interfaces.

2. The need for a framework for integration of devices, systems and technologies.

3. Comments about technology-driven changes which will influence perceptions or facilitate the achievement of the above communication and intelligence needs.

THE NEED FOR HUMAN-MACHINE INTERFACE

There is a need for research and development in the area of intelligent human-machine interfaces to support the activities of manufacturing operational control and the process control area. Human-machine interfaces can be related to

- activities of material flow, manufacturing, and process control
- data
- processing of data, flow, and processes
- communication and
- design, planning and implementation.

An ergonomic workstation should offer to the operator an efficient and simple to use machine interface. There is a need for multi-windowing, including graphics and operator guidance. Operator guidance should be based on dialogues which control access to different system control levels, data and information. A problem arises if the input knowledge from the operator is incomplete or if non-formal knowledge is used. Research on common-sense is relatively rudimentary in manufacturing, the problem areas are usually limited, thus, incomplete input knowledge may be completed via specialized dialogues. A natural human-machine interface, as proposed for the 5th-generation computer, includes the use of pictures, text, and speech (voice).

Natural human-machine interface:

At the natural human-machine interface all details related to network routing, data retrieval, operating system details, etc. should be hidden to the user. In other words, an abstraction oriented (e.g. object oriented) user-friendly human-machine interface should be used.

Some examples where such user interfaces can be applied are:

- Manufacturing task-oriented programming using cell and product information, generating control structures (graphs and sequences of elementary operations).

- Dynamic reconfiguration of cells and processes (representation of actual situations and alternatives).

- Dynamic process rescheduling.

- Decision support to the operator for alternative manufacturing operations.

- On-line manufacturing tuning requires access to on-line process information and techniques to tune the process parameter or the process structure.

- In-process knowledge acquisition.

- Predictive (preventive) maintenance.

- Manufacturing planning and scheduling.

The above-mentioned applications require strong support in terms of graphics, natural languages, algorithm and software packages, and in general access to high computational power (in future super-computing).

Graphics:

Graphics is used to support the operator with high-performance pictorial information abut the required information. Today 2-D graphics is satisfactory. There does not exist a real 3-D graphics systems supporting the presentation of solid work-pieces. Fast high performance graphics for real-time simulation of technical processes is needed.

Interactive programming of manufacturing with natural language requires graphical support related to layout, geometry, kinematics, dynamic graphs, etc. In the area of manufacturing planning and scheduling an *information trace* would be useful starting from the CAD product design (CAD-model) to the process simulation.

High performance graphics including zooming, multi-windowing, and moving graphics are required for graphical communication between the operator and the on-going process.

Natural Language:

The use of natural language is required to facilitate operator communication with complex systems. Natural language can be related to:
- Task and problem specification
- Dialogues
- Explanation
- Completion of incomplete information
- Decision support
- Planning, and more.

Access to Manufacturing Software Packages:

An intelligent human-machine interface should facilitate access to required information and material flow software. The access details should be transparent to the operator. In the future, super-computing will allow access to very complex software packages, therefore, an intelligent front-end to supercomputing is required.

THE NEED FOR A FRAMEWORK FOR FLEXIBLE INTEGRATION

Many useful tools for planning, controlling and monitoring material flow processes exist now or are being proposed. Also, many of the individual tools for design, simulation, process control, etc. are available from vendors. It is now possible to integrate a *given* selection of these into an interacting system. However, this is a difficult and expensive process in practice. Worse, the produced system is not easily expanded, modified or transferred to another application area. This is *fixed* integration.

Also, groups in manufacturing and research environments are approaching the problem from different directions:

- MAP: communications → up
- Information Technology: Information → outwards
- Systems Integrators: Sub-system by sub-system.

Expertise from different communities must be incorporated in defined areas by appropriate "implementors". It is required that elements of a material flow or manufacturing system be interchangeable with minimum disturbance to the remainder of the system.

The preceding problems have parallels in the computer hardware/software field and the multi-vendor communications field.

The requirement is for a reference model to guide the implementation and interfacing of the whole range of material flow elements and primitives. This should specify:

- Interface boundaries and services
- Functionality
- Inter-relationships ("inter-layer" operation)

But should not specify

- Implementation
- Internal operation

or

- "Intra-layer" operation

This should allow implementors of and operators of manufacturing and material flow systems to select and incorporate elements as they desire for each system and its update/reconfiguration. This approach can be viewed as flexible integration of flexible systems. The reference model/s of flexible integration would address the areas of:

- Overall integration strategy
- Design, planning and control
- Information management and access control
- In-process control and monitoring
- Implementation and evolution of systems.

Key concepts in this context are the software toolbox environment, and modular, configurable machines.

The Software Toolbox Environment

Software tools are components of a material flow system. The overall system is a large and complex software system. There is the need for a software tool system, which allows to select and to configure the required software tools. General requirements to such a toolbox system are:

- Open multi-vendor strategy
- Defined functionality of the tools
- Defined interaction and interfaces between tools
- Boundaries and constraints
- Application specific object oriented data representation necessary.

Not to be defined and standardized:

- Details of implementation
- Strategic purpose of tool
- Possible configuration of tools (open to the user).

In addition to the toolbox system configuration, systems which may be knowledge-based can be developed to support the configuration of the software modules to a complex flow scheduling and control system. With the availability of VLSI-firmware for standard manufacturing software packages, the toolbox system can be extended. Thus, manufacture planning, operational and flow control, or process control supporting workstations ("super workstations") can be realized. This requires research in the field of parallel computer architectures (i.e., transpute, networks, hypercubes), fast algorithms, and programming language. In summary, toolbox systems may be interesting for small and medium sized firms. They can use only parts of the toolbox; tools can be transferred between companies and manufacturing departments. The tools must fit into "tool holders" and should be interchangeable.

Modular Configurable Machines

The next two decades will see need for increased levels of flexible automation in both discrete and batch material flow environments. Handling, processing, assembly, packaging and inspection machines will be required to cost-effectively automate high speed, highly accurate (10^{-9}m) product realization. Such machines must fit into the logical integration architectures evolved, to facilitate flexibility (in the face of changing markets, enabling technologies, material flow and manufacturing methods) and yet serve the wide spectrum of current and future consumer needs.

The logical approach is to decentralize the machine design and creation problem, and to formulate machine building elements, or "primitives". Each primitive should be an almost indiscernible "mixture" of computer hardware/software, actuator/sensor and mechanical transmission components. Primitives will exist in hierarchical form, from low level (single degree-of-freedom, DOF elements) to multi-DOF primitives, and be embedded within a "computerized shell" supporting communication and integration services to enable aggregation of primitives. Through the principle of distributed manipulation (i.e., mechanically decoupled manipulation and synchronization) the aggregated primitives will be used in the manufacture of product families yet provide mechanical optimization. Cycle times will be shortened by an order of magnitude and accuracies increased by 10 to 100 times when compared with conventional robot manipulators. To support this interaction, synchronization and co-operation protocols and strategies should be evolved.

The evolution of machine primitives will be further facilitated by emerging technologies, such as 100 Å resolution laser tools and sensors, and materials such as composites and ceramics. Clearly, "catalog" machine primitives will be required in many guises, and novel solutions must be generated. Furthermore, research in the areas of "intelligent" tools and software must be conducted to create the "shells" by which machine primitives can be aggregated and communicate with enterprise control and information systems.

Sufficient means must be found to enable technology transfer and subsequent industrial exploitation of the research findings. In such a way specific machine primitives can properly be manufactured by machine builders and OEMS. Ultimately, machines for service industry and domestic markets will be generated.

A related concept is that such reconfigurable machines may be incorporated within learning systems, as amplified below.

Development of Learning System Capabilities

In recent research, some investigators have proposed the development of small machines which can learn from their "experience" and from the "past" to perform their actual functions more efficiently. In the meantime, research especially in the AI area has focused on machine learning methods and learning system architecture. Applying these methods to the material flow area, various learning systems can be defined.

Learning may include the human expert as teacher. It may use the physical target system as knowledge source and a knowledge base to support the desired application with the required knowledge. Learning means to add explicit knowledge into the knowledge base with the purpose to specialize or to generalize existing knowledge. Possible learning strategies range from:

- Rate Learning
- Learning by analogy
- Learning from advice/instruction
- Learning from experiments
- Learning from observation
- Learning from experience

Promising application areas of machine learning are as follows:

- Planning/scheduling
  If once a solution plan is found for a problem, it can be applied to similar problems instead of starting again from the beginning. Generalized plans can cover a well defined problem area.

- Process control (cells, devices, subsystems)
  For controlling a complex system, skill acquisition strategies may be applied to optimize the system performance with increasing time.

- Preventive maintenance
  The system learns from disturbances, errors and exceptions to optimize its effective use by generating preventive maintenance strategies.

- System configuration
  Experience from the past can be used for solving actual configuration tasks.

- For intelligent machines behavioral knowledge is required to enable the system to extend with time its skills and to react to exceptions (robots, vehicles, machine tools, etc).

- Cognitive processes
  Cognition is presently still very cumbersome. Noisy signals, fuzzy information,

incomplete reference data and incomplete knowledge about how to interpret features in context of a given task can be processed with the aid of machine learning (learning from examples, experience).

Basic research and pilot installation of learning systems will challenge the development of intelligent, flexible material flow systems.

Enabling Technologies

A further factor which the members of this research panel would bring to the attention of the wider research community is that adequate recognition must be taken of technological development fields, which may or could have significant impact on material flow. Several of these have been mentioned in the foregoing remarks. We would add two other pieces of relevant information of technology-driven change:

Laser Tools and Sensors, Optical Engineering

In the future laser tools and measuring systems will facilitate 100Å manufacture. Such tools/sensors can be fundamentally 3D in operation and can fit naturally into 3D design and product realization environments. An example application area is the etching and inspection of 3D molded PCBs: Vast increases in functionality per unit area can be facilitated and VLSI technologies utilized as an integral part of the products in which they are used. Over the next two decades, research into laser cutting of materials, laser sensing methodologies, new laser sources and beam manipulation techniques should herald the realization of optical tools which would revolutionalize manufacturing and material flow methods in certain areas.

Optical Engineering advances will also be prevalent in other relevant areas. For example, fiber optic communication media reduces restraints in regard to information handling capabilities. However, supporting studies from the electronics community will be required to extend available bandwidth on a single cable beyond 1 GHz or so.

Intelligent Actuators and Sensors

These can be considered under the wider banner of "machine primitives". Their existence will have a major impact on the evolution of distributed and integrated systems for both discrete and continuous process industries. The anticipated availability of "fieldbus" and other low level, low cost communication interfaces will be vital and provide the basis of computerized shells. Corresponding advances in fundamental actuator and sensor technologies will enhance the impact.

RESEARCH PANEL No. 3: MFS PLANNING AND CONTROL MODELS

Participants: E. Barkmeyer, L. McGinnis (Provocator), C. Moodie, D. Pape, V. Parunak (Co-Chair), F. Reubellat (Chair), M. Rosenblatt, A. Seidmann, K. Stecke, C. Sumpter

Research Problems

The main research areas discussed by this panel were

- Models for facility design and layout
- Models for planning and control of material flow systems.

Facility Design and Layout

Modeling issues to be considered would be the following:

- initial versus adaptive design and layout
- capacity requirements and planning for manufacturing resources and material handling systems
- technology selection for:
    - information acquisition and processing
    - technical components of the system, including the technical processing systems (machining, assembly . . .) and the material handling system

Planning and Control of the Technical Resources

The modeling issues to be considered would be:

- Strategies and techniques

  Analyze and compare MRP, JIT, OPT, LOP, loading control and others

  Do we need to implement pure control strategies or mixed ones according to the application context?

- Philosophy of Control

  Is the classical approach of control sufficient or do we need to use state space approach, or qualitative approach, or other alternatives?

- Exception Handling

  In the disturbed environment of a manufacturing process, how to handle exceptions like machine failures, part rework, part mix change?

- Human Factors

  A large number of MF systems being not fully automated define the place and role of people in the control structure: responsibility domain, type of supervision, task distribution between human and computer, information transfer to and from the human, and so on.

Two points were emphasized:

Distributed Control

Due to the complexity of MFS, control strategies will probably be implemented mainly through a distributed control structure. This raises the following research questions:

- architecture definition for distributed control;

- methods for coordinating individual controllers in an efficient way;

- how to manage concurrency and consistency in a distributed control structure?

Real Time Control

This is a promising research area. Important questions to be studied are goal selecting for control, data selection and monitoring for feedback, efficient methods to organize feedback loops, or reactions when needed according the systems status.

Some 'Good' Research Problems

Following is a more precise list of some "good" research problems (not prioritized):

- How can you characterize a problem instance so that you can choose the appropriate facility configuration, planning strategy, and control techniques?

- How can you characterize facility configurations, planning strategies, and control techniques so that you can tell if they are appropriate for a particular problem instance?

- What are the performance measures (or evaluation strategies) that are useful for designing and controlling material flow systems?

- **How can you estimate the performance of material handling subsystems, such as AS/RS, AGVS, lift truck fleets, etc.?**

- **How can you characterize the performance of a cell/shop with a given configuration and material flow control method?**

- **How can you describe a problem instance in a way that supports analysis/synthesis for simplification?**

- **How do you "qualify" new technology, strategy, or model for use on a class of applications?**

- **What are the asset utilization impacts of variability, flexibility, standardization, decentralization, etc.?**

Methodological Recommendations

The panel has concluded that it was not possible to make precise methodological recommendations for each of the various research problems described in the preceding paragraphs. Nevertheless, by considering the selected problems, it suggests the following general recommendations for "style of research".

- **There needs to be a balance of research which is analysis oriented (pilot studies, field studies) and that which is synthesis oriented.**

- **There needs to be some effort to develop a science/engineering base for systems analysis/structural design methods for material flow systems.**

- **There should be a portfolio of research approaches.**

- **There is a need to distinguish between research on whether a new idea *can* be used, or how it *should* be used, and the straightforward application of a new idea. Specific examples are KBS (Knowledge Based Systems), ES (Expert Systems), OOPS (Object Oriented Programming Systems), etc.**

- **Reference models are extremely valuable, not only for practice, but also for organizing, motivating, conceptualizing, and comparing research. However, it may not be a good idea to focus research only on the development of reference models.**

International Cooperation

The various ways to develop a fruitful international cooperation would be the following:

- University and lab visits

- Research articles indepth learning

- Software exchange

- Student exchange

- Joint seminars and workshops focused on material flow systems

- Joint papers/research

- Formal multilateral cooperation contracts (through national research agencies)

RESEARCH PANEL No. 4: IMPLEMENTATION AND ECONOMICS OF MFS

Participants: Y. Bozer, H. Grant (Provocator), R. Hollier
J. Solberg, A. Villa (Chair)

Preliminary Considerations

The present state-of-the art of research in manufacturing problems shows increasing efforts in designing algorithms and procedures to plan and control material handling and processing systems, as well as approaches to manage the large amount of data involved. On the other hand, accounting for implementation, one must realize that industrial managers still have a lot of difficulty in modifying their production systems and management procedures, according to newly proposed ideas. This situation is particularly evident in medium and small-scale manufacturing companies, which still represent a large part of the industrial environment of the western countries. Existence of this separation between "theoreticians" and "industrial managers" has suggested to the panel participants to concentrate their analysis on the following topic: "tools/technology for justification and incremental implementation of flexibility concepts in manufacturing companies, with particular attention to medium/small-size enterprises".

Background

Small/Medium-size Companies

It would be difficult to make a significant classification of company sizes, if not related to the product types, production efforts involved, automation possibilities, etc., relative to the dimension of the markets both of the offered products and of the machines and devices to be used. However, two possible criteria have been recognized: classify either according to personnel employed, or according to the economic dimension. Referring to the second criterion, the main characteristic of medium-size companies is their inability to support by appropriate funding large modification of their own production process, such as the implementation of a completely new department (built "ex-novo"). This situation frequently happens in mechanical production for companies with the number of employees ranging from 100 to 500. Small companies, with less that 20 to 50 employees, cannot usually plan sudden modifications of portions of their processes on an FMS scale.

Incremental Implementation

This preliminary consideration about the size of companies has pointed to an important conclusion: the necessity of accounting for incremental implementation of new processes and information technology tools. Referring to a certain desired modification of a production system, it is well known that a number of phases are necessary, ranging from a feasibility analysis and preliminary design, through the different steps of the effective implementation phase, including training of personnel. In many medium size environments, the implementation, due to funding constraints, must be split into a number of consecutive sub-phases, i.e., partial implementation phase in different times. In practice, the desired new process appears to be the final goal to be achieved through progressive modifications of some process already existing, which still must work, even if in a "non-optimal" way.

Flexibility Concepts

Obviously, the desired (optimal) performance of the final process must be stated in advance. However, it must be noted that efforts spent in defining the desired specifications of the final process are often very large. One of the reasons appears to be the fuzzy definition that the flexibility concept has in many industrial environments. To avoid confusion, panel participants have tried to summarize three flexibility concepts. In dealing with the re-organization of an existing production process, it is necessary to analyze:

- alternatives for part processing, with reference to device ability to process different types of components/products (improve the technological flexibility);

- alternatives in material handling, with reference to the ability to use different routes in order to improve the MFS flexibility;
- alternatives in rules and strategies for management of the material flow process by improving flexibility of the production management system.

It is the opinion of this panel that a careful use of the above concepts will allow an incremental implementation of new technologies and IT tools in existing industrial environments.

Economic Justification

This appears to be a crucial and difficult problem. It is a shared opinion that complex MFS implementations, e.g., of an FMS, even in large companies, are meeting some obstacles. This is owing to the inability to present and obtain both a good justification of capital investments required, and a consistent justification of the in-process costs of these new systems. Most of the time, evaluation is performed by using "old" indices (such as ROI, ROE, etc.). Such economic indices appear to be usable to analyze the convenience of high investments with respect to high variability of product market, and good flexibility of the production process. Economic justification will continue to imply evaluation of the investments' convenience. But it should also provide economic evaluation of the capability of the system to evolve in time, in terms of modifications allowed by its flexibility, and to modify the offered product types.

Needs

The above outlined concepts, even if proposed as a brief summary of some experiences of the participants, lead to the recognition of a number of needs, which are important for industrial practitioners.

a. First, they underline the necessity of using models and simulation procedures to gain a better understanding of the production environment of the MFS, in which process upgrading is required. In particular, it is of significant importance to have available simulation procedures *composed of two well-distinct blocks*, one devoted to defining the production system dynamics, the other implementing production management (PM) strategies, such as planning, control, activity scheduling, etc.

   The reason for this need is evident. By use of a "pure-process simulator", the PM block could be effectively validated, and subsequently applied to the actual production system.

   In particular, the desired "pure-process simulator" should be able to accurately describe not only the normal operation mode, but also the large variety of malfunctioning or degradated operation modes, that often represent a significant portion of the processing time.

b. The need of accurate "pure-process simulators" is directly linked to the need of a methodology to define and state desired "design-and-implementation specifications".

   More precisely, this second need can be defined as follows. In order to design and implement a new process, or an up-grading of an existing process, it is necessary:

   - to accurately define the desired "final system";

   - to obtain the best trade-off between the final process features and the desires of the end user.

It is still a problem of learning, and then describing and modeling the end user objectives; choosing technological devices and IT tools to satisfy these objectives; describing, modeling and planning the different phases to organize technological devices and IT tools in the new or upgraded system. This need is particularly evident in medium size companies, where one often finds a large "desire to change", but a low knowledge of "how to change".

c. The third need can be recognized by referring to the above mentioned flexibility concepts. When dealing with a reorganization process, the system to be implemented is always a "large-scale-system", in terms of data, variables, strategies. Then, besides the rationalization of the system in terms of its modeling, it appears mandatory to:
   - *decompose* the system into parts and inter-part connections;
   - develop individual PM modules for the different parts;
   - integrate PM modules into a PM architecture.

Two remarks must be made. First, decomposition must be oriented to obtain *simpler* sub-systems. This will allow an easier implementation of each PM module. Second, each module should have a large degree of autonomy. This will allow an incremental implementation, gradual in time, of the whole PM architecture.

d. The fourth need is brought out by both the incremental implementation concept, and the economic justification one. It often happens that a medium-size company cannot entirely support the cost of a project oriented to a significant modification of all or part of its production process. Then, it is mandatory for the system integrators to design *re-usable tools*, i.e., define methodologies, design PM modules, organize PM architecture configurations which can be applied, with small modifications, to different industrial environments.

Research Problems

All the above considerations about recognized needs suggest a number of research problems which can be approached. The panel participants suggest three interesting ones:

(1) It appears necessary to develop methodologies for economic justification of either new implementations, or upgrading of existing systems towards higher flexibility levels, i.e., justification of capital expenditures and of in-process costs. Particular attention should be devoted to methodologies able to account for qualitative, intangible benefits.

To this end, approaches based on expert systems or decision support systems concepts could be fruitful, since these tools allow for the inclusion of fuzzy as well as quantitative variables.

(2) Proposals of better representations of the different types of malfunctions that can often affect the operation modes of a MFS, are necessary. A more realistic treatment of degradated operation modes, in terms of models and algorithms, is necessary in order to allow the design of an efficient and robust PM system, one that is able to assure a good process efficiency also in failed situations. This suggestion for improved modeling tools is evidently coming from practice. The opinion of participants is that the accuracy of analytical models is not growing at the rate of technology advances in the field of MFS.

(3) The development of methodologies for the design of PM systems which allow an incremental implementation, is still very important. To all panel participants, it appears that existing methodological tools are all oriented to defining completely new PM architectures to be implemented "ex-novo". But these methodologies are weaker in dealing with problems of upgrading existing processes. *Modularity* of PM systems, *autonomy* of their modules and *integration* are still concepts to be developed, to improve the applicability and effectiveness of PM functions.

NATO ASI Series F

Including Special Programme on Sensory Systems for Robotic Control (ROB)

Vol. 1: Issues in Acoustic Signal – Image Processing and Recognition. Edited by C. H. Chen. VIII, 333 pages. 1983.

Vol. 2: Image Sequence Processing and Dynamic Scene Analysis. Edited by T. S. Huang. IX, 749 pages. 1983.

Vol. 3: Electronic Systems Effectiveness and Life Cycle Costing. Edited by J. K. Skwirzynski. XVII, 732 pages. 1983.

Vol. 4: Pictorial Data Analysis. Edited by R. M. Haralick. VIII, 468 pages. 1983.

Vol. 5: International Calibration Study of Traffic Conflict Techniques. Edited by E. Asmussen. VII, 229 pages. 1984.

Vol. 6: Information Technology and the Computer Network. Edited by K. G. Beauchamp. VIII, 271 pages. 1984.

Vol. 7: High-Speed Computation. Edited by J. S. Kowalik. IX, 441 pages. 1984.

Vol. 8: Program Transformation and Programming Environments. Report on a Workshop directed by F. L. Bauer and H. Remus. Edited by P. Pepper. XIV, 378 pages. 1984.

Vol. 9: Computer Aided Analysis and Optimization of Mechanical System Dynamics. Edited by E. J. Haug. XXII, 700 pages. 1984.

Vol. 10: Simulation and Model-Based Methodologies: An Integrative View. Edited by T. I. Ören, B. P. Zeigler, M. S. Elzas. XIII, 651 pages. 1984.

Vol. 11: Robotics and Artificial Intelligence. Edited by M. Brady, L. A. Gerhardt, H. F. Davidson. XVII, 693 pages. 1984.

Vol. 12: Combinatorial Algorithms on Words. Edited by A. Apostolico, Z. Galil. VIII, 361 pages. 1985.

Vol. 13: Logics and Models of Concurrent Systems. Edited by K. R. Apt. VIII, 498 pages. 1985.

Vol. 14: Control Flow and Data Flow: Concepts of Distributed Programming. Edited by M. Broy. VIII, 525 pages. 1985.

Vol. 15: Computational Mathematical Programming. Edited by K. Schittkowski. VIII, 451 pages. 1985.

Vol. 16: New Systems and Architectures for Automatic Speech Recognition and Synthesis. Edited by R. De Mori, C.Y. Suen. XIII, 630 pages. 1985.

Vol. 17: Fundamental Algorithms for Computer Graphics. Edited by R.A. Earnshaw. XVI, 1042 pages. 1985.

Vol. 18: Computer Architectures for Spatially Distributed Data. Edited by H. Freeman and G.G. Pieroni. VIII, 391 pages. 1985.

Vol. 19: Pictorial Information Systems in Medicine. Edited by K. H. Höhne. XII, 525 pages. 1986.

Vol. 20: Disordered Systems and Biological Organization. Edited by E. Bienenstock, F. Fogelman Soulié, G. Weisbuch. XXI, 405 pages. 1986.

Vol. 21: Intelligent Decision Support in Process Environments. Edited by E. Hollnagel, G. Mancini, D.D. Woods. XV, 524 pages. 1986.

NATO ASI Series F

Vol. 22: Software System Design Methods. The Challenge of Advanced Computing Technology. Edited by J.K. Skwirzynski. XIII, 747 pages. 1986.

Vol. 23: Designing Computer-Based Learning Materials. Edited by H. Weinstock and A. Bork. IX, 285 pages. 1986.

Vol. 24: Database Machines. Modern Trends and Applications. Edited by A.K. Sood and A.H. Qureshi. VIII, 570 pages. 1986.

Vol. 25: Pyramidal Systems for Computer Vision. Edited by V. Cantoni and S. Levialdi. VIII, 392 pages. 1986. *(ROB)*

Vol. 26: Modelling and Analysis in Arms Control. Edited by R. Avenhaus, R.K. Huber and J.D. Kettelle. VIII, 488 pages. 1986.

Vol. 27: Computer Aided Optimal Design: Structural and Mechanical Systems. Edited by C.A. Mota Soares. XIII, 1029 pages. 1987.

Vol. 28: Distributed Operating Systems. Theory und Practice. Edited by Y. Paker, J.-P. Banatre and M. Bozyiğit. X, 379 pages. 1987.

Vol. 29: Languages for Sensor-Based Control in Robotics. Edited by U. Rembold and K. Hörmann. IX, 625 pages. 1987. *(ROB)*

Vol. 30: Pattern Recognition Theory and Applications. Edited by P.A. Devijver and J. Kittler. XI, 543 pages. 1987.

Vol. 31: Decision Support Systems: Theory and Application. Edited by C.W. Holsapple and A.B. Whinston. X, 500 pages. 1987.

Vol. 32: Information Systems: Failure Analysis. Edited by J.A. Wise and A. Debons. XV, 338 pages. 1987.

Vol. 33: Machine Intelligence and Knowledge Engineering for Robotic Applications. Edited by A.K.C. Wong and A. Pugh. XIV, 486 pages. 1987. *(ROB)*

Vol. 34: Modelling, Robustness and Sensitivity Reduction in Control Systems. Edited by R.F. Curtain. IX, 492 pages. 1987.

Vol. 35: Expert Judgment and Expert Systems. Edited by J.L. Mumpower, L.D. Phillips, O. Renn and V.R.R. Uppuluri. VIII, 361 pages. 1987.

Vol. 36: Logic of Programming and Calculi of Discrete Design. Edited by M. Broy. VII, 415 pages. 1987.

Vol. 37: Dynamics of Infinite Dimensional Systems. Edited by S.-N. Chow and J.K. Hale. IX, 514 pages. 1987.

Vol. 38: Flow Control of Congested Networks. Edited by A.R. Odoni, L. Bianco and G. Szegö. XII, 355 pages. 1987.

Vol. 39: Mathematics and Computer Science in Medical Imaging. Edited by M.A. Viergever and A. Todd-Pokropek. VIII, 546 pages. 1988.

Vol. 40: Theoretical Foundations of Computer Graphics and CAD. Edited by R.A. Earnshaw. XX, 1246 pages. 1988.

Vol. 41: Neural Computers. Edited by R. Eckmiller and Ch. v. d. Malsburg. XIII, 566 pages. 1988.

NATO ASI Series F

Vol. 42: Real-Time Object Measurement and Classification. Edited by A. K. Jain. VIII, 407 pages 1988. *(ROB)*

Vol. 43: Sensors and Sensory Systems for Advanced Robots. Edited by P. Dario. XI, 597 pages. 1988. *(ROB)*

Vol. 44: Signal Processing and Pattern Recognition in Nondestructive Evaluation of Materials. Edited by C. H. Chen. VIII, 344 pages. 1988. *(ROB)*

Vol. 45: Syntactic and Structural Pattern Recognition. Edited by G. Ferraté, T. Pavlidis, A. Sanfeliu, H. Bunke. XVI, 467 pages. 1988. *(ROB)*

Vol. 46: Recent Advances in Speech Understanding and Dialog Systems. Edited by H. Niemann, M. Lang, G. Sagerer. X, 521 pages. 1988.

Vol. 47: Advanced Computing Concepts and Techniques in Control Engineering. Edited by M. J. Denham, A. J. Laub. XI, 518 pages. 1988.

Vol. 48: Mathematical Models for Decision Support. Edited by Gautam Mitra. IX, 762 pages. 1988.

Vol. 49: Computer Integrated Manufacturing. Edited by I. Burhan Turksen. VIII, 568 pages. 1988.

Vol. 50: CAD Based Programming for Sensory Robots. Edited by B. Ravani. IX, 565 pages. 1988. *(ROB)*

Vol. 51: Algorithms and Model Formulations in Mathematical Programming. Edited by Stein W. Wallace. IX, 190 pages. 1989.

Vol. 52: Sensor Devices and Systems for Robotics. Edited by Alícia Casals. IX, 362 pages. 1989. *(ROB)*

Vol. 53: Advanced Information Technologies for Industrial Material Flow Systems. Edited by S. Y. Nof and C. L. Moodie. IX, 710 pages. 1989.

Vol. 54: Reappraisal of the Efficiency of Financial Markets. Edited by R. M. C. Guimaraes, B. G. Kingsman and S. J. Taylor. X, 804 pages. 1989.

MIX
Papier aus verantwortungsvollen Quellen
Paper from responsible sources
FSC® C105338

If you have any concerns about our products,
you can contact us on
ProductSafety@springernature.com

In case Publisher is established outside the EU,
the EU authorized representative is:
Springer Nature Customer Service Center GmbH
Europaplatz 3, 69115 Heidelberg, Germany

Printed by Libri Plureos GmbH
in Hamburg, Germany